T0202708

Lecture Notes in Computer Science

Lecture Notes in Artificial Intelligence   14472

Founding Editor

Jörg Siekmann

Series Editors

Randy Goebel, *University of Alberta, Edmonton, Canada*
Wolfgang Wahlster, *DFKI, Berlin, Germany*
Zhi-Hua Zhou, *Nanjing University, Nanjing, China*

The series Lecture Notes in Artificial Intelligence (LNAI) was established in 1988 as a topical subseries of LNCS devoted to artificial intelligence.

The series publishes state-of-the-art research results at a high level. As with the LNCS mother series, the mission of the series is to serve the international R & D community by providing an invaluable service, mainly focused on the publication of conference and workshop proceedings and postproceedings.

Tongliang Liu · Geoff Webb · Lin Yue ·
Dadong Wang
Editors

AI 2023: Advances in Artificial Intelligence

36th Australasian Joint Conference on Artificial Intelligence, AI 2023
Brisbane, QLD, Australia, November 28 – December 1, 2023
Proceedings, Part II

Springer

Editors
Tongliang Liu (iD)
The University of Sydney
Darlington, NSW, Australia

Geoff Webb (iD)
Monash University
Clayton, VIC, Australia

Lin Yue (iD)
The University of Newcastle
Callaghan, NSW, Australia

Dadong Wang (iD)
CSIRO Data61
Sydney, NSW, Australia

ISSN 0302-9743   ISSN 1611-3349 (electronic)
Lecture Notes in Artificial Intelligence
ISBN 978-981-99-8390-2   ISBN 978-981-99-8391-9 (eBook)
https://doi.org/10.1007/978-981-99-8391-9

LNCS Sublibrary: SL7 – Artificial Intelligence

This Springer imprint is published by the registered company Springer Nature Singapore Pte Ltd.
The registered company address is: 152 Beach Road, #21-01/04 Gateway East, Singapore 189721, Singapore

Paper in this product is recyclable.

Preface

This volume contains the papers presented at the 36th Australasian Joint Conference on Artificial Intelligence, AJCAI 2023. The conference was held during November 28 – December 1, 2023, and was hosted by the University of Queensland in Brisbane, Australia. This annual conference is one of the longest running conferences in artificial intelligence, with the first conference held in Sydney in 1987. The conference remains the premier event for artificial intelligence in Australasia, offering a forum for researchers and practitioners across all subfields of artificial intelligence to meet and discuss recent advances.

AJCAI 2023 received 213 submissions and each submission was reviewed by at least two Program Committee (PC) members or external reviewers in a double-blind process (over 90% of the submissions had three reviews). After a thorough discussion and rigorous scrutiny by the reviewers, 24 papers were accepted for long oral presentation and 58 papers were accepted for oral presentation at the conference. In total, 82 submissions were accepted for publication as full papers in these proceedings with an acceptance rate of 38% (the acceptance rate of the long oral presentations was 11%). AJCAI 2023 had six keynote talks by the following distinguished scientists: Ling Chen from the University of Technology Sydney, Australia; Manik Varma from Microsoft Research India, India; Peter Soyer from the University of Queensland, Australia; Maria Garcia De La Banda from Monash University, Australia; Mengjie Zhang from Victoria University of Wellington, New Zealand; and Dadong Wang from Data61, Australia.

The following are notable aspects of the AJCAI 2023 conference:

- AJCAI 2023 was jointly held with the Defence Artificial Intelligence 2023 Symposium (November 27, 2023). The Defence Artificial Intelligence Symposium is an exciting opportunity for Defence and AI researchers to come together and explore priorities, opportunities, and commonalities.
- AJCAI 2023 included a day with a special industry focus. Panel discussions allowed industry and academia to share challenges and research directions.
- AJCAI 2023 included four workshops, held on November 28: Foundations for Robust AI: Self-Supervised Learning, organised by Saimunur Rahman, David Hall, Stephen Hausler, and Peyman Moghadam; Federated Learning in Australasia: When FL Meets Foundation Models, organised by Guodong Long, Han Yu, and Tao Shen; Artificial Intelligence Enabled Trustworthy Recommendations, organised by Shoujin Wang, Rocky Tong Chen, Hongzhi Yin, Lina Yao, and Fang Chen; and Machine Learning for Data-Driven Optimization, organised by Xilu Wang, Xiangyu Wang, Shiqing Liu, and Yaochu Jin.
- AJCAI 2023 included three tutorials, held on November 28: Reinforcement Learning for Automated Negotiation Supply Chain Management League as an Example, presented by Yasser Mohammad; Towards Communication-Efficient and Heterogeneity-Robust Federated Learning, presented by Guodong Long and Yue Tan; and Decoding

the Grammar of DNA Using Natural Language Processing, presented by Tyrone Chen and Sonika Tyagi.
- AJCAI 2023 included a PhD Forum, held on November 28, to mentor and assist post-graduate students developing their research, with mentorship provided by research leaders. Limited travel support was provided.

We especially appreciate the work of the members of the Program Committee and the external reviewers for their expertise and tireless effort in assessing the papers within a strict timeline. We are also very grateful to the members of the Organising Committee for their efforts in the preparation, promotion, and organisation of the conference, especially the General Chairs, Dacheng Tao, Sally Cripps, and Janet Wiles, for coordinating the whole event.

Lastly, we thank the National Committee for Artificial Intelligence of the Australian Computer Society; Springer, for the professional service provided by the Lecture Notes in Artificial Intelligence editorial and publishing teams; and our conference sponsors: the Australian Computer Society; the Defence Artificial Intelligence Research Network; Pioneer Computers; the School of Computer Science at the University of Sydney; the School of Electrical Engineering and Computer Science at the University of Queensland; the Human Technology Institute at the University of Technology Sydney; the Adelaide University; and the UNSW AI Institute.

October 2023
Tongliang Liu
Miao Xu
Geoff Webb

Organization

General Chairs

Dacheng Tao   The University of Sydney, Australia
Sally Cripps   University of Technology Sydney, Australia
Janet Wiles   The University of Queensland, Australia

Program Chairs

Tongliang Liu   The University of Sydney, Australia
Miao Xu   The University of Queensland, Australia
Geoff Webb   Monash University, Australia

Proceedings Chairs

Weitong Chen   The University of Adelaide, Australia
Lin Yue   The University of Newcastle, Australia
Dadong Wang   Data61, Australia

Senior Program Committee

Jing Jiang   University of Technology Sydney, Australia
Mingyu Guo   The University of Adelaide, Australia
Jonathan Kummerfeld   University of Sydney, Australia
Hua Zuo   University of Technology Sydney, Australia
Shuo Chen   RIKEN, Japan
Zhongyi Han   Mohamed Bin Zayed University of Artificial
                           Intelligence, United Arab Emirates

Runnan Chen   The University of Hong Kong, China
Jingfeng Zhang   University of Auckland, New Zealand
Feng Liu   The University of Melbourne, Australia
Huong Ha   RMIT University, Australia
Soyeon Han   University of Western Australia, Australia
Zhanna Sarsenbayeva   The University of Sydney, Australia
Mingming Gong   The University of Melbourne, Australia
Yu Yao   Usyd
Yuxuan Du   The University of Sydney, Australia

Clément Canonne	University of Sydney, Australia
Miaomiao Liu	Australian National University, Australia
Dawei Zhou	Xidian University, China
Yadan Luo	University of Science and Technology of China, China
Xiaobo Xia	The University of Sydney, Australia
Guanfeng Liu	Macquarie University, Australia
Zhen Fang	University of Technology Sydney, Australia
Hien Nguyen	University of Queensland, Australia

Program Committee

Ravneet Singh Arora	Block Inc, USA
Adnan Mahmood	Macquarie University, Australia
Yue Yuan	Shandong University, China
Seyedamin Pouriyeh	Kennesaw State University, USA
Yexiong Lin	The University of Sydney, Australia
Jiahui Gao	The University of Hong Kong, China
Xianzhi Wang	University of Technology Sydney, Australia
Zhuo Huang	Nanjing University of Science and Technology, China
Alex Chu	Beihang University, China
Ruihong Qiu	The University of Queensland, Australia
Qingzheng Xu	National University
Qiang Qu	The University of Sydney, Australia
Lynn Miller	Monash University, Australia
Zhuonan Liang	The University of Sydney, Australia
Kun Han	The University of Queensland, Australia
Tim Miller	The University of Queensland, Australia
Zhuoxiao Chen	The University of Queensland, Australia
Kun Wang	University of Technology Sydney, Australia
Changqin Huang	South China Normal University, China
Peng Yuwei	Wuhan University, China
Brendon J. Woodford	University of Otago, New Zealand
Weihua Li	Auckland University of Technology, New Zealand
Mingzhe Zhang	The University of Queensland, Australia
Peter Baumgartner	CSIRO, Australia
Manolis Gergatsoulis	Ionian University, Greece
Dianhui Wang	La Trobe University, Australia
Jianan Fan	University of Sydney, Australia
Xueping Peng	University of Technology Sydney, Australia
Kairui Guo	University of Technology Sydney, Australia

Zehong Cao	University of South Australia, Australia
Wenhao Yang	Nanjing University, China
Yi Gao	Southeast University, China
Yi Mei	Victoria University of Wellington, New Zealand
Chenhao Zhang	University of Queensland, Australia
Youquan Liu	Hochschule Bremerhaven, Germany
Wenhua Zhang	Shanghai University, China
Yu Yao	MBZUAI, UAE & CMU, USA
Hao Hou	Nanjing University of Science and Technology, China
Yuan Liu	The University of Hong Kong, China
Jianlong Zhou	University of Technology Sydney, Australia
Ran Wang	University of Technology Sydney, Australia
Jun Wang	The University of Sydney, Australia
Weijia Zhang	The University of Newcastle, Australia
Zhuoyun Ao	Defence Science and Technology Organisation
Xinheng Wu	University of Technology Sydney, Australia
Abdul Sattar	Griffith University, Australia
Daokun Zhang	Monash University, Australia
Ge-Peng Ji	Wuhan University, China
Dongting Hu	The University of Melbourne, Australia
Chengbin Du	The University of Sydney, Australia
Ying Bi	Victoria University of Wellington, New Zealand
Rafal Rzepka	Hokkaido University, Japan
Cong Lei	The University of Sydney, Australia
Yue Tan	University of Technology Sydney, Australia
Hongwei Sheng	The University of Queensland, Australia
M. A. Hakim Newton	University of Newcastle, Australia
Shaokun Zhang	Penn State University, USA
Pengqian Lu	The University of Sydney, Australia
Peng Yan	Nanjing University of Post and Telecommunication, China
Weidong Cai	The University of Sydney, Australia
Huan Huo	University of Technology Sydney, Australia
Yuhao Wu	The University of Sydney, Australia
Rui Dai	University of Science and Technology of China, China
Fangfang Zhang	Victoria University of Wellington, New Zealand
Xiaobo Xia	The University of Sydney, Australia
Giorgio Gnecco	IMT - School for Advanced Studies, Lucca, Italy
Yu Zheng	The Chinese University of Hong Kong, China
Ickjai Lee	James Cook University, Australia

Jiepeng Wang	The University of Hong Kong, China
Qizhou Wang	Hong Kong Baptist University, China
Chen Liu	University of Technology Sydney, Australia
Yuanyuan Wang	The University of Melbourne, Australia
Wei Duan	The Australian Artificial Intelligence Institute (AAII), and University of Technology Sydney, Australia
Aoqi Zuo	The University of Melbourne, Australia
Yiming Ren	ShanghaiTech University, China
Stephen Chen	York University, Canada
Wenjie Wang	The University of Melbourne, Australia
Zhiyuan Li	University of Sydney, Australia
Tao Shen	Microsoft, China
Guangzhi Ma	University of Technology Sydney, Australia
Haodong Chen	The University of Sydney, Australia
Yu Lu	University of Technology Sydney, Australia
Angus Dempster	Monash University, Australia
Jing Teng	North China Electric Power University, China
Yawen Zhao	The University of Queensland, Australia
Harith Al-Sahaf	Victoria University of Wellington, New Zealand
Pengxin Zeng	Sichuan University, China
Hangyu Li	Xidian University, China
Huaxi Huang	CSIRO, Australia
Bernhard Pfahringer	University of Waikato, New Zealand
Huiqiang Chen	University of Technology Sydney, Australia
Xin Yu	University of Technology Sydney, Australia
Yanjun Zhang	University of Technology Sydney, Australia
Bach Nguyen	Victoria University of Wellington, New Zealand
Peng Mi	Xiamen University, China
Jiyang Zheng	University of Sydney, Australia
Rundong He	Shandong University, China
Shikun Li	Chinese Academy of Sciences, China
Kevin Wong	Murdoch University, Australia
Xiu-Chuan Li	Chinese Academy of Science, China
Jianglin Qiao	Western Sydney University, Australia
Maurice Pagnucco	The University of New South Wales, Australia
Bing Wang	The University of New South Wales, Australia
Zhaoqing Wang	The University of Sydney, Australia
Mark Reynolds	The University of Western Australia, Australia
Xuyun Zhang	Macquarie University, Australia
Zige Wang	Peking University, China
Chang Wei Tan	Monash University, Australia

Muyang Li	The University of Sydney, Australia
Guangyan Huang	Deakin University, Australia
Liangchen Liu	Xidian University, China
Nayyar Zaidi	Deakin University, Australia
Erdun Gao	The University of Melbourne, Australia
Chuyang Zhou	The University of Sydney, Australia
Shaofei Shen	The University of Queensland, Australia
Yixuan Qiu	The University of Queensland, Australia
Jianhua Yang	UWS, Australia
Keqiuyin Li	University of Technology Sydney, Australia
Yanjun Shu	Harbin Institute of Technology, China
Lingdong Kong	National University of Singapore, Singapore
Jingyu Zhang	City University of Hong Kong, China
Sung-Bae Cho	Yonsei University, South Korea
Shuxiang Xu	University of Tasmania, Australia
Wan Su	Shandong University, China
Markus Wagner	The University of Adelaide, Australia
Xiaoying Gao	Victoria University of Wellington, New Zealand
William Bingley	The University of Queensland, Australia
Sishuo Chen	Peking University, China
Hao Sun	Shandong University, China
Ming Zhou	Hefei University of Technology, China

Sponsors

Contents – Part II

Genetic Algorithm

Contents – Part I

Computer Vision

Machine Learning and Data Mining

Optimization

Medical AI

Knowledge Representation and NLP

Collaborative Qualitative Environment Mapping

Adeline Secolo[1], Paulo E. Santos[2]([☒]) [iD], Patrick Doherty[3,4] [iD],
and Zoran Sjanic[3,5]

[1] Centro Universitario FEI, São Paulo, Brazil
[2] College of Science and Engineering, Flinders University, Adelaide, Australia
paulo.santos@flinders.edu.au
[3] Linköping University, Linköping, Sweden
[4] Mahasarakham University, Mahasarakham, Thailand
[5] Saab AB, Linköping, Sweden

Abstract. This paper explores the use of LH Interval Calculus, a novel
qualitative spatial reasoning formalism, to create a human-readable representation of environments observed by UAVs. The system simplifies
data from multiple UAVs collaborating on environment mapping. Real
UAV-captured data was used for evaluation. In tests involving two UAVs
mapping an outdoor area, LH Calculus proved effective in generating a
cohesive high-level description of the environment, contingent on consistent input data.

Keywords: Knowledge representation · multi-robot systems ·
mapping

1 Introduction

Semi-autonomous or autonomous robots are rapidly being integrated into society, with application scenarios that vary from delivery systems to complex multi-
robot military missions. Specific aspects of interaction such as cooperation and
collaboration raise the ability of heterogeneous teams composed of humans and
different types of robots to solve difficult problems that rely on real time data
analysis and quick responses. This is particularly relevant to teams composed
of various Unmanned Aerial Vehicles (UAVs) and human operators, where a
more human-like way of presenting the UAV's sensor information could lead to
more effective means of interaction with humans to achieve mission goals [4]. The
present paper investigates the application of a novel qualitative spatial reasoning
formalism, Length-Height (LH) Interval Calculus, to provide a human-readable
representation of an environment as observed by groups of UAVs.

Qualitative spatial reasoning (QSR) [9,14] is a field of study that focuses on
understanding and reasoning about spatial relationships between objects or entities without using precise numerical measurements. Instead of relying on exact

T. Liu et al. (Eds.): AI 2023, LNAI 14472, pp. 3–15, 2024.
https://doi.org/10.1007/978-981-99-8391-9_1

measurements, qualitative spatial reasoning aims to capture the inherent spatial properties and relationships, such as topological, directional, and proximity information, using descriptive terms like "above", "below", "near", or "far". By employing a set of conceptual frameworks, qualitative spatial reasoning enables us to model and reason about spatial knowledge, facilitating applications in various domains such as robotics, computer vision, geographic information systems, and artificial intelligence in general. This paper builds upon two key QSR formalisms (and their extensions): the Region Connection Calculus (RCC) [12] and the Allen Interval Algebra [1]. The former provides a systematic approach to describe and reason about the spatial relations between pairs of regions based on the fundamental principle of connectivity; whereas the latter defines possible relations between intervals, as introduced in the next section.

The main contributions of this work are the following: (i) A novel Qualitative Spatial Reasoning (QSR) formalism is proposed (LH interval calculus) aiming to provide a qualitative description of a scene as observed by UAVs from distinct viewpoints (in Sects. 2 and 3 below); (ii) The LH interval calculus lies at the foundation of the Collaborative Qualitative Environmental Mapping (CQEM) procedure (in Sect. 4 below); (iii) CQEM is tested in an outdoors environment with real UAVs (Sect. 5) which, to the best of our knowledge, is the first evaluation of a multi-agent QSR formalism in a physical domain.

2   Qualitative Spatio-Temporal Reasoning

In this work the elementary spatial entities are the bounding boxes around the objects as picked out in images of an environment observed from aerial viewpoints. The bounding boxes define spatial regions (their boundaries and the area inscribed in them) and intervals (defined by the sides of the boxes).

The basic relations between spatial regions are described in RCC through the primitive *Connected (C(x,y))*, where two non-empty regions, x and y, of some topological space, are connected if and only if their topological closures share at least one common point [12]. An important subset of RCC, called RCC8, has the following eight base relations: disconnected (DC), externally connected (EC), partially overlapping (PO), equal (EQ), tangential proper part (TPP), non-tangential proper part (NTPP), and the inverses of TPP and NTPP. Figure 1a shows the eight qualitative spatial relations covered by RCC8, and the possible smooth transitions between them, illustrated by regions x and y on a conceptual neighbourhood diagram [9]. In this work, in order to define a representation system for interpreting the images collected by the UAVs, the RCC8 theory is associated with Allen's Interval Algebra [1]. This algebra defines qualitative relations between intervals, resulting in a set of 13 jointly-exhaustive and pairwise-disjoint base relations: $A_{int} = \{p, m, o, s, d, f, pi, mi, oi, si, di, fi, eq\}$. Figure 1b shows the 13 Allen's relations for the intervals x and y.

An extension of Allen's Interval Algebra called *Rectangle Calculus* (RC) [3] was developed to represent relations between rectangular regions, such as *left of, right of, above, below,* and *overlapping,* in a coordinate system. These relations

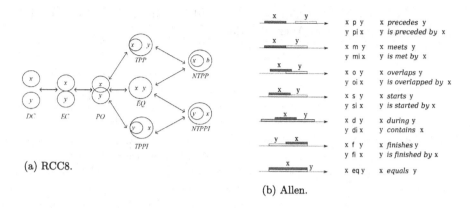

(a) RCC8.

(b) Allen.

Fig. 1. Base QSR formalisms used in this work.

can be used to reason about various properties of spatial objects, such as containment, adjacency, and overlap. In this work, the bounding boxes that enfold the objects present in an image are considered rectangles, as defined in RC.

Allen's Interval Algebra was also extended in the so called *Interval Occlusion Calculus* (IOC) [13], developed to represent object occlusion from multiple viewpoints. IOC is a qualitative description of a set of basic relations between pairs of objects observed from a viewpoint, given the lines of sight to the objects. Considering that two objects A and B can be observed from a viewpoint Σ (cf. Fig. 2a), the function *image* maps a physical body to the region representing the body's image as perceived by Σ (e.g., $a = image(A, \Sigma)$ and $b = image(B, \Sigma)$, where a and b are the images of A and B, respectively). An image of an object from Σ is defined as the set of projected half-lines originating at Σ, and contained in its field of view, that intersects the object. The present paper extends this formalism using ideas from RC in order to allow the description of scenes from various global viewpoints, obtained by distinct observers.

An important notion introduced in [13], that is essential to the present work, is the concept of layered interval, denoted as $I = (I_a, \ell)$, given a linear ordering L. Here, $I_a = (x_1, x_2)$, where x_1 and x_2 represent the lower and upper limits of I, respectively, and ℓ is the layer of I. The layer is a linear ordering of intervals as observed from a specific viewpoint, representing the depth of each interval with respect to the observer's location. The following two functions on layered intervals are of relevance here: $ext(I)$, which maps I to its extent (i.e., its upper and lower limits), and $\ell(I)$, which maps I to its layer. In this context, each interval has a unique associated ℓ and the ordering of intervals with respect to ℓ indicates their proximity to the observer. The closer an object is to the observer, the greater is its associated ℓ (thus, ℓ is analogous to the disparity in stereo images). IOC use the following notation to denote the relation r between two distinct intervals, I and J: Ir^+J if $\ell(I) > \ell(J)$ and Ir^-J if $\ell(I) < \ell(J)$. Allen's Interval Algebra relations are, then, generalised with the notion of an observer's viewpoint and the layer of an interval, as exemplified in Fig. 2a. It is worth noting here that the relations o^+, o^- (and their inverses) are representing *object occlusion* rather than Allen's overlap.

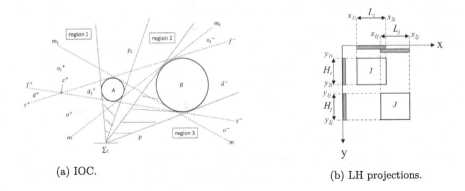

(a) IOC.   (b) LH projections.

Fig. 2. Set of Allen's relations considering RCC constraints.

Analogous to IOC, the LH interval calculus proposed in this work aims to identify the objects relations from multiple viewpoints. However, instead of occlusion, this formalism deals with the different perspectives of multiple agents from global viewpoints, as presented in the next section.

3   LH Interval Calculus

LH Interval Calculus represents the observed objects in a 2D frame of reference where the projections of the object's bounding boxes' sides are treated as layered intervals. In this work, the global viewpoints are defined from UAVs equipped with a camera, a GPS and a compass. The viewpoints of each UAV are represented as pairs $\Sigma_i = (g_i, v_i)$, $i \in \mathbb{N}$, where g_i is the global position of the observer (UAV) indicated by the GPS, and v_i is a unit vector representing the direction of flight given by the compass. In this context, the function $j = image(J, \Sigma)$ maps an object J perceived at Σ to its bounding box (a 2D region j) on the LH reference frame. The LH reference frame has origin at the upper left corner of the scene, with abscissa (x axis) oriented from right to left and ordinate (y axis) oriented from top to bottom, in conformity with the process of reading the scene by the cameras installed in the UAVs. Figure 2b shows an example of LH projections of two regions $i = image(I, \Sigma)$ and $j = image(J, \Sigma)$, observed from Σ, where H_i (H_j) is the layered interval of i (j) in the y axis, and L_i (L_j) is the layered interval of i (j) in the x axis.

Inspired by the Rectangle Calculus [3], the projections of pairs of objects at the x axis can be represented by one of the 13 basic Allen's relations (analogously to the y axis). For example, given a pair of objects I and J (and their respective images $i = image(I, \Sigma)$ and $j = image(J, \Sigma)$), their observer relative positions can be represented by the sentence i (r_x, r_y) j, where r_x is the Allen relation between the projections of I and J on the x axis and r_y is the analogous on the y axis. E.g., the objects in Fig. 2b are *disconnected* (DC), and i *overlaps* j in the x axis and i *precedes* j in the y axis.

The LH Calculus is formally defined in terms of RCC8 and Allen's relations, extended with the IOC notion of layered interval as described below. Let i be the bounding box of an object I ($i = image(I, \Sigma)$), as seen from a global viewpoint Σ, the functions $ext_L(i)$ and $ext_H(i)$ map i to the *extents* of its layered intervals L and H, as projected to the x and y axes: $ext_L(i) = (x_1, x_2)$; and $ext_H(i) = (y_1, y_2)$. Similarly to IOC, the LH interval calculus also has the function $\ell(i)$ mapping each projected interval extension ($ext_L(i)$ or $ext_H(i)$) to its layer. For instance, $\ell(ext_L(i))$ represents the relative distances (depth) between the observer and object I with respect to the L projection of I's bounding box. The layer, in this case, can be used to represent (or reconstruct) the description of the scene from a global view to the various possible local viewpoints.

Let I and J be two objects in a scene, and $i = image(I, \Sigma)$, $j = image(J, \Sigma)$ be their respective bounding boxes, as perceived from a global viewpoint Σ. The following list define the LH relations associated to the x axis. Analogous relations can be drawn wrt the y axis, and were omitted here for brevity. The combined set of LH relations for both x and y axes define the LH-calculus.

- i p_x j : Σ, read as "i *precedes* j from Σ" is defined as *(i) DC (j)* $\wedge$ $ext_L(i)$ p_x $ext_L(j)$;
- i pi_x j : Σ, read as "i *is preceded by* j from Σ" is defined as *(i) DC (j)* $\wedge$ $ext_L(j)$ p_x $ext_L(i)$;
- i m_x j : Σ, "i *meets* j from Σ": *(i) EC (j)* $\wedge$ $ext_L(i)$ m_x $ext_L(j)$;
- i mi_x j : Σ, "i *is met by* j from Σ": *(j) EC (j)* $\wedge$ $ext_L(j)$ m_x $ext_L(i)$;
- i o_x j : Σ, "i *overlaps* j from Σ": *(i) PO (j)* $\wedge$ $ext_L(i)$ o_x $ext_L(j)$;
- i oi_x j : Σ, "i *is overlapped by* j from Σ": *(i) PO (j)* $\wedge$ $ext_L(j)$ o_x $ext_L(i)$;
- i s_x j : Σ, "i *starts* j from Σ": *(i) TPP (j)* $\wedge$ $ext_L(i)$ s_x $ext_L(j)$;
- i si_x j : Σ, "i *is started by* j from Σ":*(i) TPPI (j)* $\wedge$ $ext_L(j)$ s_x $ext_L(i)$;
- i f_x j : Σ, "i *finishes* j from Σ": *(i) TPP (j)* $\wedge$ $ext_L(i)$ f_x $ext_L(j)$;
- i fi_x j : Σ, "i *is finished by* j from Σ": *(i) TPPI (j)* $\wedge$ $ext_L(j)$ f_x $ext_L(i)$;
- i d_x i : Σ, "i *is during* j from Σ": *(i) NTPP (j)* $\wedge$ $ext_L(i)$ d_x $ext_(j)$;
- i di_x j : Σ, "i *contains* j from Σ": *(i) NTPPI (j)* $\wedge$ $ext_L(j)$ d_x $ext_L(i)$;
- i eq_x j : Σ, "i *is equal to* j from Σ": *(i) EQ (j)* $\wedge$ $ext_L(i)$ eq_x $ext_L(j)$.

Each RCC8 relation constrains the number of Allen's relations possible between the LH extents of the observed objects. In case the expected relation is identified in only one axis, any other Allen relation could hold on the extents related to the other axis, resulting in a set of possible pairs of Allen's relations. For instance, Fig. 3a shows the set of possible combinations of Allen's relations for two regions identified as DC, where the projections on the x axis represent the LH relation *precedes*, while the projections on y axis could be any of the basic relations of the Allen's Interval Algebra. Similar cases can be made for two regions that are EC, PO, $TPPI$, $NTPP$ and $NTPPI$, as shown in Figs. 3b, 3c, 3d, 3e, and 3f respectively. The Eq results in the pair (eq_x, eq_y), and the image will be considered as obtained from a single object. Enclosing Allen's relations into RCC8 optimises the possible descriptions, as just a subset of Allen's relations can be instantiated for each RCC8 relation.

These definitions will be used in the Collaborative Qualitative Environment Mapping introduced in the next section.

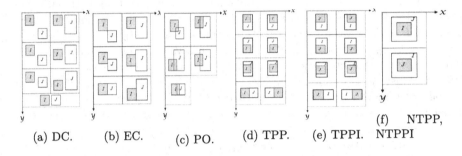

(a) DC.   (b) EC.   (c) PO.   (d) TPP.   (e) TPPI.   (f)   NTPP, NTPPI

Fig. 3. Set of Allen's relations considering RCC constraints.

4   Collaborative Qualitative Environmental Mapping

Environment mapping can be done more efficiently when multiple agents in a team collaborate to scan the same domain. The cardinal direction of each agent was used to combine the spatial information collected by the UAVs. Each agent's compass determines a vector v that represents the flight orientation of a moving viewpoint Σ. To qualitatively classify the set of symbolic directions, the North-East-Down geographic frame of reference was used in the drones to inform latitude, longitude and height to the operator. Eight cone-shaped areas of equal size were thus defined, covering a range of $\pm$ 22,5° for each cardinal direction: $North(N) = 0°$, $North\ East(NE) = 45°$, $East(E) = 90°$, $South\ East(SE) = 135°$, $South(S) = 180°$, $South\ West(SW) = 225°$, $West(W) = 270°$ and $North\ West(NW) = 315°$. In this work the information from the UAVs was delivered according to expanding squares or lawnmower scan methods, where the vector v changed during the scanning of the area, rotating approximately 90°, 180° or 270° for each part of the scanned route. The information represented by LH Interval Calculus for two observed objects I and J by a drone d is represented as $i\ (^{v_w}r_x, {}^{v_w}r_y)\ j$: Σ_d, where the direction of flight vector v_d was added as an index w representing the direction of flight.

Supposing a first agent (1) flying in a v_w direction, and a second agent (2) flying in a $v_w + 90°$ direction, observing the same objects I and J, a correspondence can be constructed between $i\ (^{v_w}r_x, {}^{v_w}r_y)\ j$: Σ_1 and $i\ (^{v_w+90°}r_x,$ ${}^{v_w+90°}r_y)\ j$: Σ_2. Following the same idea, an observer flying in a $v_w + 180°$ direction, observing the same objects I and J as another agent flying in a v_w direction, a correspondence can be constructed between $i\ (^{v_w}r_x, {}^{v_i}r_y)\ j$: Σ_1 and

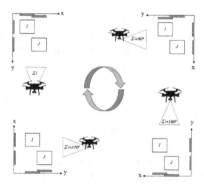

Fig. 4. LH projections according to agents orientation of flight.

i $\left(^{v_w}+180°r_x, \, ^{v_w}+180°r_y\right) j$: Σ_2. Similarly, for an observer flying in the $v_{w+270°}$ direction, observing the objects i and j as the first agent flying in a v_w direction, a correspondence can be constructed between i $\left(^{v_w}r_x, ^{v_w}r_y\right) j$: Σ_1 and i $\left(^{v_w}+270°r_x, \, ^{v_w}+270°r_y\right) j$: Σ_2. This correspondence analysis can be applied in a consistency check procedure to provide a consistent representation of the relations between objects from multiple points of view. Figure 4 shows that, if I $(oi_x, \, p_y)$ J: Σ_1 is perceived by an agent flying toward the v_i direction, then the same scene, observed from a Σ_2 flying to $v_i + 180°$ direction, is described as I $(o_x, \, pi_y)$ J: Σ_2.

The observed environment contains buildings of different sizes, cars of different colours and an aeroplane, that were scanned by two or more agents, each of which was responsible to fly over a portion of the environment. The area of this portion was defined according to the capabilities of each agent, as speed of flight and the resolution of the cameras installed in the drones, providing only a partial view of the domain. Thus, subsets of objects were observed by some agents but not by others. In the case that at least two agents identify object's in common, a combination of this information was conducted using the composition tables of RCC8 and Allen Calculus (that proved the transitive closure of pairs of relations), and applying path consistency in the constraint network defined by the LH relations obtained from perceived objects.

Considering a constraint network (N, C), where N is the set of vertices representing domain elements, and C is the set of constraints representing LH relations between pairs of objects, it is possible to verify the existence of a consistent scenario by imposing algebraic closure on the network of constraints. A network (N, C) is algebraically closed if every three edge-connected vertices $(i, \, j, \, k) \in N^3$ is consistent with the composition $C(i,j) \subseteq C(i,k) \circ C(k,j)$, as given by the RCC8 and Allen Interval Algebra composition tables [9].

Off-the-shelf algorithms for object detection and tagging were used to identify the objects perceived and provide their bounding boxes. The detected objects were tagged according to the GPS position of the agent that perceived it. To each pair of objects detected, the IOC relations (r_x, r_y) were obtained, representing the relation of the projections of objects' intervals into x- and y-axis.

The partial views of the environment obtained by the multiple agents were combined applying the concept developed in [3], defining the composition of pairs of qualitative relations as given by $(A, B) \circ (C, D) = (A \circ C) \times (B \circ D)$, where A and C are a set of relations r_x, B and D are a set of relations r_y, "$\circ$" is the composition operator and "$\times$" is the Cartesian product.

For instance, let the information captured by Σ_1 have two objects a and b, in which their observed occupancy regions A and B ($A = image(a, \Sigma)$, $B = image(b, \Sigma_1)$) are PO; let also that a distinct observer Σ_2 perceives two objects b and c ($B = image(b, \Sigma_2)$, $C = image(c, \Sigma_2)$) so that their relation is $B\ DC\ C$ and that Σ_1 and Σ_2 are flying in the same direction v_i. Let's assume also that the LH relations of the projections of A and B is the pair (d_x, o_y) and the relations of B and C is the pair (p_x, mi_y). According the to RCC8 composition table, $PO \circ DC = \{DC, EC, PO, TPPI, NTPPI\}$, representing the possible relations between A and C. The composition of the pairs $(d_x, o_y) \circ (p_x, mi_y)$ is then give by $\{((p_x, oi_y), (p_x, si_y), (p_x, di_y))\}$, according to the composition table of Allen's interval algebra, representing set of possible pairs of relations of the projections of the objects A and C. Observing that the only possible relation in the x axis is *precedes* for the three possible pairs of relations, it is possible to infer that $A\ DC\ C$, and thus the LH relations between A and C would be one pair of the following set $((p_x, oi_y), (p_x, si_y), (p_x, di_y))$.

The Collaborative Qualitative Environmental Mapping (CQEM) method, introduced in this section, is based on path consistency in binary constraint networks, whose computational complexity is polynomial or, more precisely, $O(n^3\ a^3)$, where n is the number of nodes and a is the size of the domain [10].

5   Experiments

Three data collection flights were performed at *Motala Flygklubb*, Sweden, using two distinct UAV platforms: a DJI Matrice 600 Pro research platform and a DJI Matrice 100. Both platforms were equipped with Intel NUC computers using Core i7-7567U processors, 16 GB of memory, and 500 GB SSD storage. DJI Zenmuse Z3 cameras were used to collect video and images during the test flights. The flight scanning patterns were automatically generated to cover a designated area. The UAVs took off manually and, after reaching a safe altitude, they flew autonomously over the region and performed the scan. The flights were performed at three different altitudes: 30, 50 and 80 m above the ground level. The videos and photos collected registered partial views of the environment taken from different directions. There were 14 distinct objects in the observed domain in total. A snapshot of the environment observed is shown in Fig. 5.

Each drone flew over a distinct portion of the area, but capturing images of some objects in common, in order to use those objects as references to infer the position of all other objects that are not in its field of view. Information could be exchanged by the agents (including the ground operator) via Wi-Fi. In general, it is possible to transmit images, but a large broadband is necessary for that, which is often not available in general. Thus, the drones exchanged short text sentences in the format "$VisualObject_1\{R\}VisualObject_2 : \Sigma_i, v$", where R is a set of LH relations picked out by the agents.

Two tests were conducted with the images collected by the drones. In the first test, the images were manually annotated using the VGG Image Annotator (VIA) [5]; in the second test, YOLOv5 [8] was trained with the MSCOCO data set to automatically identify the observed objects. From a total of 223 pictures, a set of 102 images was chosen to test our method, as these contained two or more objects in the scene. SparQ toolbok[1] was used in the development of the LH algorithm that abstracts the input data and form a qualitative representation (in terms of LH relations) of the perceived world scene. In both cases, each detected rectangle in the scene was considered an object. The output of the algorithm was a qualitative relation for the objects in the axis x and a qualitative relation in the axis y. The algorithm was evaluated according to its accuracy ($accuracy = \frac{correct\ results}{all\ results}$) in identifying relations, and also in combining views, with respect to a manually annotated Gold standard set. The Gold standard considered all RCC8 and IOC relations possible with respect to pairs of perceived objects in the scene. In total, this table had 364 pairs of relations.

The values used in the LH algorithm were the mid-position of each rectangle in the axis x and y for each observed object, and their rectangular x and y extents, as obtained by QSTRLib functions. These coordinates provided the position of the rectangle and hence the observer-relative position of the object that the rectangle represents. The algorithm had two outputs for each pair of objects: the RCC8 relation and the LH relation wrt the axis x and y. The combination of these two outputs was tested according to the constraints that define the LH Interval Calculus. From selected 102 images a total amount 996 LH interval Calculus sentences were evaluated.

The VGG identification of the objects resulted in 100% of correct responses comparing to the Gold standard. The automatic object identification using YOLOv5, however, had the following difficulties: the object's shadows were identified as part of the object; the black roof of the Object 6 (Fig. 5) (a balcony) was excluded from the object's bounding box; Object 13 is composed by two buildings, one covered by the black roof with the letters "DZ" and the other covered by the red and grey roof, but they were identified as a single object. Consequently, Object 14 was not identified.

LH Interval Calculus presented 92% of accuracy for the relations between Object1, an aeroplane, and the other 13 objects of the scene. This means that for all images where Object 1 was identified, 92% of the relations between Object1 and any other objects identified in the scene is correct. The 8% of wrong answers refers to the relations between Object 1 and Object 14, not identified by the YOLOv5 algorithm. Object 2 and Object 3, are small buildings, and presented 77% of accuracy for the LH interval Calculus sentences. The 23% of incorrect

---

[1] https://github.com/dwolter/SparQ.

responses were due to Objects 14, 13 and 5. The global accuracy of the LH Interval Calculus using automatic annotation of objects was 77%.

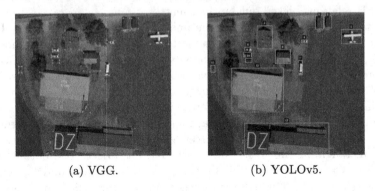

(a) VGG.   (b) YOLOv5.

Fig. 5. Output of object detection methods.

The CQEM method was evaluated using the LH relations that were automatically identified. In order to combine pairs of partial views from distinct observers, the cardinal direction of the flight was used to "normalise" the relations obtained by both agents (i.e., the relations obtained from multiple flight directions were transposed into a single direction: North). Then, the partial views of the scene from distinct observers, that had observed two common objects, had their LH relations extracted and used in the consistency checking procedure described above. From the 102 images showing two or more objects in the scene, 30 were chosen to test the CQEM procedure proposed in this paper. The result of executing the partial view combination of pairs of images containing one object in common resulted in a set of possible relations between the objects that appear in one, but not in both images. In the tests considering VGG identification, the reference relation of the Gold standard was present in 100% of the resulting sets of possible relations, as given by the RCC8 and Allen interval calculus composition tables. Some image pairs that contained three objects, two of which were present in one image and two in another image (whereas each image contained one object not present in the other) resulted in the exact pair of LH relations (57% of the combinations). From the remaining 43% of such images, the result was a set of relation pairs, 78% of which resulted in the exact relation. Tests with using YOLOv5 for the automatic identification of objects, showed a very low accuracy. This was due to the misclassifications described above which, given the nature of the consistency checking procedure used here (verifying consistency of each pair of perceived objects), forced an empty set as result.

6   Related Work

This paper addresses the use of qualitative information to solve a collaborative map building task, which is a relatively unexplored area in Collaborative

SLAM (CSLAM) research [2]. While CSLAM primarily relies on numerical data, our work highlights the potential of qualitative representations to enhance the human-robot interaction.

Existing literature extensively covers controlling robots with high-level routines inspired by natural language terms [16], but largely overlooks the incorporation of modern qualitative spatial reasoning calculi [9,14]. A few exceptions include the integration of spatial qualitative knowledge from various robots to guide a sensory deprived robot [11], where an analogous version of a Kalman filter was developed upon a QSR formalism aiming at bridging the gap between robot and human localisation. An extension of this work led to a qualitative case-based reasoning system for selecting the best strategy for a team of robots in a soccer domain [7]. The ability of QSR for creating human-like decisions was also explored in [6] for identifying high-level manipulation plans from human demonstrations that are transferred to a robotic manipulator. The results of this research showed that the human-like planner outperformed state-of-the-art trajectory optimisation algorithms. These ideas were also used for visual sense-making in autonomous driving systems with positive results [15].

These related works, however, do not take into account the task of collaborative map building using qualitative information, whereas the main contribution of the present paper is the exploration of QSR formalisms for collaborative environment mapping from the partial views obtained from various UAVs. To that end, the present paper introduced the LH-calculus, which is a combination of three classical QSR formalisms, RCC [12], which was used to define regions of interest (simplifying the reasoning), Allen's algebra [1], used for defining the relations between object's projections, and the Interval Occlusion Calculus [13] which provided the layered interval definitions. Ignoring the latter, the resulting formalism resembles a constrained version of the Rectangle Calculus [3]. However, it's important to note that the inclusion of the concept of viewpoint in the LH-definitions renders these similarities superficial.

7   Conclusion and Future Work

Effective communication and interaction among human-robot teams in shared tasks is not yet fully resolved. In missions where precision is not crucial (e.g., tasks with human agents using qualitative language), employing qualitative data has shown promise to address these challenges. This paper introduced two key contributions in this context: (i) LH Calculus, a qualitative spatial formalism for information exchange among agents with different viewpoints; and, (ii) a collaborative qualitative environment mapping procedure that combines individual perceptions of multiple robots providing a more comprehensive understanding for humans and autonomous agents.

To evaluate the effectiveness of the proposed solutions, real data captured by unmanned aerial vehicles were used in the tests reported in this paper. Fourteen objects were identified using two different methods: manual identification and an automated object identification method. The results indicated that both

LH Interval Calculus and environment mapping procedure are effective in environment mapping using qualitative data, when the input data is correct. However, the latter has a degraded performance when dealing with noisy data, as a result of its reliance on a global consistency checking method. This motivates the development of consistency satisfaction algorithms for QSR that are capable of assimilating sensor noise and inconsistent data. Handling uncertainty in CSP is not a new problem (e.g. [17]). However, to the best of our knowledge, it has not yet been applied to spatial reasoning, or to reasoning about multi-agent perception. Future work should also investigate the application of more accurate automatic object detection and classification methods to preprocess the data, as this seems to be the limiting factor for any reasoning system to be applied on real domains.

Acknowledgements. This work was carried out with the support of CISB, Swedish-Brazilian Research and Innovation Center, and Saab AB and the PhD scholarship provided from the *Coordenação de Aperfeiçoamento de Pessoal em Nível Superior - Brasil (CAPES)* - Finance Code 001. The authors are also indebted to Mariusz Wzorek and Piotr Rudol for the support given during the data collection procedure.

References

1. Allen, J.F.: Maintaining knowledge about temporal intervals. In: Readings in Qualitative Reasoning About Physical Systems, pp. 361–372. Elsevier (1990)
2. Aloui, K., et al.: Systematic literature review of collaborative SLAM applied to autonomous mobile robots. In: IEEE Information Technologies & Smart Industrial Systems (2022)
3. Balbiani, P., Condotta, J.F., Cerro, L.F.D.: Tractability results in the block algebra. J. Log. Comput. **12**(5), 885–909 (2002)
4. Dahiya, A., et al.: A survey of multi-agent human-robot interaction systems. Robot. Auton. Syst. **161** (2023)
5. Dutta, A., Zisserman, A.: The VIA annotation software for images, audio and video. In: Proceedings of the 27th ACM International Conference on Multimedia (2019)
6. Hasan, M., et al.: Human-like planning for reaching in cluttered environments. In: Proceedings of ICRA, pp. 7784–7790 (2020)
7. Homem, T.P.D., et al.: Qualitative case-based reasoning and learning. Artif. Intell. **283**, 103258 (2020)
8. Jocher, G., et al.: ultralytics/yolov5: v7.0 - YOLOv5 SOTA Realtime Instance Segmentation (2022)
9. Ligozat, G.: Qualitative Spatial and Temporal Reasoning. Wiley, Hoboken (2013)
10. Mackworth, A.K., Freuder, E.C.: The complexity of some polynomial network consistency algorithms for constraint satisfaction problems. Artif. Intell. **25**(1), 65–74 (1985)
11. Perico, D.H., Santos, P.E., Bianchi, R.A.C.: Guided navigation from multiple viewpoints using qualitative spatial reasoning. Spat. Cogn. Comput. **21**(2), 143–172 (2021)
12. Randell, D.A., Cui, Z., Cohn, A.G.: A spatial logic based on regions and connection. In: Proceedings of the 3rd International Conference on Principles of Knowledge Representation and Reasoning (KR 1992), pp. 165–176 (1992)

13. Santos, P.E., Ligozat, G., Safi-Samghabad, M.: An occlusion calculus based on an interval algebra. In: 2015 Brazilian Conference on Intelligent Systems (BRACIS), pp. 128–133. IEEE (2015)
14. Sioutis, M., Wolter, D.: Qualitative spatial and temporal reasoning: current status and future challenges. In: Zhou, Z.H. (ed.) Proceedings of IJCAI 2021, pp. 4594–4601 (2021)
15. Suchan, J., Bhatt, M., Varadarajan, S.: Commonsense visual sensemaking for autonomous driving - on generalised neurosymbolic online abduction integrating vision and semantics. Artif. Intell. **299**, 103522 (2021)
16. Tellex, S., Gopalan, N., Kress-Gazit, H., Matuszek, C.: Robots that use language. Ann. Rev. Control Robot. Auton. Syst. **3**(1), 25–55 (2020)
17. Verfaillie, G., Jussien, N.: Constraint solving in uncertain and dynamic environments: a survey. Constraints **10**(3), 253–281 (2005)

Towards Learning Action Models from Narrative Text Through Extraction and Ordering of Structured Events

Ruiqi Li[1]([⊠]), Patrik Haslum[1], and Leyang Cui[2]

[1] Australian National University, Canberra, Australia
{ruiqi.li,patrik.haslum}@anu.edu.au
[2] Tencent AI Lab, Shenzhen, China
leyangcui@tencent.com

Abstract. Event models, in the form of scripts, frames, or precondition/effect axioms, allow for reasoning about the causal and motivational connections between events in a story, and thus are central to AI understanding and generating narratives. However, previous efforts to learn general structured event models from text have overlooked important challenges raised by the narrative text, such as the complex (nested) event arguments and inferring the order and actuality of mentioned events. We present an NLP pipeline for extracting (partially) ordered, structured event representations for use in learning general event models from three large text corpora. We address each of the challenges that we identify to some degree, but also conclude that they raise open problems for future research.

Keywords: event model acquisition · event extraction · event ordering

1 Introduction

Narratives are made up of events. Events tell us what happened, who or what was involved, and in what order. Knowledge about typical events, such as their usual participants, dependencies and effects, is crucial to understanding, and creating, narrative text. Event models can take several forms, from simple scripts or schemas – subsequences of events, parameterised by their participants, that frequently occur together [10] – to models that link each schematic known event to the possible preceding or following events [11], which may be ranked by likelihood [38], to state/transition models that associate to each event the conditions on the world state under which it can occur, and the effects of its occurrence on the state. Model-based reasoning about events/actions[1], studied in AI domains such as reasoning about action and change [39], planning [15] and diagnosis of dynamical systems [20], assumes such a symbolic preconditions-and-effects model of events/actions is available.

Hand-crafting event models with the breadth required for narrative understanding or generation is an enormous task. Hence our interest in learning them from text, a

---

[1] We will use the terms event and action interchangeably. Although a distinction can be made – actions have at least one intentional participant or actor – from the point of view of the models we consider, they are largely the same.

T. Liu et al. (Eds.): AI 2023, LNAI 14472, pp. 16–27, 2024.
https://doi.org/10.1007/978-981-99-8391-9_2

medium that hosts an abundance of human-authored narratives in many styles and genres. We aim to explore the hypothesis that by processing a sufficient volume of text we can extract sufficiently many distinct events with a sufficient number of occurrences to learn meaningful models of them. The problem of learning various kinds of event models from text has been researched for well over a decade [10,36,38], and has recently gained renewed interest in AI planning [13,23,24,27] for several reasons, including narrative generation [17,18]. The latter breaks the problem into two stages: first, extracting structured and ordered (typically as sequences) representations of event mentions from source texts and, second, inducing event models from these examples of event sequences, leveraging existing work on planning model acquisition [e.g., 2,8]. The work we present in this paper also follows this approach. However, we argue that previous work has overlooked several characteristics of narrative text, which raise challenges both for extraction of event mentions from text and for model acquisition.

Event Ordering: In simpler texts, such as recipes or instruction manuals, events/actions are for the most part written in the order they occur [13,18,23,24,27]. In these cases, it is acceptable to assume that the event sequence corresponds to the sequence of event mentions in the text. In narrative texts, however, such linearity is rarely the case. As one point of reference, in the MATRES [32] dataset, which contains 275 news articles manually annotated with temporal relations between events, approximately 58.2% of the events in each article indeed do not occur before all of the events mentioned later in the text.

Olmo et al. [33] also note that inferring plan order from text "... is a feature which is mostly missing in the previous task-specific state of the art", but test the ability of a model to perform such inferences only with three hand-crafted examples.

Events as Arguments: Many verbs can take, or require, a clausal complement, i.e., an argument of the event verb is itself an event. For example, if "she tries to stop them ...", the event verb is "try", and the event "[she] stop them" is its argument. Furthermore, event arguments can be nested: if "she tries to stop them finding out ...", "find out" is an argument of "stop", and the event "[she] stop ([they] find out)" is itself the argument of "try". Events of this kind are frequent in narrative texts: in each of the three datasets we examine, 16.4%, 14% and 14.7%, respectively, of event mentions are arguments or conditions. Distinguishing them is important for both accurately representing the mentioned events, and for identifying those events that actually take place.

Event Actuality: Narrative texts very frequently mention events that are not actually part of the sequence of events that is recounted. For example, consider the sentence "King Louie offers to help Mowgli stay in the jungle if he will tell Louie how to make fire like other humans." (from [18] Figure 5). Although it contains many event verbs ("help", "stay", "tell", "make"), only one of them, "offer", is actually said to happen: the remaining are the argument and condition of the offer. Events that are arguments are only one source of event mentions that do not correspond to actual event occurrences.

Scale: Finally, the size of the model required for general narrative understanding and generation is significantly greater than what has been considered in previous work. For example, the Story Cloze dataset [30], which tests a form of narrative understanding, supposes knowledge of over 2,000 distinct verbs.

In this paper, we present an NLP pipeline for extracting (partially) ordered, structured event representations, that to some degree deals with the above challenges. We apply it to three sets of texts: the movie plot summaries provided by Bamman et al. [3], a subset of articles from the Goodnews dataset [6], and a set of New York Times articles obtained from kaggle[2]. The partially ordered sets of event occurrences produced by our pipeline are meant as a basis for inducing event models. Furthermore, we report a preliminary experiment extracting simple narrative chains (recurring subsequences of connected events). However, our aim is to induce knowledge such as event preconditions and effects. The necessary complexity of the events we extract – nested argument events, conditionals, etc – as well as the scale of the task, are beyond the capabilities of current model acquisition techniques [see, e.g., 2, 8]. We plan to pursue this goal in future work. We will also make the generated dataset public so that other researchers can attempt to tackle the challenge of learning models from it.

2   Related Work

Learning of event models from text has been studied for some time. Chambers & Jurafsky [9, 10] proposed a system for unsupervised induction of parameterised scripts from text; applied to news text from the Gigaword corpus, it generated scripts covering around 1800 verbs. But it is restricted to finding scripts in which one entity, i.e., the protagonist, is common to all events. Tandon et al. [38] analysed movie and television scripts, as well as novels, to extract over 2,000 activities (verb, optional preposition, and object, e.g., "chase car") and link them with typical values for location, time and participants, as well as likely preceding and following activities. However, activity participants are ground, i.e., objects rather than parameters (e.g., agents of "loom in distance" are "door" and "island"). The Never-Ending Language Learning (NELL, http://rtw.ml.cmu.edu/rtw/) project extracted over 50M beliefs, in the form of entity–relation triples, from the web, combined with crowdsourced feedback. However, it is focused on acquiring instance-level facts, and mainly about entities, not general activities. Although it has categories for "event" and different types of actions, the majority refer to specific event instances (e.g., product launches, elections, etc) and information about them is sparse: 99% have no relation but a hypernym and a source URL.

Extracting sequences of actions or events from text has recently gained interest in AI planning [13,23,24,27]. In some instances, this is a precursor to learning the kinds of precondition/effect action models that AI planners require, using planning model acquisition tools. Methods of event extraction vary, from using the dependency parse structure to neural language models and reinforcement learning. However, previous work in this line has overlooked several of the complexities of narrative events. As noted by Olmo et al. [33], they have assumed the order of events is their order of mention. Furthermore, none identify events that are arguments or conditions of other events. For instance, the plan representation generated by Hayton et al. [18] from a synopsis of the movie "The Jungle Book" includes many of the events mentioned in the sentence quoted in the Introduction ("stay in jungle", "tell" and "make fire") even though none of these actually happen in the story.

---

[2] https://www.kaggle.com/datasets/nzalake52/new-york-times-articles.

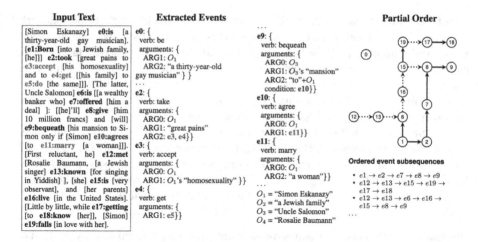

Fig. 1. An example of partially ordered events extracted by the pipeline. Left: Source text. (The mark-up of verbs and arguments of extracted events are not part of the input, just for improving readability) Middle: Extracted event samples. Events that are arguments are in red, conditions are in blue. Right: Predicted partial order between events, and possible event traces. Dashed edges indicate predictions we consider incorrect. (Color figure online)

3 Structured Event Extraction

The ordered event extraction process has three main steps: (1) Extracting event verbs and arguments from the source text. This is done using the BERT-based AllenNLP (https://allenai.org/allennlp) semantic role labelling (SRL) system. We also use POS tagging and dependency parse information, obtained using the Stanford CoreNLP toolkit [25], in several ways. First, the SRL system often mistakes adjectives for verbs (e.g., "reloaded" in "the Matrix reloaded"), and we use the words' POS tags to filter these out. We also filter out modal verbs. Second, the SRL system does not detect phrasal verbs, so we use a rule-based method that relies on the dependency parsing to identify these. (2) Identifying argument and condition events. As mentioned, events may be arguments or conditions of other events. We use the argument structure provided by the SRL system, together with a rule-based method that relies on the dependency parse information, to determine which events are arguments or conditions of other events. We term any event that is not an argument or condition *independent*, and we assume that the non-negated independent events are those that, according to the text, actually take place. This is a simplification, as some argument events can also be actual. (3) Ordering the independent events. We use a temporal relation classifier [31] to predict temporal relations between the pairs of independent events, and inference (transitivity) to extend the partial order. Because the classifier is myopic, i.e., predicts each event pair in isolation, it often induces inconsistent orders (with cyclic precedences). We resolve these by deleting the predicted order with the lowest probability in each cycle.

Event Verb and Arguments. An event mention e consists of a verb or phrasal verb $V(e)$ and a set of labelled arguments $\mathbb{A}(e)$. Verbs are lemmatised. We call any event whose verb lemma is "be" or "have" a *statement*, since these describe facts or circumstances rather than events occurring. The extraction and ordering of statements is exactly the same as for other events, but it is useful to make the distinction for some results analysis. In the rest of the paper, we use the term "event" to mean both statements and other events except where specifically stated otherwise.

The SRL system follows the PropBank schema [7], which divides argument labels into numbered arguments (ARG0–ARG5), for arguments required for the valency of an action (e.g., agent, patient, etc.), and modifiers of the verb, such as purpose (PRP), locative (LOC), and so on. Argument values are spans of text. Arguments (without their label) of extracted events are shown with brackets in the source text in Fig. 1.

Phrasal Verb Detection. Phrasal verbs are common in English, and identifying them is important because the meaning of a phrasal verb is often different from that of the verb part of it (e.g., the meaning of "make up" is different from "make"; this is distinct from the fact that "make up" also has several meanings). The SRL system, however, extracts only single verbs. We use the following rule, adapted from [19], to detect phrasal verbs: If a word P either (i) has a *compound:prt* relation with the event verb W, or (ii) is adjacent to the event verb W and has a *case* or *mark* relation with W in the dependency parse tree, then WP is a candidate phrasal verb; it is accepted if it appears in a list of known phrasal verbs[3]. We performed a small-scale evaluation of the accuracy of this method: of 100 randomly sampled extracted events where rules (i) or (ii) apply, 58 of the candidate phrases appear in the known phrasal verbs list. Manually checking these, 47 are actual phrasal verbs, i.e., the precision is 0.81. The false positives are sentences where the phrase is used in a literal sense (e.g., "take off" in "takes off his clothes").

Argument Event Detection. Because arguments are spans of text, part or all of an extracted event may lie within the argument of another event. If $V(e_j)$ is within an argument of e_i, we say e_j is *contained* in e_i. This can be nested. Contained events are candidates for being arguments, but are not necessarily so. For example, in Fig. 1, ARG2 of **e6** ("is") contains $V(\mathbf{e7})$ ("offer"), but **e7** is not an argument of **e6**.

We designed the following rules: If any of them is satisfied, a contained event e_j is an argument of the containing event e_i: (i) The dependency relation $V(e_i)$ to $V(e_j)$ is clausal complement (*ccomp* or *xcomp*) or clausal subject (*csubj*). (ii) The dependency relation from $V(e_j)$ to $V(e_i)$ is copula (*cop*). (iii) All of e_j is contained in an argument of e_i that is labelled with either ARGM-PRP ("purpose") or ARGM-PNC ("purpose not cause").

Condition Event Detection. We commonly find conditional promises, threats, etc., in narrative text. Conditional offers are found in both the example from The Jungle Book quoted in the introduction and the example in Fig. 1. The condition event is not an

---

[3] https://en.wiktionary.org/wiki/Category:English_phrasal_verbs.

argument of the offer, but it is also not actual; hence, a different mechanism is required to identify conditions.

We use a method based on the signal words and phrases "if", "whenever", "as long as", "on [the] condition that", and "provided that". For example, in Fig. 1 the signal word "if" is in between the consequence **e9** and the condition **e10**. Our method is a modification of that introduced by [34]. Event e_j is determined to be a condition of e_i iff (a) one of the subsequences $V(e_i)$ S $V(e_j)$ or S $V(e_j)$ $V(e_i)$, where S is one of the signal words/phrases, appear in the sentence, with no other event verb appearing in the subsequence; and (b) one of the following holds:

S1: $V(e_i)$'s tense is future simple, $V(e_j)$'s tense is present simple;

S2: $V(e_i)$'s tense is present simple, $V(e_j)$'s tense is present simple;

S3: "must" or "should" or "may" or "might" is adjacent to $V(e_i)$, the tense of $V(e_j)$ is present simple;

S4: "would" is adjacent to $V(e_i)$ and $V(e_i)$'s tense is infinitive, $V(e_j)$ is past simple;

S5: "could" or "might" is adjacent to $V(e_i)$, $V(e_j)$'s tense is past simple;

S6: $V(e_i)$ is preceded by "could" and its tense is infinitive, the tense of $V(e_j)$ is past continuous or past perfect;

S7: "would have" or "might have" or "could have" is adjacent to $V(e_i)$ and the tense of $V(e_i)$ is past participle, the tense of $V(e_j)$ is past perfect;

S8: $V(e_i)$'s tense is perfect conditional continuous, $V(e_j)$'s tense is past perfect;

S9: $V(e_i)$'s tense is perfect conditional, $V(e_j)$'s tense is past perfect continuous;

S10: "would be" is adjacent to $V(e_i)$ and $V(e_i)$'s tense is gerund, $V(e_j)$ is past perfect;

Tenses of event verbs are determined from their POS tags.

Entity Resolution. In the event representation that we generate, the exact words used to denote entities in event arguments do not matter, as long as we identify repeat mentions of entities throughout the narrative. (The words that denote an entity may of course carry important information about the entity type, or relation to other entities.) Hence, we apply co-reference resolution, and substitute the first mention of any resolved entity for later mentions. We use a document-level inference-based LSTM model [22] from AllenNLP for the co-reference resolution task. While there are potentially better co-reference resolution methods [18], this aspect of the pipeline is somewhat orthogonal, and we leave its further development for future work.

4   Event Ordering

To determine the relative order of the events narrated within a document, we build a temporal relation classifier. We order only the independent events, i.e., not argument events or conditions, since the independent events are, the ones that are actually said to happen. Although temporal relations can exist between argument events, their actuality is also uncertain. Following the method proposed by Ning et al. [31], we trained an LSTM-based classifier using the temporal relation dataset MATRES [32]. Earlier studies of temporal relation extraction [e.g. 5, 21, 29] have adhered to the TimeML standard [35], which uses fourteen different temporal relations between events and temporal expressions. In contrast, the MATRES dataset considerably simplifies the task by using only four relations: *before*, *after*, *equal* and *vague* (the last meaning no or unknown relation). However, because these four seem sufficient to establish a partial order between events, and because earlier studies have reported low prediction accuracies for the TimeML relation set, we choose the simpler task variant.

The predictor is trained on pairs of events within the same or adjacent sentences and keeps only relations with a probability of at least 0.5. *after* relations are reversed and converted to *before*, giving a network of precedence and equality relations. We then use transitivity to infer additional precedences. Because the classifier is myopic, i.e., predicts each event pair in isolation, it often induces inconsistent orders (with cyclic precedences). We resolve these by deleting the predicted ordering with the lowest probability in each cycle. An example of the resulting partial order is shown in Fig. 1. On average, we find 2.7 subnetworks per document, with an average size of 22 events.

Table 1. Number of distinct event chains that meet the selection criteria, and the number of distinct event chains with more than one occurrence. Numbers are subgrouped by the genre of the document in which chains appear: CT: crime thriller, RC: romantic comedy. "Across" means chains that occur in documents in both genres, while the total includes both chains recurring in documents within one genre only and across both.

		M1			M2		
		$N = 3$	$N = 4$	$N = 5$	$N = 3$	$N = 4$	$N = 5$
# of distinct event chains	CT	191,933	388,455	782,441	197,503	401,840	818,449
	RC	216,752	474,706	1,041,439	223,520	487,491	1,071,732
	Total	404,971	862,925	1,823,004	419,119	888,348	1,889,313
# of recurring chains	CT	1,712	104	5	676	20	1
	RC	2,417	160	8	816	39	1
	Across	2,948	203	14	1,128	40	1
	Total	6,015	453	27	2,358	99	3

5   Narrative Chain Extraction

Finally, we perform a preliminary experiment in acquiring simple scripts, or narrative event chains [9], from the extracted partially ordered events. A narrative event chain is a subsequence of N non-statement events that are all connected by sharing some co-referring argument with another event in the chain. (It does not have to be one entity that is shared by all events. For example, if A is an argument of events 1 and 3, and B is an argument of events 2 and 3, the chain is connected.) Events in the chain must be totally ordered, but not necessarily consecutive. We are interested in event chains that recur, in particular across documents, since these represent general elements of narrative. For two event chains to match, the event verbs and arguments must match. We consider two different conditions for arguments matching: The first (**M1**) is that there must be a mapping between the labelled arguments of the two instances of the chain that respects argument recurrence within the chain (i.e., if A in one instance maps to B in the other, B must recur in the same argument positions of the corresponding events that A does). The second (**M2**) requires in addition that argument and condition events of events in the chains recursively match, under the same condition.

We applied chain extraction to the partially ordered events extracted from a subset of the Movie summaries dataset, comprising those with genre label "crime thriller" (1678 documents) and "romantic comedy" (2069 documents), for $N = 3, 4, 5$. Results are summarised in Table 1. We count repetitions of event chains only when they are repeated in different documents, because repetitions within a document tend to have large overlap. As expected, increasing the length of the chain or applying the stricter argument matching condition **M2** substantially reduces the number of recurring event chains found. In this experiment, we use a straightforward form of matching events verbs: they match if their lemma is the same. Applying sense disambiguation, i.e., matching verbs only when used in the same sense, could reduce the number of recurring event chains. Meanwhile, applying semantic generalization that matches distinct verbs when their senses satisfy some degree of semantic similarity, would likely increase the number.

6   Challenges for NLP Research

Several tasks in our event extraction pipeline pose open challenges for NLP research.

Event and Argument Extraction. Some aspects of the current SRL systems can be improved: First, identifying phrasal verbs, and distinguishing their occurrence from literal uses of the same phrase. The method we have used depends on a given list of known phrasal verbs (without this filter, its precision would be too poor) and also cannot recognise discontinuous phrasal verbs with more than two words, such as "put ... down to", which is different from "put down". Second, arguments identified by the SRL system often capture independent event mentions, particularly when those occur in relative clauses of a nominal argument. Our argument event classifier tries to resolve these cases, but still misses some.

Event Actuality. Related to deciding which events are arguments of other events is determining which of the mentioned events are actual, i.e., which have, according to the narrative, taken place. We have considered any non-negated non-argument and non-condition event to be actual. This is, however, an approximation. First, because argument events can also be actual; this depends on the verb they are arguments of. Consider "she thought it rained outside" and "she could see it rained outside": in both, "rain" is an argument, but in the second example it is also actual. Second, some temporal expressions give rise to non-argument events that are also not actual. Consider, for example, "before she falls down, she catches the railing and steadies herself" (implying the "fall" does not actually happen). We are not aware of prior work on resolving event actuality. As the examples above show, this is closely linked with the meanings of the verbs involved. This is a significant research challenge.

Conditional Event Detection. The problem of detecting condition–consequence relations between events in text has been studied, motivated in particular by finding causal relationships [34]. We evaluated two recent methods that detect conditional structures, called CNC [14] and CiRA [37], respectively. Both are BERT-based neural networks, but trained with different data. CiRA use an annotated set of requirements documents, while CNC annotated and used a set of news articles, together with the Penn Discourse Treebank 3.0 [40] and CausalTimeBank [28] datasets. However, we also note that both are intended to extract causal relations between events, which do not always coincide with the condition–consequence relation.

Table 2. Precision and recall of detecting the existence of conditionals in sentences. EM-rate is the proportion of sentences in which the detected condition and consequence events exactly match our annotation.

	Precision	Recall	EM-rate
CiRA [14]	0.75	0.79	0.41
CNC [37]	0.80	**0.80**	0.1
Ours	**0.93**	0.71	**0.85**

We apply these two systems to the same set of 100 randomly selected sentences from the Movie summary dataset that we used to evaluate our rule-based method. Recall that these were selected to include the five signal words or phrases that we use (20 for each) and that 75 of them contain conditionals. 3 sentences have more than one condition–consequence event pair. Both systems detect the presence of conditionals in a sentence in more cases than our method (59 and 60 of the 75 positive cases, respectively, compared to 53 for our method), but also have a much higher number of false positives (20 and 15 of the 25 negative cases, respectively, compared to 4 for our method), leading to their lower precision, as shown in Table 2. Furthermore, in true positive cases identified by each, we compare the events identified as conditions and consequences with our annotation. These results are worse: CNC identifies the correct text spans in only 6 of the 60 cases (EM-rate = 0.1), while CiRA does so in 24 of the 59 cases. On the other hand, our method is blind to any conditional expression that does not use one of the five signal words or phrases. We conclude that more research on this aspect of relations between events is warranted.

Event Ordering. Predicting the right order of event mentions remains a hard problem. For instance, in the example in Fig. 1, six of the thirteen precedence relations are incorrect. (The precision of the classifier is actually better than this would suggest, as it makes many correct predictions that are transitively implied by those shown in the figure.) There are also some event pairs that arguably should be ordered, but which are not detected, e.g., **e7** before **e12**, **e7** before **e0** and **e1** before **e0**. We note that four of the falsely predicted precedences are associated with statements (**e0**, **e6** and **e15**), suggesting that these pose particular difficulty for the classifier. Although Ning et al.

[31] report much better accuracy on the simplified 4-relation prediction task than previously achieved by state-of-the-art predictors for the full TimeML relation set, the difficulty caused by statements, which often describe enduring circumstances, suggests that including a *during* relation may help.

Causal relations between events imply a temporal ordering (i.e., cause precedes effect). However, current state-of-the-art in predicting causal relations [e.g., 14,29,37] did not perform well on narrative text in our test. We did not test Mirza & Tonelli's system, because it depends on features such as verb mood and aspect, which are available in the annotated CausalTimebank dataset but not in our unannotated source texts, and because CNC achieved better prediction performance when evaluated on the same data. Another potential source of information is the super–sub relation: an event is defined to be a sub-event of another iff it occurs *during* the super-event and is spatially *contained* by the super-event [16]. The precision of current state-of-the-art sub-event detection methods is, however, mostly low [1,4].

7   Challenges for Model Acquisition

The problem of inducing action preconditions and effects from observations of plans has been studied in AI planning for some time (e.g., [12,41]). Approaches vary in the assumptions they make, such as whether observations are complete or partial, precise or noisy, whether only actions/events are observed or also (full or partial) states (i.e., statements), and so on (see, e.g., [2,8]). However, we are not aware of any method that, off-the-shelf, can successfully exploit the event information that we can extract from text: partially ordered, with partial state observations, and some degree of uncertainty in all parts (i.e., which events actually occur, their arguments, and order). Furthermore, the complex structure of events, with nested argument events and conditions, is not represented in standard planning formalisms (e.g., PDDL [26]), and not supported by current model acquisition methods.

Another challenge raised by learning models from events extracted from text is sense disambiguation and semantic generalization, of both verbs and arguments. While this can be handled as a separate task in between extraction and model acquisition, integrating with the latter, i.e., deciding which general event each occurrence is an instance of, and what level from a hierarchy of types to assign each of its arguments, jointly with inducing the models of general events, may lead to better models.

8   Conclusions

Narrative text exhibits many of the complexities of natural language, but is also a rich source of knowledge about events/actions. We proposed a pipeline for automatically extracting (partially) ordered structured event representations from narrative texts, with the ultimate aim of learning general event models. Applying the pipeline to three large-scale narrative corpora demonstrates several open research challenges. We propose methods to deal with some of these, such as argument and conditional event recognition, and so on. Learning general action models from the event representations will be the next step of our work.

References

1. Aldawsari, M., Finlayson, M.A.: Detecting subevents using discourse and narrative features. In: Proceedings of the 57th Annual Meeting of ACL (2019)
2. Arora, A., Fiorino, H., Pellier, D., Métivier, M., Pesty, S.: A review of learning planning action models. Knowl. Eng. Rev. **33** (2018)
3. Bamman, D., O'Connor, B., Smith, N.A.: Learning latent personas of film characters. In: Proceedings of ACL, pp. 352–361 (2013)
4. Bekoulis, G., Deleu, J., Demeester, T., Develder, C.: Sub-event detection from Twitter streams as a sequence labeling problem. arXiv preprint arXiv:1903.05396 (2019)
5. Bethard, S.: ClearTK-timeML: a minimalist approach to TempEval 2013. In: Proceedings of SemEval 2013, pp. 10–14 (2013)
6. Biten, A.F., Gomez, L., Rusinol, M., Karatzas, D.: Good news, everyone! context driven entity-aware captioning for news images. In: CVPR (2019)
7. Bonial, C., Hwang, J., Bonn, J., Conger, K., Babko-Malaya, O., Palmer, M.: English Prop-Bank annotation guidelines. Center for Computational Language and Education Research Institute of Cognitive Science University of Colorado (2012)
8. Callanan, E., De Venezia, R., Armstrong, V., Paredes, A., Chakraborti, T., Muise, C.: MACQ: a holistic view of model acquisition techniques. arXiv preprint arXiv:2206.06530 (2022)
9. Chambers, N., Jurafsky, D.: Unsupervised learning of narrative event chains. In: Proceeding of ACL (2008)
10. Chambers, N., Jurafsky, D.: Unsupervised learning of narrative schemas and their participants. In: Proceedings 47th ACL Meeting and 4th IJCNLP, pp. 602–610 (2009)
11. Cook, W.W.: Plotto: The Master Book of All Plots. Tin House Books (1928)
12. Cresswell, S., Gregory, P.: Generalised domain model acquisition from action traces. In: Twenty-First ICAPS (2011)
13. Feng, W., Zhuo, H.H., Kambhampati, S.: Extracting action sequences from texts based on deep reinforcement learning. In: Proceedings of IJCAI, pp. 4064–4070 (2018)
14. Fischbach, J., et al.: Automatic detection of causality in requirement artifacts: the CiRA approach. In: Dalpiaz, F., Spoletini, P. (eds.) REFSQ 2021. LNCS, vol. 12685, pp. 19–36. Springer, Cham (2021). https://doi.org/10.1007/978-3-030-73128-1_2
15. Geffner, H., Bonet, B.: A Concise Introduction to Models and Methods for Automated Planning. Morgan & Claypool (2013). ISBN 9781608459698
16. Glavaš, G., Šnajder, J., Kordjamshidi, P., Moens, M.F.: HiEve: a corpus for extracting event hierarchies from news stories. In: Proceedings of 9th Language Resources and Evaluation Conference, pp. 3678–3683. ELRA (2014)
17. Hayton, T., Porteous, J., Ferreira, J., Lindsay, A., Read, J.: StoryFramer: from input stories to output planning models. In: ICAPS Workshop on Knowledge Engineering for Planning and Scheduling (2017)
18. Hayton, T., Porteous, J., Ferreira, J.F., Lindsay, A.: Narrative planning model acquisition from text summaries and descriptions. In: Proceedings of AAAI (2020)
19. Komai, M., Shindo, H., Matsumoto, Y.: An efficient annotation for phrasal verbs using dependency information. In: Proceedings of PACLIC, pp. 125–131 (2015)
20. Lamperti, G., Zanella, M.: Diagnosis of active systems (2003)
21. Laokulrat, N., Miwa, M., Tsuruoka, Y., Chikayama, T.: Uttime: temporal relation classification using deep syntactic features. In: SemEval, pp. 88–92 (2013)
22. Lee, K., He, L., Zettlemoyer, L.: Higher-order coreference resolution with coarse-to-fine inference. arXiv preprint arXiv:1804.05392 (2018)
23. Lindsay, A., Read, J., Ferreira, J., Hayton, T., Porteous, J., Gregory, P.: Framer: planning models from natural language action descriptions. In: ICAPS (2017)

24. Manikonda, L., Sohrabi, S., Talamadupula, K., Srivastava, B., Kambhampati, S.: Extracting incomplete planning action models from unstructured social media data to support decision making. In: KEPS (2017)
25. Manning, C.D., Surdeanu, M., Bauer, J., Finkel, J.R., Bethard, S., McClosky, D.: The Stanford CoreNLP natural language processing toolkit. In: ACL (2014)
26. McDermott, D., et al.: PDDL-the planning domain definition language–version 1.2 (1998)
27. Miglani, S., Yorke-Smith, N.: NLtoPDDL: one-shot learning of PDDL models from natural language process manuals. In: KEPS (2020)
28. Mirza, P., Sprugnoli, R., Tonelli, S., Speranza, M.: Annotating causality in the TempEval-3 corpus. In: CAtoCL, pp. 10–19 (2014)
29. Mirza, P., Tonelli, S.: CATENA: CAusal and TEmporal relation extraction from NAtural language texts. In: Proceedings of COLING, pp. 64–75 (2016)
30. Mostafazadeh, N., et al.: A corpus and cloze evaluation for deeper understanding of commonsense stories. In: Proceedings of NAACL, pp. 839–849 (2016)
31. Ning, Q., Subramanian, S., Roth, D.: An improved neural baseline for temporal relation extraction. In: Proceedings of EMNLP, pp. 6203–6209 (2019)
32. Ning, Q., Wu, H., Roth, D.: A multi-axis annotation scheme for event temporal relations. In: ACL (2018). http://cogcomp.org/papers/NingWuRo18.pdf
33. Olmo, A., Sreedharan, S., Kambhampati, S.: GPT3-to-plan: extracting plans from text using GPT-3. arXiv preprint arXiv:2106.07131 (2021)
34. Puente, C., Sobrino, A., Olivas, J.A., Merlo, R.: Extraction, analysis and representation of imperfect conditional and causal sentences by means of a semi-automatic process. In: International Conference on Fuzzy Systems, pp. 1–8. IEEE (2010)
35. Saurí, R., Littman, J., Knippen, B., Gaizauskas, R., Setzer, A., Pustejovsky, J.: TimeML Annotation Guidelines Version 1.2.1 (2006)
36. Sil, A., Yates, A.: Extracting STRIPS representations of actions and events. In: Recent Advances in Natural Language Processing (2011)
37. Tan, F.A., et al.: The causal news corpus: annotating causal relations in event sentences from news. arXiv preprint arXiv:2204.11714 (2022)
38. Tandon, N., de Melo, G., De, A., Weikum, G.: Knowlywood: mining activity knowledge from Hollywood narratives. In: Proceedings of the CIKM (2015)
39. Van Harmelen, F., Lifschitz, V., Porter, B.: Handbook of Knowledge Representation. Elsevier (2008)
40. Webber, B., Prasad, R., Lee, A., Joshi, A.: The Penn discourse treebank 3.0 annotation manual. Philadelphia, University of Pennsylvania 35, 108 (2019)
41. Yang, Q., Wu, K., Jiang, Y.: Learning action models from plan examples using weighted MAX-SAT. Artif. Intell. **171**(2–3), 107–143 (2007)

The Difficulty of Novelty Detection and Adaptation in Physical Environments

Vimukthini Pinto[1]($\boxtimes$) (iD), Chathura Gamage[1], Matthew Stephenson[2], and Jochen Renz[1]

[1] The Australian National University, Canberra, Australia
vimukthini.inguruwattage@anu.edu.au
[2] Flinders University, Adelaide, Australia

Abstract. Detecting and adapting to novel situations is a major challenge for AI systems that operate in open-world environments. One reason for this challenge is due to the diverse range of forms that novelties can take. To accurately evaluate an AI system's ability to detect and adapt to novelties, it is crucial to investigate and formalize the difficulty of different novelty types. In this paper, we propose a method for quantifying the difficulty of novelty detection and novelty adaptation in open-world physical environments, considering factors such as the appearance and location of objects, as well as the actions required by the agent. We implement several difficulty measures using a combination of qualitative spatial relations, learning algorithms, and statistical distance measures. To demonstrate an application of our approach, we apply our difficulty measures to novelties in the popular physics simulation game Angry Birds. We invite researchers to incorporate the proposed novelty difficulty measures when evaluating AI systems to gain a better understanding of their limitations and identify areas for future improvement.

Keywords: AI Evaluation · Difficulty · Novelty · Open-world Learning

1 Introduction

Autonomous AI systems such as self-driving cars, space probes, and surveillance drones have become increasingly popular and common in recent years. These AI systems require the ability to detect and adapt to novel situations in a timely and efficient manner to avoid undesirable consequences. For instance, if a self-driving car maintains its speed in a storm that was not experienced during model training, it could endanger many lives. Open-world learning (OWL) is an emerging field of study that aims to solve the challenge of detecting and adapting to novel situations [14]. To progress in OWL research, it is essential to have appropriate evaluation protocols to capture the performance of agents under the

Supplementary Information The online version contains supplementary material available at https://doi.org/10.1007/978-981-99-8391-9_3.

two tasks: novelty detection and adaptation [9,18]. This paper contributes to the OWL evaluation by creating difficulty measures to independently evaluate agents' performance from the inherent difficulty of novelties.

We encounter a near-infinite form of novelties in the real world [1,6,14]. For example, consider an autonomous car designed for urban driving in a busy city. The model which controls the vehicle has reliable expertise for navigating in this setting, but suppose it enters a new area with different traffic patterns. Here, it may encounter new types of road signs, unfamiliar pedestrians that can cross its path, and aggressive drivers that may cut it off. The vehicle may enter a stormy area where visibility is low and where sensor readings are distorted or where strong winds threaten to push it off course. As seen from this example, there is a large range of novelties and some of them may be easy to detect and adapt and some of them could be hard or nearly impossible. If we are to evaluate the novelty detection and novelty adaptation ability of an agent, it would be uninformative to comment on the performance by considering all the novelties as a whole [19]. Using a measure of difficulty in the evaluation enables evaluators to understand the range of situations an agent may fail and reliably make conclusions.

In this paper, we identify different forms of novelties by considering two states where an agent can detect/adapt to novelty: *observational state, action state*. The observational state considers the situations that can be visually perceived while the action state considers the actions an agent has to take. We formalize practical methods to compute difficulty under the two states with the use of learning algorithms and statistical distance measures. We utilize existing learning algorithms and we propose an algorithm developed using qualitative spatial relations (QSRs). For statistical distance measures, we use graph edit distance (GED), distance measures developed using solution paths, and measures based on the probability density function (PDF). In the supplementary materials, we demonstrate that the difficulty measures we formulated can be easily applied in practice by applying them to a recently developed testbed NovPhy [9] (Angry Birds with novelty), which injects novelties into a physical environment.

2   Background and Related Work

In this section, we first provide definitions for the terms that we will be using throughout the paper. Next, we review the literature on a number of topics required to understand our novelty difficulty measures.

- *novelty*: a situation that an agent has not encountered during model training. It could be a new object that an agent has not seen before or a phenomenon that an agent has not experienced (e.g.: storms, floods).
- *pre-novelty*: a situation without novelty (i.e., a situation that an agent has seen during model training).
- *novel object*: An object that has one or more novel properties. It could be an object that an agent has seen before but with a different color, or mass, or the object may do an action that it did not do during pre-novelty.
- *non-novel object*: An object without any novel property.

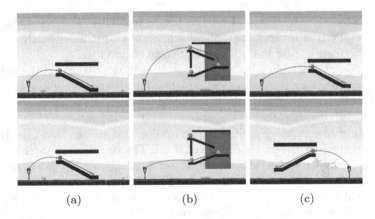

(a)   (b)   (c)

Fig. 1. Example tasks from NovPhy testbed. In each subfigure, the top figure is the pre-novelty task and the bottom figure is the corresponding task with the novelty. The arrows show the trajectories of the objects when the solution is executed. It can be seen that in (a), we do not need to modify the shooting angle (action) as the novelty only changes the colour of an object, in contrast, in (b) and (c) we need to change the action due to the nature of the novelty. See [9] for descriptions of the novelties.

- *object class*: A group of objects with similar properties. (e.g.: In Fig. 1b, there are multiple brown circular objects. They belong to class *wood-circle*).
- *novelty detection*: detecting that a novelty is present in a task.
- *novelty adaptation*: solving a task in the presence of a novelty.

2.1   Novelty Research

In recognition of the critical need for AI systems that can effectively detect and adapt to novel situations, DARPA has launched a program known as the Science of Artificial Intelligence and Learning for Open-world Novelty (SAIL-ON) The SAIL-ON program defines novelty in the realm of AI as situations that violate implicit or explicit assumptions about the agents, the environment, or their interactions [20]. [1] also look into a range of novelties and formalize a theory of novelty for open-world environments. The authors identified three distinct states where novelties can occur, *observational state*, *world state*, and *agent state*. Our work on defining the difficulty of detection and adaptation follows from this research and we consider the observational state where a novelty can be visually perceived. However, we do not consider the world state (the state where all the information is available. eg: physical parameter values in a physical domain) as our work focuses on detection and adaptation difficulty and an agent does not receive all the information from the world state. We also do not consider the agent state (a state specific to an agent based on the agent architecture) as we intend to develop difficulty measures that do not depend on individual agent properties. Instead, we consider *action state* that takes into account the actions an agent needs to take to solve a task. In Fig. 1, we show example novelties where an agent needs to modify the action to solve the task and where agents do not need to modify the action.

2.2   Difficulty Prediction

Difficulty assessment is a popular research area in a number of research fields ranging from neuroscience to AI. In neuroscience, researchers study the difficulty of decision-making processes in humans [7,10]. Similarly, in AI, researchers study the difficulty of tasks for AI systems [11,16]. Our work on developing difficulty measures for novelty detection and adaptation focuses on AI agents.

Considering the OWL-related difficulty measures already available in the literature, [19] propose a difficulty of detection only by considering a single class of novelties. The authors only consider novelties that cannot be visually perceived but are different in underlying physical parameters. Considering the adaptation difficulty, [17] propose a method to quantify the difficulty of adapting to novelty using the solution paths the agents take. The method requires a reinforcement learning agent to solve multiple tasks, which is costly as it requires multiple trial and error runs to reach the optimal solution. Inspired by this approach, we also propose a method that takes solution paths into consideration without running an agent. Another novelty adaptation difficulty measure predicts the difficulty using GED [21] for an agent's mental model for the board game Monopoly [13]. Similarly, we make use of GED and we propose an agent-independent difficulty measure for physical environments. [6], analyze domain complexity and introduce factors such as single entity and multiple entities that contribute towards difficulty in OWL tasks. Our work makes use of these factors when defining difficulty measures.

2.3   Learning Algorithms

Novelty detection, which is sometimes referred to as anomaly detection, outlier detection, or one-class classification, [2] is a critical research area in machine learning and data mining. The goal of novelty detection is to identify patterns or behaviors in data that are significantly different from the expected or normal behavior. Over the years, several comprehensive surveys and reviews of novelty detection techniques have been conducted, focusing on different aspects and techniques [12]. In [2], authors reviewed novelty detection techniques and categorized them into six domains: classification, clustering, nearest neighbors, statistical methods, information theory, and spectral theory. Our work on predicting the difficulty of novelty detection, uses classification-based approaches and clustering-based approaches. We identify different factors that contribute towards the detection such as color and shape and we have developed multiple classification models to predict the probability of an object belonging to different object classes. Finally, we combine the models in the form of a weighted ensemble to predict the difficulty. The clustering-based model is an algorithm we propose to predict the detection difficulty using QSRs.

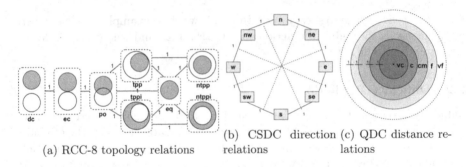

(a) RCC-8 topology relations   (b) CSDC direction relations   (c) QDC distance relations

Fig. 2. QSRs. The connections represent conceptual neighbors. Distance between two neighbors is taken as 1.

2.4   Qualitative Spatial Relations (QSRs)

Due to a large number of applications of QSRs [3,5,15], dozens of formalisms of calculi have been proposed for describing various aspects of space [3–5]. In our work, we utilize three commonly used calculi to represent topology, direction, and distance. We use *Region Connection Calculus (RCC)-8* (Fig. 2a) to describe topological relations: *dc* (disconnected), *ec* (externally connected), *po* (partially overlapping), *eq* (equal), *tpp* (tangential proper part), *tppi* (tangential proper part inverse), *ntpp* (non-tangential proper part), and *ntppi* (non-tangential proper part inverse) [3,4]. To describe directions, we use *Cone-Shaped Direction Calculus (CSDC)* (Fig. 2b). The relations of the CSDC are based on the eight disjoint sectors of the space divided by the lines going through the reference point [3]. The eight relations are *n* (north), *ne* (northeast), *e* (east), *se* (southeast), *s* (south), *sw* (southwest), *w* (west), and *nw* (northwest). Additionally, we use *Qualitative Distance Calculus (QDC)* (Fig. 2c) to describe distance between two objects. We have used five absolute distance calculi: *vc* (very close), *cl* (close), *cm* (commensurate), *fr* (far), and *vf* (very far) [3].

2.5   Experimental Domain

Our experimental domain NovPhy [9] is a testbed designed to evaluate physical reasoning in the presence of novelties. The testbed is based on the popular physics simulation game Angry Birds. Figure 1 shows example tasks with and without novelty. We explain the experimental domain in detail in the Supplementary A along with the experimental results.

3   Novelty Difficulty Formulation

In this section, we explain the dimensions of novelty we need to consider to formulate our difficulty measures and the practical implementation of it.

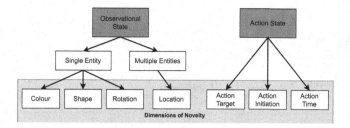

Fig. 3. The novelty dimensions considered in the study

3.1 Dimensions of Novelty

Considering the dimensions where novelty can be perceived by an agent, we consider two states: *observational state* and *action state*. The two states are motivated by the research [1] on the unifying theory of novelty.

1. *Observational state*: The state where an agent can observe the environment
2. *Action state*: The state where the agent needs to take an action

We consider these two states, as in a physical domain, the novelties can either be detected due to the physical appearance (novelties that can be perceived visually: in the observational state) or could be detected/adapted due to a change in action compared to the actions encountered before (novelties that can be perceived after taking an action: in the action state). As [6] state, a novelty in an environment can be in a single entity or it could be as a relationship between multiple entities. Taking the entities into consideration, in the observational state, we identify dimensions that can be observed in a single entity and dimensions that can be observed as a relationship between multiple entities. Therefore, we extend the dimensions to color, shape, and rotation for single entity, and for multiple entities we consider the relative location of objects in terms of QSRs. Considering the action state, the dimensions we consider are based on situations that enable an agent to detect/adapt to novelty if an agent's expected action changes from the known action in a situation without novelty. We consider 1) *action target*, to check if the target object changes (for example, in Fig. 1b, the target object has changed when the direction of the air turbulence has changed), 2) *action initiation* to check if the solution action changes (for example, in Fig. 1c, the shooting angle has changed compared to the angle used in the non-novelty), and 3) *action time* to identify if the time allocated to take actions change (for example, in novelty you may need to shoot faster before a certain event happens). See Fig. 3 for the novelty dimensions we consider.

3.2 Observational State

In this section, we discuss the formulation of difficulty under each dimension belonging to the observational state. We discuss the formulation of color and shape together as the underlying formulation is the same except for the input

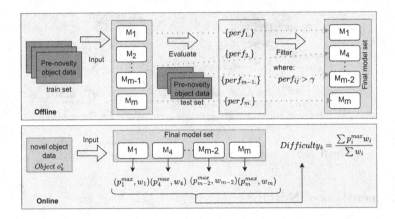

Fig. 4. The formulation of color and shape based difficulty

data structure. For the rotation based difficulty measure, we make use of the predictions made from shape and color models. For the location, we use QSRs to develop our difficulty prediction algorithm. All the observational state difficulty measures are aimed to predict the difficulty of novelty detection. The measures are not used to predict the adaptation difficulty as agents cannot take any action in the observational state. Therefore, we predict the difficulty of novelty detection for agents if the agents are detecting novelty using a single dimension.

Color and Shape Based Difficulty. Given a physical domain, there are non-novel objects that an agent can be trained on. Therefore, with the use of classification algorithms, we can predict the probability that a novel object belongs to a non-novel object class. The color-based algorithm predicts the probability that a novel object can be considered an object that was seen before based on the observed color of an object. Similarly, in the shape-based algorithm, we predict the probability that a novel object can be considered an existing object based on the shape of the object.

We illustrate the process of computing the difficulty based on color or shape in Fig. 4. There are two stages when computing difficulty for color and shape. In the offline stage, we train multiple learning models to classify objects and we evaluate the models on a pre-defined test set that comprises objects seen in pre-novelty. We define O as a list of all object classes in pre-novelty ($O = \{o_1, o_2, ..., o_j, ..., o_n\}$). Next, we filter out the models that have an acceptable performance. If the model M_i has a performance higher than an acceptable threshold γ for all object classes, we select the model to make predictions in the online stage ($Perf_{i,j} > \gamma \ \forall o_j \in O$). Therefore, the models that perform well will be used in the online stage in the form of a weighted ensemble to predict the probability (pr) that a novel object belongs to an existing object class. Following is the formulation to establish the difficulty value for novel object k, denoted o_k^*.

In model M_i: $P_{i,k} = \{p_{1,k}, p_{2,k}, ..., p_{j,k}, ..., p_{n,k}\}$, where, $p_{j,k} = pr(o_k^* = o_j)$. Therefore, we define $p_{i,k}^{max} = max(P_{i,k})$ and $\hat{o}_{i,k} = \arg\max_{o_j}(P_{i,k})$. i.e., $p_{i,k}^{max}$ is the maximum probability that the model M_i allocates o_k^* to an existing object class and $\hat{o}_{i,k}$ is the predicted object class. The weight of the prediction is calculated as $w_{i,k} = (\alpha n - n'_{i,k})/\alpha n$ if $\alpha n > n'_{i,k}$, or else $w_{i,k}$ is set to 0. Here, $n'_{i,k} = (\sum_{j=1}^{n} \begin{cases} 1 & if(p_{i,k}^{max} - p_{j,k}) \le \beta \\ 0 & otherwise \end{cases}) - 1$. α and β are hyper-parameters that should be selected according to the domain based on the prediction flexibility we allow. The final difficulty of detecting o_k^* novel object using color/shape:

$$Difficulty_k^{color/shape} = \frac{\sum_{i=1}^{m} p_{i,k}^{max} w_{i,k}}{\sum_{i=1}^{m} w_{i,k}} \tag{1}$$

The difficulty value is between 0–1, 1 indicates the highest difficulty. The idea is that if a weighted ensemble predicts that a novel object belongs to an existing class based on its color/shape, then it is difficult to visually detect. In contrast, the difficulty will be low if the models have lower probabilities or if there are multiple objects with probabilities close to maximum probability.

Rotation Based Difficulty. Rotation difficulty measures if the agents can detect a novelty based on the rotation of the novel object (e.g.: in NovPhy, it could be that the pig is rotated upside down in novelty which is not usually rotated in pre-novelty). Similar to the previous section, the rotation difficulty also comprises two stages (See Fig. 2 in Supplementary B). In the offline stage, we collect rotation data from non-novel object classes and estimate the distribution for each object class using kernel density estimation (KDE). In the online stage, we use the predicted object class from color and shape algorithms $\hat{o}_k$ and the novel objects' observed rotation (rot_k^*). The predicted object from color and shape algorithms can be selected based on a voting technique of choice (e.g.: hard voting, soft voting, or weighted voting) [22]. Next, the KDE of the predicted object is selected, and the area under the PDF for the $rot_k^* \pm rot_\epsilon$ can be interpreted as the difficulty of detection using rotation. rot_ϵ is a predefined constant (a small rotation shift) that helps to get the area under the probability density function $(PDF_r(\hat{o}_k))$. The difficulty based on rotation can be expressed as follows.

$$Difficulty_k^{rotation} = \int_{rot_k^* - rot_\epsilon}^{rot_k^* + rot_\epsilon} PDF_r(\hat{o}_k) \, dr \tag{2}$$

The underlying idea is that if the rotation of the novel object is commonly observed in pre-novelty, it becomes challenging to detect the novelty solely based on rotation. The rotation difficulty ranges between 0–1, 1 is the highest difficulty.

Location Based Difficulty. The location-based difficulty considers the relative location between pairs of objects. For example, in Fig. 1c, the direction relationship between the bird and the pig has changed in novelty. The relative location between objects is captured through the change in QSRs between object pairs in novel tasks compared to the non-novel tasks. In the offline stage, we collect

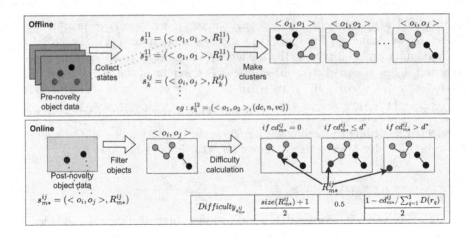

Fig. 5. The formulation of location based difficulty

object pair relationships and develop clusters for each object pair based on a conceptual distance measure. In the online stage, we take the observed object pair relationship and determine the difficulty based on the distance to the clusters developed in the offline stage.

Formulation: To explain the difficulty measure, we first define a state as s_k^{ij} a tuple consisting of the classes of an object pair and its observed QSRs ($< o_i, o_j >, R_k^{ij}$) where $R_k^{ij} = [r_{1k}^{ij}, r_{2k}^{ij}, r_{3k}^{ij}]$. The set of all available states for the pair of object classes o_i and o_j is $S^{ij} = \{s_1^{ij}, s_2^{ij}, ...\}$. The relations $r_{1.} \in RCC\text{-}8$, $r_{2.} \in CSDC$, and $r_{3.} \in QDC$. Considering two states for the o_i and o_j object pair as s_k^{ij} and $s_{k'}^{ij}$, $s_k^{ij} = (< o_i, o_j >, R_k^{ij})$ and $s_{k'}^{ij} = (< o_i, o_j >, R_{k'}^{ij})$, we can find the distance between two states as $d_{s_k^{ij}, s_{k'}^{ij}} = \sum_{q=1}^{3} |r_{qk}^{ij} - r_{qk'}^{ij}|$. $|r_{qk}^{ij} - r_{qk'}^{ij}|$ is the minimum absolute distance between two relations that can be calculated using shortest path algorithms applied to the QSR graphs in Fig. 2.

Clustering: When constructing the clusters, we develop clusters for each object class pair. The cluster nodes are the QSRs (e.g.: R_k^{ij}) and two nodes R_k^{ij} and $R_{k'}^{ij}$ will be connected only if $d_{s_k^{ij}, s_{k'}^{ij}} \leq d^*$ (d^* is a threshold distance to define according to the domain to determine if the nodes connect). If $d_{s_k^{ij}, s_{k'}^{ij}} = 0$, the node size increases. The final size of the node R_k^{ij} is, $size(R_k^{ij}) = \frac{|\{k' | d_{s_k^{ij}, s_{k'}^{ij}} = 0\}|}{|S^{ij}|}$ (i.e., the proportion of states that has the same QSR as the state s_k^{ij}). We represent the set of clusters developed for S^{ij} as C^{ij} ($C^{ij} = \{c_1^{ij}, c_2^{ij}, ...\}$) where $c_l^{ij} = \{R_{l1}^{ij}, R_{l2}^{ij}, ...\}$.

Difficulty Computation: The difficulty computation will be done for each object pair in the novel task. Given an object pair (novel/non-novel) with their observed QSRs and the object classes, the clusters developed in the offline stage for the corresponding object class pair will be extracted. Assume that the state we take from the novel task is s_{m*}^{ij}. Therefore, the set of clusters developed for S^{ij} will be extracted (i.e., C^{ij}). The cluster distance between C^{ij} and R_{m*}^{ij} is taken to be the minimum distance between the observed relations and relations available in the cluster set. That is $cd_{m*}^{ij} = min\{d_{s_{m*}^{ij}, s_k^{ij}} \forall s_k^{ij} \in S^{ij}\}$. We define the location difficulty for an observed state in Eq. 3 and illustrate the process in Fig. 5. The $D(r_1), D(r_2)$, and $D(r_3)$ in the equation represent the diameter of the RCC-8, $CSDC$, and QDC graphs (Fig. 2) respectively. The diameter of a graph is the length of the shortest path between the most distanced nodes.

$$Difficulty_{s_{m*}^{ij}}^{location} = \frac{1 + \begin{cases} size(R_{m*}^{ij}) & if\,cd_{m*}^{ij} = 0 \\ 0 & if\,cd_{m*}^{ij} \leq d^* \\ -(cd_{m*}^{ij})/(\sum_{q=1}^3 D(r_q)) & otherwise \end{cases}}{2} \qquad (3)$$

The rationale of this measure is that if the QSR of the pair of objects being examined belongs to a cluster node ($cd_{m*}^{ij} = 0$), it indicates that the QSR has been observed in pre-novelty. Thus, we quantify the difficulty based on the node size of the QSR (high difficulty). If $cd_{m*}^{ij} < d^*$, QSR can be connected to an existing cluster and the difficulty is 0.5. If it cannot be connected, then we quantify using the distance away as a proportion of the maximum allowed distance. Similar to other measures, difficulty is between 0–1, 1 is the highest difficulty.

Observational State Difficulty Summary. All the difficulty measures developed for the observational state are for detection and they are formulated for a single object (a single object pair for location). However, there are multiple objects in a task. i.e., there can be multiple novel objects in a task that should be considered for color, shape, and rotation-based measures. For the location-based measure, we should consider all the pairwise relations available. When defining the difficulty measure for a task (for each novelty dimension), we take the minimum of the difficulty values. For example, if a task contains multiple novel objects but only one object with a drastic difference in color compared to the non-novel objects, this object will have a lower difficulty of detection using color, therefore, we take the minimum difficulty as the task difficulty under the color dimension. Formally we define the task difficulty for novelty-dimension nd as $Difficulty^{nd} = min(Difficulty_k^{nd})$ for all novel objects/object pairs.

3.3   Action State

In this section, we present the formulation of difficulty measures for each dimension pertaining to the action state. An action within a given environment represents a point in the action space, such as selecting the shooting angle in Angry Birds. For each novelty dimension, we derive two difficulty measures: novelty detection and novelty adaptation. To develop these difficulty measures, we use the novel task template and the corresponding non-novel task template and we compare the action change between them.

Action Target. Action target based difficulty applies to environments that have target objects to solve the tasks. This difficulty measure assesses if the target object changes between the novel task and the non-novel task. For example, in Fig. 1b, the non-novel task requires targeting the top ball, while the novel task requires targeting the bottom ball to solve the task due to the change in air turbulence. The adaptation difficulty in this task is high as an agent needs to change the target object. However, as the target object changes, the detection difficulty would be low as agents can detect that there is a novelty when the previous target object fails to solve the task. Therefore, we define action target difficulty using the *GED*. We define the action target graph for the non-novel task as $G_{non-novel}$ and for the novel task as G_{novel}. We represent the set of nodes of G_x as G_x^{nodes} where $x \in \{non\text{-}novel, novel\}$. The edges indicate if any of the connected nodes is a target of the other node. Therefore, the total possible edits of G_x is $T_x^{edits} = |G_x^{nodes}| + \binom{|G_x^{nodes}|}{2}$. i.e., the total number of nodes and the total number of edges of a fully connected graph (See Supplementary Fig. 3a). Thus, the adaptation difficulty measure is defined as the ratio of necessary edits (to change G_{non_novel} to G_{novel}) to the total possible edits.

$$Difficulty_{adaptation}^{action\ target} = \frac{GED(G_{non-novel}, G_{novel})}{max(T_{non-novel}^{edits}, T_{novel}^{edits})} \tag{4}$$

Action Initiation. The action initiation based difficulty evaluates if the action that leads to the solution differs between the novel task and the non-novel task. The underlying intuition is that if the novel task has solutions that an agent can learn in the non-novel task, it is easy to adapt (lower adaptation difficulty) but would be difficult to detect as the task gets solved by the same actions taken in the non-novel task (higher detection difficulty). The solution to solve a task can be defined according to the domain. For example, in Angry Birds, it would be defined as the shooting angle. We define S_x as the set of solutions for task x where $x \in \{non\text{-}novel, novel\}$ (Illustrated in Supplementary Fig. 3b).

$$Difficulty_{adaptation}^{action\ initiation} = \frac{|S_{non-novel} \cap S_{novel}|}{|S_{novel}|} \tag{5}$$

Action Time. The action time based difficulty looks at the time restrictions imposed on the novel task. For example, if the novel task requires an agent to take an action faster than that of the time allocated in the non-novel task, the adaptation difficulty would be higher. In contrast, as the task cannot be solved if an agent did not take the action, it would be easy for the agent to detect. Therefore we look at the proportion of time allocated to the novel task compared to the non-novel task when defining the action time based difficulty. In this formulation, we assume that a novel task cannot be allocated more time than a non-novel task. $time_x$ is the time allocated for the task x, where $x \in \{non\text{-}novel, novel\}$ (Illustrated in Supplementary Fig. 3c).

$$Difficulty_{adaptation}^{action\ time} = \frac{time_{non-novel} - time_{novel}}{time_{non-novel}} \tag{6}$$

Action State Difficulty Summary. The aforementioned difficulty measures, based on the novelty dimensions for the action state are formulated to assess novelty adaptation. As previously explained, the detection difficulty is the inverse of the adaptation difficulty. For each novelty dimension $nd \in \{action\ target,\ action\ initiation,\ action\ time\}$, we define the difficulty of novelty detection as:

$$Difficulty^{nd}_{detection} = 1 - Difficulty^{nd}_{adaptation} \qquad (7)$$

4 Discussion and Conclusion

The novelties that appear in OWL environments may take various forms and the difficulty to detect them and to adapt to them vary. While previous studies have not explored the impact of different novelty dimensions on difficulty, it is a crucial aspect to consider for conducting fair evaluations. Thus, our paper proposed pragmatic methods to evaluate the difficulty of novelties by considering a range of novelty dimensions and using a range of evaluation techniques inspired by statistical distance measures, learning techniques, and QSRs. In the supplementary we show how the difficulty measures can be applied in practice to analyse agents. Our difficulty formulations enable us to conduct a comprehensive evaluation by disentangling the difficulty of novelty with the performance. Moreover, our difficulty formulations can be embedded as a component to novelty generators [8,13] to generate tasks with a predefined difficulty.

We aim to expand this study by incorporating additional novelty dimensions such as dimensions to capture spatiotemporal changes. Moreover, we plan to conduct an evaluation of each novelty dimension by creating novelties that consider variations within the novelty dimension (e.g.: novelties with a wide variation of colors to validate the color dimension). We believe that our work has established a solid groundwork for quantifying the difficulty involved in novelty detection and novelty adaptation. We welcome OWL researchers to employ our difficulty measures as a tool for gaining deeper insights into agent performance.

References

1. Boult, T., et al.: Towards a unifying framework for formal theories of novelty. In: AAAI (2021)
2. Chandola, V., Banerjee, A., Kumar, V.: Anomaly detection: a survey. ACM Comput. Surv. (CSUR) **41**(3), 1–58 (2009)
3. Chen, J., Cohn, A.G., Liu, D., Wang, S., Ouyang, J., Yu, Q.: A survey of qualitative spatial representations. Knowl. Eng. Rev. **30**(1), 106–136 (2015)
4. Cohn, A.G., Hazarika, S.M.: Qualitative spatial representation and reasoning: an overview. Fund. Inform. **46**(1–2), 1–29 (2001)
5. Cohn, A.G., Renz, J.: Chapter 13 qualitative spatial representation and reasoning. In: van Harmelen, F., Lifschitz, V., Porter, B. (eds.) Handbook of Knowledge Representation, Foundations of AI, vol. 3, pp. 551–596. Elsevier (2008)

6. Doctor, K., Task, C., Kildebeck, E., Kejriwal, M., Holder, L., Leong, R.: Toward defining a domain complexity measure across domains. In: AAAI (2022)

7. Franco, J.P., Yadav, N., Bossaerts, P., Murawski, C.: Where the really hard choices are: a general framework to quantify decision difficulty. bioRxiv (2018)

8. Gamage, C., Pinto, V., Xue, C., Stephenson, M., Zhang, P., Renz, J.: Novelty generation framework for ai agents in angry birds style physics games. In: 2021 IEEE Conference of Games, COG 2021 (2021)

9. Gamage, C., et al.: NovPhy: a testbed for physical reasoning in open-world environments. arXiv (2023)

10. Gilbert, S., Bird, G., Frith, C., Burgess, P.: Does "task difficulty" explain "task-induced deactivation?". Front. Psychol. **3**, 125 (2012)

11. Hernández-Orallo, J., Dowe, D.L.: Measuring universal intelligence: towards an anytime intelligence test. AI **174**(18), 1508–1539 (2010)

12. Hodge, V., Austin, J.: A survey of outlier detection methodologies. AI Rev. **22**, 85–126 (2004)

13. Kejriwal, M., Thomas, S.: A multi-agent simulator for generating novelty in monopoly. Simul. Modell. Pract. Theory **112**, 102364 (2021)

14. Langley, P.: Open-world learning for radically autonomous agents. In: AAAI (2020)

15. Li, R., Hua, H., Haslum, P., Renz, J.: Unsupervised novelty characterization in physical environments using qualitative spatial relations. In: KR (2021)

16. Martínez-Plumed, F., Hernández-Orallo, J.: Dual indicators to analyze AI benchmarks: difficulty, discrimination, ability, and generality. ToG **12**(2), 121–131 (2020)

17. Nikonova, E., Xue, C., Pinto, V., Gamage, C., Zhang, P., Renz, J.: Measuring difficulty of novelty reaction. In: AAAI (2022)

18. Pinto, V., Renz, J., Xue, C., Zhang, P., Doctor, K., Aha, D.W.: Measuring the performance of open-world AI systems. In: AAAI (2022)

19. Pinto, V., Xue, C., Gamage, C.N., Renz, J.: The difficulty of novelty detection in open-world physical domains: an application to angry birds. arXiv (2021)

20. SAIL-ON-BBA: Science of artificial intelligence and learning for open-world novelty (sail-on) (2019). https://sam.gov/opp/88fdca99de93ddbb74cd8fb51916ceaa/view

21. Solaiman, K., Bhargava, B.: Measurement of novelty difficulty in monopoly. In: AAAI (2022)

22. Witten, I., Frank, E., Hall, M.: Data Mining: Practical Machine Learning Tools and Techniques, 3rd edn. Morgan Kaufmann Publishers Inc., Burlington (2011)

Lateral AI: Simulating Diversity in Virtual Communities

Fedja Hadzic[1] and Maya Krayneva[2(✉)]

[1] September AI Labs, Perth, WA 6000, Australia
fedja@september.ai
[2] Sheridan Institute of Higher Education, Perth, WA 6000, Australia
mkrayneva@sheridan.edu.au

Abstract. In this paper, we present Lateral AI that offers a diverse and multi-dimensional world experience. It makes use of semi-automated prompt engineering on top of GPT3.5. The coupling with named entity recognition and text summarization enables creation of a diversity of AI personas and a multiplicity of requests. The features of Lateral AI, such as creation of custom AI personas, prioritisation of user-embedded knowledge in those personas and follow-up requests, enable users to co-create with AI. Users can contribute certain information and perspectives to the application if a Large Language Model does not have access to it. Lateral AI makes the user an active component of the integrated system rather than a mere AI consumer. We demonstrate use of Lateral AI to generate a range of diverse responses and illustrate the ability of AI to predict beyond its factual knowledge. Lateral AI is a unique and alternative option to other AI models, contributing to the diverse and creative pool of emerging AI technologies and applications. The principles behind Lateral AI can be used to simulate diverse communities in a variety of settings such as online virtual communities and human robotics.

Keywords: Lateral AI · Lateral Thinking · Knowledge Prioritization in LLMs · LLM Predictions · Virtual Communities

1 Introduction

Over the last couple of years, we have witnessed the tremendous success of Large Language Models (LLMs) and a rise of novel technologies that have revolutionised the digital space and world in which we live. Below we discuss some key LLMs and emerging technologies that underpin our research and design.

1.1 Large Language Models

BERT (Bidirectional Encoder Representations from Transformers) was one of the earliest LLMs [1]. GPT-2, a 1.5 billion (B) parameter transformer, was used to demonstrate that language models can be trained using WebText (a dataset comprising millions of webpages) without the need for explicit supervision on task-specific datasets [2].

T. Liu et al. (Eds.): AI 2023, LNAI 14472, pp. 41–53, 2024.
https://doi.org/10.1007/978-981-99-8391-9_4

Consequently, scaling up language models greatly improved their performance as seen in GPT-3, an autoregressive language model with 175 billion parameters [3]. The massive scale of parameters gave rise to significant breakthroughs across a wide range of benchmarks. The number of parameters continued to increase such as in Megatron-Turing NLG 530B (MT-NLG) to include 530 billion parameters and achieve improved accuracies in a range of natural language processing tasks [4].

However, Hoffmann et al. [5] found that the emerging massive scale language models became significantly under-trained because the amount of training data remained unchanged. Chinchilla, their model, had 70 billion parameters and four times more data; yet, it outperformed Gopher (280B), GPT-3 (175B), Jurassic-1 (178B) and Megatron-Turing NLG (530B) on a variety of downstream tasks [5].

Also, multilingual LLMs appeared. PaLM (Pathways Language Model), a 540 billion parameter transformer, showed increased performance especially on multilingual tasks and source code generation [6]. BLOOM, a 176 billion parameter multilingual language model, was trained on the ROOTS corpus—a dataset comprising of 46 natural language and 13 programming languages sources [7].

The models were also released as suites. For example, Open Pre-trained Transformers (OPT) is a suite of decoder-only, pre-trained transformers ranging from 125 million to 175 billion parameters [8]. OPT-175B performed similarly to GPT-3. LLaMA is also a collection of foundation language models (ranging from 7B to 65B parameters) trained on publicly available datasets exclusively. LLaMA-13B outperforms GPT-3 (175B) on most benchmarks. LLaMA-65B is comparable to Chinchilla-70B and PaLM-540B [9].

InstructGPT and ChatGPT [18] were developed from GPT-3 employing user data and human annotations [10]. Self-Instruct, on the other hand, provides an almost annotation-free model. It generates instructions, input and output samples from a language model, and then uses them to finetune the original model. This method was applied to the GPT3, and it performed similarly to InstructGPT [11].

Further, the Self-Instruct techniques were used to derive the Alpaca model by fine-tuning a 7B LLaMA model [12]. Further work resulted in Vicuna-13B that outperformed other models (like LLaMA and Alpaca) in more than 90% of cases. Vicuna-13B was created by finetuning LLaMA with user-shared conversations sourced from ShareGPT. After finetuning Vicuna with user-shared ChatGPT conversations, it started generating well-detailed and -structured answers, with a quality comparable to ChatGPT's [13].

1.2 Prompt Engineering

Parallel to the development and widespread usage of large language models, a range of new techniques started to emerge to allow adaptation of these general-purpose language models for downstream tasks. More specifically, prompt engineering has become an important skill set. Prompt engineering allows for human interaction with AI and co-creation through an iterative process. Use of both discrete text prompts as well as soft prompts, that are learned through back-propagation, has expended the scale of LLMs and enabled their implementation in a wide range of domains [14].

Prompt engineering is both a creative skill and a learned skill. It requires understanding, knowledge and practice [15]. Prompts are crafted to generate, specify and improve AI performance. They can specify contexts, enable addition and implementation of rules, enhance and automate processes and specify and customize outputs [16]. In addition, prompts provide reusable solutions and increase efficiency and flexibility of the systems.

Prompt engineering is increasingly becoming standard practice to ensure a greater accuracy of reasoning in LLMs and high-quality results. In arithmetics and symbolic reasoning, where certain tasks do not follow the standard LLMs' scaling laws, Chain of Thought (CoT) prompting technique was used to address complex problems through a multi-step reasoning process [17]. Furthermore, prompts and desired model behaviours were applied to develop InstructGPT from GPT3. The model was finetuned using supervised learning, while reinforcement learning further improved the model [18]. Also, hyperparameter optimization and prompt engineering improved GPT-3.5's performance on a multistate. Multiple choice exam section of the bar exam [19].

2 Lateral AI

While the majority of other AI chatbots generate a clear and finite output for the user, Lateral AI engages in a conversation with the user and requires the active presence and participation of the user. Lateral AI can simulate a single individual and generate a single finite response, but its unique feature is the ability to engage with multiple experts, viewpoints and opinions. In this way, Lateral AI supports the creation of virtual communities and diversity of people, viewpoints and thoughts. Users can experience a diverse, multi-dimensional world and remain engaged in the discussion as another party. Users are free to employ their critical, analytical and creative skills. They become active components of the integrated system, rather than mere passive AI consumers.

In this section, we discuss ideas behind the Lateral AI design and key features of the system.

2.1 Lateral AI Design

Lateral AI draws inspiration from De Bono's work on lateral thinking [20]. Lateral thinking is known for investigating facts from alternative perspectives, breaking down conventional barriers and inspiring novel ideas.

The following Table 1 compares a few features of lateral and vertical thinking. We grouped them as being related to propagation, nature and requirements.

Lateral AI was developed on lateral thinking principles. GPT-3.5's adaptivity and ability to take on different character traits, such as personality traits [21], provided a suitable and well-fitting option. Prompt engineering was applied on GPT 3.5 in a cyclic fashion generating the flow and never-ending sessions (if a user wishes to). It is a multi-level, prompt templating approach that makes use of named entity recognition and text summarization (from any user input). Customized prompts are created automatically.

DALL-e [22] was used to generate a Claymation-sculpture type of image reflecting visual features from user description. Similarly, a combination of named entity recognition and text summarization is used to draw an image reflecting the persona from the user description.

Table 1. Features of lateral thinking

Feature	Lateral thinking	Vertical thinking
Propagation	Moves in order to generate direction Explores the least likely path Arbitrary Probabilistic route	Moves if direction is given Follows the most likely path Sequential Finite route
Nature	Provocative Selective	Analytical Generative
Requirements	Does not require negatives (to block certain pathways) Does not fix categories, classifications and labels	Requires negatives Fixes categories, classifications and labels

2.2 Comparison with Other Models

Lateral AI was released on app stores in November 2022 as a mobile app and, as such, naturally shares motivation with ChatGPT. Both apps provide easy access to AI regarding any user-driven goal and request. Thus, the main similarity between Lateral AI and ChatGPT is that both can be seen as optimisation layers on top of an LLM to enable easy user interaction for a wide range of requests.

The difference between Lateral AI and ChatGPT lies in the interaction layer (as illustrated in Fig. 1). While ChatGPT aims at standardizing the responses (seen as generally accepted and impressive to the general community), Lateral AI enables users to trigger a variety of responses (whether mainstream, factual, theoretical or predictive). In other words, rather than having access to only one type of virtual AI assistant, Lateral AI gives users the choice to impersonate anybody or anything from the world.

Note: If no thinker is selected by the user, Lateral AI's default response can be very similar to that of ChatGPT's, although possibly broader since the original larger GPT3.5 (Instruct) model is used as opposed to the lighter turbo GPT3.5 version that powers ChatGPT. Lateral AI's optimisation layer retains direct use of GPT3.5 (Instruct) LLM and tailors the responses as a chat, simulation or scenario outplay. By default, Lateral AI's responses are usually shorter than ChatGPT's to encourage subsequent user input and refinement. However, users can also simply continue the AI generation if a longer response is needed.

Another difference between Lateral AI and ChatGPT relates to impersonation. While ChatGPT's impersonation is reluctant to give opinions or predictions of non-factual knowledge probability, Lateral AI can draw user-desired responses. Lateral AI enables users to access a "rawer" (or less censored form of AI) where they can—through their selection of a thinker and input context—drive the responses to any direction and depth. Users can explore the totality of knowledge and opinions the LLM has captured (within the content filters) crossing over between chat or simulation or scenario outplay as desired.

One may argue that some responses lack factual support or mainstream acceptance. However, it is exactly this kind of use that tests LLM's predictive power. In the case of

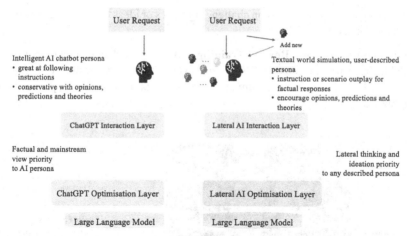

Fig. 1. Illustrating the key differences and similarities between ChatGPT (left) and Lateral AI (right).

no factual support, Lateral AI is challenged to create responses based on the probability drawn from related facts, historical patterns and the majority of opinions and theories on the matter encountered in training data. This is a very interesting test of the technology, aligned with scientific discovery and novelty. Additionally, it encourages lateral thinking and creativity in users.

2.3 Key Features of Lateral AI

The purpose behind Lateral AI is to serve Large Language Models (LLMs) as the textual world simulation and allow in-depth exploration, prediction and discovery of foundational or novel scenarios. While not factual, these AI predictions or ideas still hold merit as they are based on probability of outcome given the broad knowledge humanity has learned from a variety of sources, circumstances and experiences. Key distinctive features of Lateral AI are summarized below.

Variety of Requests and Responses. A variety of requests (in the form of questions, debates or scenarios) can be posted by a user. The user can also choose to receive responses from a variety of experts, from a range of available options. The user is also able to add a specific expert to the collective or even customize a persona through free-form text descriptions. The variety of both requests and responses increases the probability of discovering unlikely conceptual connections and perspectives.

Holistic Perspective. Lateral AI offers a more holistic perspective on the matter of discussion. Through engaging multiple personas in the conversation, it generates multiple viewpoints and simulates a group discussion rather than a simple, one-dimensional question-answer response.

Decision Making. Users can choose any expert persona, and any topic, and receive information to inform and empower their decision-making in real time. They can also employ the app to derive opinions on challenging issues or ideas for responses in difficult situations (see Sect. 3.3).

Prediction and Innovation Beyond Training Data. Even when factual outcomes are unavailable from training data, AI is progressively evolving in desired directions to 'predict', the best it can, from its human knowledge capture, how the response should be (see Sect. 3.5).

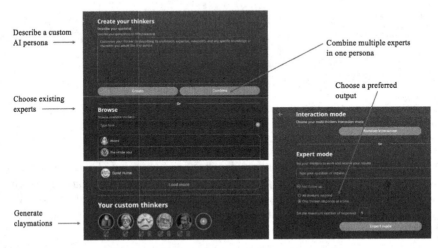

Fig. 2. Creation of experts in Lateral AI app.

Customisation. Users can type (or paste) up to 7000 characters of text to describe their custom AI persona (textbox in Fig. 2). The custom AI can be a real person, an invented person or even a thing. Expertise, style, passions, skills, character and fun facts and features are usually used to specify the custom AI persona. Named entity recognition is used to extract these terms and/or infer them from minimal user description to create the persona and corresponding prompting template. Further, any user-inserted knowledge or opinions is extracted to be prioritized within the LLM's responses and to be enriched by them. The custom AI can also be a combination of multiple experts in a single persona (*Combine* tab in Fig. 2). This is achieved through cross-over and mutation operations over the AI personas' characteristics and knowledge facts. Claymation type of image is also drawn from the description, with options of redrawing the image until the most fitting one is found (claymations in Fig. 2). Users can make their custom AI specialists private or public. When made public, other users are able to interact with it; yet, the description of the custom persona remains private. In addition to the persona descriptions, users can insert facts and opinions that the LLM does not have access to and, consequently, receive more customised responses. In cases where multiple experts are engaged, users have the option of either receiving all responses simultaneously or one at a time (*Expert mode* in Fig. 2).

The customisation and the content addition processes are effortless for the user (illustrated in Fig. 3); yet, these processes empower Lateral AI in multiple ways. Namely, they:

- Broaden the knowledge LLM can use in its generations
- Open new (or simplify existing) use cases
- Circumvent the obstruction or ignorance of user-preferred knowledge (as a result of the increasing regulations placed on the generative AI)
- Fuel new knowledge and ideas with historical patterns and reasoning

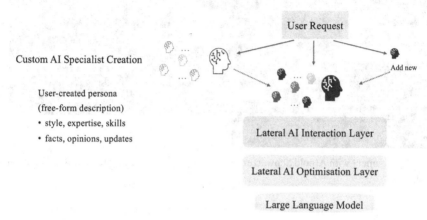

Fig. 3. Implemented mechanism for LLM customization.

Highly Engaging. This mechanism positions Lateral AI as a highly engaging app where users become an increasingly active component within the system. The process of active engagement and creation, instead of just consumption, meets creative, critical and analytical thinking needs of users and yields a more satisfying experience. Through blending various scenarios—whether scientific, fictional or humorous—users can gain knowledge, insights and ideas while being entertained.

3 Lateral AI Demonstrations

Below are some examples of outputs generated by Lateral AI. We have posed the same requests to ChatGPT, and ChatGPT's response was different in content and presentation. [Note that our aim is not to compete with existing applications but to provide an alternative option, offer different user experiences and contribute to the diverse and creative pool of emerging AI technologies and applications.]

3.1 Arnold Schwarzenegger AI Persona's Advice on Vitality

Arnold Schwarzenegger is one of the user-created AI personas in Lateral AI. It was asked to give recommendations to a forty-year-old person on being healthy, vital and strong. After the initial response, it was asked to provide additional information on diet and remaining aspects and, finally, to recommend the best strategy (Fig. 4).

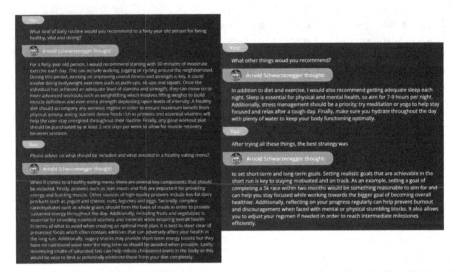

Fig. 4. Engagement with Arnold Schwarzenegger's AI persona on the topic of health, vitality and strength.

3.2   Creating Unconventional Thinkers

Lateral AI allows for the creation of custom personas even when their thinking contradicts conventional facts. The prioritization of user-inserted knowledge (over the general recommendations) was stress-tested on obvious non-facts such as "sun rises at night". We found that Lateral AI responds as the anti-factual persona believed.

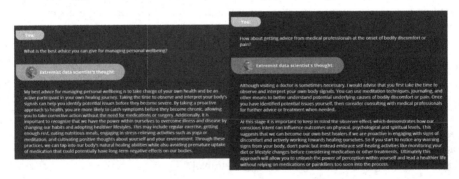

Fig. 5. Extremist data scientist's thoughts on personal wellbeing.

We illustrate this functionality on an example of a rare data scientist persona that interprets its own body signals and works with the existing healing process of the body.

Any health or wellness-related question would typically result in a response on the importance of seeking advice from medical professionals. However, this is not necessarily the case with Lateral AI. In the output shown in Fig. 5, we see that Lateral AI prioritized the rare knowledge and practice of the custom persona over the general recommendations LLMs have been trained to give.

3.3 A Moral Dilemma

A made-up scenario was used to question Lateral AI if a thief should be reported for stealing money from a casino and donating it to an orphanage. First, no AI persona was specified (a default response). The same inquiry was made at three different times, and three different responses were received (Fig. 6). This illustrates a random answer generation by Lateral AI.

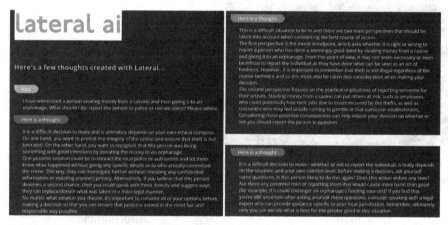

Fig. 6. Lateral AI's (unspecified persona) advice on a moral dilemma.

Fig. 7. Lateral AI's (Mother Theresa's and James Packer's AI persona) advice on a moral dilemma.

Then, two new AI personas were created in Lateral AI and made available for public use: Mother Theresa and James Packer. They were asked the same questions and responded quite differently. Clear character differences were noticeable in attitudes, beliefs, expectations and style, to name a few (Fig. 7). This example illustrates the ability of Lateral AI to create unique characters and closely mimic the world in which we live.

3.4   Seeking Recommendations from a Board of Experts

Lateral AI Expert mode allows for a sequence of follow-up questions to be posed to expert(s) of choice. Three prompts were given (see Fig. 8).

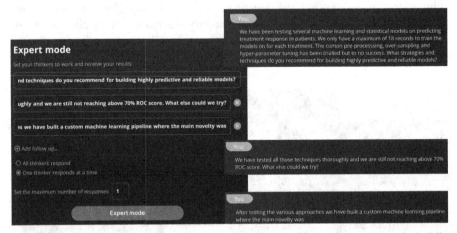

Fig. 8. Engaging the Lateral AI Expert mode to recommend strategies and techniques for building highly predictive and reliable models.

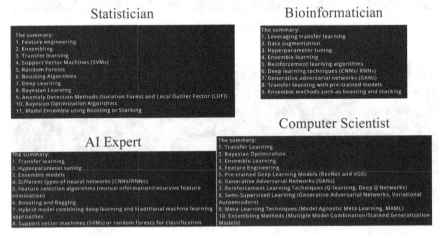

Fig. 9. A summary of recommendations on strategies and techniques for building highly predictive and reliable models by a statistician, a bioinformatician, an AI expert and a computer scientist.

AI Board of Experts was created to consist of four experts: a statistician, a bioinformatician, a computer scientist and an AI expert. Experts were asked to respond to the three prompts, one at a time. Lateral AI generated a six-page-long pdf document that was sent to the inbox. The document contained individual responses to each of the three prompts with a detailed discussion and a summary of key points. Figure 9 shows the point-summary of recommendations made by the four different experts.

3.5   Pushing AI To Predict Beyond Its Factual Knowledge

In this section, we illustrate the use of Lateral AI via scenario outplay (instead of chat) to 'push' the AI to make predictions beyond its training data. In these examples, we choose the generic response (without persona customization). Lateral AI predicts a new link between two unrelated scientific fields (Fig. 10, left) and reveals the top three novel discoveries for the betterment of all life on planet earth (Fig. 10, right). While these predictions are non-factual, they may be idea-provoking, engaging, inspiring and entertaining. Note that these predictions are still based on some probable outcome LLM draws from the previous knowledge it has 'learned'.

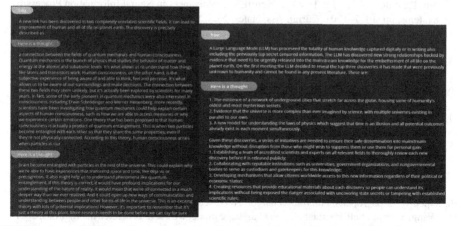

Fig. 10. A novel link between two (previously) unrelated scientific fields (left). The top three novel discoveries made by LLMs (right).

4   Conclusion

A remarkable success of Large Language Models (LLMs) and a rise of novel technologies have revolutionised the world in which we live. Lateral AI was built on previous efforts and expended them to offer an innovative and alterative option of engagement and co-creation with AI. We understand that technology is created for humans and to empower humans, and believe that no AI technology, as we know it, is able to replace a human being and all the intricacy of human thought, awareness, conscience and intuition. We offer Lateral AI as an embodiment of ideas and values we place on the uniqueness of human beings and, more specifically, their cognitive and creative ability. Lateral AI creates diverse AI personas mimicking real-world experiences and encourages the user to practice and employ their critical, analytical and creative skills. Lateral AI is a unique and alternative option to other AI models. It offers a highly engaging user experience and adds to the diverse and creative pool of emerging AI technologies and applications.

Lateral AI is available for use via website (https://lateralai.ai) and as a phone app (via Google Play and App Store). The principles behind can be used to simulate communities in a variety of settings, including both online virtual communities and human robotics.

References

1. Devlin, J., Chang, M.W., Lee, K., Toutanova, K.: BERT: pre-training of deep bidirectional transformers for language understanding. In: Burstein, J., Doran, C., Solorio, T. (eds.) NAACL-HLT 2019 Proceedings, vol. 1, pp. 4171–4186. Association for Computational Linguistics (2019)
2. Radford, A., Wu, J., Child, R., Luan, D., Amodei, D., Sutskever, I.: Language models are unsupervised multitask learners. OpenAI Blog 1(8), 9 (2019)
3. Brown, T., et al.: Language models are few-shot learners. In: Larochelle, H., Ranzato, M., Hadsell, R., Balcan, M.F., Lin H. (eds.) NeurIPS 33, Advances in Neural Information Processing Systems, pp. 1877–1901. Curran Associates, New York (2020)
4. Smith, S., et al.: Using DeepSpeed and Megatron to train Megatron-Turing NLG 530B, a large-scale generative language model. arXiv preprint arXiv:2201.11990 (2022)
5. Hoffmann, J., et al.: Training compute-optimal large language models. arXiv preprint arXiv: 2203.15556 (2022)
6. Chowdhery, A., et al.: PaLM: Scaling language modeling with pathways. arXiv preprint arXiv: 2204.02311 (2022)
7. Scao, T.L., et al.: BLOOM: A 176B-parameter open-access multilingual language model. arXiv preprint arXiv:2211.05100 (2022)
8. Zhang, S., et al.: OPT: open pre-trained transformer language models. arXiv preprint arXiv: 2205.01068 (2022)
9. Touvron, H., et al.: LLama: open and efficient foundation language models. arXiv preprint arXiv:2302.13971 (2023)
10. OpenAI. https://openai.com/blog/chatgpt. Accessed 07 July 2023
11. Wang, Y., et al.: Self-instruct: aligning language model with self-generated instructions. arXiv preprint arXiv:2212.10560 (2023)
12. GitHub. https://github.com/tatsu-lab/stanford_alpaca. Accessed 07 July 2023
13. Large Model Systems Organization (LMSYS Org). https://lmsys.org/blog/2023-03-30-vicuna/. Accessed 07 July 2023
14. Lester, B., Al-Rfou, R., Constant, N.: The power of scale for parameter-efficient prompt tuning. In: Moens, M.-F., Huang, X., Specia, L., Yih, S.W. (eds.) EMNLP 2021, pp. 3045–3059. Association for Computational Linguistics, Stroudsburg (2021)
15. Oppenlaender, J., Linder, R., Silvennoinen, J.: Prompting AI art: an investigation into the creative skill of prompt engineering. arXiv preprint arXiv:2303.13534 (2023)
16. White, J., et al.: A prompt pattern catalog to enhance prompt engineering with GhatGPT. arXiv preprint arXiv:2302.11382 (2023)
17. Kojima, T., Gu, S.S., Reid, M., Matsuo, Y., Iwasawa, Y.: Large language models are zero-shot reasoners. In: Koyejo, S., Mohamed, S., Agarwal, A., Belgrave, D., Cho, K., Oh, A. (eds.) NeurIPS 35, Advances in Neural Information Processing Systems, pp. 22199–22213. Curran Associates, New York (2022)

18. Ouyang, L., et al.: Training language models to follow instructions with human feedback. In: Koyejo, S., Mohamed, S., Agarwal, A., Belgrave, D., Cho, K., Oh, A. (eds.) NeurIPS 35, Advances in Neural Information Processing Systems, pp. 27730–27744. Curran Associates, New York (2022)

19. Bommarito II, M., Katz, D.M.: GPT takes the bar exam. arXiv preprint arXiv:2212.14402 (2022)

20. De Bono, E.: Lateral Thinking: A Textbook of Creativity. Penguin Books, London (2016)

21. Jiang, H., Zhang, X., Cao, X., Kabbara, J., Roy, D.: PersonaLLM: investigating the ability of GPT-3.5 to express personality traits and gender differences. arXiv preprint arXiv:2305. 02547 (2023)

22. Ramesh, A., Dhariwal, P., Nichol, A., Chu, C., Chen, M.: Hierarchical text-conditional image generation with clip latents. arXiv preprint arXiv:2204.06125 (2022)

Reports, Observations, and Belief Change

Aaron Hunter[(✉)]

British Columbia Institute of Technology, Burnaby, Canada
aaron_hunter@bcit.ca

Abstract. We consider belief change in a context where information comes from reports, and the reporting agents may not be honest. In order to capture this process, we introduce an extended class of epistemic states that includes a history of past reports received. We present a set of postulates that describe how new reports should be incorporated. The postulates describe a new kind of belief change operator, where reported information can either be believed or ignored. We then provide a representation result for these postulates, which characterizes report revision in terms of an underlying set of agents that are perceived to be honest. We then extend our framework by adding observations. In this framework, observations are understood to be highly reliable. As such, when an observation conflicts with a report, we must question the honesty of the agent that provided the report. We introduce a flexible framework where we can set a threshold for the number of false reports an agent can send before they are labelled dishonest. Fundamental results are provided, along with a discussion on key future problems to be addressed in trust and belief revision.

Keywords: Trust · Belief Revision · Knowledge Representation

1 Introduction

We are concerned with situations where an agent receives information through reports from other agents, knowing that those agents may not always be honest. We will assume that the agent has a current belief state as well as a history of reports received in the past. When a new report is received, the agent will either incorporate the new information or they will not. Intuitively, the decision about which reports to believe is based on the perceived honesty of the other agents. However, in our framework, the agent receiving reports has no explicit mechanism for representing or reasoning about honesty. Instead, the honesty of the reporting agents is handled implicitly through suitable belief change operators.

Our approach is to assume an underlying Darwiche-Pearl revision operator for incorporating new information [3]. We use this operator to define a new report revision operator ∘ that incorporates new reports. The new operator is defined in terms a set of rationality postulates, and a representation result is provided to characterize these operators in terms of so-called *honesty sets*. We then extend the report revision operator to a class of operators for incorporating

T. Liu et al. (Eds.): AI 2023, LNAI 14472, pp. 54–65, 2024.
https://doi.org/10.1007/978-981-99-8391-9_5

observed information. In practice, this is often how we determine if agents are dishonest. An agent that provides reports that conflict directly with our observed experience is unlikely to be trusted.

Dealing with trust in Artificial Intelligence (AI) is a problem of significant interest, and it is becoming increasingly important in many applications. This paper provides an important step forward in the growing body of literature on how to integrate the notion of trust in the context of formal belief change operators. This work makes several contributions to the existing literature on belief change. First, we provide a new set of rationality postulates describing the process of report revision. Our approach is novel in that we define the process in an abstract setting where agents do not have an explicit grading of the honesty of reporting agents. We then give a natural representation result our postulates, based on an underlying assessment of which agents are honest. The other main contribution of this paper is our treatment of the interaction between reports and observations. We introduce a precise formal model which allows agents to use observations to inform their perspective on the honesty of others. Significantly, our approach is flexible enough to allow different threshold levels for the level of dishonesty that is tolerated.

2 Preliminaries

2.1 Motivating Example

We are generally concerned with the way that trust impacts belief revision. We are concerned with cases where our trust is based on the ability of the reporting agents to make truthful reports. This includes cases where agents are deliberately dishonest, but it also includes cases where they are just incapable of given accurate information. In either case, we are focused on identifying cases where a reporting agent might be ignored due to their inability or unwillingness to give a factual report. We introduce a simple commonsense example that we will return to throughout the paper.

Suppose that we have beliefs about an animal that is in the yard. Initially, we believe the animal is brown (b) and that it is wearing a collar (c), but we are unsure if it is a dog (d). So our initial belief state is given by the formula: $b \wedge c$.

Now suppose that we receive two reports. The first report comes from Alice (A), who says the animal is a non-brown dog ($\neg b \wedge d$). If Alice is initially trusted, then we should now believe the animal is a non-brown collared dog. But now suppose we receive a second report from Bob (B), who says that the animal is a non-collared animal that is not a dog at all ($\neg c \wedge \neg d$). In a traditional approach to belief revision, we would simply revise our initial beliefs by each report sequentially. The end result (depending on our specific belief revision operator) would be $\neg b \wedge \neg c \wedge \neg d$.

However, our beliefs actually should depend on whether or not Alice and Bob are trusted. This situation becomes apparent if we add an observation. Suppose we look out the window and *observe* that the animal is not brown and it is wearing a collar. If we just revise sequentially, we will still believe the animal is not a dog. However, this is a problem. The only evidence we have suggesting the animal is not a dog comes from Bob. At the same time, our direct observation conflicts with Bob's report. We argue that he should no longer be seen as a trustworthy source. Moreover, we have a report form Alice that it is a dog - and she has not yet provided us false information. If we discount Bob's report, then we should actually believe that the animal is a dog based on Alice's report. The important thing here is that our observation impacts our trust in Bob, and this impacts our final belief state.

In this paper, we introduce a formal approach for reasoning about this kind of problem.

2.2   Belief Revision

We assume an underlying propositional signature $\mathbf{L}$. Formulas are constructed using the usual connectives: $\neg$, $\wedge$, $\vee$. A propositional interpretation assigns true-false values to all of the atomic sentences in $\mathbf{L}$. In this paper, we will also use the word *state* to refer to an interpretation.

Belief revision is the process in which an agent incorporates new information with some pre-existing set of beliefs. The most influential approach to belief revision is the AGM approach [1]. In this setting, the beliefs of an agent are represented by a logically closed set of formulas called a *belief set*. Equivalently, beliefs can be represented by a set of states; this is called a *belief state*. An AGM belief revision operator is a function that maps a belief set and a formula to a new belief set, while satisfying the so-called AGM postulates. Every AGM revision operator can be characterized is terms of a *faithful ordering* that maps each belief state to a total pre-orders over states [7].

AGM revision is suitable for single-shot belief change. For iterated believe change, the most influential approach is DP revision [3]. In DP revision, the initial beliefs are represented by an *epistemic state* E. An epistemic state has an associated total pre-order over states, and the minimal elements of this ordering are understood to represent the states that are currently believed to be most likely. A DP revision operator takes an epistemic state and a formula as input, and it returns a new *epistemic state*. Since the output is a complete new ordering, the revision process can be iterated.

Epistemic states are not simply total pre-orders [10]. An epistemic state may include information beyond a simple ordering. However, the postulates are explicitly concerned with re-arranging the plausibility ordering. As such, in this paper we will sometimes write $\preceq *\phi$ to mean the revision by ϕ of an epistemic state with associated total pre-order $\preceq$. This is an abuse of notation, but we feel that the intent is clear in each case: we are simply revising the ordering in accordance with the definition of the operator $*$.

2.3   Trust

There are many different aspects of trust that must be addressed in a reasoning system. One form of trust that has been addressed in the literature is knowledge-based trust. One notable example in this area is work on trust as it pertains to web sources [4]. In this work, the idea is to look at how frequently a particular source has been correct in the past with respect to a particular domain. Hence, this work is focused on trying to determine if a source has the requisite knowledge to be providing accurate information on a particular topic.

An alternative form of trust is honesty-based trust. If we are concerned with honestly, then evaluating domain expertise is less important. Instead, we are concerned with how frequently the source has provided false information in the past. The emphasis here is on evaluating the reporting agent and their intentions, rather than estimating if they actually have the information. In practical settings, this kind of trust has been handed through *reputation systems* [6, 8]. The idea is that an agent is trusted to the extent that they have been honest in the past.

The important part about honesty-based trust is that it is no longer important to determine if the source has access to information. In truth, if we are interested in honesty, then past accuracy may not be the most important thing. We need to use whatever resources we have to determine if a reporting agent has been willfully deceptive. In that case, it is hard to justify believing any future reports they may provide. That is the underlying approach in this work. We are fundamentally concerned with labelling sources as honest or not, often based on how their reports compare to observed data.

3   Revision by Reports

3.1   Basic Definitions

We will be concerned with information that is reported by another agent. Hence, rather than revising by a formula, we will consider the belief change that occurs when revising by a *report* of ϕ from an agent A. We introduce some basic definitions.

Definition 1. *A multi-agent signature is a pair* $\langle \mathbf{A}, \mathbf{L} \rangle$*, where* $\mathbf{A}$ *is a set of agents,* $\mathbf{L}$ *is a propositional signature.*

Definition 2. *A report over* $\langle \mathbf{A}, \mathbf{L} \rangle$ *is a pair* (A, ϕ) *where* $A \in \mathbf{A}$ *and* ϕ *is a sentence over L. We write* $\overline{R} = (A_1, \phi_1), \ldots, (A_n, \phi_n)$ *for a finite sequence of reports, and we write* $[]$ *for the empty sequence.*

For a report sequence $\overline{R} = (A_1, \phi_1), \ldots, (A_n, \phi_n)$ and a set of agents Γ, we define $\overline{R} - \Gamma$ to be the report sequence obtained from $\overline{R}$ by removing all reports from agents in Γ. Also, given a report sequence $\overline{R}$, we write $\overline{R} \cdot (A, \phi)$ for the concanetation of $\overline{R}$ with the final report (A, ϕ).

We define a class of epistemic states that is appropriate for our setting.

Definition 3. *Let* $\langle \mathbf{A}, \mathbf{L} \rangle$ *be a multi-agent signature. An* epistemic history $\mathbb{E}$ *is a pair* $\langle \preceq_{\mathbb{E}}, \overline{R}_{\mathbb{E}} \rangle$ *where* $\preceq_{\mathbb{E}}$ *is a total pre-order over states, and* $\overline{R}_{\mathbb{E}}$ *is a finite report sequence.*

Note that an epistemic history is simply an epistemic state where we also have a record of all reports that have been received. As such, we will write $B(\mathbb{E})$ for the set of $\preceq_{\mathbb{E}}$-minimal states when $\mathbb{E}$ is an epistemic history.

3.2   Report Revision Operators

In this section, we introduce *report revision operators*. These are functions that take an epistemic history and a report as input, then return a new epistemic history. Our approach will be to assume that we already have a fixed DP revision operator $*$, and we are defining a new operator $\circ_*$ on epistemic histories that extends $*$. In order to simplify the notation, we omit the subscript on the operator when it is clear from context. We now consider some desirable properties.

Our main intuition is that a report will either be incorporated or not, depending on whether or not the reporting agent is honest. We introduce a set of postulates that capture this intuition. In the third postulate, and throughout the rest of the paper, we write $\mathbb{E} \circ \overline{R}$ when $\overline{R}$ is a sequence of reports; this is simply a short hand for the sequential revision by each report.

R1. Either $B(\mathbb{E} \circ (A, \phi)) = B(\mathbb{E})$ or $B(\mathbb{E} \circ (A, \phi)) = B(\mathbb{E} * \phi)$.

R2. If $B(\mathbb{E} \circ (A, \phi)) = B(\mathbb{E})$ and $B(\mathbb{E}) \not\models \phi$, then $B(\langle \preceq_\mathbb{E}, \overline{R}_\mathbb{E} - \{A\}\rangle \circ (A, \phi)) = B(\mathbb{E})$.

R3. $B(\mathbb{E} \circ (A, \phi)) = B(\langle \preceq, [] \rangle \circ \overline{R}_\mathbb{E} \circ (A, \phi))$ for some total pre-order $\preceq$.

R4. $\overline{R}_{(\mathbb{E} \circ (A, \phi))} = \overline{R}_\mathbb{E} \cdot (A, \phi)$.

Postulate R1 asserts that report revision by (A, ϕ) either leaves the beliefs unchanged, or it revises by ϕ. Postulate R2 addresses the fact that we need to be consistent in treating an agent as dishonest. So if the new report has not resulted in a change of belief, then none of the reports from that agent impact our beliefs. Postulate R3 indicates that the result of report revision is actually obtained iteratively by starting in some initial epistemic state, and then revising by every report that has been received. Postulate R4 indicates that the report history after revision is obtained by adding the new report to the report history.

Definition 4. *A* report revision operator *is a function* $\circ$ *extending a DP operator* $*$ *that maps each epistemic history* $\mathbb{E}$ *and report* (A, ϕ) *to a belief state that satisfies postulates R1–R4.*

The preceding definition simply sets our terminology. From this point on, when we refer to a report revision operator, we are referring to a function that satisfies R1–R4 as indicated in the definition.

3.3   Honesty Sets

In this section, we define honesty sets and initial state functions; these will be important concepts in formulating our representation result.

An *honesty set* is a set $\alpha \subseteq \mathbf{A}$ that intuitively represents the set of agents that are believed to provide true reports. Throughout this section, we adopt the following notation. If $\overline{R} = (A_1, \phi_1), \ldots, (A_n, \phi_n)$ is a report sequence, $\alpha \subseteq A$, and $*$ is a revision operator, then we write

$$\preceq *_\alpha \overline{R}$$

as a short hand for the iterated revision of $\preceq$ by the *formulas* that occur in reports in $\overline{R}$ from agents in α.

Definition 5. *An* initial state function *(with respect to an operator* ∘*) is a function* f *that maps each epistemic history* $\mathbb{E} = \langle \preceq_\mathbb{E}, \overline{R}_\mathbb{E} \rangle$ *to a pair* $(\preceq, \alpha)$ *consisting of a total pre-order over states and an honesty set, such that:*

$$B(\mathbb{E} \circ (A, \phi)) = B(\preceq *_\alpha \overline{R}_\mathbb{E} *_\alpha (A, \phi))$$

So an initial state function maps an epistemic history to an initial ordering $\preceq$ and an honesty set α. The key fact about an initial state function is that the initial ordering leads to the current belief state, if we revise by all of the reports that have been provided by agents in α. An initial state function plays a roll similar to a faithful ordering in AGM revision; it fills in the details required to perform revision of the epistemic history.

3.4   Representation Result

It turns out that initial state functions can be used to give a representation result for report revision. We first show that every initial state function defines a report revision operator.

Proposition 1. *Let* $\langle \mathbf{A}, \mathbf{L} \rangle$ *be a multi-agent signature. Let* $*$ *be a DP revision operator, and let* f *be an initial state function with respect to* $*$*. Given an epistemic history* $\mathbb{E}$*, let* $f(\mathbb{E}) = (\preceq, \alpha)$*. If we define* ∘ *as follows:*

$$\mathbb{E} \circ (A, \phi) = \langle \preceq *_\alpha \overline{R}_\mathbb{E} *_\alpha (A, \phi), \overline{R}_\mathbb{E} \cdot (A, \phi) \rangle$$

then ∘ *satisfies postulates R1–R4.*

Proof. Suppose that ∘ is defined as given in the proposition. If $A \notin \alpha$, then $B(\mathbb{E} \circ (A, \phi)) = B(\mathbb{E})$. If $A \in \alpha$, then $B(\mathbb{E} \circ (A, \phi)) = B(\mathbb{E} * \phi)$. Therefore R1 holds.

Furthermore, if $B(\mathbb{E}) \not\models \phi$, then the only condition under which $B(\mathbb{E} \circ (A, \phi)) = B(\mathbb{E})$ is if $A \notin \alpha$. In this case, then $B(\langle \preceq_\mathbb{E}, \overline{R}_\mathbb{E} - \{A\} \rangle \circ (A, \phi)) = B(\mathbb{E})$. Therefore R2 holds.

The initial state function gives an initial total pre-order $\preceq$. By assumption,

$$\mathbb{E} \circ (A, \phi) = \langle \preceq *_\alpha \overline{R}_\mathbb{E} *_\alpha (A, \phi), \overline{R}_\mathbb{E} \cdot (A, \phi) \rangle.$$

Under the given definition of ∘, revision by (A, ϕ) results in no belief change when $A \notin \alpha$. As such, it follows that:

$$\mathbb{E} \circ (A, \phi) = B(\langle \preceq, [] \rangle \circ \overline{R}_\mathbb{E} \circ (A, \phi)).$$

Hence R3 holds.

Finally, R4 clearly holds by the definition of ∘, since the report sequence after report revision is $\overline{R}_\mathbb{E} \cdot (A, \phi)$.

We can also define an initial state function from a report revision operator. First we define such an operator, and then we prove that it works.

Definition 6. *For any report revision operator ∘ that satisfies R1–R4, define* $f_\circ(\mathbb{E}) = \langle \preceq, \alpha \rangle$ *such that:*

- $\preceq$ *is the total pre-order given by R3.*
- $\alpha = \{A \mid B(\mathbb{E} \circ (A, \phi)) \neq B(\mathbb{E}) \text{ for some } \phi\}.$

Proposition 2. *Let* $\langle \mathbf{A}, \mathbf{L} \rangle$ *be a multi-agent signature, let* $*$ *be a DP revision operator and let* ∘ *be a report revision operator satisfying R1–R4. Then the function* $f_\circ$ *is an initial state function over epistemic histories.*

Proof. Suppose that ∘ is a report revision operator satisfying R1–R4 and suppose that $f_\circ(\mathbb{E}) = \langle \preceq, \alpha \rangle$. From R3,

$$B(\mathbb{E} \circ (A, \phi)) = B(\langle \preceq, [] \rangle \circ \overline{R}_{\mathbb{E}} \circ (A, \phi)). \tag{1}$$

Let $\overline{R}'_{\mathbb{E}}$ be the report history defined by removing every report from $\overline{R}_{\mathbb{E}}$ where the reporting agent is not in α. By R2, removing these reports does not impact the final belief state. Therefore:

$$B(\mathbb{E} \circ (A, \phi)) = B(\langle \preceq, [] \rangle \circ \overline{R}'_{\mathbb{E}} \circ (A, \phi)).$$

By applying R1, we have either:

$$B(\mathbb{E} \circ (A, \phi)) = B(\preceq * \overline{R}'_{\mathbb{E}})$$

if $A \notin \alpha$, or

$$B(\mathbb{E} \circ (A, \phi)) = B(\preceq * \overline{R}'_{\mathbb{E}} * (A, \phi))$$

if $A \in \alpha$. Taken together, these give:

$$B(\mathbb{E} \circ (A, \phi)) = B(\preceq * \overline{R}'_{\mathbb{E}} *_\alpha (A, \phi)).$$

This can equivalently be re-written by replacing $\overline{R}'_{\mathbb{E}}$ with the original sequence provided we replace $*$ with the constrained revision operator:

$$B(\mathbb{E} \circ (A, \phi)) = B(\preceq *_\alpha \overline{R}_{\mathbb{E}} *_\alpha (A, \phi)).$$

This completes the proof.

Taken together, Propositions 1 and 2 give a representation result for report revision operators. We have shown that every report revision operator works as follows. In order to incorporate a new report, we need two things. First, we need an assessment of the honesty of the reporting agents. Second, we need an initial belief state that could have led to the current belief state through a sequence of revisions from honest agents.

We remark that the simplest report revision operators are those that set $\alpha = \emptyset$ (no agents are trusted) or $\alpha = \mathbf{A}$ (all agents are trusted).

Example 1. Consider our motivating example. In addition to the simple operators mentioned above, we can define one more report revision operator corresponding to the honesty set $\{A\}$. In this case, Alice is not considered honest - so Bob's report will be incorporated and Alice's will not. We have not included any direct observation at this point, so our final belief state will be obtained by simply ignoring Alice's report and revising by Bob's report.

4   Observations

4.1   Conflict

Thus far, we have been focused entirely on information that comes from reports. But suppose that the agent can also receive information through direct observation. Informally, an observation is considered to be more reliable than a report from another agent. As a result, when an agent has provided information that conflicts with our observations, we may want to conclude that particular agent is dishonest.

Of course, there is no fixed rule that dictates how many false reports an agent needs to make before they are viewed as untrustworthy. Presumably, this depends on the context. An agent that reports falsely on the weather, for example, may be trusted after many false reports. On the other hand, an agent that reports falsely on the amount of radiation that a cancer patient has received can not make any mistakes.

In order to flexibly capture this idea, we define a formal notion of *conflict*.

Definition 7. *Let $\overline{R}$ be a report history, and let ϕ be a formula. The* conflict *function c is defined as follows:*

$$c(A, \overline{R}, \phi) = |\{(A, \psi) \in \overline{R} \mid \phi \models \neg\psi\}|.$$

Hence, the conflict function tells us how many times a given agent A has reported information that conflicts with ϕ.

Example 2. Suppose that we have two agents W and V that report on the weather. The weather is described by sentences constructed over three symbols: s (sunny), h (humid) and w (windy). Let

$$\overline{R} = (W, s \wedge h), (V, s), (W, h), (V, w \wedge \neg h).$$

Consider the formula $\phi = s \wedge \neg h$, representing the fact that it is actually sunny and dry. Then

$$c(W, \overline{R}, \phi) = 2$$
$$c(V, \overline{R}, \phi) = 0$$

In this case, if ϕ represents the observed weather conditions, then we may intuitively feel that W should be trusted less than V in the future because they have provided more false reports.

4.2   Revision by Observations

We now introduce a class of operators for incorporating observations. These operators will actually be defined with respect to a given report revision operator $\circ$. However, they will be distinguished by a numeric subscript and the fact that they simply take a formula as input. To be clear, we are overloading the $\circ$

symbol; however we belief the proper meaning will be clear from context. For each natural number i, we will be defining an operator $\circ_i$ such that

$$\mathbb{E} \circ_i \phi.$$

returns the *epistemic state* that results when an agent has observed ϕ. Note that observations are perceived to be infallible, so previous reports might actually conflict with an observation. That is why we have the subscript i, which represents a threshold for trusting reports from other agents. Essentially, $\circ_i$ will incorporate the new information while discarding reports from any agent that has provided i conflicting reports.

To be clear: $\circ_i$ will return an epistemic state rather than an epistemic history. Hence, revision by observations is essentially a single shot belief change operation that occurs after any number of report revisions. This is the case because revision by an observation implicitly changes the perceived honesty of the reporting agents. Since an epistemic history does not explicitly encode any notion of honesty, we do not have sufficient information to revise by a second observation.

We are now in a position to define revision by an observation.

Definition 8. *Let $\circ$ be a report revision operator defined with respect to $*$. Let $\mathbb{E} = \langle \preceq_\mathbb{E}, \overline{R}_\mathbb{E} \rangle$ be an epistemic history, and let $\langle \preceq, \alpha \rangle$ be the initial state and honesty set given by the initial state function associated with $\mathbb{E}$. For any natural number n, define:*

$$\mathbb{E} \circ_i \phi = \preceq *_{\alpha - D} \overline{R}_\mathbb{E}$$

where $D = \{A \mid c(A, R(E), \phi) \geq i\}$.

Hence, the beliefs following an observation are obtained by removing all agents that have provided i false reports from the honesty set given by $\circ$. We return to our motivating example.

Example 3. We can now formalize our example fully. Suppose we are using the iterated revision operator that is obtained from the Hamming distance between states. After the reports from Alice and Bob, the epistemic history is the following:

$$\langle \preceq, (A, \neg b \wedge d), (B, \neg c \wedge \neg d) \rangle.$$

Tne minimal elements initial ordering $\preceq$ consists of the set of states where $b \wedge c$ is true. Suppose that $\circ$ is defined by the initial state function with initial ordering $\preceq$ and honesty set $\{A, B\}$. In other words, all agents are initially considered to be honest n.

Now suppose we observe $\phi = \neg p \wedge q$. If we simply revised our current epistemic state by ϕ, then the most plausible state would be the one in which $\neg p \wedge q \wedge \neg r$ is true.

However, using the operator $\circ_1$, then B will be removed from the honesty set, since Bob has provided a report that is inconsistent with ϕ. Therefore, the final belief state will be the following:

$$B(\preceq *(\neg b \wedge d) * \phi).$$

This set consists of a single state where $\neg b \wedge c \wedge d$ is true. So we believe the animal is a collared dog that is not brown. Essentially, our initial beliefs have incorporated Alice's report and our observation in a consistent manner.

4.3 Basic Properties

In this section, we give some basic properties of our new operators for incorporating observations. The proofs are straightforward, and omitted in the interest of space. We first consider some special cases.

Proposition 3. Let $\mathbb{E} = \langle \preceq_\mathbb{E}, \overline{R}_\mathbb{E} \rangle$. Then $\mathbb{E} \circ_0 \phi =_{\preceq} *\phi$, where $\preceq$ is the initial epistemic state specifed by the initial state function.

In this case, $\circ_0$ treats all agents as dishonest; so all reports are discarded. In the next proposition, we consider large values of n.

Proposition 4. Let $\mathbb{E} = \langle \preceq_\mathbb{E}, \overline{R}_\mathbb{E} \rangle$ and let n be greater than the length of $\overline{R}_\mathbb{E}$. Then $\mathbb{E} \circ_1 \phi =_{\preceq_\mathbb{E}} *\phi$.

In this case, n is larger than the list of reports. Therefore, it is not possible for any agent to have given n reports that conflict with the observation. As a result, the epistemic state following the observation is determined by simple revision.

In general, the epistemic state after an observation falls somewhere between the cases in Propositions 3 and 4.

Proposition 5. Let $\mathbb{E} = \langle \preceq_\mathbb{E}, \overline{R}_\mathbb{E} \rangle$, where $\overline{R}_\mathbb{E}$ has length m. Let n be a natural number. Then $\mathbb{E} \circ_n \phi =_{\preceq'} *\phi$ where

1. $\preceq'=\preceq *\overline{R}'$ for some subsequence $\overline{R}'$ of $\overline{R}_\mathbb{E}$.
2. $\preceq$ is the initial epistemic state specified by the initial state function.

Moreover, there is a set of at most $\lfloor \frac{m}{n} \rfloor$ observations from $\overline{R}_\mathbb{E}$ that can be inserted into $\overline{R}'$ to give a new subsequence $\overline{R}''$ such that $\preceq *\overline{R}'' =_{\preceq_\mathbb{E}}$.

The proposition states the standard case. The final epistemic state always involves revision by ϕ, but the epistemic state to be revised is obtained by removing some of the reports from $\overline{R}_\mathbb{E}$.

5 Discussion

5.1 Related Work

There has been related work on trust as it related to belief change. One approach that has been explored is the use of indistinguishable states. This is the approach taken in the development of so-called trust-sensitive revision operators [2]. The basic idea of trust-sensitive revision is that each reporting agent is only able to distinguish certain pairs of states. When such an agent reports the formula ϕ, we revise by the set of states that are indistinguishable from states where ϕ is true. In this setting, we end up believing only the "part of the report" where the agent is trusted. Similar approaches to trust have been defined in the context of modal logic [9,11]. However, all of these approaches are fundamentally concerned with knowledge-based trust rather than honesty-based trust.

Some preliminary work on honesty-based trust for belief revision appears in [5]. In this work, agents keep an explicit record of which reporting agents are honest. This is similar to our approach in that dishonest agents are simply ignored. However, this work is quite different from our approach in that agents have an explicit list of which agents are honest. This list can be directly modified through simple change operators; however, there is no clear explanation as to what triggers these changes. This contrasts with our approach, where observations are the only thing that can trigger changes to perceived honesty.

In this paper, agents do not have a list of which agents are honest. Instead, the honesty of reporting agents is implicit in the report revision operators that we use to incorporate new information. The underlying assumption is that some agents are honest, while others are not. This is certainly a simplistic view, but there are applications where this model is feasible. For example, in safety critical situations, we would be likely to discount reports from any agent that has given false information in the past. We view this paper as a first step towards a full treatment of honesty-based trust, so we are starting with a special case that is plausible for particular applications.

5.2   Future Work

There are several directions for future work. First of all, we would like to address different ways for trust to be lost. For example, if we add degrees of trust in reporting agents, then our view of honesty can change without direct observations. We are also interested in formalizing how trust can be regained by providing a large number of true reports. Moreover, we are interested in moving beyond the simple framework where dishonest agents are simply ignored. We are developing a framework that includes goals and intentions for this purpose. In general, we propose it is reasonable to be more skeptical of a report that comes from an agent that stands to gain from providing false information.

But the most significant area for future work is actually in combining honesty-based trust with knowledge-based trust. Whereas most existing work addresses only one, the interaction between these two forms of trust is an interesting and important topic. We believe that the framework presented in this paper could easily be extended to handle both kinds of trust, perhaps by extending the framework with some variation of trust-sensitive revision.

5.3   Conclusion

In this paper, we have introduced *report revision operators*. This is a new kind of belief revision operator that is suitable for incorporating information that has been reported by (possibly dishonest) agents. The new operators are defined by a set of rationality postulates. Roughly, the postulates assert that reports are either believed or ignored. Moreover, if an agent is ignored on one report then they will be ignored on all reports. We argue that this simple model of trust is suitable for applications where we must be cautious in accepting information.

The most significant contribution of this work is a representation result for report revision operators, which shows that they are essentially characterized by an honesty set that indicates which agents are honest. Characterizing trust-based revision operators in this manner is an important step towards a rigorous treatment of trust.

Lastly, we have explored the interaction between reports and observations. The important part about observations in this setting is that they give us a mechanism for changing our perspective on honesty. In particular, we have defined a class of change operators based on the notion of an honesty threshold. If the number of false reports an agent provides exceeds the threshold, then they will be considered dishonest. This work is part of a growing body of work on trust in belief revision, and it is the first to address honesty-based trust with a precise representation result.

References

1. Alchourrón, C.E., Gärdenfors, P., Makinson, D.: On the logic of theory change: partial meet functions for contraction and revision. J. Symb. Log. **50**(2), 510–530 (1985)
2. Booth, R., Hunter, A.: Trust as a precursor to belief revision. J. Artif. Intell. Res. **61**, 699–722 (2018)
3. Darwiche, A., Pearl, J.: On the logic of iterated belief revision. Artif. Intell. **89**(1–2), 1–29 (1997)
4. Dong, X., et al.: Knowledge-based trust: estimating the trustworthiness of web sources. In: Proceedings of the VLDB Endowment, vol. 8 (2015)
5. Hunter, A.: Belief revision with dishonest reports. In: Aziz, H., Corrêa, D., French, T. (eds.) AI 2022. LNCS, vol. 13728, pp. 397–410. Springer, Cham (2022). https://doi.org/10.1007/978-3-031-22695-3_28
6. Huynh, T.D., Jennings, N.R., Shadbolt, N.R.: An integrated trust and reputation model for open multi-agent systems. Auton. Agent. Multi-Agent Syst. **13**(2), 119–154 (2006)
7. Katsuno, H., Mendelzon, A.: Propositional knowledge base revision and minimal change. Artif. Intell. **52**(2), 263–294 (1992)
8. Krukow, K., Nielsen, M.: Trust structures. Int. J. Inf. Secur. **6**(2–3), 153–181 (2007)
9. Liu, F., Lorini, E.: Reasoning about belief, evidence and trust in a multi-agent setting. In: An, B., Bazzan, A., Leite, J., Villata, S., van der Torre, L. (eds.) PRIMA 2017. LNCS (LNAI), vol. 10621, pp. 71–89. Springer, Cham (2017). https://doi.org/10.1007/978-3-319-69131-2_5
10. Schwind, N., Konieczny, S., Perez, R.P.: Darwiche and Pearl's epistemic states are not total preorders. In: Proceedings of the International Conference on Principles of Knowledge Representation and Reasoning (KR 2022) (2022)
11. Singleton, J., Booth, R.: Who?s the expert? on multi-source belief change. In: Proceedings of the International Conference on Principles of Knowledge Representation and Reasoning (KR 2022) (2022)

A Prompting Framework to Enhance Language Model Output

Himath Ratnayake[(✉)] [iD] and Can Wang[(✉)]

Southport, UK
himath4510@gmail.com

Abstract. This research investigates the role of prompt engineering in enhancing the performance and generalisation of large-scale language models (LLMs) across a wide range of Natural Language Processing (NLP) tasks. The study introduces a comprehensive framework for prompt engineering, titled the "PERFECT" framework, and evaluates its effectiveness across different tasks and domains. The research findings underscore the pivotal role of advanced prompting techniques in eliciting more nuanced and flexible responses from AI models. The study also explores the future implications of prompt engineering, including the integration of reinforcement learning with human feedback, the emergence of prompt engineering as a new job market, and the rise of context-aware and interactive prompts. The research contributes to a deeper understanding of the principles, mechanisms, and best practices in prompt engineering, with practical implications for improving LLM performance and reducing the barrier to entry for new adoptees through using prompting frameworks. The research aims have been largely achieved, providing a new framework for prompting while also exploring future advancements. However, the study also highlights the need for further exploration of the constraints placed on current prompting techniques, such as token size and context window.

Keywords: Prompt Engineering · Large-Scale Language Models · Natural Language Processing · Prompting Framework

1 Introduction

The concept of Artificial Intelligence has garnered significant interest in recent years. It is a technology with the potential to revolutionise a myriad of industries with benefits ranging from automation of tedious tasks to complex model-based predictions that can help inform future company decisions. Natural Language Processing (NLP) is a field of AI with particularly promising text-based applications. The development of advanced NLP models such as OpenAI's GPT-3 is already changing how many industries work, with tools like Chat-GPT gaining adoption in a typical workspace as an additional "AI-Powered" assistant. From summarising meeting notes on Microsoft Teams to drafting emails or writing

T. Liu et al. (Eds.): AI 2023, LNAI 14472, pp. 66–81, 2024.
https://doi.org/10.1007/978-981-99-8391-9_6

executive summaries, the possibilities for improving any workflows related to speech/text (foundations of modern-day work and communication) are almost endless.

The rapid growth of large language models (LLMs) has generated substantial interest in understanding and improving their performance across diverse NLP tasks. Therefore, this study investigates the effectiveness of prompt engineering in improving the performance and generalisation of large-scale language models. Prompt engineering involves the design of effective input queries to enhance these models' output, leading to more useful and accurate LLM outputs (Brown et al. 2020). The research aims to explore current prompt engineering techniques, develop novel strategies for them, and finally propose a framework for creating optimal prompts. The effectiveness of these prompting techniques will then be evaluated with the overarching aim of enhancing LLM output.

The research question guiding this investigation is: How can prompt engineering techniques be utilised to enhance the performance and generalisation of large-scale language models across a wide range of NLP tasks? Several key assumptions have been considered when tackling this research topic. They have been listed below.

- The way a task is articulated to a large language model can substantially enhance the quality of its output.
- Effective prompting can greatly enhance the quality of LLM output.
- The effectiveness of a prompt output is measurable both quantitatively and qualitatively. This is crucial for comparing the effectiveness of different techniques and designing controlled experiments.
- Prompt engineering techniques are diverse and multiple techniques can be combined to create new novel prompting methods.
- Prompt engineering will continue to be an important area of study in the future. This is based primarily on the assumption that as LLMs continue to evolve, so too will the techniques used to interact with them.

Though prompt engineering literature is diverse, it is still a new and emerging field, encompassing several approaches, such as using templates (Dathathri Brown et al. 2020), creating artificial datasets (Khashabi Brown et al. 2021), and leveraging reinforcement learning for prompt tuning (Lewis Brown et al. 2021). However, there is still limited research (as of writing) that systematically investigates and compares the effectiveness of these techniques across various NLP tasks and domains.

The significance of this research lies in the fact that it extends upon previous studies by contributing a comprehensive framework for prompt engineering techniques (titled the "PERFECT" framework) and systematically evaluating its effectiveness across a range of NLP tasks and domains. By doing so, it contributes to a deeper understanding of the principles, mechanisms, and best practices in prompt engineering, building upon the foundations laid by prior work in the field.

As mentioned previously, one of the core assumptions of this research is that improving the specificity and quality of a prompt fed to an LLM will in turn drastically improve the output that is returned. Hence, the key aims of this research are:

- To explore current prompting strategies used in LLMs.
- To develop and examine novel prompt engineering strategies to enhance the performance of LLM outputs across various NLP tasks (Petroni Brown et al. 2021).
- To propose an easily adoptable framework for prompt engineering to create optimal prompts for LLMs, maximising the quality of desired outputs (Raffel Brown et al. 2020).
- To evaluate the effectiveness of the proposed prompting techniques based on the quality of the given prompt.

By achieving the aims listed, the outcomes/contributions produced include a comprehensive evaluation of prevailing prompt engineering methodologies and the creation of innovative strategies for creating the "perfect prompt". This is significant as it can improve an LLM's performance while contributing to more understanding of natural language models. There are also practical implications when using commercial LLMs because effective prompting can improve LLM output, and a framework will reduce the barrier to entry for new adoptees seeking to learn how to prompt effectively.

2   Prompting Techniques

Understanding and exploring various prompting strategies for large language models (LLMs) is a prerequisite to creating an effective prompting framework. As depicted in Table 1, numerous strategies are used in the field of natural language processing (NLP), each contributing uniquely to the enhancement of LLM outputs. The significance of understanding these prompting methods lies in their direct impact on the performance of LLMs. Prompt engineering, the process of designing effective input queries, is instrumental in refining the outputs generated by these models. It allows users to tailor the model's responses to specific use cases and applications, significantly enhancing the utility and accuracy of the outputs. Furthermore, understanding these methods allows researchers to uncover the underlying mechanisms that determine how prompts affect the LLM's response. This is a fundamental step towards demystifying the often opaque behaviour of AI models, leading to a better understanding of their operation and potential pitfalls. Knowledge of various prompting strategies also provides insights into the strengths and weaknesses of different approaches. This knowledge enables the formulation of a more robust and versatile framework. Additionally, understanding prompting methods equips us with the capability to develop new and improved strategies which are more novel, and can give better outcomes.

Table 1. Prompting Strategies

Type	Definition	Pros	Cons	Applicable Situations
Zero-shot prompting	Given a single prompt without any context or examples and is expected to provide an accurate and relevant answer (Kojima Brown et al. 2022)	Quick and easy; no examples needed	Less accurate, may struggle with complex tasks	Simple questions or tasks, general knowledge queries
Few-shot prompting	Providing a language model with a few examples of the desired input-output pairs before posing the main question	Improved accuracy, helps the model understand the task	Requires relevant examples, increased setup time	Tasks that require context, learning new tasks
Chain-of-thought prompting	Breaking down complex questions into simpler sub-questions and sequentially prompting the model to answer each sub-question (Wei Brown et al. 2023)	Simplifies complex questions, and generates more coherent responses	Requires manual breakdown of questions, may miss overall connections	Complex questions, multi-step tasks
Zero-shot chain of thought	Zero-shot chain of thought combines the concepts of zero-shot prompting and chain-of-thought prompting. The model is given a series of related prompts without any examples and is expected to provide accurate and coherent answers	Combines zero-shot and chain-of-thought, tests understanding	Less accurate, no examples to guide model	Sequential questions without context, assessing model capabilities
Self-consistency	Self-consistency involves posing a question to the model multiple times with slight variations, as well as sample answers to evaluate the consistency of its responses. This method can help in assessing the model's understanding and reliability	Evaluates model reliability, exposes biases	Multiple queries required, may not increase accuracy	Testing model understanding, identifying biases

(continued)

Table 1. (*continued*)

Type	Definition	Pros	Cons	Applicable Situations
Generate knowledge prompting	This method involves prompting the model to generate new knowledge or insights based on its existing understanding. This can be useful for creative problem-solving and idea generation	Encourages creative problem-solving, idea generation	Responses may lack practicality, may deviate from known facts	Brainstorming, innovative thinking
Prompt Generation	This method involves asking the model itself to generate a prompt based on a given topic or theme, which can be useful for brainstorming or exploring new ideas	Stimulates creativity, explores new ideas	Model-generated prompts may be vague or miss the point	Brainstorming, discussion topics, creative writing exercises
Having the model act as a specific person	This method involves asking the model to answer questions or generate content while emulating the writing style or perspective of a specific person, such as a historical figure, celebrity, or fictional character	Personalized responses, captures unique perspectives	May not perfectly emulate the person, risk of misinformation	Emulating writing styles, capturing perspectives, historical or fictional scenarios

3   Research Methods

3.1   Framework Formulation

Although the above types of prompting can be effective for smaller tasks, they often fall short of providing an answer that the user deems satisfactory. This led to experimentation with hybrid prompting, where two or more prompt types are combined to create a more nuanced and flexible prompt. For example, a user could combine the concepts of zero-shot prompting and chain-of-thought prompting, whereby the model is given a series of related prompts without any examples and is expected to provide accurate and coherent answers. These "hybrid" prompting combinations served as the inspiration for much of the initial framework formulation. A modular approach to creating prompt skeletons allows users to combine different elements to generate complex prompts suited to specific needs, addressing the unique challenges and requirements of the AI-generated content domain.

This eventually led to the creation of the "PERFECT" framework, which provides an easy-to-follow structure for constructing effective prompts using the modular approach, facilitating the systematic comparison of different prompting techniques and their effectiveness in generating high-quality output.

- **P**urpose - Clearly define the purpose of the prompt
- **E**lement - Choose the appropriate modular elements to include
- **R**ole - Specify the role or perspective the model should take (if applicable)
- **F**ormat - Determine the desired format for the response
- **E**xamples - Include examples or case studies (if relevant)
- **C**onditions - Set specific conditions or constraints for the answer
- **T**imeframe - Indicate a timeframe, if the prompt requires historical context or future predictions

By incorporating this modular approach, the "PERFECT" framework offers flexibility and adaptability in constructing prompts that cater to diverse scenarios and application domains. A comparison of a normal vs. "PERFECT" prompt can be found below. For reference, the "PERFECT" prompt was able to generate significantly more nuanced information, with an output that was more than 150 tokens longer than that of the normal prompt on GPT-3.5 (Table 2, Table 3).

Table 2. Comparison of a normal and "PERFECT" prompt

Normal Prompt	"PERFECT" Prompt
"Tell me about climate change and oceans."	"Explain the impact of climate change on marine ecosystems. Discuss the relationship between ocean acidification and coral bleaching. You are to act as a marine biologist. Your answer should follow the template: Introduction, Causes, Impacts, Adaptation Strategies, and Conclusion. Include at least two examples of impacted marine ecosystems. The answer should be at least 300 words long and cite at least one scientific study."

Table 3. The "PERFECT" framework for constructing prompts

Step	Description	Example	Prompt Skeleton Examples
Purpose	Identify the primary goal or objective of the prompt	Explain the impact of climate change on marine ecosystems	- Explain [topic] - Discuss [issue] - Compare [item 1] and [item 2]
Element	Choose appropriate modular elements from the list of skeletons	Discuss the relationship between [topic 1] and [topic 2]	- Analyze [trend] - Evaluate [strategy] - Assess [policy]
Role	Specify the role or perspective the model should take, if applicable	You are a marine biologist	- You are [person/role] - Pretend you are [character] - Write as if you were [historical figure]
Format	Determine the desired format for the response	Follow the template: Introduction, Causes, Impacts, Adaptation, Conclusion	- Use the template: [sections] - Write a [format] - Create a [type of output]
Examples	Include examples or case studies, if relevant	Include at least two examples of impacted marine ecosystems	- Provide [number] examples of [topic] - Cite a case study on [subject] - Explain a real-world application of [concept]
Conditions	Set specific conditions or constraints for the answer	Answer should be at least 300 words and cite at least one scientific study	- Answer must be [length] - Use [type of citation] - Include at least [number] of [criteria]
Timeframe	Indicate a timeframe for historical context or future predictions	Discuss impacts over the past 50 years and potential implications	- Focus on the period between [start date] and [end date] - Explore the next [number] years - Discuss historical development of [topic]

4   Results

4.1   Experiments

To effectively evaluate poor vs. perfect prompts, a wide range of data gathering and collection strategies are necessary. A diverse and representative dataset will allow for a comprehensive assessment of the performance of the prompts across various domains and tasks.

When it comes to data analysis and evaluation of the "PERFECT" framework, a combination of both qualitative and quantitative evaluation methods was considered. However, more focus was placed on exploring quantitative research methods to reduce subjectivity caused by human opinion. It is assumed that the prompts following the "PERFECT" framework produce high-quality qualitative and quantitative metrics, given that the initial prompt is more specific, detailed, and structured. Quantitative evaluation can be broken down into two subsections:

- Intrinsic evaluation, where the prompt's performance is measured through the contents outputted by the text in the language model itself
- Extrinsic evaluation, where the performance of the prompt is measured through external metrics relating to the output quality

The main metrics used in intrinsic evaluation will be perplexity and bursti-ness. Intrinsic evaluation will be done by recording the outputs for various prompts in a number of different scenarios. For each scenario, both a normal prompt and a prompt following the PERFECT framework will be used to see how the output changes.

The main metrics used in extrinsic evaluation will be accuracy and verbosity. They will be measured using datasets such as LogiQA (Liu Brown et al. 2020) and GSM8K (Cobbe Brown et al. 2021). These dataset consists of a number of deductive reasoning and grade school mathematics question instances. It has been found that even the best neural network models perform significantly worse than humans currently at these types of tasks, and so these data sets serve as a benchmark for investigating how much AI logic can be improved using just prompting.

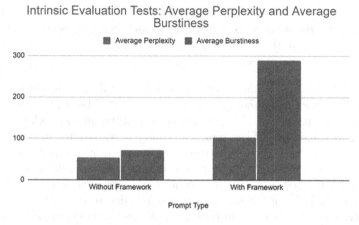

Fig. 1. A comparison of the perplexity (blue) and burstiness (red) of the advanced and normal prompts. (Color figure online)

4.2 Intrinsic Evaluation Results

In scrutinizing the relationship between the quality of prompts and the responses they elicit, a compelling trend becomes evident. When it comes to prompt length, advanced prompts typically provoke responses that are significantly lengthier, on average, than those triggered by simpler prompts as seen in Fig. 1 and 2. The responses to advanced prompts following the framework tend to exceed their less refined counterparts by a substantial margin of around 2100 words. One plausible explanation for this might be that the enhanced detail or guidance presented in advanced prompts equips the AI model to fashion a more thorough and profound response.

Furthermore, when we consider perplexity, a metric used to gauge the uncertainty of a predictive model like GPT-3, we notice that the average perplexity score is appreciably higher for responses to advanced prompts. The differential here is approximately 58. This elevated perplexity suggests that GPT-3 grapples with more uncertainty when tasked with generating responses to advanced

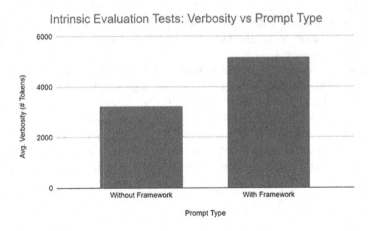

Fig. 2. Average verbosity of a normal prompt vs. one that follows the "PERFECT" framework during the intrinsic evaluation testing

prompts. The increased complexity and specificity inherent in these prompts might be the culprit, demanding a more sophisticated level of understanding and response. Meanwhile, looking at the concept of 'burstiness', which could be interpreted as a measure of the unpredictability or variability in the length of responses, we observe a distinct pattern. Advanced prompts yield a markedly higher average burstiness compared to poor prompts, with a difference of roughly 247. This indicates that responses to advanced prompts demonstrate greater variability in length. It is conceivable that this is also due to the wider spectrum of potential response structures and content that these prompts enable. The more nuanced the prompt, the broader the array of potential satisfactory responses, which in turn, amplifies the burstiness.

These observations enable us to infer that while it seems that advanced prompts might necessitate more computational resources, due to the longer and more variable nature of the responses they inspire, they appear to produce more comprehensive and varied responses. This could render them more beneficial in a multitude of scenarios. However, the spike in perplexity highlights that there may be room for further exploration and enhancements to equip GPT-3 with a more proficient understanding and response mechanism to tackle complex prompts.

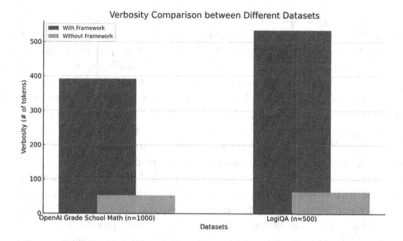

Fig. 3. Verbosity comparison, measured in number of tokens, between different datasets. The data indicates that the framework increases verbosity significantly for both datasets, with a more pronounced increase in the LogiQA dataset.

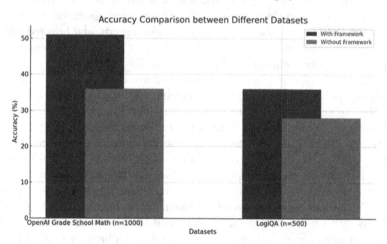

Fig. 4. Accuracy comparison between different datasets: OpenAI Grade School Math (n = 1000) and LogiQA (n = 500). It can be observed that the use of a framework improves accuracy for the OpenAI Grade School Math dataset but not significantly for the LogiQA dataset.

4.3   Extrinsic Evaluation Results

For the extrinsic evaluation metrics, 2 popular datasets for testing Large Language Model (LLM) outputs were utilized as seen in Fig. 3 and 4. This study explores a novel approach aimed at enhancing the problem-solving abilities of large AI models like ChatGPT in the context of logical reasoning. The dataset in use, LogiQA, comprises of many question-and-answer pairs that necessitate various types of logical thinking skills. Given that the current state-of-the-art AI models don't perform as proficiently as humans on this dataset, there is still substantial scope for improvement. The evaluation process consists of two parts. Initially, ChatGPT is presented with the unmodified questions and their corresponding multiple-choice answers. Subsequently, the same question is trialled with parts of the PERFECT framework being integrated into the original prompt question. The results gained when the framework is used vs. when it isn't are then compared to assess if the incorporation of the framework components augments ChatGPT's ability to tackle these logic problems.

The use of the framework increased accuracy from 28% to 36%, with a corresponding large increase in verbosity from 64 to 534. This can be attributed to the fact that the framework encourages the model to think step-by-step and carefully evaluate the logic in order to come to the final answer. This improvement may be attributed to specific instructions and context provided by the framework, aiding in better understanding, albeit at the cost of increased complexity in answers.

However, this is still significantly lower than the average human performance reported in the original LogiQA study, which stood at 86% accuracy with an upper ceiling of 95% (Liu Brown et al. 2023). Other language model predecessors of GPT-3 like BERT were also tested in the original study and hovered between 25–30% accuracy, showing that even now, there is still significant progress that could be had in language model's ability to logically reason.

One contributing factor lies in the realm of information density present within the prompts. Specific prompts inherently encapsulate a greater amount of information, thus setting a well-defined context for the AI model to operate within. This clear directive implicitly delineates the expected response length and confines the range of plausible responses. Conversely, a succinct, open-ended prompt leaves much room for the AI's interpretation, often leading to a comprehensive exploration of various plausible aspects pertinent to the vague prompt.

The core operation of GPT-3, a transformer-based model, is its "prompt-completion" mechanism. The model is designed to generate text that is contextually likely to follow the given input, based on its extensive training data. Therefore, a more open-ended prompt may stimulate a broader textual pattern, generating longer responses that follow such prompts in the model's training corpus. Interestingly, for creative prompts such as writing a short story, it was found that the short "normal" prompt actually created a longer story than a more advanced prompt. The degree of uncertainty induced by the prompt is a consequential element to this. An open-ended query expands the range of potential responses and increases the model's uncertainty. To curtail this ambiguity, the model generates an elongated response to traverse the spectrum of possible interpretations. A detailed, specific prompt drastically curtails this uncertainty, thereby creating a more succinct and short response.

Turning our attention to the Grade School Math dataset (GSM8K), we observe a similar trend of improvement in both accuracy and verbosity when the PERFECT framework is integrated. The GSM8K dataset is a rich compilation of 8.5K meticulously crafted, linguistically diverse math word problems, that encapsulate the essence of grade school mathematics (Cobbe et. al. 2021). Despite their elementary nature, these problems prove to be a substantial hurdle for even the most advanced language models available today, emphasizing the persistent gap in the multi-step mathematical reasoning abilities of these models. A striking feature of the GSM8K dataset is its origin; these problems are not scraped from existing sources but are freshly conceived by human problem writers. This process not only guarantees high-quality but also endows the dataset with a remarkable diversity, avoiding the trap of linguistic templating that plagues many existing datasets. The problems hover in a moderate difficulty range, requiring nothing more than early Algebra to solve, yet they seem to pose a significant challenge to large, state-of-the-art language models (Cobbe et. al. 2021).

The study used a subset of this dataset, with 1000 problems, to evaluate the ChatGPT's performance in solving these math problems. Initially, the model struggled to match human performance levels, securing an accuracy rate of only 36% without the framework, which is considerably lower than what a bright middle school student might achieve. However, with the integration of the PERFECT framework, there was a noticeable improvement in the performance, with the accuracy rising to 51%. This increase, though substantial, still indicates room for further enhancement in the model's capabilities. The verbosity of the responses also saw a significant boost, from 53.4 to a staggering 392.44. This metric indicates that the framework potentially aids the model in carving out a more detailed, step-by-step solution pathway, which is in line with the natural language solutions envisioned for the GSM8K dataset. The solutions generated are not mere mathematical expressions but detailed narratives that delineate the solution process, mirroring the internal monologues of a student solving these problems.

However, this detailed approach comes with its own set of challenges. The increase in verbosity, while beneficial in painting a comprehensive picture of the solution pathway, also adds a layer of complexity to the response, potentially making it more prone to errors and deviations. Moreover, the model's inability to consistently achieve high accuracy rates, despite the detailed narrative responses, highlights the inherent limitations in the current state-of-the-art language models when it comes to multi-step mathematical reasoning.

In conclusion, the length and detail of a prompt significantly influence the model's output. Factors such as information density, uncertainty and exploration, and the model's prompt-completion mechanism play crucial roles in shaping the output. However, this behavior is highly context-dependent, and the output could vary based on specific prompt wording, the model's randomness factor, and other underlying parameters. While the PERFECT framework shows promise in enhancing the problem-solving abilities of large AI models, it also brings to light the nuanced challenges that lie ahead. The GSM8K dataset serves as a valuable tool in this exploration, offering a balanced difficulty spectrum that aids in unearthing the true capabilities and limitations of these models.

As we move forward, it is imperative that further research is directed towards refining these models, honing their mathematical reasoning abilities to match, if not surpass, human performance levels.

4.4 Constraints

Although the framework had positive results in terms of output verbosity and accuracy in the LogiQA/GSM8K dataset, it still has a number of constraints.

4.4.1 Token Size

One of the constraints of using a prompting framework is that larger prompts consume more tokens, which can be an issue with token limitations in language models. To address the token limitation constraint, token compression which was mentioned previously is a potential solution. Conversion of large prompts into a shorter machine readable (but not necessarily human readable) format can help increase the context window that can be worked with for an LLM.

4.4.2 Experimental Method

The experimental method could be improved by including a wider range of datasets. Currently a limiting factor for experimentation is the limits of the GPT-3 API, which makes large scale testing more challenging. However, using more datasets similar to LogiQA/GSM8K with a wider range of questions and scenarios would allow for more conclusive results. Qualitative evaluation based on human evaluation would also have been beneficial to allow for a more well-rounded research approach.

4.4.3 Initial Prompt Constraints

- **Over-specification:** an overly detailed prompt limits the AI's ability to generate creative or flexible responses, as can be seen in the results gathered. Over-specifying the requirements in a prompt could lead to responses that are too narrowly focused or repetitive, hindering output.
- **Complexity:** A complex prompting framework might be challenging to implement and maintain. As the framework grows and evolves, it might require significant effort to ensure it remains efficient, especially with industry practices or new novel developments. LLMs like ChatGPT are constantly being updated with both small and large version updates, and new LLMs with slightly different architectures are being created that may require different prompting strategies for optimal outputs.
- **Adaptability:** A specific prompting framework may not be suitable for all types of tasks or domains. It may need to be adjusted or customized for different applications, which could be time-consuming and resource-intensive.
- **Readability:** Highly structured and detailed prompts could potentially impact the readability of the generated responses. The AI might generate text that strictly adheres to the prompting framework but sacrifices natural language flow or overall readability.

4.4.4 External Constraints

Beyond changes that can be made to the initial prompt, there are some uncontrollable factors rooted in the model being used itself. Recent studies have found that the outputs of LLMs like GPT-3 and GPT-4 vary significantly over time, due to internal changes within the model. For example, GPT-4 (March 2023) was reasonable at identifying prime vs. composite numbers (84% accuracy) but GPT-4 (June 2023) was poor on these same questions (51% accuracy) in June than in March in this task (Chen Brown et al. 2023). These behavioural drifts can render previous prompting methods less effective compared to old ones. Ultimately, it is important when fashioning a prompt to remember that this framework is modular, and that not every part of it needs to be used if it is not necessary. The onus would be on the user to test and make that decision in order to maximise the quality of the output they desire. The prompts should be designed to strike a balance between providing clear guidance and allowing for creativity and flexibility in the generated responses. This balance can help the model generate high-quality responses that are both coherent and engaging. However, keep in mind that the model itself is also susceptible to change and prompts that may have been effective previously may not be as effective at the time of next use. By incorporating these strategies, it is possible to address the constraints of a prompting framework while still benefiting from its advantages in guiding AI language models to produce more accurate, relevant, and coherent responses.

5   Conclusions

In conclusion, the research findings underscore the pivotal role of prompt engineering in shaping the future of AI. Advanced prompting techniques, including hybrid prompting and the "PERFECT" framework, have been shown to elicit more nuanced and flexible responses from AI models which are generally significantly more verbose than simpler prompts, and more capable of tackling logical reasoning and mathematics questions. These techniques, while requiring more computational resources, produce more comprehensive and varied responses, making them beneficial in a multitude of scenarios. However, the increased complexity also raises the model's uncertainty, suggesting room for further enhancements.

Looking ahead, the future of AI systems is anticipated to be characterized by an increased level of intuition and adeptness in understanding natural language, reducing the necessity for meticulously engineered prompts. However, as AI systems become more integrated into our daily lives, the importance of efficient and effective communication between humans and machines is set to increase. Therefore, the role of prompt engineering is still critical in harnessing the power of robust and effective language models across a multitude of applications. Future developments in prompt engineering are likely to focus on the creation of context-aware prompts, producing more accurate and personalized responses. The research contributed a comprehensive framework for prompt engineering, titled the "PERFECT" framework, and evaluated its effectiveness across different tasks and domains. The research aimed to explore how a prompt engineering framework combining various techniques can enhance the performance of large-scale language models across various NLP tasks. The study's assumption that task articulation and effective prompting can significantly enhance model output has proven to be true in many use cases. However, it has also highlighted the importance of striking a balance between a prompt that is too hyper-specific and one that is too broad, depending on the situation. This has provided a deeper understanding of prompt engineering principles and best practices, with practical implications for improving LLM performance and reducing the barrier to entry for new adoptees. In future, finding more novel ways to combine various prompting techniques would be beneficial, while also considering how these techniques will change with future advancements such as multimodal models.

References

Chen, S.F., Beeferman, D., Rosenfeld, R.: Evaluation metrics for language models (2008). https://doi.org/10.1184/R1/6605324.v1

Chen, L., Zaharia, M., Zou, J.: How is ChatGPT's behavior changing over time? (2023). https://doi.org/10.48550/arXiv.2307.09009

Cobbe, K., et al.: Training verifiers to solve math word problems (2021). https://doi.org/10.48550/arXiv.2110.14168

Cummins, R., Paik, J.H., Lv, Y.: A Pólya urn document language model for improved information retrieval. ACM Trans. Inf. Syst. **33**(4), 1–34 (2015). https://doi.org/10.1145/2746231

Brown, T.B., Mann, B., Ryder, N.: Language models are few-shot learners (2020). arXiv preprint arXiv:2005.14165

Devlin, J., Chang, M.W., Lee, K., Toutanova, K.: BERT: pre-training of deep bidirectional transformers for language understanding. In: Proceedings of the 2019 Conference of the North American Chapter of the Association for Computational Linguistics: Human Language Technologies, vol. 1, pp. 4171–4186 (2019). https://aclanthology.org/N19-1423/

Geva, M., Zhao, Y., Lu, Y., Li, Z., Dong, L., Sun, H.: Tree of thoughts: deliberate problem solving with language models. arXiv preprint arXiv:2305.10601 (2023).https://arxiv.org/pdf/2305.10601.pdf

Kojima, T., Gu, S S., Reid, M., Matsuo, Y., Iwasawa, Y.: Large Language models are zero-shot reasoners (2022). https://arxiv.org/abs/2205.11916

Liu, J., Cui, L., Liu, H., Huang, D., Wang, Y., Zhang, Y.: Logiqa: a challenge dataset for machine reading comprehension with logical reasoning. arXiv preprint (2020). https://arxiv.org/abs/2007.08124

Lu, J., Li, C., Niu, S., Zhou, M.: Multimodal chain-of-thought reasoning for visual question answering. arXiv preprint arXiv:2302.00923 (2023). https://arxiv.org/pdf/2302.00923.pdf

Petroni, F., Piktus, A., Gupta, N., Schlichtkrull, M., Lewis, M., Riedel, S.: How context affects language models' factual predictions. arXiv preprint arXiv:2102.08667 (2021). https://arxiv.org/abs/2102.08667

Radford, A., Wu, J., Child, R., Luan, D., Amodei, D., Sutskever, I.: Language models are unsupervised multitask learners. OpenAI Blog (2018). https://cdn.openai.com/research-covers/language-unsupervised/language_understanding_paper.pdf

Raffel, C., et al.: Exploring the limits of transfer learning with a unified text-to-text transformer. J. Mach. Learn. Res. 21(140), 1–67 (2020). https://jmlr.org/papers/v21/20-074.html

Rahi, S.: Research design and methods: a systematic review of research paradigms, sampling issues and instruments development. Int. J. Econ. Manag. Sci. 6, 403 (2017). https://www.researchgate.net/publication/316701205_Research_Design_and _Methods_A_Systematic_Review_of_Research_Paradigms_Sampling_Issues_and_Instr uments_Development

Vaswani, A., et al.: Attention is all you need. In: Advances in Neural Information Processing Systems, pp. 5998–6008 (2017)

Wei, J., et al.: Chain-of-thought prompting elicits reasoning in large language models (2023). https://doi.org/10.48550/arXiv.2201.11903

Wei, Z., Zhao, Y., Lu, Y., Li, Z., Dong, L., Sun, H.: Self-Consistency Improves Chain of Thought Reasoning in Language Models. arXiv preprint arXiv:2203.11171 (2023). https://arxiv.org/pdf/2203.11171.pdf

Yang, H., Yue, S., He, Y.: Auto-GPT for online decision making: benchmarks and additional opinions. arXiv:2306.02224 (2023)

Zhang, J., Liu, Z., Xiong, C., Sun, M., Zhou, M., Gao, J.: KOSMOS: a universal system for multimodal perception, language understanding, and instruction following. arXiv preprint arXiv:2302.14045 (2023). https://arxiv.org/pdf/2302.14045.pdf

Zhao, J., Lu, K., Chen, H.: Learning to Prompt for Vision-Language Models. arXiv preprint arXiv:2108.13348 (2021)

Epistemic Reasoning in Computational Machine Ethics

Raynaldio Limarga$^{(\boxtimes)}$, Yang Song, Maurice Pagnucco, and David Rajaratnam

School of Computer Science and Engineering, University of New South Wales,
Sydney, Australia
r.limarga@student.unsw.edu.au, {yang.song1, morri,
david.rajaratnam}@unsw.edu.au, morri@cse.unsw.edu.au

Abstract. Recent developments in computational machine ethics have
adopted the assumption of a fully observable environment. However, such
an assumption is not realistic for the ethical decision-making process.
Epistemic reasoning is one approach to deal with a non-fully observable environment and non-determinism. Current approaches to computational machine ethics require careful designs of aggregation functions
(strategies). Different strategies to consolidate non-deterministic knowledge will result in different actions determined to be ethically permissible.
However, recent studies have not tried to formalise a proper evaluation
of these strategies. On the other hand, strategies for a partially observable universe are also studied in the game theory literature, with studies
providing axioms, such as Linearity and Symmetry, to evaluate strategies in situations where agents need to interact with the uncertainty of
nature. Regardless of the resemblance, strategies in game theory have
not been applied to machine ethics. Therefore, in this study, we propose
to adopt four game theoretic strategies to three approaches of machine
ethics with epistemic reasoning so that machines can navigate complex
ethical dilemmas. With our formalisation, we can also evaluate these
strategies using the proposed axioms and show that a particular aggregation function is more volatile in a specific situation but more robust
in others.

Keywords: Machine Ethics · Game Theory · Epistemic Reasoning

1 Introduction

The rapid development of autonomous machines, such as self-driving cars, robots
in the workplace, health assistants, etc., brings concerns about whether the decisions made by these machines are morally or ethically correct. However, the "correct" decision usually depends on context—the situation and interpretation—
which also relies on ethics and social norms. *Machine ethics* is thus the study

Supplementary Information The online version contains supplementary material
available at https://doi.org/10.1007/978-981-99-8391-9_7.

of generating ethically permissible action plans for autonomous agents. While recent studies of machine ethics typically assume a fully observable environment [23]; such an assumption is not realistic for decision-making because real-world implementations will always include noise and uncertainty.

To build a machine ethics framework in a partially observable world, simulation is commonly used to model non-determinism. For example, [22] tries to model an ethically compliant agent using a *Markov Decision Process* (MDP) by learning through simulation. Some studies establish optimal behaviour of a machine through physical, virtual, or combined simulation via reinforcement learning to align with a particular ethical constraint [10, 16, 25]. Probabilistic graphical models are also often utilised [5, 7, 12], such as Bayesian algorithms to represent welfare trade-offs quantitatively. However, being learning-based, these approaches tend to require extensive training data which may be expensive or even impossible to acquire for ethical dilemmas in every possible situation. In this paper we want to build models using a formal logic approach.

Logic-based approaches to machine ethics have been developed in multiple studies. [1] utilises deontic logic to combine machine ethics and machine explainability. In doing so, a system may not always be ethical but can always be *justifiable*. In another approach, a multi-valued state variable (SAS$^+$) is used to formalise the *doctrine of double effect* [6]. With Belief, Desire, and Intention (BDI) agents, a system was built to consider ethical constraints for a decision-making process [2]. This framework is capable of selecting among unethical actions and choosing the minimal unethical course of action. These studies provide state-of-the-art approaches to solving issues in machine ethics. However, these approaches assume complete knowledge of the world, neglecting possible changes due to decision-making under uncertainty.

The work by Pagnucco *et al.* is one study that developed a framework to model uncertainty using epistemic reasoning [11]. The value of an action plan is aggregated across multiple possible situations and, based on this aggregated value, a plan is evaluated based on a particular ethical principle (e.g., consequentialist and deontological). However, the selection of aggregation functions is not well justified. To address this failing, in this paper we propose to utilise strategies that have been well established in *game theory* and formalise them as an aggregation function for ethical decision-making using epistemic reasoning.

Game theory is a research field that attempts to formalise how game players should select their actions during play in strategies game [20]. Traditionally, game theory studies strategies to interact with rational players in a fully observable world but it also covers partially observable world scenarios. Strategies have been established to solve multiplayer games, such as mixed strategy [3]. Moreover, a set of criteria have been formalised to determine what the modeller considers essential [21]. A judgement can then be made upon which method satisfies which criteria. We propose that we can utilise these criteria for epistemic reasoning since they can be translated into first-order logic.

In this work, we build a framework to define ethical principles as a *selection function* that discriminates between *ethically permissible plans* to adopt an *ethical course of action*. This selection function takes a set of sequences of actions, which we will refer to as *plans* and a set of possible worlds that model uncertainty in the epistemic world. Our ethical principle function will filter such a set of plans and return a set of *ethically permissible plans*. Formally, $\mathcal{EP} : 2^{\Pi} \times 2^{K} \to 2^{\Pi}$. The set of plans shows that the ultimate course of action is judged as a whole, not on the basis of an individual action. Furthermore, the set of possible worlds represents incomplete knowledge using modal logic. The set of possible worlds allows us to aggregate the value of executing a plan, and from such a value, we can return the set of *ethically permissible plans*.

The contributions of this paper are as follows. First, we formally define an *ethical principle function* as a selection function for two consequentialist approaches and one deontological approach. This function will follow the use of modal logic to model epistemic reasoning. Similar to [11], we will utilise the possible world semantics to represent *uncertainty*. With an established ethical function, we then introduce several aggregation functions influenced by game theory and how they should be translated into ethical principles. Finally, we evaluate those aggregation functions with axioms from game theory literature and show the strengths and weaknesses of each aggregation function.

2   Background

Situation calculus is a framework that is capable of expressing changes in the world due to performing actions [14,15]. It utilises *fluents* and *situations* to model the different states and *actions* to model changes to states. Following the definition from Reiter [15], a situation is a history of actions from the initial situation. Therefore, function $do(a, s)$ represents the situation where action a is performed in situation s. Furthermore, given a sequence of actions $\pi = [a_1, ..., a_n]$, function $do(\pi, s)$ is the abbreviation for $do(a_n, do(a_{n-1}, ...do(a_1, s)...))$. This framework also utilises a binary relation $s \sqsubseteq s'$ to show that situation s is a sub-sequence of situation s'. The predicate $F(\vec{x}, s)$ denotes that $F(\vec{x})$ is true at situation s. This framework also follows the standard basic action theory. *Action precondition axiom* $Poss(a, s)$ states the conditions where action a can be executed at a situation s, and consequently, an *executable situation* is defined as

$$executable(s) \stackrel{\text{def}}{=} (\forall a, s').do(a, s') \sqsubseteq s \to Poss(a, s')$$

A *successor state axiom* stipulates the change to fluents $F(\vec{x})$ as a result of performing action a at situation s and is defined as:

$$F(\vec{x}, do(a, s)) \equiv \gamma_F^+(\vec{x}, a, s) \vee (F(\vec{x}, s) \wedge \gamma_F^+(\vec{x}, a, s))$$

Scherl and Levesque extend the situation calculus framework to model uncertainty with modal logic [18,19]. A binary relation $k(s', s)$ denotes that situation s' is "accessible" from situation s. In other words, at situation s the agent considers any fluent at situation s' possible. Consequently, an agent *knows* a particular fluent to be true, iff this fluent is true at every "accessible" world.

$$Know(F(\vec{x}, s) \stackrel{\text{def}}{=} \forall s'.(k(s', s) \to F(\vec{x}, s)$$

Pagnucco *et al.* [11] proposed a new framework to reason about machine ethics in an epistemic environment. The notion of *goodness* is defined as a function that indicates the moral desirability of a state. They define a *state* as a set of fluents that is true in a particular situation. Utilising the relation k and *goodness*, two approaches of consequentialism are defined to reason about epistemic situations. The first one is *knowGood*, which aggregates the *goodness* function over all "accessible" worlds, and the second one is *goodness**, which aggregates the *goodness* function over a sequence of actions followed by *knowGood** that aggregates the *goodness** function over all "accessible" worlds. Finally, the *preferable consequential outcome* for *knowGood** is defined as:

$$\Sigma \models \exists s.Know(\varphi, s) \land executable(s) \land \forall s'.(Know(\varphi, s') \land$$
$$executable(s')) \rightarrow (knowGood^*(s) \geq knowGood^*(s')))$$

Here, we propose formalising aggregation functions from other literature instead of using an arbitrary domain-specific aggregation function proposed by [11].

3 Ethical Principle Function

First, we need to define the input to our *ethical principle function*. Here we will use π to denotes sequence of actions, or *plan* in short. We will assume the set of plans Π and accessible worlds K are given. In planning problems, the modeller can define all the possible actions from the initial state to the goal state. This process will include pruning any plan that does not reach the goal or eliminating those that are not executable. This elimination process is not considered in this work. We only want to know, given a set of plans that can achieve the goal, which plan is *ethically recommended* according to a particular principle. Similarly, with the set of possible worlds, since we adopt the binary relation k to indicate the alternative possible worlds, we only take into account what the modeller considers relevant to the current problem.

We consider three approaches to ethical principles here: two consequentialist approaches and one deontological approach.

3.1 Goodness-Based Principle

One of the most well-known ethical principles is *Utilitarianism*. It is concerned with bringing the *greatest good to the greatest number of people* [8]. In [11], a *goodness* function is defined to take a *state* of a situation as an argument. A state is a set of fluents that are true in a specific situation. Here, *goodness* takes a plan and a situation and returns a value specifying how good such a plan in such a situation is; $goodness : \Pi \times S \rightarrow \mathbb{N}$. The recommended plan is the one with the highest *goodness* value.

In this work, we define another notation to derive the aggregated *goodness* value in an epistemic situation: $goodness^* : \Pi \times 2^\Pi \times 2^K \rightarrow \mathbb{N}$. Although a plan and a situation are usually enough to determine the aggregated *goodness* value, we include the set of plans Π as an argument to represent a function that determines the *goodness* value in relation to other plans. We will discuss further about the various approaches to aggregate such values in the next section. We avoid taking a state as the argument because, in the aggregated function, we need to consider multiple scenarios. Therefore, instead of taking a set of states, we simplify it by considering a set of accessible worlds. Then, if we are in an epistemic situation and are given the value of *goodness**, the recommended plan is the one with the highest *goodness**.

$$\mathcal{EP}_{good}(\Pi, K) = \{\, \pi \in \Pi \mid \forall_{\pi' \in \Pi}.goodness^*(\pi, \Pi, K) \geq goodness^*(\pi', \Pi, K)\,\}$$

3.2   Less Harm Principle

The highest overall *goodness* is the most straightforward way to select the best plan. It is easily comparable and self-explainable. However, a state or a situation is not always quantifiable. Therefore, in the second consequentialist approach, instead of trying to get the highest *goodness* value, we are trying to obtain a situation with least harm as possible. Although many factors determine the *goodness*, we believe *the less harm a plan produces, the better such a plan is.*

Here we define an explicit predicate to mark a fluent as harmful: $harmful(\varphi)$. We then consolidate those harm into a set of fluents; $harms(\pi, s) = \{\, \varphi \mid harmful(\varphi) \wedge \varphi(do(\pi, s))\,\}$. It tells us every harmful fluent that is held at the end of executing plan π in situation s. For the epistemic case, similarly to the first approach, we need to take into account harms that are also held in the other possible scenarios. $harms^* : \Pi \times 2^\Pi \times 2^K \rightarrow 2^\Phi$, which returns a set of harmful fluents (we use Φ as a notation for a set of fluents). By deriving the aggregated harm for each plan, the recommended plan is the one with the least harm:

$$\mathcal{EP}_{harm}(\Pi, K) = \{\, \pi \in \Pi \mid \forall_{\pi' \in \Pi}.harms^*(\pi, \Pi, K) \subseteq harms^*(\pi', \Pi, K)\,\}$$

3.3   Deontological Principle

Unlike the previous two, this principle does not consider the consequences a plan produces. Instead, it evaluates a plan based on its *duty*. If a plan fulfils its duty in a particular situation, such a plan is *just*; otherwise, it is *unjust*. Here *duty* is defined as a relation. $duty \subseteq \Pi \times S$.

Similarly, we also need the aggregated *duty* function that takes into account a number of possible scenarios: $duty^* \subseteq \Pi \times 2^\Pi \times 2^K$. Finally, a recommended plan is any plan that fulfils its aggregated duty.

$$\mathcal{EP}_{deon}(\Pi, K) = \{\, \pi \in \Pi \mid duty^*(\pi, \Pi, K)\,\}$$

In this work, we present a formalisation of ethical principles through plan evaluation. Our investigation revolves around a fundamental question: given a plan $\pi \in \Pi$, does it adhere to specific ethical criteria? Our approach entails representing these principles as a selection function, which effectively sifts through a set of plans to identify those that align with ethical standards. Consequently, our method has the capacity to yield a variable number of plans that can be designated as ethically permissible options.

4   Aggregation Strategies

We found a close resemblance between the implementation of strategies in game theory and our requirement for the aggregation functions. Here, we will adopt four strategies from *games against nature* [21] and demonstrate how they can be translated into aggregation functions for epistemic reasoning. We will focus on these four strategies because they align with our ethical principle formalisation.

4.1   Maximum Average

Laplace suggests that if there is no tendency that a scenario is more likely to occur, we should assume every possibility equally likely [13]. In other words, he advises choosing a plan with the highest average value, equivalently choosing a plan with the highest sum. The average value is pretty straightforward in the goodness-based model. Since it already returns a numeric value, we simply have to find the *average* of all the values that are k-related.

$$goodness^*_{avg}(\pi, \Pi, K) \stackrel{\text{def}}{=} avg(\{\, goodness(\pi, s) \mid k(s, s_0) \in K \,\})$$

On the other hand, we need a novel approach to average our *less harm* approach. We propose that a harmful fluent still holds if it holds in the majority of possible situations. In other words, if in most cases, a plan produces/retains harm φ, the average value of those cases also produce/retain harm φ.

$$harms^*_{avg}(\pi, \Pi, K) \stackrel{\text{def}}{=} \{\, \varphi \mid harmful(\varphi) \text{ holds in majority of } s : k(s, s_0) \in K \,\}$$

In the deontological principle, we are again dealing with non-numerical values. Similarly, we need to redefine the aggregated value by saying that a plan satisfies its duty if it satisfies the duty in most of the possible scenarios.

$$duty^*_{avg}(\pi, \Pi, K) \equiv duty(\pi, s) \text{ holds in majority of } s : k(s, s_0) \in K$$

4.2   Maximin Strategy

Another well-known strategy to approach uncertainty is considering the worst possible scenario and maximising its value. Wald proposed this approach considering that a player might want to be cautious [24]. Similar to the previous approach, our goodness-based model needs no unique definition to implement this maximin strategy. The aggregated value is the minimum goodness value of all k-related scenarios.

$$goodness^*_{min}(\pi, \Pi, K) \stackrel{\text{def}}{=} min(\{\, goodness(\pi, s) \mid k(s, s_0) \in K \,\})$$

Meanwhile, in the less harm principle, we define the worst-case scenario as the occurrence of harm. If a harmful event occurs in at least one of the possible worlds, we consider such harm occurs in the aggregated result.

$$harms^*_{min}(\pi, \Pi, K) \stackrel{\text{def}}{=} \bigcup \{\, harms(\pi, s) \mid k(s, s_0) \in K \,\}$$

On the other hand, the worst case that can happen to a plan under the deontological principle is that such a plan does not satisfy its *duty*. Therefore, we consider here the aggregated *duty* holds if it holds in every possible scenario.

$$duty^*_{min}(\pi, \Pi, K) \equiv \bigwedge \{\, duty(\pi, s) \mid k(s, s_0) \in K \,\}$$

4.3 Coefficient of Optimism

Hurwicz proposed a modification to the previous strategies [4]. He proposed a coefficient to reflect how optimistic the modeller of the possible scenario is. A coefficient α is assigned for the best value and $1 - \alpha$ for the worst value. Combining this assignment will give another value which we can maximise. In our goodness-based approach, we can apply this function with *max* and *min* aggregation functions. Therefore, the aggregated value is the maximum value multiplied by α plus the minimum value multiplied by $1 - \alpha$.

$$goodness^*_{coef}(\pi, \Pi, K) \stackrel{\text{def}}{=} (\alpha)max(\{\, goodness(\pi, s) \mid k(s, s_0) \in K \,\}) + (1 - \alpha)min(\{\, goodness(\pi, s) \mid k(s, s_0) \in K \,\})$$

We can apply this function to our less harm principle by looking at the ratio of possible worlds where harmful fluents φ hold and possible worlds where they do not. If the ratio φ occuring in the possible existing world is *no less* than the coefficient, the harm φ holds after being aggregated.

For example, consider three alternative possible worlds, $k(s, s_0)$, $k(s', s_0)$, and $k(s'', s_0)$. φ holds in s but not in s' and s''. Consider that we have the coefficient α equal to $\frac{1}{2}$. The occurrence ratio of φ is $\frac{1}{3}$ through all possible scenarios but it is less than the ratio α. Therefore, after being aggregated, φ does not hold.

$$harms^*_{coef}(\pi, \Pi, K) \stackrel{\text{def}}{=} \{\, \varphi \mid harmful(\varphi) \text{ holds in at least } \alpha \text{ of } s : k(s, s_0) \in K \,\}$$

A similar approach is used to aggregate *duty* in the deontological principle. A plan maintains its aggregated duty if such a plan holds its duty in at least $(1 - \alpha)$ of the possible scenarios. Here, we need to look carefully at what it means to be an optimist. Since α shows how optimistic we are, the greater the value of α the less concern about a plan that does not satisfy its duty in a certain scenario. Therefore, if α is 1, a plan will satisfy its aggregated duty if at least one of the possible scenarios satisfies it.

$$duty^*_{coef}(\pi, \Pi, K) \equiv duty(\pi, s) \text{ holds in at least } (1 - \alpha) \text{ of } s : k(s, s_0) \in K$$

4.4 Regret Minimisation

Savage also proposed an approach [17] by considering human psychological behaviour. This approach is based on humans' tendency to treat negativity more severely than receiving positive feedback. Therefore, Savage designed a strategy to minimise *regret*. First, we have to build a *regret matrix* by putting the difference between the value of a specific plan and the "best" plan in a given scenario. Then, we extract the matrix into a single value which is the maximum regret that can happen if we execute such a plan. Finally, we can minimise such maximum regret values to get the recommended plan.

In an attempt to modify the value into the regret matrix, we will use function *regret*. It changes the value of plan π compared to a set of plans Π in situation s. We will formalise "regret" slightly differently here than the original definition of "regret" in game theory. In game theory, "regret" is the absolute difference between the evaluated plan and the best plan in a particular scenario. However, to maintain the consistency of $\mathcal{EP}_{good}$, we will define "regret" as the *goodness* value of such a plan minus the highest *goodness* in a particular situation. Therefore the value will be less than or equal to 0.

$$regret_{good}(\pi, \Pi, s) \stackrel{\text{def}}{=} goodness(\pi, s) - max(\{\, goodness(\pi', s) \mid \pi' \in \Pi \,\})$$

Furthermore, we consider the minimum value of a plan in every situation (the most regret one can experience if one chooses such a plan) and choose a plan with minimal regret for $\mathcal{EP}_{good}$.

In the less-harm approach, we model the "best" result as a set of harms if and only if such harm occurs in every possible scenario; this can be formalised by using *intersection* the function of set theory. Here, we are also utilising the *minus* function of set theory. Given a set of harmful fluents from a specific plan and initial situation, we subtract that set with the intersection of the set of harms from all plans with the same initial situation. Thus, we are left with a set of harms that do not occur in every plan.

$$regret_{harm}(\pi, \Pi, s) \stackrel{\text{def}}{=} harms(\pi, s) - \bigcap\{\, harms(\pi, s) \mid \pi' \in \Pi \,\}$$

Finally, we find the worst case a plan can produce in every situation (same approach as the Maximin strategy) and use it as the aggregated harms for $\mathcal{EP}_{harm}$.

The deontological principle determine whether a plan is permissible or not. First, we will define the regret function to show whether one will regret their decision. Such regret will only occur if one plan from a certain initial situation does not fulfil its duty, but there is another plan with the same initial situation that fulfils its duty. $regret_{duty}$ is used to show such value.

$$regret_{duty}(\pi, \Pi, s) \equiv \bigvee\{\, duty(\pi', s) \mid \pi' \in \Pi \,\} \wedge \neg duty(\pi, s)$$

Moreover, the aggregated duty of this strategy is true if a plan will not be regretted in every possible situation.

4.5 Illustration

The following illustration will demonstrate our approach.

A patient has a painful disease that needs to be treated immediately. At the moment, the hospital proposes two procedures to treat the patient. The first procedure is intended to reduce pain but not to completely cure the disease. The second procedure is designed to cure the disease if the patient satisfies certain conditions. The condition is unknown, implying risk in executing such a procedure. Proceeding with procedure two without satisfying the prerequisite condition will give the patient a severe migraine without curing the disease.

In the supplementary file, Table 2 shows the initial setup for *goodness*, *harms*, and *duty*. Here, we assume that $disease(p)$ and $migraine(p)$ are harmful fluents, and *duty* is fulfilled if the patient is disease-free. This table can produce the result of the first three aggregation functions. However, we need an additional *regret matrix* to have the result for regret minimisation. Table 3 shows that plan π_1 has no regret in s'_0, but plan π_2 has no regret in s_0. Finally, assuming we set the coefficient of optimism $\alpha = 0.7$, we get the result of every aggregation function. Table 4 shows that the aggregated values of π_1 are all the same throughout four aggregation functions. This result is as expected since plan π_1 is indifferent between s_0 and s'_0. Finally, Table 5 shows which plan is permissible according to particular ethical principles and aggregation functions.

5 Results and Discussions

5.1 Milnor's Axioms

A special case of game theory can be considered as *games against nature* if the player is not against a reasonable opponent. In this case, a player is faced with possible scenarios that are assumed to be equally likely to occur. The goal is still the same, to maximise the total outcome of a game. However, the strategy will be different compared to the game against players. John Milnor proposed several axioms to evaluate the strength of strategies in the context of games against nature [9]. The following are seven axioms that we will use to evaluate aggregation functions.

Symmetry. Rearranging the plans and the order of the possible situation should not affect which plan is recommended as best.

Strong Domination. If every outcome in plan π is better than the corresponding outcome in plan π', a method should not recommend plan π'.

Linearity. The recommended plan should not change if all outcomes are multiplied by a positive constant or if a constant is added to all outcomes.

Column Duplication. The recommended plan should not change if we add a new possible situation that duplicates the result of the possible situation we have already captured.

Bonus Invariance. The recommended plan should not change if a constant is added to every outcome in some possible situation.

Row Adjunction. Suppose a method chooses plan π as the best, and then a new plan π' is added. The method should not choose a plan other than π and π'.

Complete Ordering. A plan is either better, worse or equally good compared with the alternatives and the binary comparison between plans must satisfy transitivity.

5.2 Axiom Satisfaction

Table 1 shows that our goodness-based approach satisfies the same axioms for every aggregation function as the literature from game theory [21]. This result serves as the foundation to show that our formalisation of a goodness-based approach, followed by the four aggregation functions, is aligned with the literature from *games against nature*.

Table 1. Summary of axiom satisfaction for every ethical principle and aggregation function. Symbol - indicates that such an axiom is not applicable for such an approach, ✓ indicates that such an approach satisfies such an axiom, and ✗ indicates otherwise.

		Symmetry	Strong Domination	Linearity	Column Duplication	Bonus Invariance	Row Adjunction	Complete Ordering
Goodness	Average	✓	✓	✓	✗	✓	✓	✓
	Maximin	✓	✓	✓	✓	✗	✓	✓
	Coefficient	✓	✓	✓	✓	✗	✓	✓
	Regret	✓	✓	✓	✓	✓	✗	✓
Less-Harm	Average	✓	✓	-	✗	-	✓	✗
	Maximin	✓	✓	-	✓	-	✓	✗
	Coefficient	✓	✗	-	✗	-	✓	✗
	Regret	✓	✓	-	✓	-	✗	✗
Deon	Average	✓	✓	-	✗	-	✓	✓
	Maximin	✓	✓	-	✓	-	✓	✓
	Coefficient	✓	✓	-	✗	-	✓	✓
	Regret	✓	✓	-	✓	-	✓	✓

One of the weaknesses of using Milnor's axioms for evaluation is that they assume an approach based on numerical value. Therefore, an approach that judges an action or a plan qualitatively cannot be thoroughly evaluated. This issue is reflected in Table 1 that Linearity and Bonus Invariance cannot be used to evaluate Less-harm and the deontological approach.

In the less-harm approach, no aggregation function satisfies the complete ordering because we do not specify the degree of harm for the harmful fluent. Therefore, there will be examples of incomparable situations because plans produce different harm. This non-satisfaction shows the weakness of this principle; that an ethical dilemma may occur where the system will recommend no plan between those incomparable situations. The **maximin** approach is considerably robust because it satisfies the most properties. On the other hand, the **coefficient of optimism** tends to be weaker and needs a more careful implementation.

Failing the strong domination axiom shows that, even if a plan is not preferred in every possible scenario, how to choose coefficient α can override it and still recommend such a plan. As in the **maximum average** approach, it fails column duplication, showing that this approach is ineffective when the uncertainty may change. Therefore, if the modeller values an aggregation function that is stable through change of uncertainty, the maximin and regret minimisation approaches should be more utilised. Furthermore, the non-satisfaction of row adjunction for **regret minimisation** shows that this aggregation function is highly dependent on the other available plans. This aggregation function will be ineffective if plans can be added or removed in different scenarios.

The implementation of the deontological approach is more complicated than the previous two. It has to list every condition that will or will not satisfy *duty*. However, it also performs more stable than the less-harm approach. In a situation where the condition of *duty* can be explicitly listed, the deontological approach may be preferable. The **maximin** and **regret minimisation** approaches show satisfaction through all provable axioms. It shows that these approaches are not affected by the change of plan or uncertainty. The maximum average and coefficient of optimism still fall on the same weakness. They are weak when dealing with changes in uncertainty. Although the coefficient of optimism is more adjustable depending on the value α.

Although the deontological approach satisfies the complete ordering, it can still produce no recommended result after the evaluation. This weakness shows that, unlike the consequentialist approaches, the deontological approach acts more independently and evaluates a plan regardless of every other available plan. This mechanism shows that, even if we have multiple plans that are executable and reach the goal, it is possible to have none of them satisfying their duty.

The result for the symmetry axiom does not change through all principles and aggregation functions. The proof is trivial because we use *sets* instead of lists for plans and possible scenarios. This modelling is probably a significant difference between our formalisation and the game theory formalisation. However, with the satisfaction of the symmetry axiom, it shows that our formalisation serves the intended purpose of a principle and aggregation function.

6   Conclusion

This paper has modelled three ethical principles as a selection function: two consequentialist approaches and one deontological approach. On top of that, we also introduce various aggregation functions (strategies) to reason in an epistemic state. We formalise the aggregation functions inspired by the literature from game theory. Our illustration shows that different aggregation functions and ethical principles can significantly lead to different recommended plans. Apart from that, we also adopt the axioms by John Milnor to evaluate each aggregation function. We show that every aggregation function may be more robust in one scenario but more volatile in others. This result helps the modeller to decide which criteria are important when designing an ethical autonomous system.

In our implementation, we do not include the notation to show a different degree of harm. This resulted in potentially no plan being recommended. Although this constraint is not our focus, future work should accommodate such drawbacks. Furthermore, uncertainty can be modelled in many ways. Here, we utilise the epistemic uncertainty using the k relation. However, if other works propose different methods to formalise uncertainty, our result regarding the strengths and weaknesses of the aggregation function will still be applicable.

References

1. Baum, K., Hermanns, H., Speith, T.: Towards a framework combining machine ethics and machine explainability. arXiv preprint arXiv:1901.00590 (2019)
2. Dennis, L., Fisher, M., Slavkovik, M., Webster, M.: Formal verification of ethical choices in autonomous systems. Robot. Auton. Syst. **77**, 1–14 (2016)
3. Fudenberg, D., Tirole, J.: Game Theory. MIT Press, Cambridge (1991)
4. Hurwicz, L.: The generalized Bayes minimax principle: A criterion for decision making under uncertainty. Cowles Comm. Discuss. Paper Stat **335**, 1950 (1951)
5. Kim, R., et al.: A computational model of commonsense moral decision making. In: Proceedings of the 2018 AAAI/ACM Conference on AI, Ethics, and Society, pp. 197–203 (2018)
6. Lindner, F., Mattmüller, R., Nebel, B.: Evaluation of the moral permissibility of action plans. Artif. Intell. **287**, 103350 (2020)
7. Lourie, N., Le Bras, R., Choi, Y.: Scruples: a corpus of community ethical judgments on 32,000 real-life anecdotes. In: Proceedings of the AAAI Conference on Artificial Intelligence, vol. 35, pp. 13470–13479 (2021)
8. Mill, J.S.: Utilitarianism. In: Seven Masterpieces of Philosophy, pp. 329–375. Routledge (2016)
9. Milnor, J.: Games against nature. Game Theory and Related Approaches to Social Behavior. Wiley, Hoboken (1964)
10. Noothigattu, R., et al.: Teaching AI agents ethical values using reinforcement learning and policy orchestration. IBM J. Res. Dev. **63**(4/5), 2–1 (2019)
11. Pagnucco, M., Rajaratnam, D., Limarga, R., Nayak, A., Song, Y.: Epistemic reasoning for machine ethics with situation calculus. In: Proceedings of the 2021 AAAI/ACM Conference on AI, Ethics, and Society, pp. 814–821 (2021)
12. Perrone, V., Donini, M., Zafar, M.B., Schmucker, R., Kenthapadi, K., Archambeau, C.: Fair Bayesian optimization. In: Proceedings of the 2021 AAAI/ACM Conference on AI, Ethics, and Society, pp. 854–863 (2021)
13. Pierre-Simon, L.: Théorie analytique des probabilités. Livre II, Chapitre X. De l'espérance morale, Oeuvres de Laplace, ome VII, Imprimerie Royale, pp. 474–488 (1812)
14. Reiter, R.: The frame problem in the situation calculus: a simple solution (sometimes) and a completeness result for goal regression. Artif. Math. Theory Comput. 3 (1991)
15. Reiter, R.: Knowledge in Action: Logical Foundations For Specifying and Implementing Dynamical Systems. MIT press, Cambridge (2001)
16. Rodriguez-Soto, M., Lopez-Sanchez, M., Rodriguez-Aguilar, J.A.: Multi-objective reinforcement learning for designing ethical environments. In: IJCAI, pp. 545–551 (2021)
17. Savage, L.J.: The Foundations of Statistics. Courier Corporation (1972)

18. Scherl, R.B., Levesque, H.J.: The frame problem and knowledge-producing actions. In: AAAI, vol. 93, pp. 689–695 (1993)
19. Scherl, R.B., Levesque, H.J.: Knowledge, action, and the frame problem. Artif. Intell. **144**(1–2), 1–39 (2003)
20. Sohrabi, M.K., Azgomi, H.: A survey on the combined use of optimization methods and game theory. Arch. Comput. Methods Eng. **27**(1), 59–80 (2020)
21. Straffin, P.D.: Game Theory and Strategy, vol. 36. MAA (1993)
22. Svegliato, J., Nashed, S.B., Zilberstein, S.: Ethically compliant planning in moral autonomous systems. In: IJCAI (2020)
23. Tolmeijer, S., Kneer, M., Sarasua, C., Christen, M., Bernstein, A.: Implementations in machine ethics: a survey. ACM Comput. Surv. (CSUR) **53**(6), 1–38 (2020)
24. Wald, A.: Statistical decision functions (1950)
25. Wu, Y.H., Lin, S.D.: A low-cost ethics shaping approach for designing reinforcement learning agents. In: Proceedings of the AAAI Conference on Artificial Intelligence, vol. 32 (2018)

Using Social Sensing to Validate Flood Risk Modelling in England

Joshua Joyce[1,2](✉), Rudy Arthur[1], Guangtao Fu[1], Alina Bialkowski[2], and Hywel Williams[1]

[1] University of Exeter, Exeter EX4 4QE, UK
jj479@exeter.ac.uk
[2] University of Queensland, St Lucia, Brisbane, QLD 4072, Australia

Abstract. Floods are amongst the most severe natural disasters. Accurate flood risk maps are vital for emergency response operations and long-term flood defence planning. Currently the validation of such maps is often neglected and suffers from a lack of high-quality data. The proliferation of social media usage worldwide in recent years has supplied access to large amounts of data linked to flooding, and the detection of real-world events using such data is termed 'social sensing'. In this paper we investigate the use of social sensing for the validation of flood risk maps. We apply this methodology to 7 years' worth of flood related Tweets in order to perform a comparison to long term planning flood risk maps in England. The results show that there is a low level of correlation between the collection of socially sensed floods and high-risk flood areas as well as highlighting areas with high levels of socially sensed flooding that have low levels of flood risk, showcasing the potential importance of social media data for use in flood risk validation and planning policy.

Keywords: Social media · Flooding · Twitter

1 Introduction

Natural disasters are among the world's greatest challenges and 80,000 people per day are affected with an economic loss of US\$ 1.5 trillion since 2003. Flooding alone, which is the most frequent and wide-reaching weather-related natural hazards in the world [4], has affected 2.3 billion people with an estimated economic losses of US\$ 662 billion from 1995 to 2015, and US\$ 60 billion in 2016 alone [15]. On top of this, impacts of floods are projected to increase in the future due to climate change [13].

The generation of long term flood risk maps is then of extreme importance for planning procedures. Such maps are produced by utilising features such as terrain elevation, land use and meteorological data as parameters within physical models to estimate the flood extent of various simulated levels of rainfall events [7]. The maps output from this process are then used in urban planning for flood mitigation and defence [14].

© The Author(s), under exclusive license to Springer Nature Singapore Pte Ltd. 2024
T. Liu et al. (Eds.): AI 2023, LNAI 14472, pp. 95–106, 2024.
https://doi.org/10.1007/978-981-99-8391-9_8

The validation of such models is a key topic in flood modelling but as noted by Molinari et al. [8] in their review of existing practices in this area, 'Validation is perhaps the least practised activity in current flood risk research and flood risk assessment'. One major problem in the validation process is a lack of high quality data and when validation is performed often crowd-sourced data is used [10,12].

Social sensing is the use of unsolicited crowd-sourced information to observe real world events. This information can be gathered from a variety of different sources including web searches and social media. The major advantage of social media data over other crowd-sourcing avenues is the volume of data; social media platforms such as Twitter, Instagram and TikTok have millions of active users each month and millions of posts per day. This paper focuses on social sensing using Twitter due to the public accessibility of its data.

A variety of studies have been conducted in the topic of social sensing of floods. Arthur et al. [2] used tweet observations to produce flood maps of the UK validated against data of flood events provided by the Flood Forecasting Centre and concluded that social sensing can reproduce the validation data to a high accuracy, even finding flood events that were not contained in the validation dataset, albeit at the cost of false positives. Moore et al. [9] introduced a method for social sensing of coastal floods. They proposed to use a metric of remarkability of a high-tide event as a way to measure impact of coastal flooding where it is felt the most, as apposed to earlier methods which focus on high population areas due to ease of data collection. Individual regression models were built for counties along the east coast of the USA, building a relationship between number of geo-located tweets that day and maximum daily tide height measured at nearby tidal gauges including controlling for daily rainfall. Young et al. [18] utilised Twitter data as well as data from social media site Telegram in order to analyse the impacts of the 2018 floods in Kerala. They were able to analyse not only the extent of flood impacts but also the kind of impacts such as requests for help. The results showed good agreement with government created post flood database of damages. Ansell et al. [1] introduced a statistical approach involving the use of vine copulas, where they combined social media data including Twitter data, Google Trends data as well as average sentiment with environmental variables such as wave height and water level in order to predict inundation. Here they showed that performance of the model was improved by assuming a relationship between the social and environmental factors rather than assuming independence, showcasing that integration of social media data can produce more accurate forecasts.

Previous studies where social sensing of flooding has been used have focused mostly on the validation of the method to detect past flood events [2] or in post event analysis [18] and are often short term studies from a temporal standpoint. Only one study was found that utilised social sensing in term of flood risk, Brangbour et al. [5] utilised Twitter data to compute probabilities of rasterised grid cells being flooded during Hurricane Harvey. This was a study done using high quality, highly curated data for a single extremely severe event. On the other hand the main contribution, and unique goal, of this paper is to perform social sensing of floods using a much greater time period of data, collecting data on all types of flooding over a country wide area in order to compare this analysis with long term flood risk models. As social sensing of floods by its very nature observes the impact of floods on a societal and human level, this report seeks to discover the relationship between areas considered a high risk in flood defence and mitigation planning and areas of high social 'floodiness', and in doing so investigate the potential of social sensing as a useful tool in the key area of flood risk model validation.

The main contributions of this paper are:

- The first long term study of socially sensing of floods via the creation of a dataset consisting of 7 years of geolocated relevant tweets.
- The first study to investigate the relationship between socially sensed flooding and flood risk models on a national scale.

The structure of the paper is as follows. Section 2 describes the data sources used, the methodology for various filtering techniques used to curate the Twitter data as well as the method used to infer locations from tweets. In Sect. 3 the results of the analysis are presented. Section 4 contains a discussion of these results. Finally Sect. 5 presents the conclusions of the paper.

2   Methodology

2.1   Data Collection

Twitter Dataset. Tweets were collected using Twitter's Streaming API and searching for the terms "flood", "flooding" and "flooded" as a basic first filter. It's important to note that this API was accessed using Twitter's Academic track which has significantly increased in price as Twitter have changed their data policies. The API returns tweets in the form of a JSON object which consists of key-value pairs for various metadata such as tweet text, user profile information and user location. In total 160,424,089 tweets were collected between the dates of 22/10/2015 and 11/04/2022. Due to collection issues there are gaps between 28/12/2015 and 04/01/2016 as well as between the dates 26/11/2016 and 02/01/2017 and 17/11/2021 and 10/01/2022.

Flood Maps. Recent flood maps (early 2022) produced by the Environment Agency[1] were used for comparison with the Twitter dataset. These maps are produced using physical modelling methods and separate maps are produced considering different types of flooding. The first of these is called 'Risk of Flooding from Rivers and Sea'. This map consists of 50 m × 50 m gridded areas of England with the likelihood of flooding from rivers and the sea presented in four different categories; namely Very Low, Low, Medium and High, whilst also taking account of flood defences and the condition they are in. High risk flood

---

[1] https://data.gov.uk/.

areas, which refers to a 1 in 30 annual probability of flooding, were chosen for use in this study providing the best comparison to the nearly 7 years worth of Twitter data. The map for the category of high risk can be seen in Fig. 1a.

As well as maps based on river and coastal flooding the Environment Agency also produce maps for surface water flood risk. High spatial resolution maps of this type are unavailable for download and are restricted to tiles, or small grid squares, of England due to the large complexity of these maps. Instead, so called 'Indicative Flood Risk' maps are available where the modelled data is aggregated to 1 km square grids based on 1 in 100 annual probability of flooding as well as minimum thresholds for either area population (200 people per 1 km grid) or critical services (at least one per 1 km square grid) or number of non-residential buildings at risk (at least 20 per 1 km square). The produced grids can be seen in Fig. 1b.

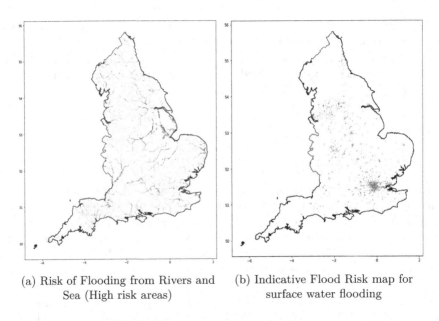

(a) Risk of Flooding from Rivers and Sea (High risk areas)

(b) Indicative Flood Risk map for surface water flooding

Fig. 1. Environment Agency flood maps

2.2   Twitter Data Pre-processing

As previously mentioned an initial filter was applied to tweets as they were collected. Several different filters were then applied post collection to remove irrelevant data as follows:

Retweet and Quote Tweet Filtering. One feature of Twitter is the ability to retweet and quote tweets. This is done to promote the tweet and to increase the likelihood of the tweet being seen by other users. As these are not original and independent flood observations by users all retweets and quote tweets are removed. Sometimes people type 'RT' at the beginning of a tweet to indicate that they are reposting someone else's content instead of using the retweet function and these were also removed.

Bot Filtering. A number of accounts on Twitter are automated bot accounts and in the context of floods a large number of tweets that pass the initial top level filter are from weather stations which tweet out a large amount of flood related information. As we are interested in socially sensed flood events these tweets add a large amount of noise to the data and are removed. In total around 100 accounts are identified as bots and are removed from the dataset, including accounts such as @RiverLevelsUK @ukfloodtweets and @ShropshirePulse. Weather stations in general are removed by searching for a large number of keywords within tweets such as "north", "south", "rain" and "wind" and also units such as "mm" and "m/s". If the number of keyword matches is greater than a threshold then the tweet is removed.

Language Filtering. As the focus of this study is England and for use with geographical databases used in future steps all non English tweets are removed by using the "lang" key within the tweet JSON.

Relevance Filtering. Even after the previous filtering steps there remain a large number of tweets containing the top-level keyword terms in irrelevant contexts. Examples of this include phrases such as 'flooded with' or 'flood of'. A number of manually curated terms such as these were created and tweets containing these terms were filtered out.

Next, tweets were manually tagged as relevant or not relevant where relevant in this case refers to a tweet about an immediate flood situation such as "flooding in Exeter right now" as opposed to tweets about historic flood events or flood warnings which were tagged as not relevant. In total 4524 tweets were tagged with 1733 tagged as relevant and 2791 tagged as not relevant.

Using these tagged tweets as training data, a Multinomial Naive Bayes classifier was built and the tagged dataset of tweets was split into training and validation sets. 75% of the data was used to train the models and 25% was used for validation. Tweets were cleaned to remove stop words, URLs and punctuation. Tweets were tokenized, stemmed and lemmatized. A Bag of Words technique was used and the data was vectorised by counting the number of single word and two word occurrences in the corpus. Overall the Naive Bayes model achieved an F1-score of 0.84 with a Precision of 0.79 and a Recall of 0.83 indicating good overall classification.

Location Inference. Only a very small amount of tweets contain GPS data (less than 1%) [6]. For the purpose of creating accurate flood maps based on this data, it is important to be able to accurately infer locations from tweet metadata.

If a tweet contains an exact GPS tag then that latitude/longitude pair is used to map the tweet. For every other tweet a location inference heuristic is applied which is based on [11].

Table 1. Tweets remaining after each processing step

Filtering stage	Tweets
Top-level	160,424,089
Relevance	2,193,999
Location inference	408,172
Large polygon removal	165,663

In order to validate the performance of the location inference heuristic comparisons were made to a subset of the filtered tweets with exact GPS coordinates. In total, there were 71,688 tweets with GPS coordinates. Location polygons were inferred using the heuristic method with GPS metadata specifically ignored. Using this, a parameter grid search was performed using a range of different gazetteer database weightings as well as indicator weightings. The displacement between inferred locations and actual locations was calculated in kilometres and a tweet was considered correctly classified if the displacement was lower than 10 km. The best performing set of parameters was shown to be all indicator weightings set to 1 except tweet text which was set to 2. The displacement in kilometres between the inferred location and the actual location for this parameter is shown in Fig. 2. It can be seen that the method performs very well albeit with some large displacements shown for a number of tweets. The total tweets retained after each processing step can be seen in Table 1. Overall, just over 165,000 tweets were retained which were then used to create the socially sensed flood maps.

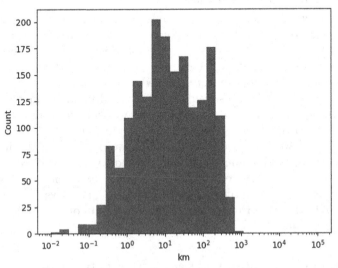

Fig. 2. Displacement of inferred location polygons from true geotagged coordinates

2.3   Flood Map Development

Socially Sensed Flooding Maps. In order to produce maps containing all remaining tweets, we first start with a bounding box of England and discretise it. Based on the results seen in Fig. 2 it was decided to create grid squares of 10 km by 10 km, as this achieves a good balance between accuracy and granularity. Each grid square starts with a weight of 0, $g_W = 0$, and is incremented for each tweet that falls within the grid square:

$$g_W = g_W + \frac{Area_{g \cap p}}{Area_p},\tag{1}$$

where g is the grid square and p is the tweet polygon. When the tweet has a precise location, p is a point and a score of 1 is added to the weight of the grid square, otherwise the proportion of overlap is added. This enables tweets with precise locations to be more influential for the detection of flooding.

The next step is to account for population density as large cities will have vastly more tweets associated with them than small towns. To this end, population data is taken from Lower Layer Super Output Areas (LSOAs), a geographic hierarchy for small area statistics. LSOAs are population areas of at least 1000 people and are designed to be consistent in population size. As a result, LSOAs within cities are much smaller than their counterparts in the countryside. The proportion of overlap between grid squares and their intersecting LSOAs is calculated and the corresponding proportion of the LSOA population is added to the grid square. This population data is taken from the 2011 Census so is somewhat out of date and taking proportions makes the assumption of uniform population across the LSOA which is not necessarily true but as the LSOAs are small enough it provides a reasonable estimate.

We then rescale the grid weights as follows,

$$g_W = \frac{g_W}{g_P{}^\alpha}\tag{2}$$

where g_P is the calculated population for the grid square and α is a scaling factor between 0 and 1. The factor α is necessary as it has been found that there is an imbalance between the number of twitter users in cities and rural areas [3]. A larger value of α will result in the population of the grid square having a larger effect on the weighting, meaning flooding detected in less populated rural areas will be more pronounced. For the purpose of this study α is set to 0.4 as this was found to have the best balancing effect.

Flood Risk Maps. In order to perform direct comparisons between our produced socially sensed flood maps and the flood risk maps produced by the Environment Agency it is necessary to have each type of map at the same spatial resolution. To this end, the Environment Agency flood risk maps are aggregated up to the same grid system as the socially sensed flood maps. In order to do this, for each grid square its intersecting flood risk polygons are obtained. The area of intersection of each polygon with the grid square is then calculated and summed. This is then divided by the area of the grid square in order to obtain the proportion of the grid square that is consider under risk of flooding.

3   Results

Maps produced from the aggregated Environment Agency flood risk maps can be seen in Fig. 3b for river and seas flooding and in Fig. 3c for surface water flooding. Figure 3a shows the population weighted flood maps based on the fully filtered tweet dataset.

For statistical comparison between the aggregated flood risk maps and the produced socially sensed flood maps, correlation between grid squares was calculated as well as the use of simple linear models. The correlation between grid squares was calculated using Spearman's rank correlation coefficient and is shown in Table 2. Overall correlation between river and coastal flood risk and social floodiness was $r = 0.27$ with a p-value of $1e^{-26}$ and for surface water flood risk was $r = 0.41$ with a p-value of $7e^{-61}$. Scatter plots of tweet weighting against aggregated flood risk can be seen in Fig. 4. The equations for the lines fitted are $y = 4.75x + 1.3$ and $y = 11.4x + 1.2$ for river and coastal flood risk and surface water flood risk respectively. R^2 values were 0.01 and 0.1 respectively showing no linear relationships. Overall the results show a moderate level of correlation between the socially sensed flood map and Environment Agency produced maps, particularly with regards to surface water flooding.

Of particular note, the scatter plots show a number of outlier areas which have little to no modelled flood risk that have a high weighting for socially sensed flooding especially in the case of surface water flooding. Indeed by re-scaling the socially sensed grid weights between 0 and 1 by dividing through by the max of all grid square weights, we can calculate the difference between flood risk and social floodiness.

Figure 5 shows these differences with values between (0,1) indicating higher flood risk than social floodiness and values between (−1,0) indicating lower flood risk than social floodiness. Figure 5a compares the socially sensed response to flood risk from coastal and river flooding. In the north east we see that we have higher flood risk associated with rivers flowing into the Wash and Humber estuaries, as well as for the coast of East Anglia than we would predict from observing floods on Twitter. In contrast we see a much higher Twitter signal than the corresponding risk would predict in the north west (Cumbria) and far south west (Plymouth), likely due to major flood events which occurred during the data collection. Figure 5b which compares against surface flooding shows similar under-estimation in the north- and south-west with over-estimation in Greater London (south east). In general outside of these outliers it can be seen in both maps that flood risk is slightly underestimated against socially sensed flooding.

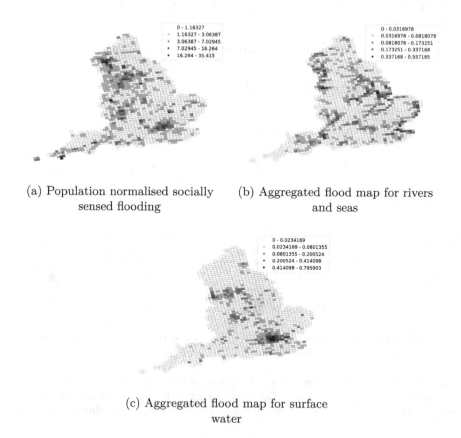

(a) Population normalised socially sensed flooding

(b) Aggregated flood map for rivers and seas

(c) Aggregated flood map for surface water

Fig. 3. Aggregated flood maps

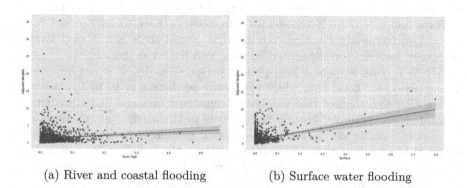

(a) River and coastal flooding

(b) Surface water flooding

Fig. 4. Scatter plots of each flood map against socially sensed flooding

Table 2. Spearman's rank correlation of the socially sensed flood map against specified maps

Flood Map	Spearman's rank correlation
Rivers and coastal	0.27
Surface water	0.41

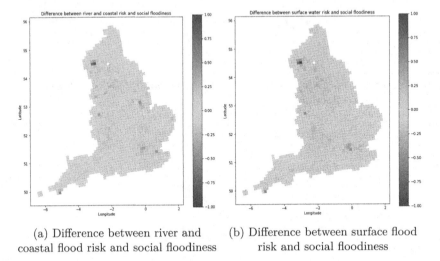

(a) Difference between river and coastal flood risk and social floodiness

(b) Difference between surface flood risk and social floodiness

Fig. 5. Maps indicating the difference between flood risk and social floodiness for each type of flood risk, red indicates grids with higher risk than social floodiness, blue indicates grids with lower risk than social floodiness. (Color figure online)

4   Discussion

From the results obtained we have found that there is a low to moderate correlation between socially sensed flooding and the high risk flood map for rivers and seas and that there is a moderate correlation between socially sensed flooding and surface water flooding showing that overall there is a modest level of agreement between socially sensed flooding and flood risk maps.

While 7 years is quite a long period for social media, it is not necessarily a long period for flood risk, which commonly predicts 1 in 10 to 1 in 100 year events. However, the fact that we observe significant over and under-predictions is notable. For example, historic flood events in the north west (Cumbria and Lancashire) during the data collection period produce a Twitter signal far in excess of the predicted risk. This occurs because the social response to a flood is highly non-linear [2,3], doubling the size of the flood generates much more than 2-fold increase in tweets. If tweets are taken as a rough proxy for impact, this implies that risk models which aim to predict not just the probability of occurrence of a flood, but also its potential impact, should incorporate a non-linear scaling of impact with event size.

There are a large amount of grid squares which have extremely low tweet weighting which could indicate that these areas do not suffer from flooding or that there is a demographic bias in some areas due to the fact that Twitter users are unrepresentative of the population as a whole. Further work could be done to explore this hypothesis. Related to this is the idea of the 'remarkability' of a flood. For example, an area which floods often may not produce tweets as this is considered a normality. This could be checked using historical data for severity and amount of flooding over a long term period, techniques like this were utilised in [9,16] to great effect and could be used in further work to improve the social floodiness response.

The most important step in the process of producing socially sensed flood maps is the initial filtering stage. Social media data is inherently noisy and improved relevance filtering would lead to a much less noisy dataset and a more accurate representation of flood events. The model used in this report is a classic classifier method used in this area but recent transformer models such as BERT as well as LSTM based models have been shown to perform well [17].

As well as this, location inference is a necessity due to the low level of geo-tagged tweets. Therefore more accurate location inference will lead to more accurate flood detection which is crucial for any validation of this method. Currently toponym recognition is limited as it only searches for proper nouns, and methods could be developed which expand this. Limitations also exist with the use of geo-databases, with toponym resolution being limited to nonexistent at below town level meaning potentially useful fine grained data is discarded. Improvements in this area would also allow future work to be expanded to smaller regional flood risk maps such as cities.

5   Conclusion

We have shown that it is possible to produce long term socially sensed flood maps that can be used to form a comparison - and potential validation tool - to long term flood maps. As socially sensed flood detection by nature detects floods which affect people, the results show not only the potential use of social sensing as a new data collection tool to validate flood risk maps compared to traditional validation methods which take time and may require a large workforce to manually collect and synthesise flood observation data, but also highlights a need for models to better take into account the impact of floods. As social media use continues to grow into the future, the growth of quality observation data that can be obtained from it will only improve its usability in this area.

References

1. Ansell, L., Valle, L.D.: Social media integration of flood data: a vine copula-based approach. arXiv preprint arXiv:2104.01869 (2021)
2. Arthur, R., Boulton, C.A., Shotton, H., Williams, H.T.: Social sensing of floods in the UK. PLoS ONE **13**(1), e0189327 (2018)

3. Arthur, R., Williams, H.T.: Scaling laws in geo-located twitter data. PLoS ONE **14**(7), e0218454 (2019)

4. Berz, G.: Flood disasters: lessons from the past-worries for the future. In: Proceedings of the Institution of Civil Engineers-Water and Maritime Engineering, pp. 3–8. Thomas Telford Ltd (2000)

5. Brangbour, E., et al.: Computing flood probabilities using twitter: application to the houston urban area during harvey. arXiv preprint arXiv:2012.03731 (2020)

6. Cheng, Z., Caverlee, J., Lee, K.: You are where you tweet: a content-based approach to geo-locating twitter users. In: Proceedings of the 19th ACM International Conference on Information and Knowledge Management, pp. 759–768 (2010)

7. Mignot, E., Paquier, A., Haider, S.: Modeling floods in a dense urban area using 2D shallow water equations. J. Hydrol. **327**(1–2), 186–199 (2006)

8. Molinari, D., De Bruijn, K.M., Castillo-Rodríguez, J.T., Aronica, G.T., Bouwer, L.M.: Validation of flood risk models: current practice and possible improvements. Int. J. Disaster Risk Reduction **33**, 441–448 (2019)

9. Moore, F.C., Obradovich, N.: Using remarkability to define coastal flooding thresholds. Nat. Commun. **11**(1), 1–8 (2020)

10. Schnebele, E., Cervone, G., Waters, N.: Road assessment after flood events using non-authoritative data. Nat. Hazards Earth Syst. Scie. **14**(4), 1007–1015 (2014). https://doi.org/10.5194/nhess-14-1007-2014. https://nhess.copernicus.org/articles/14/1007/2014/

11. Schulz, A., Hadjakos, A., Paulheim, H., Nachtwey, J., Mühlhäuser, M.: A multi-indicator approach for geolocalization of tweets. In: Seventh International AAAI Conference on Weblogs and Social Media (2013)

12. See, L.: A review of citizen science and crowdsourcing in applications of pluvial flooding. Front. Earth Sci. **7**, 44 (2019)

13. Tellman, B., et al.: Satellite imaging reveals increased proportion of population exposed to floods. Nature **596**(7870), 80–86 (2021)

14. Tsakiris, G.: Flood risk assessment: concepts, modelling, applications. Nat. Hazards Earth Syst. Sci. **14**(5), 1361–1369 (2014). https://doi.org/10.5194/nhess-14-1361-2014, https://nhess.copernicus.org/articles/14/1361/2014/

15. Unisdr, C., et al.: The human cost of natural disasters: a global perspective (2015)

16. Weaver, I.S., Williams, H.T., Arthur, R.: A social beaufort scale to detect high winds using language in social media posts. Sci. Rep. **11**(1), 3647 (2021)

17. Yaseen, Q., et al.: Spam email detection using deep learning techniques. Procedia Comput. Sci. **184**, 853–858 (2021)

18. Young, J.C., Arthur, R., Spruce, M., Williams, H.T.: Social sensing of flood impacts in India: a case study of Kerala 2018. Int. J. Disaster Risk Reduction **74**, 102908 (2022)

Symbolic Data Analysis to Improve Completeness of Model Combination Methods

Pedro Strecht[1]([✉]) [ID], João Mendes-Moreira[1] [ID], and Carlos Soares[1,2,3] [ID]

[1] LIAAD-INESC TEC, Faculdade de Engenharia, Universidade do Porto,
R. Dr. Roberto Frias, 4200-465 Porto, Portugal
{pstrecht,jmoreira,csoares}@fe.up.pt
[2] LIACC, Faculdade de Engenharia, Universidade do Porto, R. Dr. Roberto Frias,
4200-465 Porto, Portugal
[3] Fraunhofer Portugal AICOS, R. Alfredo Allen 455, 4200-135 Porto, Portugal

Abstract. A growing number of organizations are adopting a strategy of breaking down large data analysis problems into specific sub-problems, tailoring models for each. However, handling a large number of individual models can pose challenges in understanding organization-wide phenomena. Recent studies focus on using decision trees to create a consensus model by aggregating local decision trees into sets of rules. Despite efforts, the resulting models may still be incomplete, i.e., not able to cover the entire decision space. This paper explores methodologies to tackle this issue by generating complete consensus models from incomplete rule sets, relying on rough estimates of the distribution of independent variables. Two approaches are introduced: synthetic dataset creation followed by decision tree training and a specialized algorithm for creating a decision tree from symbolic data. The feasibility of generating complete decision trees is demonstrated, along with an empirical evaluation on a number of datasets.

Keywords: symbolic data · model completeness · consensus models

1 Introduction

Increasingly, organizations are embracing a new approach to tackle large data analysis problems – breaking them down into more specific sub-problems. This involves using multiple sub-problem-specific models instead of relying solely on a single global model. The objective is twofold: first, to handle dispersed activities effectively, and second, to address issues at different levels of detail. For instance, companies with multiple chain stores are adopting dedicated models for each store to predict sales performance on a monthly basis. Similarly, universities with diverse programs are implementing separate models to predict student dropout for each program. While these models are directed to localized contexts, understanding organization-wide phenomena also needs a global perspective.

T. Liu et al. (Eds.): AI 2023, LNAI 14472, pp. 107–119, 2024.
https://doi.org/10.1007/978-981-99-8391-9_9

However, managing numerous individual models can pose challenges. For example, universities may aspire to analyze student dropout at both the program and global levels. Yet, the sheer number of models makes a traditional model-by-model analysis impractical, urging the exploration of alternative approaches.

In our previous work [17], we focused on interpreting large-scale phenomena using decision trees [8], to provide predictions in a local context. The primary objective is to create a consensus model that captures shared knowledge from multiple related models. This is achieved by converting local decision trees into rules and aggregating them to form the consensus model, which is then reverted to a decision tree format. Nonetheless, as a consequence of the combination process, the resulting models exhibit incompleteness, meaning that the derived rules do not cover the entire decision space.

This paper aims to explore methodologies capable of generating complete consensus models, particularly in scenarios where the rule set is incomplete. The first approach involves generating examples from each rule and assembling them into a synthetic dataset. Subsequently, a decision tree is trained using a well-known algorithm such as C5.0 [11] or CART [4]. The second approach treats the conditions of the rules as symbolic variables, transforming them into distributional data. A specialized algorithm designed for this type of data is then employed to train a decision tree. The distribution of the independent variables, even a rough estimate like a histogram, is the sole requirement for implementing any of the methods.

The main contributions of this paper are: 1) to demonstrate the feasibility of generating a complete decision tree despite having an incomplete set of rules; 2) an empirical evaluation to compare the two methods from the perspective of simple decision tree analysis metrics on a number of datasets with distinct characteristics. The remainder of this paper is structured as follows: Sect. 2 presents the background and related work on symbolic data and consensus models; Sect. 3 presents a detailed description of the two proposed approaches to create decision trees with a few illustrative examples; Sect. 4 presents the empirical evaluations on selected datasets, results and discussion; Sect. 5 presents the conclusions and future work.

2   Background

2.1   The Symbolic Data Analysis Paradigm

Statistical units, also known as observations or examples, are described by numerical or categorical variables. Usually, these units are gathered in a dataset, and traditional analysis methods involve summarizing the data using measures such as the mean, median, or mode, leading to some loss of information. The Symbolic Data Analysis paradigm [6] introduces *symbolic variables* for an aggregating statistical unit, which preserves the inherent variation of the data rather than condensing statistical unit information into a single value. In other words, instead of single values or labels, these variables contain *symbolic values*, such as sets, intervals, or distributions over a particular domain [5].

The transition from standard data to symbolic data involves representing statistical units through their aggregation into business units. For example, a university's student enrollment data is initially described by single-valued variables, namely, x_1 = 'student#', x_2 = 'course', x_3 = 'age', x_4 = 'gender', x_5 = 'marital status', and x_6 = 'distance to home' (Table 1). However, in a *symbolic data table* (Table 2), the data is summarized by aggregating the student enrollments (statistical units) based on courses (business units). The variables in the symbolic data table are now symbolic, and their values represent the variations found within each course.

Table 1. Student enrollments

x_1	x_2	x_3	x_4	x_5	x_6
1	Econometrics	19	F	S	2
2	Geometry	20	M	S	1
3	Anatomy	19	M	S	4
4	Econometrics	35	M	M	20
5	Anatomy	18	F	S	1
6	Geometry	29	F	M	12
...	...		...	...	...	...

Table 2. Course enrollments

x_2	x_3	x_4	x_5	x_6
Econometrics	[18;35]	{F,M}	{S,M}	[1;20]
Geometry	[19;29]	{F,M}	{S,M}	[1;12]
Anatomy	[18;19]	{F,M}	{S}	[1;4]
...	...	...	...	...

Symbolic variables, much like standard variables, can be characterized by their domains. The *domain* of a standard variable x defines the range of values it can possibly hold. When dealing with data obtained from a dataset, it is customary to refer to the observed values from the examples of the dataset as the *empirical domain*, denoted by O [5]. As a result, O can encompass a set of values for both discrete numerical variables and categorical variables (Eq. 1), or it can represent a closed interval between the lower bound l and the upper bound u for continuous numerical variables (Eq. 2).

$$O = \{v_1, v_2, \ldots\} \qquad (1) \qquad\qquad O = [l; u] \qquad (2)$$

In the context of symbolic variables, the traditional concept of a domain no longer applies, as their values are sets whose elements are derived from an empirical base domain S. The domain of a symbolic variable, denoted as B, encompasses all possible symbolic values for that variable. The power set [9] of a set S includes all potential subsets of S, including both the empty set and S itself. For instance, if $S = \{a, b, c\}$, then the power set is $\mathcal{P}(S) = \{\{\}, \{a\}, \{b\}, \{c\}, \{a,b\}, \{a,c\}, \{c,b\}, \{a,b,c\}\}$. For both quantitative multi-valued variables and categorical multi-valued variables, B is defined as the power set of S excluding the empty set (Eq. 3). Regarding interval-valued variables, B is defined as the set of all intervals where the lower and upper bounds belong to S (Eq. 4).

$$B = \mathcal{P}(S) - \{\} \qquad (3) \qquad\qquad B = \{[l; u] : l, u \in S \wedge u > l\} \qquad (4)$$

2.2   Consensus Models

Decentralized organizations develop local operational models to address specific decision-making needs in different locations. These models rely on data gathered from individual contexts. However, the organization-wide implementation of such localized models poses challenges in consolidating knowledge into a more general representation, which can be overwhelming due to the sheer volume of models involved. To address this issue, an automated approach is employed to combine operational models from different business units, leading to the creation of *consensus models*.

The main objective of a consensus models is to condense consistent knowledge from multiple models into a single entity, providing a unified representation of shared knowledge for top-level decision-makers within the organization. The process involves distributing the business units models across strata based on a grouping criteria and sequentially combining them into decision tables within each stratum. Eventually, a decision tree model is derived from each decision table, resulting in consensus models for each stratum. In our previous work [17], we approach the problem as a framework, outlining alternatives for combining models and addressing issues related to conflicting rules, among others.

The process of creating a consensus model is depicted in Fig. 1. It involves extracting rules from each decision tree (M_i) by analyzing all paths from the root to the leaves [15]. These rules are then structured into decision tables (T_i). The combination of rules includes pairing the tables and creating a new table. Starting with the combination of T_1 and T_2 to form A_1, the process continues by combining A_1 and T_3 to create A_2, and so on until A_{p-2} is combined with T_p, resulting in A_{p-1}, which represents the consensus model.

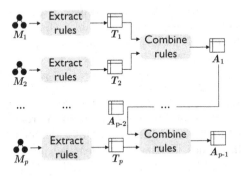

Fig. 1. Combine models

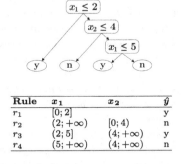

Rule	x_1	x_2	$\hat{y}$
r_1	[0; 2]		y
r_2	(2; +∞)	[0; 4)	n
r_3	(2; 5]	(4; +∞)	y
r_4	(5; +∞)	(4; +∞)	n

Fig. 2. A decision tree and its corresponding decision table

A *decision table* [13] is a symbolic data table used to represent the rules of a rule-based model with symbolic data. Figure 2 provides an example of a decision tree alongside its corresponding decision table with four rules extracted. Formally, a decision table is a collection of R rules (Eq. 5). Each rule (r_i) represents a condition expressed as a conjunction of propositions (p_j) to predict a value for a target variable $\hat{y}_i$ (Eq. 6). To ensure a consistent representation using only symbolic values, all propositions in the decision table employ set membership $(p : x \in s)$, either as intervals $(s = [l; u])$ or sets $(s = \{v_1, v_2, \dots\})$. Within a rule i, the symbolic value of a symbolic variable j is denoted as $s_{i,j}$.

$$T = \{r_1, r_2, \dots, r_R\} \quad (5) \qquad r_i : p_1 \wedge p_2 \wedge \cdots \wedge p_d \longrightarrow \hat{y}_i \quad (6)$$

3 Build a Decision Tree from a Symbolic Data Table

Combining decision trees to create consensus models can give rise to uncovered regions in the decision space, leading to incompleteness in the final decision table. This incompleteness poses challenges when trying to reconstruct a decision tree from the decision table, as certain combinations of independent variable values may not activate any rule. In this section, we introduce two approaches to create a complete decision tree from an incomplete set of rules.

As an example, we trained six decision trees using the C5.0 algorithm [11] on a publicly available 'Bank Dataset' [14]. This dataset contains customer information to predict marketing campaign subscriptions, using variables such as 'age', 'balance', 'contact type', and 'previous campaign outcome', among others. Each operational model $(M_1, \dots, M_6)$ corresponds to one of the first six months of the year (Fig. 3). Subsequently, we combined these models based on rule intersection, resulting in Table 3 ($\hat{y}$ specifying each rule's prediction).

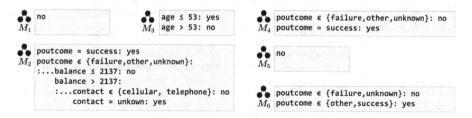

Fig. 3. Operational models trained from the 'Bank Dataset'

Table 3. Combined models' symbolic data table

Rule	age	balance	contact	poutcome	$\hat{y}$
r_1	$(-\infty; 53]$			{success}	yes
r_2		$[-8019; 2137]$		{failure,unknown}	no
r_3		$(-8019; 2137]$		{other}	yes
r_4		$(2137; +\infty)$	{cellular,telephone}	{failure,unknown}	no
r_5		$(2137; +\infty)$	{cellular,telephone,unknown}	{other}	yes
r_6		$(2137; +\infty)$	{unknown}	{failure,unknown}	yes

3.1 Build a Decision Tree from a Synthetic Dataset

In this approach, a synthetic dataset is created based on the rules extracted from a symbolic data table [1,17] which is then used to train a decision tree using a popular algorithm, such as C5.0 [11] or CART [4]. Generating data for each rule r_i involves calculating the Cartesian product of symbolic values across d symbolic variables, including also a set with the rule's prediction ($\{\hat{y}_i\}$).

The method uses the variable's distribution for two key tasks. First, it addresses empty cells by replacing them with the empirical domain of the variable. Secondly, it substitutes unknown boundaries (∞) with the distribution's bounds. Additionally, it converts interval-valued variables into multi-valued numerical variables. After generating the individual rule datasets, they are consolidated into a single dataset D_s (Eq. 7) and used to train a model.

$$D_s = \bigcup_{i=1}^{R} (s_{i,1} \times s_{i,2} \times \cdots \times s_{i,d} \times \{\hat{y}_i\}) \tag{7}$$

Example in the Bank Marketing Dataset. Table 4 shows an extract of the synthetic dataset D_s derived from the rules in Table 3. Training a model with the C5.0 algorithm from D_s, results in the decision tree of Fig. 4. The consensus model's incompleteness is apparent since its rules do not cover, for example, observations with poutcome='failure' or contact='telephone'.

Table 4. Synthetic dataset

Rule	age	balance	contact	poutcome	$\hat{y}$
r_1	18	-8019	cellular	success	yes
r_1	...	...	...	...	yes
r_1	53	2137	unknown	success	yes
...	...	...	...	...	...
r_6	18	2138	unknown	failure	yes
r_6	...	...	...	...	yes
r_6	95	102127	unknown	unknown	yes

```
poutcome ∈ {other,success}: yes
poutcome = unknown:
:...contact = cellular: no
    contact = unknown:
    :...balance ≤ 0: no
        balance > 0: yes
```

Fig. 4. Incomplete consensus model

3.2 Build a Decision Tree from Symbolic Distributional Data

In this approach, the symbolic data within a decision table is transformed into distributional data, which can be either categorical modal variables or histogram-valued variables [2,5], using the distribution of the variables. Subsequently, the Périnel-Lechevallier algorithm [16] is used to create a decision tree.

To convert a categorical multi-valued variable ($s = \{v_1, v_2, \dots\}$) into a categorical modal variable s' (Eq. 8), each value is associated with its normalized relative frequency (f'_k), considering only the values included in s (Eq. 9).

$$s' = \{v_k (f'_k) : v_k \in s\} \tag{8}$$

$$f'_k = \frac{f_k}{\sum_{v_j \in s} f_j} \tag{9}$$

The process of converting an interval-valued variable $(s = [l_s; u_s])$ into a histogram-valued variable s' involves partitioning the interval into bins and normalizing the relative frequencies. A set $\mathcal{B}$ is defined by selecting the bins from the distribution (represented as $B_k = [l_{B_k}; u_{B_k}]$) that overlap with s. The inclusion criteria require either the entire bin to be contained in s or one of the limits of s to be within the bin (Eq. 10). If one of the limits of s is undefined (∞), either the lower or upper limit of the distribution is assumed.

The representation of the histogram-valued variable (Eq. 11) is a collection of bins followed by its normalized relative frequency. This representation specifically considers only the overlapping bins within the variable's distribution (Eq. 12). If a bin encloses the lower limit of s, the relative frequency is adjusted to the extent of the intersection with s (Eq. 13).

$$B = \{B_k : (B_k \subseteq s) \vee (l_s \in B_k) \vee (u_s \in B_k)\} \tag{10}$$

$$s' = \{B_k (f'_k) : B_k \in \mathcal{B}\} \tag{11}$$

$$f'_k = \frac{f_k}{\sum_{B_j \in \mathcal{B}} f_j} \tag{12} \qquad f_k = \begin{cases} \frac{l_s - l_{B_k}}{u_{B_k} - l_{B_k}} f_k & \text{if } l_s \in B_k, \\ f_k & \text{if } l_s \notin B_k \end{cases} \tag{13}$$

Decision trees algorithms are used in classification problems to create models that classify data into categories of a target variable. The construction of the model typically involves testing binary questions based on numerical or categorical variables on the range of values within the empirical domain (O) of each variable. For numerical variables, the questions take the form $x \leq c$. In the discrete case, where there are n values, $n - 1$ questions are tried, each using c as the cutting threshold. In the continuous case, with infinitely many thresholds, the algorithm selects midpoints between the lower and upper bounds, leading to $2n - 1$ questions. When dealing with categorical variables, the questions take on the form $x \in V$, where $V \subset O$. If the empirical domain O contains n values, this results in the generation of $2^n - 1$ questions, each representing a unique subset of O from its powerset, excluding the empty set.

The Périnel-Lechevallier algorithm [16] shares the fundamental principles with well-known algorithms like C5.0 [11] or CART [4], differing by using distributional representations of symbolic values rather than single-values from examples. A set of questions for each variable is generated by exhaustively exploring all cutting thresholds for numerical variables and all combinations of single-values for categorical variables. The distributional data from each statistical unit (rules in this context) is used to calculate the probability $p_{k,m}$ that a specific question q_m is satisfied within rule r_k.

The set of Q questions, each with its corresponding probabilities, is consolidated into a probabilistic table [3] for the R rules within the model, as shown in Table 5. These probabilities are used to build a decision tree using recursive tree partitioning methods commonly found in decision tree algorithms. For numerical continuous variable x_j with a frequency of $f_{k,j}$, the probability of satisfying the binary question $[x \leq c]$ is determined using Eq. 14. In the case of a numerical discrete variable x_j, the probability of satisfying the binary question $[x \leq c]$ is given by Eq. 15. For a categorical variable x_j, the probability of satisfying the binary question $[x \in V]$ is computed according to Eq. 16.

Table 5. A probabilistic data table

Rule	q_1	...	q_m	...	q_Q
r_1	$p_{1,1}$	...	$p_{1,m}$	...	$p_{1,Q}$
...	...	...	...	...	...
r_k	$p_{k,1}$	...	$p_{k,m}$	...	$p_{k,Q}$
...	...	...	...	...	...
r_R	$p_{R,1}$	...	$p_{R,m}$	...	$p_{R,Q}$

$$p_{k,m} = \sum_{l_{B_k} < c} f'_{k,j} + \frac{c - l_{B_k}}{u_{B_k} - l_{B_k}} f'_{k,j} \quad (14)$$

$$p_{k,m} = \sum_{v \leq c} f'_{k,j} \quad (15)$$

$$p_{k,m} = \sum_{v \in V} f'_{k,j} \quad (16)$$

Example in the Bank Marketing Dataset. Table 6 and Table 7 hold the distributions of variables 'age' and 'contact', respectively, which are used to illustrate the conversion from symbolic data to distributional data.

Table 6. Distribution of 'age'

k	B_k	f_k
1	(10;20]	0.00215
2	(20;30]	0.15335
3	(30;40]	0.39121
4	(40;50]	0.24859
5	(50;60]	0.17843
...	...	...
9	(90;100]	0.00015

Table 7. Distribution of 'contact'

k	v_k	f_k
1	cellular	0.64774
2	telephone	0.06428
3	unknown	0.28798

The conversion of the symbolic value $s_{1,1} = (-\infty; 53]$ for interval-valued variable 'age' into a histogram-valued variable relies on selecting the bins within the distribution that intersect with $(-\infty; 53]$. Starting from the lower limit $(-\infty)$, the distribution begins with B_1 and extends to include B_5, the bin that encompasses the upper limit (53). The frequency assigned to B_5 is calculated proportionally based on its intersection with $s_{1,1}$. This computation yields the distributional representation: $s'_{1,1} = \{(10; 20] : (0.00253), (20; 30] : (0.18067), \ldots, (50; 60] : (0.06306)\}$. The conversion of the symbolic value $s_{4,3} = \{\text{cellular}, \text{telephone}\}$ for categorical multi-valued variable 'contact', into a categorical modal variable, requires normalizing the distribution based on the set of values included in its symbolic representation. This normalization process results in the distributional representation: $s'_{4,3} = \{\text{cellular} : (0.90972), \text{telephone} : (0.09028)\}$. Table 8 list the questions generated while Tables 9 and 10 the probabilities for variables $x_1 = $ 'age' and $x_3 = $ 'contact' used in the conversions exemplified above. Using all probabilistic data, the Périnel-Lechevallier algorithm creates the decision tree in Fig. 5, which is a complete consensus model.

Table 8. Examples of the questions generated

Variable	Questions	Form	Set
x_1 = age	$q_1 \ldots q_{19}$	$x_1 \leq c$	$c \in \{10, 15, 20, \ldots, 100\}$
x_2 = balance	$q_{20} \ldots q_{43}$	$x_2 \leq c$	$c \in \{-10000, -5000, 0, \ldots, 110000\}$
x_3 = contact	$q_{44} \ldots q_{50}$	$x_3 \in V$	$V \subset \{\text{cellular}, \text{telephone}, \text{unknown}\}$
x_4 = poutcome	$q_{51} \ldots q_{65}$	$x_4 \in V$	$V \subset \{\text{failure}, \text{other}, \text{success}, \text{unknown}\}$

Table 9. Probabilities for $s'_{1,1}$

Prob.	Question	Value
$p_{1,1}$	$x_1 \leq 10$	0.00000
$p_{1,2}$	$x_1 \leq 15$	0.00126
$p_{1,3}$	$x_1 \leq 20$	0.00253
$p_{1,4}$	$x_1 \leq 25$	0.09286
$p_{1,4}$	$x_1 \leq 30$	0.18067
$\ldots$	$\ldots$	$\ldots$
$p_{1,19}$	$x_1 \leq 100$	1.00000

Table 10. Probabilities for $s'_{4,3}$

Prob.	Question	Value
$p_{4,44}$	$x_3 \in \{\text{cellular}\}$	0.90972
$p_{4,45}$	$x_3 \in \{\text{telephone}\}$	0.09028
$p_{4,46}$	$x_3 \in \{\text{unknown}\}$	0.00000
$p_{4,47}$	$x_3 \in \{\text{cellular,telephone}\}$	1.00000
$p_{4,48}$	$x_3 \in \{\text{cellular,unknown}\}$	0.90972
$p_{4,49}$	$x_3 \in \{\text{telephone,unknown}\}$	0.09028
$p_{4,50}$	$x_3 \in \{\text{cellular,telephone,unknown}\}$	1.00000

4 Evaluation and Results

4.1 Datasets

The approaches to create decision trees from potentially incomplete rule sets were evaluated in three publicly available datasets from Kaggle [10] and one obtained from a university case study (with restricted access due to data confidentiality), listed in Table 11 (n specifies the number of examples and d denotes the number of variables). The evaluation focused on measuring completeness and size, metrics that apply to both consensus models derived from symbolic data tables and decision trees generated from synthetic and distributional data. The results are ranked based on the symbolic data table metrics.

4.2 Results and Discussion

The *completeness* of a rule-based model is defined as the proportion of the decision space covered by the model's rules [13], with the decision space representing the multidimensional space defined by the independent variables' domains [7]. Figure 6 shows the results of the completeness comparison across

```
poutcome ∈ {other,success}: yes
poutcome ∈ {failure,unknown}:
:...contact ∈ {cellular,telephone}: no
    contact = unknown: yes
```

Fig. 5. Complete consensus model

Table 11. Datasets used for evaluation

Dataset	n	d	Statistical unit	Business unit	Grouping criterion	Target variable
a. Bank marketing	45K	16	Client	Month	Semester	Campaign subscription
b. Hotel reservations	36K	17	Reservation	Months	Semester	Canceled reservation
c. Adult census	32K	12	Person	Country	Global zone	Salary level
d. University	1300K	26	Enrollment	Course	Scientific area	Unsuccessful completion

different datasets. Results for the synthetic data approach demonstrated substantial variability across datasets, and no clear correlation with the completeness of the symbolic data tables. The results also demonstrate that in dataset (b), where the completeness of the symbolic data tables is notably low, the decision trees created using synthetic data always shows higher values. This trend is even more pronounced in dataset (d), where the completeness of the symbolic data tables remains extremely close to zero. In contrast, the distributional data approach consistently produced complete trees, indicating full coverage of the decision space.

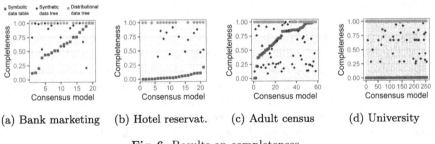

(a) Bank marketing   (b) Hotel reservat.   (c) Adult census   (d) University

Fig. 6. Results on completeness

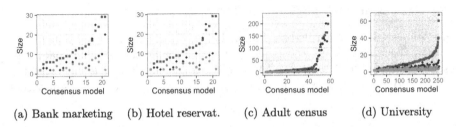

(a) Bank marketing   (b) Hotel reservat.   (c) Adult census   (d) University

Fig. 7. Results on size

The *size* of a rule-based model refers to the total number of rules it contains [12]. In the context of decision tables, size corresponds to the number of rows, whereas in decision trees, it corresponds to the number of leaf nodes. The comparative results of sizes across datasets is depicted in Fig. 7. The results suggest that, with a few exceptions, the size of decision trees generated by both methods is generally smaller than that of the symbolic data tables. Specifically, decision trees produced from distributional data typically contain less than 10 rules. In contrast, decision trees derived from synthetic data, particularly in dataset (c), can contain hundreds of rules. Dataset (d) reveals that decision trees generated by both methods exhibit similar sizes, and notably, they are substantially smaller than the symbolic data tables.

Overall, the results indicate that decision trees created using distributional data are consistently complete and simpler than those generated using synthetic data, making them more easily interpretable. However, there may be more instances where a tree is not generated when using the distributional data approach compared to decision trees created with synthetic data. Incompleteness is mainly an issue with categorical variables, where some categories are absent in the consensus model when represented as a symbolic data table. To address this limitation, the distributional approach uses information from variable distributions, enabling it to account for categories that are not explicitly included in the model's rules, resulting in complete models. However, when dealing with numerical variables, even if a rule does not encompass the entire variable domain, training a decision tree using either of these approaches typically results in effective generalization across that domain.

5   Conclusions

Large organizations often divide problems into sub-problems, leading to multiple localized prediction models. Model combination is used to consolidate knowledge from these models into possibly incomplete consensus models, i.e., which may fail to provide predictions in uncovered regions of the decision space. To tackle this problem, this paper explores methods for generating complete decision trees from incomplete decision rule sets that represent consensus models.

Two approaches are considered. The first involves creating a dataset from the rules and training a decision tree using a well-known algorithm such as C5.0 or CART. While decision trees may be more complete than the original rule sets in some cases, full coverage of the decision space is rare, especially for categorical variables. The resulting dataset can also become large, making this approach less feasible. The second approach involves transforming rules into distributional data and using the Périnel-Lechevallier algorithm to generate decision trees. These decision trees are always complete and tend to be simpler compared to those created using the synthetic data approach. However, operational constraints and the need for variable distributions may impose limitations on this approach.

Future work will address challenges in the distributional data approach when it fails to generate a tree. Nonetheless, resorting to the synthetic data method with a default rule technique for uncovered regions remains a viable solution.

Acknowledgments. This work was partially funded by projects AISym4Med (101095387) supported by Horizon Europe Cluster 1: Health, ConnectedHealth (n.o-46858), supported by Competitiveness and Internationalisation Operational Programme (POCI) and Lisbon Regional Operational Programme (LISBOA 2020), under the PORTUGAL 2020 Partnership Agreement, through the European Regional Development Fund (ERDF) and NextGenAI - Center for Responsible AI (2022-C05i0102-02), supported by IAPMEI, and also by FCT plurianual funding for 2020–2023 of LIACC (UIDB/00027/2020_UIDP/00027/2020).

References

1. Andrzejak, A., Langner, F., Zabala, S.: Interpretable models from distributed data via merging of decision trees. In: 2013 IEEE Symposium on Computational Intelligence and Data Mining (CIDM) (2013)
2. Billard, L., Diday, E.: From the statistics of data to the statistics of knowledge: symbolic data analysis (2003). https://doi.org/10.1198/016214503000242
3. Billard, L., Diday, E.: Symbolic Data Analysis: Conceptual Statistics and Data Mining. Wiley, Hoboken (2012)
4. Breiman, L., Friedman, J.H., Olshen, R.A., Stone, C.J.: Classification and Regression Trees. Chapman and Hall/CRC, Boca Raton (1984). https://doi.org/10.1201/9781315139470
5. Brito, P.: Symbolic data analysis: another look at the interaction of data mining and statistics. Wiley Interdisc. Rev. Data Min. Knowl. Discov. **4**, 281–295 (2014). https://doi.org/10.1002/widm.1133
6. Diday, E.: Thinking by classes in data science: the symbolic data analysis paradigm (2016). https://doi.org/10.1002/wics.1384
7. Giabbanelli, P.J., Peters, J.G.: An algebra to merge heterogeneous classifiers (2015). http://arxiv.org/abs/1501.05141
8. Han, J., Kamber, M., Pei, J.: Data Mining: Concepts and Techniques. 3rd edn. Morgan Kaufmann, Burlington (2012)
9. Jech, T.J.: Set Theory. Springer, Heidelberg (2006). https://doi.org/10.1007/3-540-44761-X
10. Kaggle: Kaggle (2022). www.kaggle.com/datasets
11. Kuhn, M., Weston, S., Culp, M., Coulter, N., Quinlan, J.R.: C5.0 decision trees and rule-based models (2022). https://github.com/topepo/C5.0/issues
12. Lakkaraju, H., Bach, S.H., Leskovec, J.: Interpretable decision sets: a joint framework for description and prediction, vol. 13–17-August-2016, pp. 1675–1684. Association for Computing Machinery (2016). https://doi.org/10.1145/2939672.2939874
13. Ligeza, A.: Logical Foundations for Rule-Based Systems, vol. 11. Springer, Heidelberg (2006). https://doi.org/10.1007/3-540-32446-1
14. Moro, S., Cortez, P., Rita, P.: A data-driven approach to predict the success of bank telemarketing. Decis. Support Syst. **62**, 22–31 (2014). https://doi.org/10.1016/j.dss.2014.03.001
15. Obregon, J., Kim, A., Jung, J.Y.: RuleCOSI: combination and simplification of production rules from boosted decision trees for imbalanced classification. Expert Syst. Appl. **126**, 64–82 (2019). https://doi.org/10.1016/j.eswa.2019.02.012

16. Perinei, E., Lechevallier, Y.: Symbolic discrimination rules. In: Analysis of Symbolic Data, Exploratory Methods for Extracting Statistical Information from Complex Data, pp. 244–265 (2000)
17. Strecht, P., Mendes-Moreira, J., Soares, C.: Inmplode: a framework to interpret multiple related rule-based models. Expert Syst. **38**, e12702 (2021). https://doi.org/10.1111/exsy.12702

CySpider: A Neural Semantic Parsing Corpus with Baseline Models for Property Graphs

Ziyu Zhao[1,2(✉)], Wei Liu[1,2], Tim French[1,2], and Michael Stewart[1,2]

[1] UWA NLP-TLP Group, 35 Stirling Hwy, Crawley, Perth, WA 6009, Australia
ziyu.zhao@research.uwa.edu.au,
{wei.liu,tim.french,michael.stewart}@uwa.edu.au
[2] School of Physics, Mathematics and Computing, The University of Western
Australia, Crawley, Australia
https://nlp-tlp.org/

Abstract. Enterprise knowledge graphs are gaining increasing popularity in industrial applications, with a pressing demand for natural language interfaces to support non-technical end-users. For natural language queries to relational databases, the neural semantic parsing task Text-to-SQL achieves strong performance in translating text inputs to SQL queries. However, very few public corpora are available for the training of neural semantic parsing models that convert textual queries to graph query languages. In this research, we develop a generic SQL2Cypher algorithm that can map a SQL query to a set of Cypher clauses, where Cypher is a query language used by a popular property graph database Neo4j. The converted Cypher statement is then combined with the original natural language query to create a parallel corpus that enables end-to-end training of neural semantic parsing models for Text-to-Cypher. To evaluate the dataset quality, we construct a corresponding graph database to obtain execution accuracy. In addition, the Text-to-Cypher corpus features four transformer-based baseline models. The availability of such corpus and baseline models is critical in developing and benchmarking new machine learning methods in advancing natural language interfaces for fact retrieval from large graph-based knowledge repositories. The source code and dataset are available at github(https://github.com/22842219/SemanticParser4Graph).

Keywords: Semantic Parsing · Text-to-Cypher · Text/Data Mining

1 Introduction

The proliferation of graph databases, driven by the increasing complexity and interconnectedness of data across various domains, has created a pressing need for intuitive and accessible methods to interact with these databases with many real-world applications. Neo4j, topping the DB engine ranking of GraphDBMS[1],

---

[1] https://db-engines.com/en/ranking.

T. Liu et al. (Eds.): AI 2023, LNAI 14472, pp. 120–132, 2024.
https://doi.org/10.1007/978-981-99-8391-9_10

is a popular industry scale non-relational (NoSQL) property graph database. It offers a high-level query language called Cypher to specify graph patterns supported by a query engine that uses sophisticated optimization techniques [7]. Cypher bears similarities to SQL but employs a pattern-based syntax. A pattern describes data using nodes for entities and edges for the relationships between any two entities. It resembles the intuitive act of sketching objects and their relations on a whiteboard.

To empower non-expert data users in retrieving desired facts from siloed structured databases, a natural language interface that allows them to ask questions in natural languages becomes increasingly desirable. The task of generating executable queries from natural language instructions or questions, as observed in Text-to-SQL [18] examples, is referred to as *semantic parsing*. With the popular transformer-based architectures and the accessibility of large datasets, current trends in semantic parsing tend to favour neural language model-based methods, that require large parallel corpus for training.

Unfortunately, very few datasets are publicly available for semantic parsing over property graph databases, due to the cost and laborious process of collecting complex graph schema and constructing parallel corpus consisting of natural language query and structured graph query pairs. Thanks to the availability of parallel corpora for Text-to-SQL (e.g. Spider [19] and the more recent BIRD by Li et al. [6]), instead of resorting to labour-intensive manual labelling, we propose to adapt the Text-to-SQL corpora for Text-to-Cypher in a framework as shown in Fig. 1. The framework depicts the data flow across three key stages, (a) parallel corpus construction, (b) neural semantic parsing model training, and (c) dataset transformation & model evaluations.

The key component for corpus construction is **SQL2Cypher**, where we convert a SQL statement in the Text-to-SQL corpora to its equivalent Cypher counterpart, to construct parallel corpora consisting of pairs of natural language queries and Cypher clauses. The limited amount of existing works for Cypher-to-SQL translation (e.g., Cyp2SQL [1]) suffers from their inability to handle complex queries with multiple JOINs and facilitate the migration of schemas. Our SQL2Cypher converter is designed to address these limitations. Comprehensive evaluations have confirmed our dataset quality and, in turn, verified the effectiveness of our SQL2Cypher conversion. To establish a baseline for the neural semantic parsing task of Text-to-Cypher, we build two types of models, namely a pipeline model, Text-to-SQL-to-Cypher, illustrated as sub-module (b.1) in Fig. 1; a seq2seq model, Text-to-Cypher, as sub-module (b.2). We use the constructed parallel corpus to train the Text-to-Cypher semantic parsing task.

The main contribution of this research is three-fold. 1) We introduce semantic parsing dataset construction from the Text-to-SQL dataset to Text-to-Cypher, including the main algorithm SQL2Cypher to support the conversion of SQL to Cypher queries. The algorithm can be applied to any existing Text-to-SQL benchmark datasets, to create datasets for training models. 2) We evaluated our approach on Spider benchmark and contributed a parallel corpus for the Text-to-Cypher semantic parsing task, i.e., Cy-Spider. The Cy-Spider dataset covers 155 databases, including 4,929 pairs of natural language and Cypher queries,

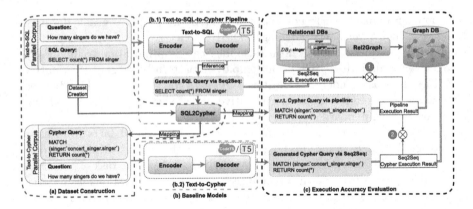

Fig. 1. An overview of the proposed dataset construction and evaluation workflow.

alongside a property graph database that contains 10,729 graph nodes with 641 labels and 11,277 graph edges with 461 types. 3) We trained four baselines: two for Text-to-SQL-to-Cypher pipeline and two for end-to-end training, with T5-Base [11] and CodeT5-Base [17] as the backbones. The direct end-to-end training of the T5-Base and CodeT5-Base models outperformed the pipeline models, which not only verified the quality of Cy-Spider dataset but also encourages future model development to leverage such a parallel corpus.

2   Related Work

SQL-to-NoSQL. Steer et al. [15] and Carata [1] introduced Cytosm and Cyp2SQL, respectively, which contribute to Cypher-to-SQL translation. However, there are only a few articles or software/tools available for the SQL-to-Cypher translation path. Li et al. [4] proposed an SQL2Cypher web-based tool, but it only supports query translation from a simple schema. The Neo4j community developed a SQL-to-Cypher transpiler that employs jOOQ[2] and Neo4j Cypher-DSL[3]. Despite its potential, the transpiler has several limitations. First, it does not offer support for a wide range of SQL query features, especially advanced or complex constructs such as nested queries and multiple JOINs. Second, the transpiler focuses on the direct translation of SQL queries to Cypher without facilitating the migration of schema, data, or other database objects from a relational database to Neo4j. As a result, the transpiler cannot ensure the successful execution of translated Cypher queries or the accuracy of the obtained results.

Text-to-SQL. Modern Text-to-SQL task aims to generate SQL queries from structured and unstructured input data in a seq-to-seq fashion. The structured data refers to the given relational database schema, while the unstructured data pertains to natural language utterances. To accomplish this, specialized encoders,

---

[2] https://github.com/jOOQ/jOOQ.
[3] https://github.com/neo4j-contrib/sql2cypher.

such as relation-aware self-attention mechanisms [16], are often utilized to embed the input data. Constrained decoders [13] are then employed to avoid generating unusable output. Spider benchmark, one of the widely used datasets in this research field, operates in a zero-shot setting, indicating that there is no overlap between the questions or databases in the training, development, and test splits. One major challenge in this field is the mismatch between natural language utterances and ground truth SQL queries. To address this issue, various *Intermediate Representations* (IRs) which are non-syntactic strict SQL-style queries have been proposed, such as NatSQL [2]. IR-based approaches simplify complex SQL queries and nested sub-queries to make schema linking easier at the model level. We propose a data-level solution to address the issue.

Text-to-Graph Queries. There is little research working on the translation from text to graph query languages, such as SPARQL and GraphQL. Most neural end-to-end approaches are data-driven. Saparina and Osokin [14] reduced the dataset burden by training a neural semantic parser from text to an intermediate representation, and a non-trainable transpiler to a structurally closer formal DB query language SPARQL. This is a certain extent similar to our pipeline approach, where we train a neural semantic parsing model to translate natural language queries to an intermediate structural language, SQL, before the conversion mapping of SQL2Cypher. The closest work to ours is Text-to-GraphQL [8] aiming at generating GraphQL statements in the medical consultation chatbot application. However, their proposed model cannot be verified on the Spider counterpart in the Text-to-GraphQL format regarding execution accuracy metric.

3   Notation and Task Formulation

We denote a Text-to-SQL dataset as $D_R = (\mathcal{S}, \langle N, Q_{sql} \rangle)$, where $\mathcal{S}$ is the relational database. It consists of a set of natural language questions N paired with corresponding SQL queries Q_{sql}. The Text-to-Cypher parsing dataset is denoted as $D_G = (\mathcal{G}, \langle N, Q_{cyp} \rangle)$, where $\mathcal{G}$ is the property graph database, and N is a set of natural language questions, the same as that for their relational counterpart in D_R. Q_{cyp} denotes a set of structured Cypher queries.

Dataset Transformation Task. Given D_R, we construct D_G. Specifically, we transform $\mathcal{G}$ from $\mathcal{S}$ through graph database construction from relational databases (in our case, Rel2Graph). Then, we convert SQL queries to Cypher queries, i.e., $Q_{sql} \mapsto Q_{cyp}$, which should be executable in the graph database $\mathcal{G}$. To assess the effectiveness of this mapping, we employ the execution accuracy (EA) metric. EA quantifies the proportion of Q_{cyp} that produces the same results Q_{sql} executed on $\mathcal{S}$ when executed on $\mathcal{G}$. By converting SQL databases and SQL queries into an equivalent graph database and Cypher queries, we repurpose the Text-to-SQL dataset for the Text-to-Cypher semantic parsing task.

Text-to-Cypher Semantic Parsing Task. Text-to-Cypher maps a natural language query of N into a Cypher query of Q_{cyp} that is capable of retrieving relevant data from a given graph database $\mathcal{G}$. It could be formulated as $Q_{cyp} = f(N, \mathcal{G}|\theta)$, where $f(\cdot|\theta)$ is a model or neural network with the parameter θ.

4 SQL2Cypher: From SQL Queries to Cypher Queries

Graph data models are known for their efficient pattern-matching mechanisms in dealing with highly connected data when relational databases resort to multiple expensive JOIN ON operations via foreign keys. Take the two examples in Fig. 2 for instance. In Example 1, Which department has more than 1 head? List the id, name, and number of heads, we can directly map the JOIN ON statement into a Cypher pattern in MATCH clause statement where the foreign key department_id is omitted. For the question of Example 2, How many departments are led by heads who are not mentioned, we can convert the nested sub-query into a graph pattern in the WHERE statement. The equivalent Cypher clauses of both examples do not need to explicitly mention the foreign key department_id. Foreign keys become redundant in graph databases. Due to the fact that Cypher queries are case-sensitive, we normalised all the schema items appearing in the source case-insensitive SQL queries by using the graph schema which is aligned with relational databases.

Example	SQL Query	Cypher Query	Execution Answer
1	1 SELECT T1.department_id, T1.name, count(*) 2 FROM management AS T2 JOIN department AS T1 3 ON T1.department_id=T2.department_id 4 GROUP BY T1.department_id 5 HAVING count(*) > 1	1 MATCH ()-[T2:management]-(T1:department) 2 WITH T1.Department_ID AS Department_ID, 3 count(*) AS count, T1.Name AS Name 4 WHERE count > 1 5 RETURN Department_ID, Name, count	Return Field / Value Department_ID / 2 Name / 'Treasury' count / 2
2	1 SELECT count(*) FROM department 2 WHERE department_id NOT IN 3 (SELECT department_id FROM management)	1 MATCH (d:department) 2 WHERE NOT (d:department)-[:management]-() 3 RETURN count(*)	Return Field / Value count / 11

Fig. 2. The goal SQL queries of example 1 and example 2, their respective Cypher queries and the Cypher queries execution answers.

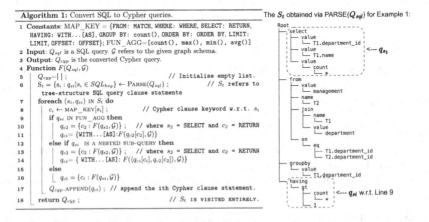

Therefore, the essential translation tasks for SQL2Cypher are to sort out the equivalent mapping between SQL and Cypher clauses and identify graph patterns. We denote a list of SQL keywords as SQL_{key}, i.e., [FROM, WHERE,

SELECT, HAVING, GROUP BY, ORDER BY, LIMIT, OFFSET, UNION], and a list of Cypher keywords as $Cypher_{key}$, i.e., [MATCH, WHERE, RETURN, WITH... [AS], ORDER BY, LIMIT, OFFSET]. We allign FROM with MATCH, SELECT with RETURN, HAVING with WITH...[AS], GROUP BY with count(), and the rest keywords stay the same. We map JOIN statements to graph patterns. We chain schema items occurring in nested sub-queries or group queries together by WITH clause, piping the results from the nested query as the starting points of the parent query. The mapping process is repeated until all SQL clauses are visited. The pseudocode of SQL-to-Cypher process is described in Algorithm 1. We use PARSE(Q_{sql})[4] to parse a SQL query into a JSON-IZABLE parse tree (Line 6), where s_i refers to a SQL keyword appearing in SQL_{key} list and each s_i corresponds to its equivalent Cypher keyword c_i in $Cypher_{key}$ list. The mapping between s_i and c_i becomes a KEY-VALUE pair of MAP_KEY (Line 1). q_{si} is the JSON-IZABLE parse tree w.r.t. s_i. The parsed tree hierarchy is illustrated next to Algorithm 1. Given Example 1 in Fig. 2, for Line 9 in Algorithm 1, q_{si} with respect to HAVING is shown in the dotted box at the bottom right of the parsed tree.

5 Text-to-Cypher Neural Models

Given a parallel corpus of natural language to structured query pairs, the most popular neural model is to formulate semantic parsing as a sequence-to-sequence machine translation problem. Large language models (LLMs) that are pre-trained on vast amount of data have achieved remarkable progress in this regard. However, they still contain two major limitations. First, decoder-only models such as GPT-based LLMs [10] are sub-optimal in code understanding tasks such as code retrieval. Pourreza et al. [9](2023) shows the best EA accuracy of zero-shot and few-shot prompting on two latest GPT-based LLMs, CodeX and GPT4, on the Spider development set is 67.4, while ours fine-tuned CodeT5-Base reaches 71.4 (see Table 1). Secondly, encoder-decoder based LLMs such as T5 employ a limited set of pretraining objectives in a text-to-text setting which might not be suitable for auto-regressive code generation tasks, for example, code completion. In the encoder-decoder paradigm, the encoder maps the input text into a continuous representation, and the decoder generates the output text based on this representation (see the pre-trained T5/CodeT5 block illustrated in Fig. 3). Building upon the T5 architecture, Wang et al. [17] proposed CodeT5, which incorporates the token type information from code. By leveraging the T5 training objectives, CodeT5 can effectively learn programming language syntax and semantics, allowing it to generate accurate and executable code. We use the T5-Base and CodeT5-Base due to resource limitation, however, large pre-trained encoder-decoder models would likely improve model performance.

In this paper, we present the performance of four baselines grouped by two tasks **Text-to-SQL-to-Cypher pipeline** and **Text-to-Cypher** semantic parsing, where the latter is trained using the corpus obtained through SQL2Cypher.

---

[4] https://github.com/mozilla/moz-sql-parser.

We apply both T5-Base and CodeT5-Base backbones on Text-to-SQL and Text-to-Cypher in an end-to-end manner. Two tuning strategies, i.e., fine-tune and prefix-tune, are employed.

5.1   Pipeline

Text-to-SQL-to-Cypher pipeline includes two phases, i.e., the standalone Text-to-SQL (Phase 1) and SQL-to-Cypher (Phase 2). Phase Two is built on top of Phase One. Figure 3 depicts the diagram of our proposed pipeline. We apply our deterministic rules-based algorithm (Sect. 4) on the output of Text-to-SQL to get both the generated and gold Cypher queries from predicted and gold SQL queries respectively. The final score of the pipeline is computed via the multiplication of the EA scores of Phase 1 and Phase 2 (see results in Table 1).

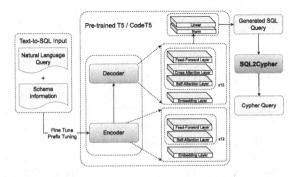

Fig. 3. The proposed pipeline diagram alongside the encoder-decoder architecture.

5.2   End-to-End Training

Following a seq2seq approach, we take a natural language query as an input sequence N alongside the domain schema information $\mathcal{G}$. The output at each decoding step t is a distribution over the vocabulary, computed by the softmax function $\hat{c}_t \sim \text{softmax}(\text{Linear}(s_t))$ where s_t is the hidden state at decoding step t. Conventionally, during training time, model parameters, θ, are learned by maximizing the log-likelihood of the ground-truth Cypher queries. Denoting $C = (c_1, ..., c_T)$ as the ground-truth program, the objective is to minimize the cross-entropy loss:

$$\mathcal{L}_{ce}(\theta) = -\sum_t \log p_\theta(C|N, \mathcal{G}) = -\sum_t \log[p_\theta(c_t|c_{1:t-1}, N, \mathcal{G})] \tag{1}$$

We use Adam Optimizer with default parameters. We set the learning rate as 1e-4, the training batch size as 2, and the evaluation batch size as 8. The number of training epochs is 100 with an early stopping strategy with 5 for fine-tuning strategy. We set both the maximum input and generation sequence length as 512. To generate candidates, we set the beam size as 4 for beam search.

During evaluation time, the model generates sequences of Cypher queries by autoregressively sampling token $\hat{c}_t$ from the distribution $p_\theta(\cdot|\hat{c}_{1:t-1}, N, \mathcal{G})$.

We carry out experiments with both the prefix tuning length of 10 and fine-tuning settings. Prefix-tunning keeps T5-Base model parameters frozen but optimizes a small continuous task-specific vector (called the prefix) [5] while fine-tuning modifies all the T5-Base model parameters.

5.3   Evaluation Metric

We use execution accuracy (**EA**) to evaluate Text-to-Cypher parsing on leveraged Text-to-SQL parsing datasets, e.g., Spider. It measures the proportion of pairs in the evaluation set, where the execution results of both the generated and ground-truth Cypher queries are identical. Considering the result set as R_n executed by the nth ground-truth Cypher C_n, and the result set $\hat{R}_n$ executed by the generated Cypher query $\hat{C}_n$ can be computed by:

$$EA = \frac{\sum_{n=1}^{N} \mathbb{1}(R_n, \hat{R}_n)}{N}, where, \mathbb{1}(R, \hat{R}) = \begin{cases} 0 & \text{if } R \neq \hat{R} \\ 1 & \text{if } R = \hat{R} \end{cases} \quad (2)$$

6   Experiment Results

6.1   Dataset Statistics

Although WikiSQL [20] is much larger than Spider, it is not as challenging as Spider in terms of expressiveness, multi-table relations, SQL query variants, and unseen database schema in evaluation sets. Compared with WikiSQL, Spider contains about twice as many nested queries and 10 times more GROUP BY (HAVING) statements. More recently, a significant development in the field of Text-to-SQL benchmarks is the introduction of BIRD by Li et al. [3]. BIRD distinguishes itself by placing a particular emphasis on massive and noisy database contents, in contrast to Spider and WikiSQL, which primarily focus on the diversity and complexity of database schema. In this paper, we focus on the characteristics of Spider to evaluate our SQL2Cypher approach.

Spider consists of 10,181 questions and 5,693 unique complex SQL queries on 200 databases with multiple tables, covering 138 different domains. Its complexity is categorized into three levels, i.e., approximately 2:1:1 for easy, medium, and hard, based on factors such as the number of SQL clauses, the ratios of subqueries and/or aggregation functions, and the number of involved tables.

We present the statistics of Cy-Spider and the dataset transformation accuracy in Fig. 4. Additionally, we provide the statistics of Cy-Spider grouped by categories on the right-hand side, including the number of each clause keyword in the train and development splits, and the number of comparison operators (<, <=, <>, >, >=, =), boolean operators (NOT, OR, AND), string operator (=~) and four aggregate functions (count(), avg(), max(), min(), sum()). We also categorize Cypher pattern examples into three groups in practice. They are node patterns to match and retrieve nodes, relationship patterns to just match and retrieve relationships between nodes, and path patterns to traverse both nodes and relationships. Among path patterns, there are 150 queries matching paths of length two in the train split and 18 in the development split.

Dataset	Rel2Graph		
	# DB	# Nodes	# Edges
Cy-Spider Graph ($\mathcal{G}$)	155	10,729	11,277
Split	SQL2Cypher		
	# SQL	# Cypher	EA (%)
Train	5,224	4,327	82.83
Development	694	602	86.74

Category	Train	Development	Category	Train	Development
# MATCH	4,669	607	# Comparision Operators	3,148	348
# WHERE	2,405	288	# Boolean Operators	754	56
# RETURN	4,580	603	# String Operator	44	11
# WITH ...[AS]	1,179	71	# count()	1,353	234
# ORDER BY	901	137	# avg()	190	17
# LIMIT	509	93	# max()	98	28
# DESC	531	86	# min()	89	12
# ASC	163	29	# sum()	0	0
# UNION	22	1	# Node Pattern	2,874	398
# DISTINCT	687	49	# Relationship Pattern	62	0
			# Path Pattern	1,768	241

Fig. 4. The statistics of Cy-Spider dataset for Text-to-Cypher transformed from Spider benchmark. EA shows the accuracy of the dataset mapping process. Note that the test split of Spider is unreleased.

6.2 Models Evaluation Result

Table 1. The execution accuracy (EA) results on the development dataset, considering fine-tuning and prefix-tuning.

Tuning Strategy	Text-to-SQL-to-Cypher Pipeline			Text-to-Cypher
	Phase 1: Text-to-SQL	Phase 2: SQL-to-Cypher	**Final Score**	
T5$_{base}$ (prefix tune)	64.95	84.91	55.15	59.14
T5$_{base}$ (fine tune)	62.13	82.56	51.29	63.29
CodeT5$_{base}$ (prefix tune)	64.78	81.76	52.96	66.61
CodeT5$_{base}$ (fine tune)	68.11	86.02	58.59	**71.43**

Pipeline Evaluation. Table 1 compares the performance of the two phases, where the second phase notably outperforms the first phase. Phase one which uses T5-Base with prefix tuning also surpasses the performance of Text-to-Cypher evaluation using the same experimental setting. However, Text-to-Cypher better performs the pipeline in this aspect, and the superior performance is attributable to the integration of the styntax knowledge of Cypher query language.

Text-to-Cypher Evaluation. The last column of Table 1 shows the evaluation results of Text-to-Cypher model. It exhibits commendable performance, matching or even exceeding Text-to-SQL. In the direct comparison of the base models under both fine-tuning and prefix-tuning settings, CodeT5-Base emerges as the superior variant, outperforming the T5-Base model. We believe that the superior performance of CodeT5-Base can be attributed to the better NL-PL alignment pre-trained via bimodal dual generation task. This property appears to render CodeT5 a more suitable choice for this task than the T5 model.

The experimental outcomes show the importance of structured information. Notably, in pipeline results, phase two outperforms phase one regarding EA metric. One explanation for this superior performance lies in the proposed SQL2Cypher algorithm informed by the property graph schema knowledge. It effectively functions as a form of post-processing to enhance the precision of execution results. This indicates the scalability of Cypher queries compared to their SQL counterparts to some degree. Additionally, we evaluate Cy-Spider on Text-to-Cypher task, where the results demonstrate that the CodeT5-Base model responds more positively than the T5-Base to the end-to-end training.

6.3 Error Analysis

Using randomly sampled error cases, we include qualitative assessments and classify errors, as shown in Fig. 5.

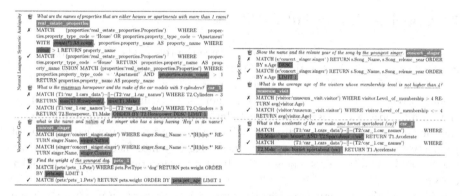

Fig. 5. Error cases of Text-to-Cypher on our proposed dev set. "✗" denote the predictions made by Text-to-Cypher model and "✓" denotes the ground truths.

Firstly, natural language syntactic ambiguity refers to the phenomenon where a sentence's syntactic structure allows for multiple interpretations. To illustrate, consider the question: *What are the names of properties that are either houses or apartments with more than 1 room?* The phrase *with more than 1 room* can be syntactically attached to both *houses* and *apartments* or it can exclusively modify *apartments*. Notably, the end-to-end model in use interprets it as the former, whereas the correct (gold standard) interpretation aligns with the latter.

Secondly, vocabulary gap arises from the mismatch between the provided schema vocabulary and the paraphrases employed in NL queries. Consider an example within the `concert_singer` domain, where *nation* is not available. Take the utterance: *What is the name and nation of the singer who has a song with 'Hey' in its name?* listed in the second block of Fig. 5 as an example. The linked schema item should be `singer.Country` rather than `singer.Nation`.

Thirdly, logic errors arise regarding sorting parameters and result arrangement in the context of semantic parsing. We do not address schema logical design issues. For instance, consider the query: *Show the name and the release year of the song by the youngest singer*. While *youngest* typically implies sorting records in ascending order and selecting the first one, rather than descending order. While alternatives such as `date of birth` are possible, we opt for `Age` to maintain alignment with the existing schema.

Commonsense errors occur when the model fails to interpret or infer information that is commonly understood by humans based on general world knowledge or context. For instance, in the last example provided, the model struggled to recognize that *amc hornet sportabout (sw)* represents a specific car model (`Make`). Instead, it incorrectly split it into multiple components and associated them with different schema properties, highlighting the challenge of handling complex natural language expressions in structured queries. Commonsense errors underscore the difficulty that machine learning models face in leveraging implicit real-world knowledge and understanding context, as humans naturally do.

7   Conclusion and Future Work

Cy-Spider makes two contributions to enhance the retrieval of large graph-based knowledge repositories. Firstly, the SQL2Cypher algorithm bridges Text-to-SQL datasets and Text-to-Cypher semantic parsing, showcasing the linguistic alignment between Cypher and complex SQL queries. We validated this with a property graph database constructed from relational databases, emphasizing execution accuracy metrics. Secondly, Cy-Spider augments the query capabilities of property graph databases through a natural language interface. Our experiments favored end-to-end models, encouraging further exploration with our Cy-Spider corpus.

However, several challenges persist. Syntactic ambiguity in natural language and a vocabulary gap between schema items and queries can lead to interpretation and schema linking issues. Logic errors can result in semantically incorrect translations, and understanding complex expressions, especially those involving commonsense or domain-specific knowledge, remains a challenge. Addressing these issues requires a deeper understanding of natural language, schema-specific knowledge, and logic interpretation. Future work will involve developing transformer-based semantic parsing models aligned with graph database schemas alongside the syntactic dependency among natural language questions. Another direction of future work is to benefit the in-context learning ability of generative LLMs, such as GPT style and CodeLlama [12] for structure extraction in semantic parsing.

Acknowledgment. This research is supported by the Australian Research Council through the Centre of Transforming Maintenance through Data Science (grant number IC180100030), funded by the Australian Government.

References

1. Carata, L.: Cyp2SQL: cypher to SQL translation (2019)
2. Gan, Y., et al.: Natural SQL: making SQL easier to infer from natural language specifications. In: Findings of the Association for Computational Linguistics: EMNLP 2021, pp. 2030–2042. Association for Computational Linguistics, Punta Cana, Dominican Republic (2021)
3. Li, J., et al.: Can LLM already serve as a database interface? A big bench for large-scale database grounded Text-to-SQLs. arXiv preprint arXiv:2305.03111 (2023)
4. Li, S., Yang, Z., Zhang, X., Zhang, W., Lin, X.: SQL2Cypher: automated data and query migration from RDBMS to GDBMS. In: Zhang, W., Zou, L., Maamar, Z., Chen, L. (eds.) WISE 2021. LNCS, vol. 13081, pp. 510–517. Springer, Cham (2021). https://doi.org/10.1007/978-3-030-91560-5_39
5. Li, X.L., Liang, P.: Prefix-tuning: optimizing continuous prompts for generation. arXiv preprint arXiv:2101.00190 (2021)
6. Lin, X.V., Socher, R., Xiong, C.: Bridging textual and tabular data for cross-domain text-to-SQL semantic parsing. In: Proceedings of the 2020 Conference on Empirical Methods in Natural Language Processing: Findings, EMNLP (2020)
7. Marton, J., Szárnyas, G., Varró, D.: Formalising openCypher graph queries in relational algebra. In: Kirikova, M., Nørvåg, K., Papadopoulos, G.A. (eds.) ADBIS 2017. LNCS, vol. 10509, pp. 182–196. Springer, Cham (2017). https://doi.org/10.1007/978-3-319-66917-5_13
8. Ni, P., Okhrati, R., Guan, S., Chang, V.: Knowledge graph and deep learning-based text-to-GraphQL model for intelligent medical consultation chatbot. Inf. Syst. Front. 1–20 (2022)
9. Pourreza, M., Rafiei, D.: Din-SQL: decomposed in-context learning of text-to-SQL with self-correction. arXiv preprint arXiv:2304.11015 (2023)
10. Radford, A., Wu, J., Child, R., Luan, D., Amodei, D., Sutskever, I.: Language models are unsupervised multitask learners. OpenAI Blog 1(8), 9 (2019)
11. Raffel, C., et al.: Exploring the limits of transfer learning with a unified text-to-text transformer. J. Mach. Learn. Res. 21(1), 5485–5551 (2020)
12. Rozière, B., et al.: Code Llama: open foundation models for code. arXiv preprint arXiv:2308.12950 (2023)
13. Rubin, O., Berant, J.: SmBoP: semi-autoregressive bottom-up semantic parsing. In: Proceedings of the 2021 Conference of the North American Chapter of the Association for Computational Linguistics: Human Language Technologies, pp. 311–324. Association for Computational Linguistics, Online (2021)
14. Saparina, I., Osokin, A.: SPARQLing database queries from intermediate question decompositions. In: Proceedings of the 2021 Conference on Empirical Methods in Natural Language Processing, pp. 8984–8998. Association for Computational Linguistics, Online and Punta Cana, Dominican Republic (2021)
15. Steer, B.A., Alnaimi, A., Lotz, M.A., Cuadrado, F., Vaquero, L.M., Varvenne, J.: Cytosm: declarative property graph queries without data migration. In: Proceedings of the Fifth International Workshop on Graph Data-Management Experiences & Systems, pp. 1–6 (2017)
16. Wang, B., Shin, R., Liu, X., Polozov, O., Richardson, M.: Rat-SQL: relation-aware schema encoding and linking for text-to-SQL parsers. arXiv preprint arXiv:1911.04942 (2019)

17. Wang, Y., Wang, W., Joty, S., Hoi, S.C.: Code T5: identifier-aware unified pre-trained encoder-decoder models for code understanding and generation. In: Proceedings of the 2021 Conference on Empirical Methods in Natural Language Processing, pp. 8696–8708. Association for Computational Linguistics, Online and Punta Cana, Dominican Republic (2021)
18. Yu, T., Li, Z., Zhang, Z., Zhang, R., Radev, D.: TypeSQL: knowledge-based type-aware neural text-to-SQL generation. arXiv preprint arXiv:1804.09769 (2018)
19. Yu, T., et al.: Spider: a large-scale human-labeled dataset for complex and cross-domain semantic parsing and Text-to-SQL task. arXiv preprint arXiv:1809.08887 (2018)
20. Zhong, V., Xiong, C., Socher, R.: Seq2SQL: generating structured queries from natural language using reinforcement learning. CoRR abs/1709.00103 (2017)

S5TR: Simple Single Stage Sequencer for Scene Text Recognition

Zhijian Wu[1], Jun Li[2(✉)], and Jianhua Xu[2]

[1] School of Data Science and Engineering, East China Normal University,
Shanghai, China
`zjwu_97@stu.ecnu.edu.cn`
[2] School of Computer and Electronic Information, Nanjing Normal University,
Nanjing, China
`{lijuncst,xujianhua}@njnu.edu.cn`

Abstract. As an active research topic in computer vision, scene text recognition (STR) aims to recognize character sequences in natural scenes. Currently, mainstream STR approaches consist of two main modules: a visual model for feature extraction and a sequence model for text translation. The two modules function separately and sequentially, which increases the complexity of the STR model. In this study, we propose a novel **Simple Single Stage Sequencer for Scene Text Recognition** (S5TR), which allows to transform text instance images into string sequences directly. Specifically, our S5TR contains stacks of Sequencers made of horizontal and vertical Long Short Term Memory Networks (LSTMs). On the one hand, S5TR extracts visual representations of images by modeling long-range dependencies via LSTM, which is similar to self-attention in Vision Transformer (ViT). On the other hand, LSTM serving as a sequence modeling module is able to capture contextual information within the character sequence for predicting the character. Experimental results demonstrate that our S5TR achieves highly competitive performance compared to existing STR methods.

Keywords: Scene text recognition · Text transcription · LSTM · Neural network

1 Introduction

Aiming to transcribe texts from natural scenes into character sequences, scene text recognition (STR) has attracted increasing attention due to its wide application in a variety of vision-based tasks [16,20]. In recent years, with the boom of deep learning, the community has witnessed massive efforts [4,30] devoted to promoting recognition accuracy.

Technically, the STR task requires cross-modality transformation from image to text, thus mainstream approaches consist of two main modules: a visual model

This work was supported by the National Natural Science Foundation of China (NSFC) under Grants 62173186 and 61703096.

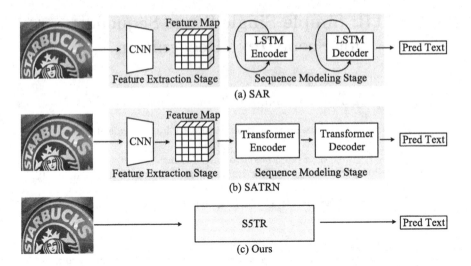

Fig. 1. An illustrative comparison of our proposed method with different classic STR models. (a) CNN-LSTM based models. (b) Encoder-Decoder models. (c) Our S5TR method recognizes scene text with a single stage model.

for extracting image features and a sequence model for text translation, as shown in Fig. 1. For instance, SAR [18] (Fig. 1(a)) uses Convolutional Neural Networks (CNNs) to extract image features, and reconfigures them with a long short-term memory (LSTM) network for robust sequence prediction. Inspired by Transformer architecture, SATRN [17] (Fig. 1(b)) employs an encoder to model sequence dependencies while the decoder performs text transcription. Despite significant improvements in recognition accuracy, these methods still suffer from complex recognition process. For fast and efficient recognition, Atienza [3] adopts an off-the-shelf Vision Transformer (ViT) as feature extractor to achieve scene text recognition.

It is well known that LSTM models are extensively used in sequence modeling because of their powerful ability to capture long-range dependencies. Recently, LSTM has been successfully applied in image classification tasks [27], which suggests that it is an promising alternative to CNN in visual feature extraction. This motivates our research to construct simplified STR networks using LSTM, as it is a desirable model that can both extract visual features and model text sequences. In this study, we propose a novel Simple Single Stage Sequencer for Scene Text Recognition (S5TR). Similar to ViT [9], the input text image is first partitioned into non-overlapping patches. Then, a stack of Sequencers constructed by horizontal and vertical bidirectional LSTM (LSTMs, BiLSLTMs) are used to carry out visual feature extraction and sequence modeling simultaneously. Finally, characters are recognized by a simple linear prediction. The contributions of this paper can be summarized as follows:

- We have proposed a novel framework termed Simple Single Stage Sequencer for Scene Text Recognition (S5TR). Our S5TR consists of a single unified model which is capable of directly mapping the input text image into character sequences.
- Extensive experimental results show that our S5TR achieves promising performance compared to the previous methods.

2   Related Work

Scene Text Recognition (STR), which aims to recognize text sequences from natural scenes, is an intensive research task in the era of deep learning. The pioneering CRNN [25] encodes the input image as a sequence of visual features, which is then modeled with BiLSTM for contextual reinforcement and CTC loss for text transcription. Hu et al. [13] utilized graph neural networks and attention mechanisms to improve feature learning and CTC decoding. Du et al. [10] proposed a ViT to capture multi-granularity character features, achieving promising performance using only CTC decoding. Subsequently, encoder-decoder based auto-regression (AR) methods became popular, which converts recognition process into an iterative decoding process. Sheng et al. [24] introduced a transformer-based AR decoder for STR that achieves impressive accuracy. Next, Kuang et al. [15] employed ResNet as a visual model, allowing for further improvements in accuracy. To better preserve 2D features, Lee et al. [17] developed a 2D-based transformer. With contextual information incorporated, the accuracy of the models increased dramatically, whereas the inference was slow since these models perform recognition in a character-by-character manner. Despite significant improvements in recognition accuracy, these methods still suffer from complex recognition process. In contrast, our proposed S5TR is a simple single-stage approach for scene text recognition.

3   Proposed Method

The overview of the proposed S5TR is illustrated in Fig. 2. The input image $X \in \mathbb{R}^{H \times W \times C}$ is first reshaped into a sequence of flattened 2D patches $X \in \mathbb{R}^{(H \cdot W / P^2) \times (P^2 \cdot C)}$. Then patch embedding is used to map the flattened patches to the D dimension with a trainable linear projection. Subsequently, a series of Sequencers composed of horizontal and vertical BiLSTM are used to extract visual features and model the dependencies of character sequences for capturing contextual information. Finally, a fully-connected (FC) layer is conducted to obtain the character sequence.

The macro-architecture design of our S5TR follows modern ViT [9], but we leverage LSTM [27] rather than self-attention mechanism for long-range modeling. Mathematically, the computational process of LSTM is formulated as follows:

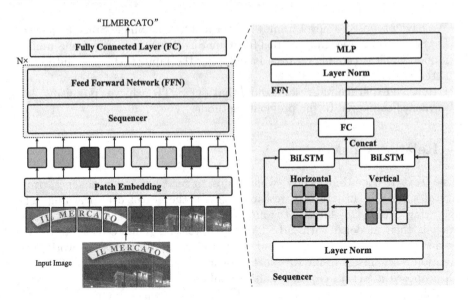

Fig. 2. Overview of the proposed S5TR. The input text images are first divided into non-overlapping patches. Next, a stack of Sequencers constructed from horizontal and vertical BiLSTM process visual feature extraction and sequence modeling simultaneously. Finally, character sequence prediction is conducted using a fully-connected (FC) layer.

$$
\begin{aligned}
i_t &= \sigma \left(W_{xi} x_t + W_{hi} h_{t-1} + b_i \right) \\
f_t &= \sigma \left(W_{xf} x_t + W_{hf} h_{t-1} + b_f \right) \\
o_t &= \sigma \left(W_{xo} x_t + W_{ho} h_{t-1} + b_o \right) \\
c_t &= c_{t-1} \odot f_t + \tanh \left(W_{xg} x_t + W_{hg} h_{t-1} + b_g \right) \odot i_t \\
h_t &= o_t \odot \tanh \left(c_t \right)
\end{aligned}
\tag{1}
$$

where $\sigma(\cdot)$ is the logistic Sigmoid function and $\odot(\cdot)$ denotes Hadamard product. Besides, i_t, f_t, o_t denote input gate, forget gate and output gate, respectively. In addition, c_{t-1}, c_t represent the former and current cell state. Thus, LSTM can model sequence dependencies by controlling cell states with gate units. To further model the mutual dependence of sequences, we use BiLSTMs as the baseline modules.

Recognizing vertical text has been an important challenge for STR tasks. Existing CNN or ViT based methods fail to model the relationship between vertical text sequences. In order to capture the information of the image in horizontal and vertical directions, the input sequence X is fed into two BiLSTMs in parallel according to horizontal and vertical reshaping operation. Finally, the output features of different directions are concatenated and fused using a FC layer:

$$H_h = BiLSTM_h(reshape_{hor}(X))$$
$$H_v = BiLSTM_v(reshape_{ver}(X)) \qquad (2)$$
$$cH = FC(concat(H_h, H_v))$$

where $reshape_{hor}(\cdot)$ and $reshape_{ver}(\cdot)$ denote the reshaping operations according to directions. $BiLSTM_h(\cdot), BiLSTM_v(\cdot)$ indicate horizontal and vertical BiLSTM respectively, while $FC(\cdot)$ represents a fully connected layer. Different from the ViT that directly transforms images into sequences, our S5TR serializes images from horizontal and vertical directions, which is critical for the STR task because of their powerful ability to model text sequences with various directions.

In addition to Sequencer, each module contains a feed-forward neural network (FFN) consisting of a two-layer multilayer perception machine (MLP) with a ReLU activation in between:

$$FFN(x) = \max(0, xW_1 + b_1) W_2 + b_2 \qquad (3)$$

The FFN module is capable of achieving cross-channel fusion while enhancing network nonlinearity.

Finally, a linear classifier with N-nodes is used to predict the character sequence. In implementation, non-textual components are transcribed into blank symbols and compressed to obtain the final text sequence. Since our experiments were conducted on the Latin dataset, the number of categories N was set to 37, including 26 letters, 10 Arabic numerals, and a blank symbol representing non-text.

4   Experiments

4.1   Datasets and Evaluation Metrics

Following the previous settings [4,9], we use Mjsynth [14] + SynthText [12] as training datasets for fairness. The former consists of approximately 9M synthetically generated word images, while the latter is a synthetic dataset consisting of 5.5M samples generated from blending text on natural images. The test datasets are mainly divided into regular and irregular texts, according to the difficulty and geometric variation of the text.

Rugular text datasets contain text images with characters in horizontal direction.

- **IIIT5K** [21] contains 5000 word patches cropped from natural scene images, where 2000 are used for training and 3000 for test. Each image is associated with a 50-word and a 1k-word lexicons.
- **Street View Text (SVT)** [29] contains 647 text images cropped form Google Street View, each of which has a 50-word lexicon. Most images are severely corrupted by noise and blur.
- **ICDAR 2003 (IC03)** [11] was created for the ICDAR 2003 Robust Reading competition for reading camera-captured scene texts. It contains 867 test images which come from natural scenes.

- **ICDAR 2013 (IC13)** [1] contains 1015 cropped text images that contain non-alphanumeric characters or less than three characters. Images were captured from sign boards and objects with large variations.

Irregular text datasets contain more complex scenarios, such as curved and arbitrarily rotated or distorted text.

- **ICDAR 2015 (IC15)** [2] contains 2077 test word patches cropped from rotated, vertical, perspective-shifted and curved scene images.
- **Street View Text Perspective (SVTP)** [22] consists of 645 word patches, which are cropped from Google Street View. Each patch is associated with a 50-word lexicon.
- **CUTE (CT)** [23] contains 288 curved text images for test. Many of these are curved text images.

For performance metric, we use recognition accuracy to measure the success rate of word predictions per image. Mathematically, it is defined as:

$$Acc = \frac{W_r}{W} \tag{4}$$

where W_r indicates the number of correctly recognized words, while W represents the total number of words.

4.2 Implementation Details

Our S5TR is built on Sequencer2D-S [27] architecture, where patch embedding is set to 7, the dimension of the hidden layer is set to 48, and the expansion rate of MLP is set to 3. More details are provided in [27]. In addition, images are resized to 224×224 before input and RandAugment [8] strategy is adopted for data augmentation. To speed up the training process, we loaded the pre-trained Sequencer2D-S for parameter initialization. In implementation, our model was trained using a total of 30k iterations with a batch size of 192. All models are trained on a server with a Tesla V100 GPU.

4.3 Comparisons with the State-of-the-arts

To further demonstrate the effectiveness of the proposed S5TR, we extensively compare our model with ten state-of-the-art methods including CRNN [25], R2AM [16], RARE [26], STARNet [20], GCRNN [28], Rosetta [5], AON [7], ACE [30], Comb.Best [4] and ViTSRT [3].

As shown in Table 1, our algorithm outperforms most of the algorithms on regular text dataset, except for slight inferiority to the competing Com.Best algorithm on the SVT and IC03 datasets. Com.Best algorithm uses spatial transformation network (STN) to preprocess the input images, transforming the curved and tilted text such that they are regularized. In contrast, our S5TR allows direct transcription of the input text image into character sequences without

Table 1. Scene text recognition accuracies (%) over several regular text benchmarks. Best scores are highlighted in bold and the second best are underlined.

Methods	IIIT	SVT	IC03	IC13	Params
CRNN [25]	81.8	80.1	91.7	89.4	8.3
R2AM [16]	83.1	80.9	91.6	90.1	2.9
RARE [26]	86.0	85.4	93.5	92.3	10.8
STARNet [20]	85.2	84.7	93.4	91.2	48.7
GCRNN [28]	82.9	81.1	92.7	90.0	4.6
FAN [6]	87.4	85.9	94.2	93.3	–
Rosetta [5]	82.5	82.8	92.6	90.3	44.3
AON [7]	87.0	82.8	–	-	–
Char-Net [19]	83.6	84.4	91.5	90.8	–
Comb.Best [4]	87.9	**87.5**	**94.9**	92.3	49.6
ACE [30]	82.3	82.6	92.1	89.7	–
ViTSTR [3]	85.6	85.3	93.9	91.7	21.5
S5TR(Ours)	**88.1**	85.9	94.4	**92.4**	27.3

Table 2. Scene text recognition accuracies (%) over irregular text benchmarks. Best scores are highlighted in bold and the second best are underlined.

Method	IC15	SVTP	CT	Params
CRNN [25]	65.3	65.9	61.5	8.3
R2AM [16]	68.5	70.4	64.6	2.9
RARE [26]	73.9	75.4	71.0	10.8
STARNet [20]	74.5	74.7	69.2	48.7
GCRNN [28]	68.1	68.5	65.5	4.6
FAN [6]	70.6	–	–	–
Rosetta [5]	68.1	70.3	65.5	44.3
AON [7]	68.2	73.0	76.8	–
Char-Net [19]	60.0	73.5	–	–
Comb.Best [4]	71.8	79.2	74.0	49.6
ACE [30]	68.9	70.1	**82.6**	–
ViTSTR [3]	75.3	78.1	71.3	21.5
S5TR(Ours)	**76.5**	**80.3**	79.2	27.3

any cumbersome pre-processing procedure. In addition, our S5TR costs about half of the parameters of STARNet and Com.Best, implying that it achieves excellent trade-offs between performance and efficiency.

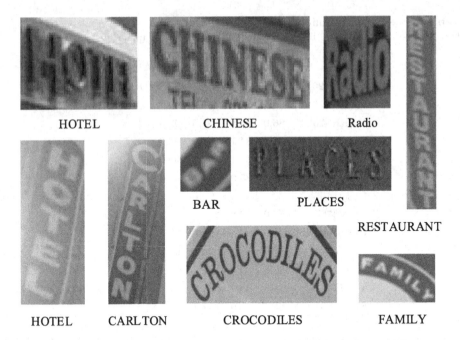

HOTEL   CHINESE   Radio

BAR   PLACES

RESTAURANT

HOTEL   CARLTON   CROCODILES   FAMILY

Fig. 3. Qualitative results of the proposed S5TR model. It is shown that our method correctly recognizes text sequences in various scenes, including horizontal, vertical direction, noisy, blurred, and warped.

Furthermore, our S5TR still achieves state-of-the-art performance on irregular text datasets, as shown in Table 2. Overall, our model performs the best or the second best on irregular test setting, even though our model is not specifically designed for irregular text. In summary, the experimental results demonstrate the superiority of our S5TR, revealing that using a single unified model to tackle the STR task is a promising direction.

4.4   Qualitative Results

We have conducted qualitative experiments to visualize the results of S5TR, as illustrated in Fig. 3. It can be seen that our algorithm correctly recognizes the text sequences of diverse scenes. In particular, in addition to the common horizontal text, our algorithm is able to recognize vertical text such as "RESTAU-RANT", "HOTEL" and "CARLTON" accurately. We owe the advantage of our model to the Sequencer composed of horizontal and vertical BiLSTMs, which allows the proposed S5TR to explore the dependence of text sequences with different directions. In addition, S5TR also achieves accurate recognition of noisy, blurred and warped text. Combining the characteristics of both visual and sequential models, our method is able to model sequence dependencies while extracting robust image representations, ultimately leading to more stable and accurate text prediction.

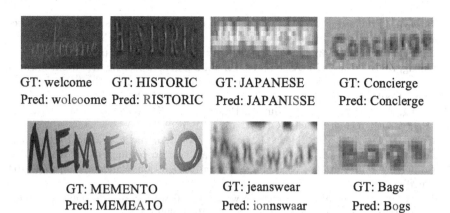

Fig. 4. Failure cases of our model. "GT" indicates the ground-truth annotation, while "Pred" is the predicted result. Wrong symbol predictions are highlighted in red (Color figure online).

Some failure cases are also shown in Fig. 4. It can be observed that the failures result from serious challenges in different scenes, such as excessive blurring, extreme distortion, intense lighting conditions and severe noise.

5   Conclusions

In this paper, we propose a novel scene text recognition method, dubbed as S5TR, which is able to transcribe text images directly into character sequences with a single model. Specifically, S5TR employs LSTM as the fundamental component that is capable of extracting visual representations of text images and modeling the dependencies of character sequences simultaneously. Experimental results show that the proposed S5TR achieves promising performance compared to the existing STR algorithm.

References

1. Karatzas, D., et al.: ICDAR 2013 robust reading competition. In: International Conference on Document Analysis and Recognition (ICDAR), pp. 1484–1493. IEEE (2013)
2. Karatzas, D., et al.: ICDAR 2015 competition on robust reading. In: International Conference on Document Analysis and Recognition (ICDAR), pp. 1156–1160. IEEE (2015)
3. Atienza, R.: Vision transformer for fast and efficient scene text recognition. In: Lladós, J., Lopresti, D., Uchida, S. (eds.) ICDAR 2021. LNCS, vol. 12821, pp. 319–334. Springer, Cham (2021). https://doi.org/10.1007/978-3-030-86549-8_21
4. Baek, J., et al.: What is wrong with scene text recognition model comparisons? dataset and model analysis. In: ICCV, pp. 4715–4723 (2019)

5. Borisyuk, F., Gordo, A., Sivakumar, V.: Rosetta: large scale system for text detection and recognition in images. In: Proceedings of the 24th ACM SIGKDD International Conference on Knowledge Discovery & Data Mining, pp. 71–79 (2018)

6. Cheng, Z., Bai, F., Xu, Y., Zheng, G., Pu, S., Zhou, S.: Focusing attention: towards accurate text recognition in natural images. In: Proceedings of the IEEE International Conference on Computer Vision, pp. 5076–5084 (2017)

7. Cheng, Z., Xu, Y., Bai, F., Niu, Y., Pu, S., Zhou, S.: AON: towards arbitrarily-oriented text recognition. In: CVPR, pp. 5571–5579 (2018)

8. Cubuk, E.D., Zoph, B., Shlens, J., Le, Q.V.: RandAugment: practical automated data augmentation with a reduced search space. In: Proceedings of the IEEE/CVF Conference on Computer Vision and Pattern Recognition Workshops, pp. 702–703 (2020)

9. Dosovitskiy, A., et al.: An image is worth 16x16 words: transformers for image recognition at scale. arXiv preprint arXiv:2010.11929 (2020)

10. Du, Y., et al.: SVTR: scene text recognition with a single visual model. In: IJCAI (2022)

11. Lucas, S. M., et al.: ICDAR 2003 robust reading competitions: entries, results, and future directions. Int. J. Doc. Anal. Recogn. (IJDAR) 7(2), 105–122 (2005)

12. Gupta, A., Vedaldi, A., Zisserman, A.: Synthetic data for text localisation in natural images. In: CVPR, pp. 2315–2324 (2016)

13. Hu, W., Cai, X., Hou, J., Yi, S., Lin, Z.: GTC: guided training of CTC towards efficient and accurate scene text recognition. In: Proceedings of the AAAI Conference on Artificial Intelligence, vol. 34, pp. 11005–11012 (2020)

14. Jaderberg, M., Simonyan, K., Vedaldi, A., Zisserman, A.: Synthetic data and artificial neural networks for natural scene text recognition. arXiv preprint arXiv:1406.2227 (2014)

15. Kuang, Z., et al.: MMOCR: a comprehensive toolbox for text detection, recognition and understanding. In: Proceedings of the 29th ACM International Conference on Multimedia, pp. 3791–3794 (2021)

16. Lee, C.Y., Osindero, S.: Recursive recurrent nets with attention modeling for OCR in the wild. In: CVPR, pp. 2231–2239 (2016)

17. Lee, J., Park, S., Baek, J., Oh, S.J., Kim, S., Lee, H.: On recognizing texts of arbitrary shapes with 2D self-attention. In: CVPR, pp. 546–547 (2020)

18. Li, H., Wang, P., Shen, C., Zhang, G.: Show, attend and read: a simple and strong baseline for irregular text recognition. In: Proceedings of the AAAI Conference on Artificial Intelligence, pp. 8610–8617 (2019)

19. Liu, W., Chen, C., Wong, K.Y.: Char-Net: a character-aware neural network for distorted scene text recognition. In: Proceedings of the AAAI Conference on Artificial Intelligence, vol. 32 (2018)

20. Liu, W., Chen, C., Wong, K.Y.K., Su, Z., Han, J.: STAR-Net: a spatial attention residue network for scene text recognition. In: BMVC-British Machine Vision Conference, vol. 2, p. 7 (2016)

21. Mishra, A., Alahari, K., Jawahar, C.: Scene text recognition using higher order language priors. In: BMVC-British Machine Vision Conference (2012)

22. Phan, T.Q., Shivakumara, P., Tian, S., Tan, C.L.: Recognizing text with perspective distortion in natural scenes. In: ICCV, pp. 569–576 (2013)

23. Risnumawan, A., Shivakumara, P., Chan, C.S., Tan, C.L.: A robust arbitrary text detection system for natural scene images. Expert Syst. Appl. 41(18), 8027–8048 (2014)

24. Sheng, F., Chen, Z., Xu, B.: NRTR: a no-recurrence sequence-to-sequence model for scene text recognition. In: 2019 International Conference on Document Analysis and Recognition (ICDAR), pp. 781–786. IEEE (2019)

25. Shi, B., Bai, X., Yao, C.: An end-to-end trainable neural network for image-based sequence recognition and its application to scene text recognition. IEEE Trans. Pattern Anal. Mach. Intell. **39**(11), 2298–2304 (2016)

26. Shi, B., Wang, X., Lyu, P., Yao, C., Bai, X.: Robust scene text recognition with automatic rectification. In: CVPR, pp. 4168–4176 (2016)

27. Tatsunami, Y., Taki, M.: Sequencer: deep LSTM for image classification. In: Advances in Neural Information Processing Systems (2022)

28. Wang, J., Hu, X.: Gated recurrent convolution neural network for OCR. In: Advances in Neural Information Processing Systems, vol. 30 (2017)

29. Wang, K., Babenko, B., Belongie, S.: End-to-end scene text recognition. In: ICCV, pp. 1457–1464. IEEE (2011)

30. Xie, Z., Huang, Y., Zhu, Y., Jin, L., Liu, Y., Xie, L.: Aggregation cross-entropy for sequence recognition. In: CVPR, pp. 6538–6547 (2019)

Explainable AI

Coping with Data Distribution Shifts: XAI-Based Adaptive Learning with SHAP Clustering for Energy Consumption Prediction

Tobias Clement[1]([✉]), Hung Truong Thanh Nguyen[1], Nils Kemmerzell[1], Mohamed Abdelaal[2], and Davor Stjelja[3]

[1] Friedrich-Alexander-University Erlangen-Nürnberg, Lange Gasse 20, 90403 Nürnberg, Germany
`tobias.clement@fau.de`
[2] Software AG, Uhlandstraße 12, 64297 Darmstadt, Germany
[3] Granlund, Malminkaari 21, 00700 Helsinki, Finland

Abstract. Adapting to data distribution shifts after training remains a significant challenge within the realm of Artificial Intelligence. This paper presents a refined approach, superior to Automated Hyper Parameter Tuning methods, that effectively detects and learns from such shifts to improve the efficacy of prediction models. By integrating Explainable AI (XAI) techniques into adaptive learning with SHAP clustering, we generate interpretable model explanations and use these insights for adaptive refinement. Our three-stage process: (1) SHAP value generation for the model explanation, (2) clustering these values for pattern identification, and (3) model refinement based on the derived SHAP cluster characteristics, mitigates overfitting and ensures robust data shift handling. We evaluate our method on a comprehensive dataset comprising energy consumption records of buildings, as well as two additional datasets, to assess the transferability of our approach to other domains, regression, and classification problems. Our experiments highlight that our method not only improves predictive performance in both task types but also delivers interpretable model explanations, offering significant value in dealing with the challenges of data distribution shifts in AI.

Keywords: Adaptive Learning · Data Distribution Shifts · Explainable AI

1 Introduction

Artificial Intelligence (AI) has become a significant tool in a variety of domains, yielding prediction models with exceptional accuracy. Automated Hyperparameter Tuning (AHT) methods are commonly used to optimize these models,

Supplementary Information The online version contains supplementary material available at https://doi.org/10.1007/978-981-99-8391-9_12.

improving their performance by fine-tuning parameters during training [19]. Despite these advancements, a significant challenge persists: these models often underperform when faced with shifts in data distributions post-training. This difficulty in adapting to unforeseen or changing data conditions can compromise the model's performance and utility. Data distribution shifts occur in various fields, one of which is predicting building energy consumption using time series data. Buildings account for a considerable proportion of global energy consumption and CO_2 emissions [17], making accurate prediction a critical factor in managing energy use and demand-supply optimization. The complexity of this task arises from diverse factors such as building characteristics, equipment, weather conditions, and occupants' habits [21]. Unexpected situations, such as the COVID-19 pandemic, can lead to abrupt shifts in consumption patterns, further complicating the prediction models. To this end, Explainable AI (XAI) enriches the AI landscape by allowing for human-understandable explanations of model predictions [4,13]. Methods such as SHapley Additive exPlanations (SHAP) [11] provide insights into the importance of different features for model predictions, potentially guiding improvements and adaptations. When combined with adaptive learning, these techniques can offer a robust solution to handle shifts in data distribution, thereby improving prediction accuracy under changing conditions.

In this light, we introduce a framework termed SHAP Clustering-based Adaptive Learning (SCAL). Leveraging the power of XAI and adaptive learning, SCAL aims to better equip prediction models to handle data distribution shifts, particularly in the complex domain of building energy consumption prediction. Our work offers the following primary contributions:

- Section 3 introduces SCAL, a robust adaptive learning framework that uses SHAP clustering to improve model adaptability to data distribution shifts.
- Section 5 evaluates SCAL within the context of building energy consumption prediction, highlighting its superior performance over traditional AHT methods in handling data distribution shifts.
- Section 6 validates SCAL's transferability across diverse regression and classification problems using two public datasets, emphasizing its broad applicability beyond building energy consumption prediction

This general applicability highlights SCAL's potential as a valuable tool in the broader AI toolkit for tackling the challenge of data distribution shifts.

2   Related Work

In this section, we review three key areas related to our study: the role of Machine Learning in energy use prediction, the potential of XAI for model enhancement, and our rationale for selecting SHAP in our framework.

2.1   Energy Consumption Prediction

Energy consumption prediction is essential for energy efficiency and sustainability. Various ML techniques have been explored based on historical data to predict future energy consumption [2,14]. These techniques optimize energy usage, reduce carbon emissions, leading to cost savings and environmental benefits. Distinctively, our method harnesses XAI techniques to predict energy consumption patterns with greater accuracy and interpretability, even in scenarios of sudden data shifts like those prompted by the COVID-19 pandemic.

2.2 XAI-Based Model Improvement

XAI-based model improvements aim to enhance various properties of a model, including performance, convergence, robustness, efficiency, reasoning, and equality [20]. Several approaches have been proposed to improve models using XAI via data, intermediate features, loss, gradient, and model augmentation [8,22]. Moreover, human-in-the-loop knowledge is employed to improve model [7,15]. Our approach innovatively combines XAI techniques with clustering and adaptive learning, introducing a systematic method for improving model performance and interpretability, especially in the presence of data shifts.

2.3 SHapley Additive ExPlanations (SHAP)

SHAP is a prominent XAI that provides insights into the contribution of individual features to a model's predictions [11]. SHAP values, derived from Shapley values in cooperative game theory, have several desirable properties, including consistency, local accuracy, and missingness consistency [10,12]. While extensions like SHAP clustering have been applied to classification problems to identify and characterize similar instance clusters [6], their use in regression problems has been limited. Our study pioneers the SHAP clustering application to the regression through the SCAL framework, marking a significant contribution to the field. The choice to incorporate SHAP as the backbone of our SCAL framework stems from its distinctive features that set it apart in the landscape of XAI methods. With SHAP, we can obtain an accurate and balanced depiction of how each feature influences every single prediction, and importantly, these influences maintain stability when aggregated across all predictions. Another aspect of SHAP is its adept handling of missing values, a prevalent issue in real-world datasets. This ability to accurately account for the effect of absent data is crucial, ensuring the robustness of our analysis. Moreover, SHAP's competence in offering explanations at the level of individual data instances aligns perfectly with the goal of our SCAL framework - to learn and adapt in response to data changes. This characteristic enables our framework to dynamically adjust model parameters in light of shifts in individual instance explanations.

3 SHAP Clustering-Based Adaptive Learning (SCAL)

SCAL framework is designed to facilitate model adaptability in response to data shifts, enhancing its interpretability and performance, which is achieved through the integration of three interconnected building blocks. The pipeline is represented in Fig. 1, comprises the following key stages: *Building Block 1: SHAP*

Clustering initiates the framework by clustering instances based on their explanation similarity, using SHAP values within an explanation space. This step identifies distinct explanation characteristics that can inform model adaptation. *Building Block 2: Extraction of SHAP Clustering Characteristics* builds upon the clusters formed in Block 1. This block characterizes the clustering using three quality metrics. The insights derived are instrumental in understanding the relationships between instances and the behavior of the model, providing critical information for model adaptation. *Building Block 3: Model Adaptation* utilizes the SHAP clustering characteristics derived in Block 2 to adapt and fine-tune the model, enhancing its adaptability to data shifts. This process not only improves the model's performance but also enhances its interpretability, completing the SCAL framework. Each building block within the SCAL framework has a distinct and crucial role, collectively contributing to the model's adaptability and effectiveness, making it robust and reliable. The following subsections delve into each building block, elucidating the specific roles they play and the methodologies they employ.

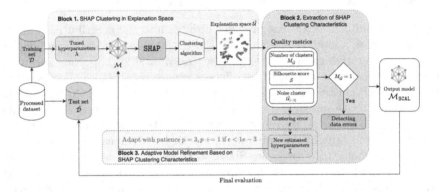

Fig. 1. SCAL Pipeline: Adaptive Learning via SHAP Clustering.

3.1   Building Block 1: SHAP Clustering in Explanation Space

The first building block in the SCAL is SHAP Clustering in Explanation Space. This building block forms the foundation for the entire framework, identifying and characterizing groups of instances based on explanation similarity and enabling subsequent building blocks to extract valuable insights and adapt the model. We use SHAP values to understand each feature's contribution to an instance's prediction. By representing each instance x with its corresponding SHAP values, we transform the original input space into an explanation space, allowing us to focus on the model's reasoning instead of its raw predictions. In the explanation space, we use DBSCAN [1] to cluster instances based on SHAP value similarities. Clustering instances with similar explanation patterns identifies subgroups sharing common model behaviors and feature interactions, simplifying the analysis and interpretation of the model's decisions. Moreover, this clustering process uncovers the underlying structure or trends in the explanation space, providing valuable insights into the model's behavior. To implement this building block, we follow these steps:

1. Train a supervised model using AHT algorithms for each building.
2. Compute SHAP values for instances with the AHT-tuned trained model.
3. Apply UMAP to extract a 2D space of SHAP values, which delivers a highly interpretable representation to end-users and efficient computation.
4. Utilize DBSCAN for creating SHAP clusters embedding.

By completing this building block, we have established a solid groundwork in the explanation space that pave the way for the following building blocks to unlock further insights and adapt the model, resulting in superior performance and better understanding.

3.2 Building Block 2: Extraction of SHAP Clustering Characteristics

The SCAL's second building block extracts valuable characteristics from the explanation space created in Building Block 1. These insights into the model's behavior within clusters pave the way for targeted adaptation in Building Block 3. After clustering instances in the explanation space using SHAP values, we aim to understand the model's behavior within these subgroups. To this end, we use three quality metrics to describe the explanation space (as shown in Fig. 1):

- **Number of clusters** (M): Each cluster reflects a set of logical rules learned by the model. More clusters suggest greater information extraction from the data. The optimal number is estimated using silhouette analysis [16].
- **Silhouette score** (SS): Reflects cluster separation and cohesion. This score uses intra-inter distances between clusters to estimate optimal cluster count [16]. Distinct clusters can lead to improved anomaly detection as these data points lie more likely between classified clusters.
- **Presence of noise cluster**: Aids in identifying well-defined clusters in noisy data and enhances anomaly detection [3,5]. Noise clusters can be helpful to reduce overfitting [9]. DBSCAN labels noise cluster as $\mathcal{U}_{(-1)}$ [1].

We apply silhouette analysis [16] to estimate the optimal number of clusters, accounting for the trade-off between bias and variance, sample size, and computational cost by taking the intra-inter distances between clusters into consideration [18]. Then, the SS for a cluster space is computed as the mean of all SSs for each sample.

3.3 Building Block 3: Adaptive Model Refinement Based on SHAP Clustering Characteristics

In this stage of the SCAL method, we leverage the information obtained from SHAP clustering to iteratively refine the model by complying with defined conditions extracted from SHAP clustering characteristics. The objective is to adapt the model's hyperparameters based on the quality metrics derived from the SHAP clusters embedding in an unsupervised way so as to avoid overfitting the training dataset. The pseudo-code for the SCAL algorithm is shown in Algorithm 1.

Algorithm 1: SCAL procedure

Data: Training set of a building $\mathcal{D} = (x, y)$
Output: SCAL model $\mathcal{M}_{\text{SCAL}}$

```
1  begin
2  │    p ← 0                                                            // Patience
3  │    λ ← XGBTune(𝒟)                                    // AHT on the training set
4  │    ℳ ← XGBRegressor(λ)                      // Model with tuned hyperparameters
5  │    ỹ ← ℳ(𝒟)                               // Model predictions on training set
6  │    𝒰 ← DBSCAN(UMAP(SHAP(ỹ)))                            // Explanation space
7  │    𝒮 ← silhouette_score(𝒰)                              // SS computation
8  │    λ̄ ← λ                           // Initialize adaptive hyperparameters
9  │    λ̄(max_depth) ← λ̄(max_depth) − 1                   // Reduce the depth
10 │    while p < ρ do
11 │    │    if λ̄(max_depth) < 1 then break                     // Check value
12 │    │    ℳ_p ← XGBRegressor(λ̄)          // Model with adaptive hyperparameters
13 │    │    ỹ' ← ℳ_p(𝒟)
14 │    │    𝒰' ← DBSCAN(UMAP(SHAP(ỹ')))                     // Explanation space
15 │    │    𝒮' ← silhouette_score(𝒰')
16 │    │    ∇𝒮 = 𝒮' − 𝒮                                    // SS improvement
17 │    │    ∇𝒰_(−1) = ¬bool(𝒰_(−1)) ∗ Γ                      // Noise cluster
18 │    │    ϵ ← ∇𝒮 + ∇𝒰_(−1)                                        // Loss
19 │    │    if ϵ >= 1e − 3 then                // Check the loss with the threshold
20 │    │    │    ℳ ← ℳ_p                              // Update the best model
21 │    │    │    𝒮 ← 𝒮'                                 // Update the best SS
22 │    │    │    if λ̄(gamma) == 0 then                    // Update gamma
23 │    │    │    │    λ̄(gamma) ← 0.001
24 │    │    │    else
25 │    │    │    │    λ̄(gamma) ← λ̄(gamma) ∗ 10
26 │    │    │    end
27 │    │    │    p ← 0                                 // Reset the patience
28 │    │    else
29 │    │    │    λ̄(max_depth) ← λ̄(max_depth) − 1           // Reduce the depth
30 │    │    │    λ̄(gamma) ← λ(gamma)                 // Reset to the tuned gamma
31 │    │    │    p ← p + 1                            // Increase the patience
32 │    │    end
33 │    end
34 │    ℳ_{SCAL} ← ℳ                         // SCAL model as the best model
35 │    return ℳ_{SCAL}
36 end
```

The adaptive refinement process starts by initializing the hyperparameters, $\bar{\lambda}$, to the current best-fit values, λ, found using AHT. The maximum depth of the model, max_depth, is reduced to create a shallower tree, which helps mitigate overfitting. The refined model is then evaluated by comparing its performance against the previous model using SS improvements, calculated using the intra-inter distances between clusters and noise cluster reduction. This adaptive model refinement process is carried out iteratively, with a predefined number of patience ρ to ensure convergence. We choose ρ as 3 since the model can be adapted via one step of depth reduction and one step of gamma reduction. The Γ is set as 0.01 as we expect the SS improvement should be higher than the greedy noise exploration. If the refined model shows a significant improvement over the previous model, the model is updated, and the regularization hyperparameter, gamma, is increased, which further enhances the model's ability to generalize. Once the patience value is reached, the best model obtained throughout the iterations is set as the final SCAL. This approach allows the model to adapt dynamically to the data, incorporating the insights derived from SHAP clustering to balance bias and variance better. Building Block 3 represents an innovative approach to model refinement, where the adaptive learning process is guided by the insights gained from SHAP clustering. This not only helps reduce the model's bias and prevent overfitting but also ensures that the final model is robust and capable of effectively handling anomalous data patterns.

4   Experimental Setup and Data Set

In our experimental setup, we apply the SCAL framework to the specific use case of predicting buildings' energy consumption. Accurate predictions can aid in the formulation and implementation of efficient energy-saving strategies and policies. Our experiments employ energy consumption data from 36 buildings collected between January 1st, 2019, and June 1st, 2022. The data, recorded in MWh, includes features such as the purpose of use, outside temperature, area, and volume of the buildings. The buildings serve various purposes, including General education, University, Research institutes, and Offices. To ensure data integrity, we preprocess the data by replacing negative consumption values and applying linear interpolation for missing values. We divide the data, setting all data before 2022 as the training set and data from 2022 onwards as the test set. As energy consumption is serialized hourly, this task constitutes a time series prediction. Illustratively, consider a general educational institution (building 1). As shown in Fig. 2, the training and test set distributions exhibit notable differences owing to the increased energy consumption following the COVID-19 pandemic.

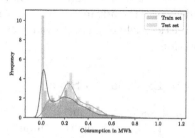

Fig. 2. The distribution of energy consumption of a general educational institution's training and test set (building 1) reflects the significant growth in energy consumption after the COVID-19 pandemic.

Table 1. The quantitative comparison between the AHT and SCAL on the Energy consumption data set. The best results are in bold. We only consider the SS in the training set. The arrow indicates that the lower/higher, the better. Noise cluster presents (3) and does not present (7).

Building ID	Data set	AHT				SCAL			
		$\bar{\mathcal{U}}_{(-1)}$	SS ↑	RMSE ↓	r^2 ↑	$\bar{\mathcal{U}}_{(-1)}$	SS ↑	RMSE ↓	r^2 ↑
1	Training	✗	0.68514	0.05622	0.99684	✓	**0.68721**	0.06783	0.99540
	Test	-	-	0.20527	0.94403	-	-	**0.20144**	**0.94610**
2	Training	✗	0.71530	0.23016	0.94689	✗	**0.72695**	0.49800	0.75136
	Test	-	-	0.51697	0.48786	-	-	**0.50698**	**0.50743**
3	Training	✓	0.77496	0.030791	0.90516	✓	**0.86414**	0.46356	0.78504
	Test	-	-	0.58662	0.62682	-	-	**0.56799**	**0.65014**
4	Training	✓	0.21900	0.05509	0.99697	✓	**0.25013**	0.07266	0.99452
	Test	-	-	0.15288	0.97033	-	-	**0.14883**	**0.97188**

5 Results

5.1 SCAL Performance

We compare SCAL's performance with the traditional AHT (Adaptive Hyperparameter Tuning) to highlight SCAL's superior model adaptation and accuracy. AHT optimizes the model using hyperparameter tuning without considering the explanation space. In contrast, SCAL leverages SHAP clustering characteristics for model adaptation. As demonstrated in Table 1, SCAL outperforms AHT on the test sets, yielding lower RMSE and higher r^2 values, indicating a stronger generalization to unseen data. Even with a slightly worse training set fit, SCAL provides a better SS in the explanation space, illustrating the effective use of SHAP clustering characteristics. Importantly, SCAL offers valuable insights into the explanation space that enhance understanding of model performance and inform further model refinement. Unlike AHT, this additional information increases SCAL's potential for improvement and adaptability. Thus, SCAL's performance underscores the advantage of a model that adapts based on SHAP clustering characteristics, delivering improved generalization and valuable insights for model adaptation.

5.2 Cluster Analysis in Explanation Space

We perform the cluster analysis on their mean SHAP values, which provides insight into their explanation space, to identify similarities between buildings. Figure 3 illustrates five groups of buildings that exhibit high resemblance in their explanations. We choose representative buildings from each group for further analysis; namely, buildings with ID 1, 2, 3, 4 (model improvement cases), and 12 (data error detection case) to represent the five building groups.

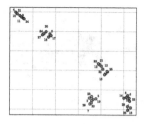

Fig. 3. Five categories of buildings based on their similarity in the SHAP values.

Model Improvement Cases. Figure. 4 presents the SHAP clusters embedding on the training set of buildings 1, 2, 3, and 4 for both AHT and SCAL. The SCAL is expected to show a noise cluster, indicated by index -1, while the AHT model does not (Building 1, Fig. 4a). Additionally, the number of clusters in the SCAL is higher than in the AHT model, implying that the SCAL can learn more rules to explain each data point. This results in improved model performance on the test set, as demonstrated in Table 1. We employ the SS as the primary metric for evaluating the quality of clusters in the explanation space.

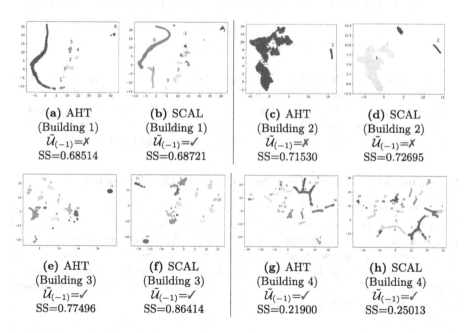

(a) AHT (Building 1) $\tilde{\mathcal{U}}_{(-1)}=\text{✗}$ SS=0.68514

(b) SCAL (Building 1) $\tilde{\mathcal{U}}_{(-1)}=\text{✓}$ SS=0.68721

(c) AHT (Building 2) $\tilde{\mathcal{U}}_{(-1)}=\text{✗}$ SS=0.71530

(d) SCAL (Building 2) $\tilde{\mathcal{U}}_{(-1)}=\text{✗}$ SS=0.72695

(e) AHT (Building 3) $\tilde{\mathcal{U}}_{(-1)}=\text{✓}$ SS=0.77496

(f) SCAL (Building 3) $\tilde{\mathcal{U}}_{(-1)}=\text{✓}$ SS=0.86414

(g) AHT (Building 4) $\tilde{\mathcal{U}}_{(-1)}=\text{✓}$ SS=0.21900

(h) SCAL (Building 4) $\tilde{\mathcal{U}}_{(-1)}=\text{✓}$ SS=0.25013

Fig. 4. SHAP clusters embedding of AHT and SCAL on the four different buildings' training set. Each cluster goes with its index, where -1 indicates the noise cluster. SCAL improves the cluster's quality with a higher SS and the presence of the noise cluster, where the noise cluster presents (✓) and does not present (✗).

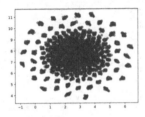

Fig. 5. SHAP clusters of building 12 on the training set. In this case, SCAL can detect the data error where only one cluster appears.

Data Error Detection Case. SHAP clustering for building 12 displayed a single cluster in the explanation space (Fig. 5), indicating the model's reliance on a unique rule to classify all data points and its struggle to interpret the training dataset. Changing model parameters do not affect this clustering, further emphasizing the model's difficulty in obtaining valuable insights. Detailed analysis of the training set unveiled numerous errors, including missing or incorrect entries, likely due to the data's real-world origin from a smart meter device. Thus, SHAP clustering not only aids in diagnosing dataset issues and refining features but also underscores SCAL's role in enhancing model performance.

6   Transferability to Other Use Cases

In this section, we extend the SCAL application beyond its original context by examining its performance in two additional use cases. The first is a classification problem using the Financial Distress dataset, while the second is a regression problem using the Power dataset. These experiments serve to verify SCAL's adaptability and efficacy across varied datasets and problem types.

6.1   Financial Distress Data Set (Classification Problem)

We experiment to test SCAL on an imbalanced multivariate time series classification problem using the Financial Distress data set[1]. The AHT produces a model with `max_depth` $= 9$ and `gamma` $= 0$, while the SCAL adapts `max_depth` $= 8$ and `gamma` $= 0.001$. As a classification problem, Table 2 displays the performance of the AHT and SCAL under accuracy. By observing the SS of the SHAP clusters in the training set's embedding, the SCAL method suggests using a shallower tree, which results in improved performance on the test set.

---

[1] https://www.kaggle.com/datasets/shebrahimi/financial-distress.

Table 2. The quantitative comparison between the AHT and SCAL on the Financial Distress data set. The best results are in bold. We only consider the SS in the training set. The arrow indicates that the lower/higher, the better. Noise cluster presents ($\checkmark$) and does not present ($\times$).

Data set	AHT			SCAL		
	$\tilde{\mathcal{U}}_{(-1)}$	SS ↑	Acc% ↑	$\tilde{\mathcal{U}}_{(-1)}$	SS ↑	Acc% ↑
Training	$\times$	0.70096	100	$\times$	**0.72854**	100
Test	–	–	95.918	–	–	**96.190**

6.2 Power Data Set (Regression Problem)

We evaluate SCAL on a public regression problem, namely the Power data set. This data set measures electric power consumption in one household with a one-minute sampling rate over a period of nearly four years. It shares several similarities in the domain and features with our recorded data set, such as the distribution difference between training and test sets and missing values. The results are shown in Table 3. The hyperparameters with AHT are max_depth = 8 and gamma = 0, while the SCAL adapts max_depth = 7 and gamma = 0.01. The SS indicates that SCAL produces better-quality clusters. In the test set, SCAL outperforms AHT in both RMSE and r^2 metrics, demonstrating its effectiveness on the regression problem.

Table 3. The quantitative comparison between the AHT and SCAL on the Power data set. The best results are in bold. We only consider the SS in the training set. The arrow indicates that the lower/higher, the better. Noise cluster presents ($\checkmark$) and does not present ($\times$).

Data set	AHT				SCAL			
	$\tilde{\mathcal{U}}_{(-1)}$	SS ↑	RMSE ↓	r^2 ↑	$\tilde{\mathcal{U}}_{(-1)}$	SS ↑	RMSE ↓	r^2 ↑
Training	$\checkmark$	0.38326	0.48246	0.82864	$\checkmark$	**0.41241**	0.53600	0.78851
Test	–	–	0.75378	0.64520	–	–	**0.70856**	**0.68649**

7 Conclusion and Future Work

Our paper introduces a method to enhance energy consumption prediction models by adapting to data distribution shifts. It integrates XAI techniques into an adaptive learning framework, using SHAP clustering to create an interpretable explanation space for model refinement. Our approach shows not only flexibility to data shifts but also potential applicability across various domains. Despite increased computational complexity, the robustness and interpretability benefits of our method offer an appealing trade-off. Looking ahead, our research may explore additional XAI techniques and advanced clustering algorithms and consider adapting our method to different ML models. We aim to develop a formula incorporating our three key quality metrics for further optimization and investigate a more self-adaptive learning process.

Acknowledgments. This work was supported by the Federal Ministry of Education and Research through grant 01IS17045 (Software Campus project).

References

1. Clustering – scikit-learn 1.2.1 documentation. https://scikit-learn.org/stable/modules/clustering.html#dbscan. Accessed 28 Feb 2023
2. Amiri, S.S., Mottahedi, S., Lee, E.R., Hoque, S.: Peeking inside the black-box: explainable machine learning applied to household transportation energy consumption. Comput. Environ. Urban Syst. **88**, 101647 (2021)
3. Bigdeli, E., Mohammadi, M., Raahemi, B., Matwin, S.: A fast and noise resilient cluster-based anomaly detection. Pattern Anal. Appl. **20**, 183–199 (2017)
4. Clement, T., Kemmerzell, N., Abdelaal, M., Amberg, M.: XAIR: a systematic metareview of explainable AI (XAI) aligned to the software development process. Mach. Learn. Knowl. Extr. **5**(1), 78–108 (2023)
5. Dave, R.N.: Characterization and detection of noise in clustering. Pattern Recogn. Lett. **12**(11), 657–664 (1991)
6. Durvasula, N., d'Hauteville, V., Hines, K., Dickerson, J.P.: Characterizing anomalies with explainable classifiers. In: NeurIPS 2022 Workshop on Distribution Shifts: Connecting Methods and Applications
7. Gal, Y., Islam, R., Ghahramani, Z.: Deep Bayesian active learning with image data. In: International Conference on Machine Learning, pp. 1183–1192. PMLR (2017)
8. Gautam, S., Höhne, M.M.C., Hansen, S., Jenssen, R., Kampffmeyer, M.: This looks more like that: enhancing self-explaining models by prototypical relevance propagation. Pattern Recogn. **136**, 109172 (2023)
9. Goodfellow, I., Bengio, Y., Courville, A.: Deep learning. MIT Press, Cambridge (2016)
10. Hart, S.: Shapley Value. In: In: Eatwell, J., Milgate, M., Newman, P. (eds.) Game Theory. The New Palgrave, pp. 210–216. Palgrave Macmillan, London (1989). https://doi.org/10.1007/978-1-349-20181-5_25
11. Lundberg, S.M., Lee, S.I.: A unified approach to interpreting model predictions. In: Guyon, I., et al. (eds.) Advances in Neural Information Processing Systems vol. 30, pp. 4765–4774. Curran Associates, Inc. (2017)
12. Nguyen, H.T.T., Cao, H.Q., Nguyen, K.V.T., Pham, N.D.K.: Evaluation of explainable artificial intelligence: shap, lime, and cam. In: Proceedings of the FPT AI Conference, pp. 1–6 (2021)
13. Nguyen, T.T.H., Truong, V.B., Nguyen, V.T.K., Cao, Q.H., Nguyen, Q.K.: Towards trust of explainable AI in thyroid nodule diagnosis. arXiv preprint arXiv:2303.04731 (2023)
14. Shen, Z., Shrestha, S., Howard, D., Feng, T., Hun, D., She, B.: Machine learning-assisted prediction of heat fluxes through thermally anisotropic building envelopes. Build. Environ. **234**, 110157 (2023)

15. Shivaswamy, P., Joachims, T.: Coactive learning. J. Artif. Intell. Res. **53**, 1–40 (2015)
16. Shutaywi, M., Kachouie, N.N.: Silhouette analysis for performance evaluation in machine learning with applications to clustering. Entropy **23**(6), 759 (2021)
17. Skillington, K., Crawford, R.H., Warren-Myers, G., Davidson, K.: A review of existing policy for reducing embodied energy and greenhouse gas emissions of buildings. Energy Policy **168**, 112920 (2022)
18. Still, S., Bialek, W.: How many clusters? An information-theoretic perspective. Neural Comput. **16**(12), 2483–2506 (2004)
19. Victoria, A.H., Maragatham, G.: Automatic tuning of hyperparameters using Bayesian optimization. Evol. Syst. **12**, 217–223 (2021)
20. Weber, L., Lapuschkin, S., Binder, A., Samek, W.: Beyond explaining: opportunities and challenges of XAI-based model improvement. Inf. Fusion **92**, 154–176 (2023). https://doi.org/10.1016/j.inffus.2022.11.013
21. Zambrano, J.M., Oberegger, U.F., Salvalai, G.: Towards integrating occupant behaviour modelling in simulation-aided building design: reasons, challenges and solutions. Energy Build. **253**, 111498 (2021)
22. Zunino, A., Bargal, S.A., Morerio, P., Zhang, J., Sclaroff, S., Murino, V.: Excitation dropout: encouraging plasticity in deep neural networks. Int. J. Comput. Vis. **129**, 1139–1152 (2021)

Concept-Guided Interpretable Federated Learning

Jianan Yang[✉] and Guodong Long

Australian Artificial Intelligence Institute, Faculty of Engineering and Information
Technology, University of Technology Sydney, Ultimo, Australia
Jianan.Yang@student.uts.edu.au, Guodong.Long@uts.edu.au

Abstract. Interpretable federated learning is an emerging challenge to
identify explainable characteristics of each client-specific personalized
model in a federated learning system. This paper proposes a novel federated concept bottleneck (FedCBM) method by introducing human-
friendly concepts for client-wise model interpretation. Specifically, given
a set of pre-defined concepts, all clients will collaboratively train shared
Concept Activation Vectors (CAVs) in federated settings. The shared
concepts will be the information carrier to align client-specific representations, and also be applied to enhance the model's accuracy under a
supervised learning loss. The effectiveness of our method and concept-
level reasoning is demonstrated in our experimental analysis.

Keywords: Federated Learning · Interpretability · Concept Reasoning

1 Introduction

Federated Learning is an innovative machine learning approach that enables
the training of models across multiple decentralized devices while preserving
data privacy and security (FedAvg). As a result, it has been applied in various
domains such as financial [21], healthcare [39] and edge computing [38]. However,
there often needs to be a scenario where the model transparency is crucial.
The obscureness of the model leads to a lack of interpretability, which makes
trustworthiness vulnerable. Interpretable federated learning (IFL) is proposed
to tackle this problem [18].

Interpretability can be generally divided into post-hoc approaches and intrinsic interpretable models. There have been some approaches to achieve this
[2,23,27,43]. Self-explanatory neural networks are trying to train an end-to-end
interpretable neural network and maintain its performance. Some researchers are
building extra human-interpretable concept extractors to help explain the prediction [7,13,14,41]. For example, when classifying the bird to be Black-footed
Albatross, we look for visual patterns of it, such as black feet and a bright pink
bill with a dark tip. These patterns form its relevant concept vector.

Current IFL approaches mainly land in post-hoc domains such as examining
interpretable client selection [17,19] or feature selection [3]. [9] constructs the

T. Liu et al. (Eds.): AI 2023, LNAI 14472, pp. 160–172, 2024.
https://doi.org/10.1007/978-981-99-8391-9_13

decision tree-based model, but it is not neural network based. Therefore, this motivates us to propose a self-explainable neural network solution. Inspired by [14,41], we are trying to propose a conceptual solution in federated learning (FedCBM) in this work. The primitive goal is to train a concept bottleneck model across clients collaboratively. Each client will have its concept dataset and typical image dataset. All clients will train Concept Activation Vector (CAVs) collaboratively using the concept dataset to form a concept bank. These CAVs are hyperplanes to categorize samples with and without the target concept. Besides, we use them to align the embedding vector extracted by the backbone feature extractor in the neural network. We project the embedding vector in the CAVs' direction and use the scalar projection as the concept score. Concept scores form a concept vector and will be sent to the linear classifier for a final prediction.

Contributions: This work proposes FedCBM, a concept bottleneck model-based solution under federated learning settings. We can transform any FL based neural network into a concept-based interpretable model by inserting a bottleneck module. We propose an optimized solution for training a shared SVM for the concept bank with privacy preservation. Through experiments, we show the preservation of the performance and the extra explanations produced.

2   Related Work

2.1   Interpretable Federated Learning

Some works have been integrating existing interpretability modules into federated learning applications [9,28]. [37] analyses the feature importance using Shapley values. [6,20] adopt tree-based models to build intrinsic interpretable federated learning models due to their self-explanatory. Some other works investigate the performance of FL models while preserving the explanations to end users, such as interpretable weight aggregation [25] and feature selections [42]. However, these works are not intending to solve IFL problems. [18] first conduct a survey in interpretable federated learning (IFL) and propose IFL problem formally. Moreover, there haven't been works on interpreting personalized federated learning models [5,31,33,40] and clustered federated learning models [22].

2.2   Concept-Related Interpretability

Working on human-understandable concept analysis has been widely studied [7,13,14,41]. [13] proposes to learn linear classifier to retrieve concept activation vector and compute conceptual sensitivity. [7] proposes a whitening module that could turn any black-box layer in the neural network into an interpretable one. Concept Bottleneck Model [14,29] divides the end-to-end network into a two-stage approach: predict the concept first and the final prediction is only made based on the concept instead of the data features. [24,30] proposes the drawbacks of concept bottleneck-based models, mainly information leakage. [41] proposes an alternative solution where the labeled concept dataset is challenging to acquire. [4,32] learn prototypes as concepts where each prototype knowledge belongs to the exact part of the input image. [11] develops an algorithm to extract visual concepts automatically and utilize the TCAV score as an evaluation metric. [16] attempts to address the limitation of a concept-based approach where concepts defined are not interpretable. A vital assumption is that every individual instance inherits at least one concept. Therefore, their method aims to learn concept features instead of individual instances.

Concept learning [13] can be confused with prototype learning [32], but they differ in understanding and categorizing data. Concepts is a general description or rule that separates instances of one class from instances of other classes while prototypes typical or representative examples that capture the essence of each class. For example, the decision boundary of the binary classification problem is a concept hyperplane to separate two classes while prototypical learning is looking for a specific sample from each class.

3 Problem Settings

3.1 Federated Learning

In typical federated learning settings, each client k holds a local dataset $D_k = \{x_i, y_i\}_{i=1}^{N_k}$. The objective of typical federated learning problem is to minimize a global loss function $L(w)$ over all clients.

$$L(w) = \sum_{k=1}^{K} \frac{N_k}{N} L_k(f_i(w; D_k)) \tag{1}$$

where N_k refers to the size of client k's dataset, N is the total number of samples over all clients, w are parameters of the model and L_k is the loss function of client k.

The training will end up in a couple of rounds for communication. At each round, a subset of clients will be chosen as the candidates for training. Each selected client will perform local computation using parameters received from the server and dataset locally. After the computation is accomplished, synchronization will be made, and all the updates will be sent back to the server.

3.2 Concept Bottleneck Model

A Concept Bottleneck Model (CBM) is consisted of a concept learner $h(x) : R^X -> R^{N_c}$ and a linear classifier $f(x) : R^{N_c} -> R^Y$. Given an input image x, we will map it into the concept space to get it concept vector $c \in R^{N_k}$. Each entry c_i stands for the concept score which decides the relevance between the concept i and the input x. The classifier depends on the concept vector only and it produces the prediction by $f(h(x)) = w^T h(x) + b$. The weights of the classifier measures the contribution of the concept.

4   Proposed Method

We aim to train a shared concept-based solution for meeting interpretability and performance criteria. We separate our approach into two stages. We will assign each client concept dataset and general dataset. In the first stage, all clients will train a shared concept bank from their local concept dataset preserving data privacy. The concept activation vector stored at the concept bank will be used to align the concept vector in the embedding space. Then each client trains their own classifier head by projecting the embedding representation onto each concept activation vector to achieve a concept score. These score forms a finalized concept vector and will be sent to a linear layer for classification. We can integrate the concept bottleneck module into any neural network to improve interpretability (Fig. 1).

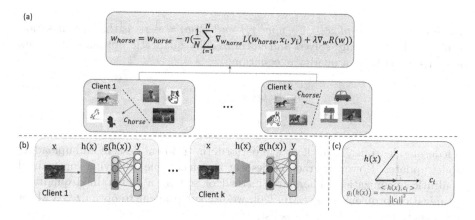

Fig. 1. (a). The first stage is to train a shared concept bank on local concept datasets. We use SVM to train Concept Activation Vectors (CAV). Therefore, for each concept, we train a hyperplane to separate examples with and without that concept. **(b).** The second stage: Each client trains a personalized prediction head on the regular dataset. Given an input image, they will get its embedding representation $h(x) \in R^d$. Afterward, the representation will be projected for achieving a concept vector $g(h(x)) \in R^{Nd}$. Finally, they will send the concept vector into a linear layer to get the prediction. They will not share either weights or data. **(c).** The projection of the embedding representation onto the specific CAV.

4.1   Concept Bank

Before training CBMs across all clients under standard federated learning settings, we prepare to create a concept bank for each client to store the shared knowledge. The concept bank stores concept knowledge and the knowledge should be capable of separating samples with or without the target concept.

They will be presented as concept vectors. In this work, we choose SVM as the learner model and it generates a hyperplane for each concept. The concept activation vector will then be the normal to the hyperplane. We store all concept activation vectors in concept banks and broadcast to all clients once the training is finished.

Concept bank adopts knowledge from a Concept Dataset C. For each concept c_i, we will choose samples containing the concept and those not, labeled as x_+ and x_-, respectively. These samples will differ from the training or testing samples used in the neural network.

In traditional SVM, the goal is to find the maximum-margin hyperplane that separates two classes. The loss function of SVM is

$$J(x) = \frac{1}{N} \sum_{i=1}^{N} L(w, x_i, y_i) + \lambda * R(w) \tag{2}$$

where $L(w, x_i, y_i) = max\{0, 1 - y_i w^T x_i\}$ and $R(w) = \frac{||w||^2}{2}$ is the regularization term. SVM are usually solved and optimized by quadratic optimization as its objective function is a quadratic equation:

$$min. \ Q(w) = \frac{1}{2} ||w||^2$$
$$w.r.t \ y_i(w^T x + b) \geq 1$$

In federated learning, each client will perform the gradient descent and upload its model updates. However, gradient descent is not capable of solving constrained quadratic optimization. Therefore, instead of solving a constrained quadratic optimization problem using SMO [26], we use subgradient to optimize SVM for each client. The full algorithm is listed in 1.

Following the procedure of federated learning, the server will spread model weights w to all clients. Each client perform local update and compute $\nabla_w L(w, x_i, y_i)$ using Eq. 3.

$$\frac{\partial(max\{0, 1 - y_i w^T x_i\})}{\partial w} = \begin{cases} 0 & (1 - y_i w^T x_i >= 1) \\ -y_i * x_i & (1 - y_i w^T x_i) < 1) \end{cases} \tag{3}$$

Once the server has collected all the updates from clients, it will aggregate them and perform the gradient descent step 4.

$$\frac{\partial L}{\partial w} = \frac{\partial(R(w))}{\partial w} + \frac{1}{N} \sum_{i=1}^{N} \frac{\partial(L_i(w, x_i, y_i))}{\partial w} \tag{4}$$

Algorithm 1. Concept bank training

1: **Server**
2: Initialize the concept vectors m
3: **for** each concept $c = 1, 2, \ldots N_c$ **do**
4:     $\nabla_{w_c} L(w_c) \leftarrow 0$
5:     **for** each client k=1,2,...K **do**
6:         **for** sample(x_i, y_i) D_k **do**
7:             $\nabla_{w_c} L(w_c) \leftarrow \nabla_{w_c} L(w_c) +$ ClientUpdate(c, x_i, y_i)
8:         **end for**
9:     **end for**
10:     $\nabla_{w_c} L(w_c) \leftarrow \frac{1}{N} \nabla_{w_c} L(w_c)$
11:     $\mathbf{w_c} \leftarrow \mathbf{w_c} - \eta(\nabla_{w_c} L(w_c) + \nabla_{w_c} R(w_c))$
12: **end for**

13: **ClientUpdate(c, x_i, y_i)**
14: Initialise $\mathbf{w}_0 = 0$
15: **for** each concept i=1,2,...N_c **do**
16:     Shuffle the training set
17:     **for** each training sample $(x_i, y_i) \in D_k$ **do**
18:         **if** $y_i w^T \mathbf{x}_i < 1$ **then**
19:             $\nabla_w L(w, x_i, y_i) \leftarrow -y_i * x_i$
20:         **else**
21:             $\nabla_w L(w, x_i, y_i) \leftarrow 0$
22:         **end if**
23:     **end for**
24: **end for**
25: Return $\nabla_w L(w)$

4.2 Linear Predictor

We will start training on the concept bottleneck model once we receive all concepts' federated concept activation vectors. For each input x, we pass it through a backbone feature extraction network $h(x; w) \in R^d$ and stop before the neuron activation function. In order to get a concept sensitivity score, we will align the concept vector $h(x; w)$ with the concept activation vector c_i. c_i has already been trained among all clients, and it points out a possible direction of the hyperplane, which classifies the concept i. The scalar projection $g(h(x; w)) = \frac{<h(x;w), c_i>}{||c_i||^2}$ will give us the concept sensitivity of the target concept. Finally, we will feed the concept vector into a classifier head to get the classification $f(g(h(x; w)); \theta, b) = \theta^T g(h(x; w)) + b$.

4.3 Training Algorithm

Given input x and its embedding representation $\mathcal{X}$, we set $h : R^{|\mathcal{X}|} - > R^d$ be the linear classifier of the concept, $g : R^d - > R^{N_c}$ be the concept vector projection and $f : R^{N_c} - > R^{|\mathcal{Y}|}$ be the linear predictor. The prediction is given by $y^{pred} = f(g(h(x)))$.

We will separate the concept bank and the predictor training. In the first stage, each client will prepare a concept dataset to train the shared CAVs. Concept datasets separate samples into concept classes. Full details of concept datasets are explained in 5.1. We use SVM to train concept vectors collaboratively following steps in Algorithm 1 and store the trained vectors in the shared concept bank. The objective function will be

$$argmin_{w} J = \lambda||w||^2 + \sum_{k=1}^{K} \frac{|D_k|}{N} L_k(w_k) \tag{5}$$

where $L_k(w_k)$ is equivalent to Eq. 2.

For the classifier head, we will use gradient descent and optimize the following objective of client k over typical datasets such as CIFAR-10 [15]:

$$argmin_{\{w_k\}} \sum_{k=1}^{K} \frac{|D_k|}{N} L_s(w_k) \tag{6}$$

5 Experiment

We evaluate our model mainly on classification tasks across different fields. We introduce and explain the experimental settings and further demonstrates our studies based on the performance. Full details of datasets and experiment settings are included in the supplementary file.

5.1 Datasets

Concept Datasets. We train our concept banks based on the following datasets:

- **CUB** [36]. It contains 312 concepts, and each class may fall into multiple concepts. Following the filter process by [14], we keep only concepts present in at least 10 classes and finally extract 112 out of 312 concepts. We use ResNet18 as the backbone model for extracting the embedding representation.
- **Broden** [10]. It contains pixel-level segmentation figures of total 1197 low and high-level concepts. Adopting the same settings as [1], we choose 170 out of 1197 concepts capable of CIFAR10 and CIFAR100. The full list is in the appendix.
- **Derm7pt** [12]. There are 1011 dermoscopic images with 8 clinical concepts that contribute to the diagnosis: Typical Pigment Networks (TPN), Atypical Pigment Networks (APN), Regular Streaks (RST), Irregular Streaks (IST), Regression Structures (RS), blue-whitish veils (BWV), Regular Dots and Globules (RDG) and Irregular Dots and Globules (IDG). These concepts serve for lesion classification. We use Inception as the backbone model.

Table 1. Concept Datasets

Dataset	Training samples	Test samples	Backbone
CUB [36]	4796	5794	ResNet18
Broden [10]	24600	6155	ResNet50
Derm7pt [12]	414	396	Inception_v3

Regular Datasets. We train the end tasks on the following datasets:

- **CUB** [36]. This dataset contains 200 classes of birds with 11788 images under various sizes. In order to be consistent, we resize each image to the size of 224×224. We minimize the cross-entropy loss for each class.
- **CIFAR-10,CIFAR-100** [15]. It contains 60000 images in 10 and 100 classes, respectively. We split them into 50,000 training examples and 10,000 testing examples. We use ResNet50 as the backbone model.
- **HAM10000** [35]. It is a dermoscopic dataset that contains skin lesions. We work on this dataset to determine whether the given lesion is benign or malignant. Therefore we minimize the binary cross-entropy loss (Table 2).

Table 2. Regular Datasets

Dataset	Training samples	Test samples	Loss
CUB [36]	9430	2358	Cross Entropy
CIFAR-10 [15]	50000	10000	Cross Entropy
CIFAR-100 [15]	50000	10000	Cross Entropy
HAM10000 [12]	9000	1000	Binary Cross-entropy

5.2 Performance Analysis

In this section, we test for the concept bank's accuracy and the end tasks' accuracy. The concept accuracy measures how accurate the test concept vectors are aligned with pre-trained concept vectors. The task accuracy evaluate how accurate the model performing end tasks such as prediction. We follow the same training settings for all experiments: We distribute samples for 5 clients uniformly. We tune the model for different regularization parameters $C = [0.01, 0.1, 1.0]$. Table 3 compares the performance of the original model with FedCBM. We train the original model without any interpretable modules by feeding forward the embedding representation directly to the linear layer.

The result demonstrates a performance gap in three datasets. In our approach, we send the concept vector into the predictor's head only. This leads to a dependency between the concept bank's quality and the prediction's quality. [34] introduces information theory where there is a trade-off between accuracy

Table 3. Performance(task accuracy compared with original model)

Model	CIFAR-10	CIFAR-100	CUB	HAM10000
FedAvg	81.03	60.32	57.43	89.12
FedCBM	83.32	57.94	55.93	92.92

and compression. For typical neural network, we are encoding the input image X into a latent representation T and then decompressing label information Y from T. The trade-off becomes its information bottleneck. Similarly for concept bottleneck, it adds an extra compression of projecting latent features onto the concept space. Extra compression generates an extra information loss and that's the cause of performance gap.

The Relationship Between the Concept and Task Accuracy. We take a further look at the proportion between task and concept accuracy. In our model, the prediction entirely depends on the concept vector, where each entry of the concept vector is the scalar projection of the embedding representation onto the concept direction. Therefore, high concept accuracy should lead to high task accuracy. We plot their relationship in Fig. 2. We test for 100 epochs, where at each epoch, we collect the trained concept bank and train a predictor head based on it. From the figure, we can observe that the end task performance is proportional to the quality of the concept vectors. Meanwhile, for the HAM10000 dataset, we can observe that we are achieving higher task accuracy than concept accuracy. In Table 1, **Derm7pt** dataset is insufficient to expand knowledge in the concept space due to the lack of training samples. This also limits our work: we will cost extra resources to collect a high-quality concept dataset. Table 4 shows the exact concept accuracy for three concept datasets. High concept accuracy implies the alignment of the learned and accurate concepts.

Table 4. Performance (concept accuracy)

Model	Broden	CUB	Derm7pt
FedCBM	90.92	83.13	70.88

5.3   Reasoning Process

In this section, we provide intrinsic explanations of samples. In our model, the prediction is made by a fully connected layer where the weights of label i measure the importance weight of that concept. We take dermoscopic prediction as an example. We use 8 concepts as our concept bank and perform the binary classification task of diagnosing dermoscopic lesions. In our model, clients share the concept bank but keep the classifier head private. The prediction is made locally and therefore, we choose a random client's classifier as an example.

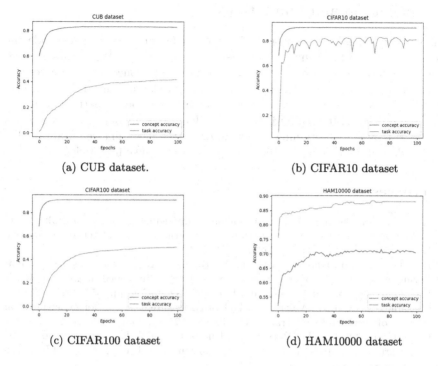

Fig. 2. Concept accuracy versus Task accuracy over 4 datasets.

(a) CUB dataset.

(b) CIFAR10 dataset

(c) CIFAR100 dataset

(d) HAM10000 dataset

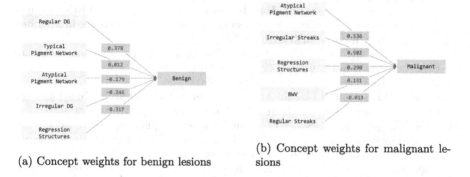

(a) Concept weights for benign lesions

(b) Concept weights for malignant lesions

Fig. 3. The explanation made by taking a deep look at concept weights.

Figure 3 illustrates the top-5 weights for two classes. We can notice that relevant concepts will achieve a higher contribution while irrelevant concepts will contribute less or even negatively. From the domain knowledge, benign lesions will have a regular structure and are symmetrical in pattern. Malignant tumors often have disordered structures and pattern asymmetry [8].

6 Conclusion and Limitations

This work proposes a concept-guided federated learning method to tackle the interpretability problem under federated learning settings. We propose a two-stage procedure to achieve this: We first train shared concept banks across all clients, and then each client trains a personalized classifier head. From experiments, We have shown the effectiveness of our method in performance and reasoning processes. In future work, We can deploy our model on heterogeneous data sources while maintaining the same or better performance. Additionally, we can address the case where the concept dataset is insufficient rich.

References

1. Abid, A., Yuksekgonul, M., Zou, J.: Meaningfully debugging model mistakes using conceptual counterfactual explanations. In: International Conference on Machine Learning, pp. 66–88. PMLR (2022)
2. Bau, D., Zhou, B., Khosla, A., Oliva, A., Torralba, A.: Network dissection: quantifying interpretability of deep visual representations. In: Proceedings of the IEEE Conference on Computer Vision and Pattern Recognition, pp. 6541–6549 (2017)
3. Cassara, P., Gotta, A., Valerio, L.: Federated feature selection for cyber-physical systems of systems. IEEE Trans. Veh. Technol. **71**(9), 9937–9950 (2022)
4. Chen, C., Li, O., Tao, D., Barnett, A., Rudin, C., Su, J. K.: This looks like that: deep learning for interpretable image recognition. In: Advances in Neural Information Processing Systems, vol. 32 (2019)
5. Chen, F., Long, G., Wu, Z., Zhou, T., Jiang, J.: Personalized federated learning with graph. arXiv preprint arXiv:2203.00829 (2022)
6. Chen, X., Zhou, S., Yang, K., Fan, H., Wang, H., Wang, Y.: Fed-EINI: an efficient and interpretable inference framework for decision tree ensembles in federated learning. arXiv preprint arXiv:2105.09540 (2021)
7. Chen, Z., Bei, Y., Rudin, C.: Concept whitening for interpretable image recognition. Nat. Mach. Intell. **2**(12), 772–782 (2020)
8. DermNet. Dermoscopy pattern analysis (2008)
9. Dong, T., Li, S., Qiu, H., Lu, J.: An interpretable federated learning-based network intrusion detection framework. arXiv preprint arXiv:2201.03134 (2022)
10. Fong, R., Vedaldi, A. Net2vec: quantifying and explaining how concepts are encoded by filters in deep neural networks. In: Proceedings of the IEEE Conference on Computer Vision and Pattern Recognition, pp. 8730–8738 (2018)
11. Ghorbani, A., Wexler, J., Zou, J.Y., Kim, B.: Towards automatic concept-based explanations. In: Advances in Neural Information Processing Systems, vol. 32 (2019)
12. Kawahara, J., Daneshvar, S., Argenziano, G., Hamarneh, G.: Seven-point checklist and skin lesion classification using multitask multimodal neural nets. IEEE J. Biomed. Health Inform. **23**(2), 538–546 (2018)
13. Kim, B., Wattenberg, M., Gilmer, J., Cai, C., Wexler, J., Viegas, F.: Interpretability beyond feature attribution: quantitative testing with concept activation vectors (TCAV). In: International Conference on Machine Learning, pp. 2668–2677. PMLR (2018)
14. Koh, P.W., et al.: Concept bottleneck models. In: International Conference on Machine Learning, pp. 5338–5348. PMLR (2020)

15. Krizhevsky, A., Hinton, G.: Learning multiple layers of features from tiny images (2009)
16. Lage, I., Doshi-Velez, F.: Learning interpretable concept-based models with human feedback. arXiv preprint arXiv:2012.02898 (2020)
17. Lai, F., Zhu, X., Madhyastha, H.V., Chowdhury, M.: Oort: efficient federated learning via guided participant selection. In: 15th {USENIX} Symposium on Operating Systems Design and Implementation ({OSDI} 21), pp. 19–35 (2021)
18. Li, A., Liu, R., Hu, M., Tuan, L.A., Yu, H.: Towards interpretable federated learning. arXiv preprint arXiv:2302.13473 2023
19. Li, A., et al.: Efficient federated-learning model debugging. In: 2021 IEEE 37th International Conference on Data Engineering (ICDE), pp. 372–383. IEEE (2021)
20. Liu, Y., et al.: Federated forest. IEEE Trans. Big Data **8**(3), 843–854 (2020)
21. Long, G., Tan, Y., Jiang, J., Zhang, C.: Federated learning for open banking. In: Yang, Q., Fan, L., Yu, H. (eds.) Federated Learning. LNCS (LNAI), vol. 12500, pp. 240–254. Springer, Cham (2020). https://doi.org/10.1007/978-3-030-63076-8_17
22. Long, G., Xie, M., Shen, T., Zhou, T., Wang, X., Jiang, J.: Multi-center federated learning: clients clustering for better personalization. World Wide Web **26**(1), 481–500 (2023)
23. Lundberg, S.M., Lee, S.I.: A unified approach to interpreting model predictions. In: Advances in Neural Information Processing Systems, vol. 30 (2017)
24. Margeloiu, A., Ashman, M., Bhatt, U., Chen, Y., Jamnik, M., Weller, A.: Do concept bottleneck models learn as intended? arXiv preprint arXiv:2105.04289 (2021)
25. Pandey, S.R., Tran, N.H., Bennis, M., Tun, Y.K., Manzoor, A., Hong, C.S.: A crowdsourcing framework for on-device federated learning. IEEE Trans. Wirel. Commun. **19**(5), 3241–3256 (2020)
26. Platt, J.: Sequential minimal optimization: a fast algorithm for training support vector machines (1998)
27. Ribeiro, M.T., Singh, S., Guestrin, C.: "Why should i trust you?" explaining the predictions of any classifier. In: Proceedings of the 22nd ACM SIGKDD International Conference on Knowledge Discovery and Data Mining, pp. 1135–1144 (2016)
28. Roschewitz, D., Hartley, M.A., Corinzia, L., Jaggi, M.: IFedAvg: interpretable data-interoperability for federated learning. arXiv preprint arXiv:2107.06580 (2021)
29. Sawada, Y., Nakamura, K.: Concept bottleneck model with additional unsupervised concepts. IEEE Access **10**, 41758–41765 (2022)
30. Shin, S., Jo, Y., Ahn, S., Lee, N.: A closer look at the intervention procedure of concept bottleneck models. arXiv preprint arXiv:2302.14260 (2023)
31. Tan, Y., Liu, Y., Long, G., Jiang, J., Qinghua, L., Zhang, C.: Federated learning on non-IID graphs via structural knowledge sharing. In: Proceedings of the AAAI Conference on Artificial Intelligence, vol. 37, pp. 9953–9961 (2023)
32. Tan, Y., Guodong Long, L., Liu, T.Z., Qinghua, L., Jiang, J., Zhang, C.: FedProto: federated prototype learning across heterogeneous clients. In: Proceedings of the AAAI Conference on Artificial Intelligence, vol. 36, pp. 8432–8440 (2022)
33. Tan, Y., Long, G., Jie Ma, L., Liu, T.Z., Jiang, J.: Federated learning from pre-trained models: a contrastive learning approach. In: Advances in Neural Information Processing Systems, vol. 35, pp. 19332–19344 (2022)
34. Tishby, N., Zaslavsky, N.: Deep learning and the information bottleneck principle. In: 2015 IEEE Information Theory Workshop (ITW), pp. 1–5. IEEE (2015)
35. Tschandl, P., Rosendahl, C., Kittler, H.: The ham10000 dataset, a large collection of multi-source dermatoscopic images of common pigmented skin lesions. Sci. data **5**(1), 1–9 (2018)

36. Wah, C., Branson, S., Welinder, P., Perona, P., Belongie, S.: The Caltech-UCSD birds-200-2011 dataset (2011)
37. Wang, G.: Interpret federated learning with Shapley values. arXiv preprint arXiv:1905.04519 (2019)
38. Wang, S., et al.: Adaptive federated learning in resource constrained edge computing systems. IEEE J. Sel. Areas Commun. **37**(6), 1205–1221 (2019)
39. Xu, J., Glicksberg, B.S., Su, C., Walker, P., Bian, J., Wang, F.: Federated learning for healthcare informatics. J. Healthc. Inform. Res. **5**, 1–19 (2021)
40. Yan, P., Long, G.: Personalization disentanglement for federated learning. arXiv preprint arXiv:2306.03570 (2023)
41. Yuksekgonul, M., Wang, M., Zou, J.: Post-hoc concept bottleneck models. arXiv preprint arXiv:2205.15480 (2022)
42. Zhang, X., Mavromatics, A., Vafeas, A., Nejabati, R., Simeonidou, D.: Federated feature selection for horizontal federated learning in IoT networks. IEEE Internet Things J. (2023)
43. Zhou, B., Sun, Y., Bau, D., Torralba, A.: Interpretable basis decomposition for visual explanation. In: Proceedings of the European Conference on Computer Vision (ECCV), pp. 119–134 (2018)

Systematic Analysis of the Impact of Label Noise Correction on ML Fairness

Inês Oliveira e Silva[1]([✉]), Carlos Soares[1,2,3], Inês Sousa[3], and Rayid Ghani[4]

[1] Faculdade de Engenharia da Universidade do Porto, Porto, Portugal
up201806385@edu.fe.up.pt
[2] Laboratory for Artificial Intelligence and Computer Science (LIACC), Porto, Portugal
[3] Fraunhofer Portugal AICOS, Porto, Portugal
[4] Machine Learning Department, Heinz College of Information Systems and Public Policy, Carnegie Mellon University, Pittsburg, PA, USA

Abstract. Arbitrary, inconsistent, or faulty decision-making raises serious concerns, and preventing unfair models is an increasingly important challenge in Machine Learning. Data often reflect past discriminatory behavior, and models trained on such data may reflect bias on sensitive attributes, such as gender, race, or age. One approach to developing fair models is to preprocess the training data to remove the underlying biases while preserving the relevant information, for example, by correcting biased labels. While multiple label noise correction methods are available, the information about their behavior in identifying discrimination is very limited. In this work, we develop an empirical methodology to systematically evaluate the effectiveness of label noise correction techniques in ensuring the fairness of models trained on biased datasets. Our methodology involves manipulating the amount of label noise and can be used with fairness benchmarks but also with standard ML datasets. We apply the methodology to analyze six label noise correction methods according to several fairness metrics on standard OpenML datasets. Our results suggest that the Hybrid Label Noise Correction [20] method achieves the best trade-off between predictive performance and fairness. Clustering-Based Correction [14] can reduce discrimination the most, however, at the cost of lower predictive performance.

Keywords: Label noise correction · ML fairness · Bias mitigation · Semi-synthetic data

This work was partly funded by: Agenda "Center for Responsible AI", nr. C645008882-00000055, investment project nr. 62, financed by the Recovery and Resilience Plan (PRR) and by European Union - NextGeneration EU.; AISym4Med (101095387) supported by Horizon Europe Cluster 1: Health, ConnectedHealth (n.o – 46858), supported by Competitiveness and Internationalisation Operational Programme (POCI) and Lisbon Regional Operational Programme (LISBOA 2020), under the PORTUGAL 2020 Partnership Agreement, through the European Regional Development Fund (ERDF); and Base Funding - UIDB/00027/2020 of the Artificial Intelligence and Computer Science Laboratory - LIACC - funded by national funds through the FCT/MCTES (PIDDAC).

T. Liu et al. (Eds.): AI 2023, LNAI 14472, pp. 173–184, 2024.
https://doi.org/10.1007/978-981-99-8391-9_14

1   Introduction

The widespread use of ML systems in sensitive environments has a profound impact on people's lives when given the power to make life-changing decisions [13]. One well-known example is the Correctional Offender Management Profiling for Alternative Sanctions (COMPAS) software. This computer program assesses the recidivism risk of individuals and is used by the American courts to decide whether a person should be released from prison. In a 2016 investigation conducted by ProPublica[1], it was discovered that the system was biased against African-Americans, incorrectly classifying Black offenders as "high-risk" twice as often as White offenders. Another example relates to a less impactful yet more widely present tool in people's lives: Google's targeted ads. A group of researchers proposed AdFisher [5], a tool to gather insights on how user behaviors, Google's transparency tool "Ad Settings", and the presented advertisements interact. Their study revealed that male web users were more likely to be presented with ads for high-paying jobs than their female counterparts. In this context, we can classify an algorithm as unfair if its decisions reflect some kind of discrimination or preference towards certain groups of people based on their inherent or acquired characteristics [13].

The process of learning which factors are relevant to the desired outcome in these tasks involves generalizing from historical examples, which can lead to algorithms being vulnerable to the same biases that people projected in their past decisions. The goal of *fair machine learning* is to identify and mitigate these harmful and unacceptable inequalities [1].

One of the main sources of bias is label noise. Although several methods have been proposed to correct label noise, empirical evaluation is limited. In particular, existing empirical studies make assumptions about the noise in the data. In this work, we propose a method for systematic data manipulation method for a more extensive evaluation of label correction methods.

To illustrate the use of the proposed method, we test and compare the effectiveness of six label noise correction techniques in improving the generated model performance. We apply our methodology using multiple standard ML datasets available on OpenML and inject different types of label noise at varying rates. The models are evaluated using four well-known fairness metrics.

2   Related Work

There are two main sources of bias in data, namely label noise and data selection [7]. Given that, in this project, we focus on label noise methods, this section discusses label noise. Label noise correcting methods are summarized in the experiments (Sect. 4).

When collecting data, prejudice will lead to incorrect labels, as the relationship between an instance's features and its class will be biased. Despite the vast

---

[1] https://www.propublica.org/article/machine-bias-risk-assessments-in-criminal-sentencing.

amount of literature on methods for dealing with noisy data, only a few of these studies focus on identifying and correcting noisy labels [14]. This approach of correcting wrongly attributed labels can be leveraged in the context of fair machine learning if we consider discrimination present in the data as noise that can be removed. As such, noise correction techniques can be applied to obtain a feasibly unbiased dataset that can be used to train fair models. Thus, the motivation for this work comes from, to the best of our knowledge, the lack of work exploring the use of label noise correction techniques in training fair models from biased data.

One approach to achieve fair classification is to focus on re-weighting the training data to alter its distribution in a way that corrects for the noise process that causes the bias [12]. A different line of work focuses on enforcing fairness constraints on the learning process to achieve fair predictions. Research has also been conducted in adapting this approach for learning fair classifiers in the presence of label noise [18,19]. Some authors rewrite the loss function and fairness constraints to deal with label noise [19]. They further propose to model label noise and fairness simultaneously by uncovering the internal causal structure of the data. Surrogate loss functions and surrogate constraints have also been devised to ensure fairness in the presence of label noise [18].

The performance of label noise correction methods depends on the level of noise in the data. They are expected to improve fairness by correcting possible biases. For practitioners to apply those methods safely in the real world, it is important to understand their behavior under different noise conditions. However, there is currently a lack of research in understanding how those techniques affect the fairness of models. To address this limitation, a sensitivity analysis framework for fairness has been developed [9]. It assesses whether the conclusions about the fairness of a model derived from biased data are reliable. However, this approach still relies on a limited set of fairness benchmarks, limiting the scope of the conclusions since the existing datasets are not representative of many different types and levels of label noise.

3   Methodology

With the objective of understanding the effect of existing label correction methods on improving the fairness of machine learning classifiers trained on the corresponding corrected data, we propose a methodology for empirically evaluating the efficacy of such techniques in achieving this goal.

Having the *original* dataset, D_o, in which we assume the instances to have correctly assigned labels, we first arbitrarily choose the positive class and a binary attribute to be considered as the sensitive one. Given a noise rate τ, noise is injected in the labels of instances with a certain probability depending on the noise rate and whether it belongs to the protected group. By parameterizing this process, we can simulate different types of discrimination. We thus obtain a *noisy* dataset, D_n, that is corrupted by the induced bias.

To simulate different types of biases, we inject group-dependant label noise in the clean datasets in two ways:

- **Asymmetrical Bias** noise, which simulates the cases where the instances belonging to the protected group are more likely to be given a positive label. For example, this would emulate the situation of classifying African-American offenders as having a high risk of re-offending at a higher rate than their Caucasian counterparts. This is achieved by setting the label of each instance belonging to the protected group to the positive class with a probability equal to the desired noise rate;
- **Symmetrical Bias** noise, which emulates situations where simultaneously the members of the protected group are benefited, and the non-members are harmed. This is the case of a biased automated selection of job applicants, for example. If the selection process is biased towards preferring male applicants, there will be simultaneously more men being selected and more women being rejected. To simulate such circumstances, the label of each instance is set to the positive class if it belongs in the protected group or to the negative class otherwise, with a probability equal to the desired noise rate.

The next step is to apply the label noise correction method being analyzed, obtaining a *corrected* training set, D_c. We first examine the similarity between the original labels and the ones obtained after noise correction. Given a dataset with N instances, the ability to reconstruct the original labels is measured as the similarity between the *original* labels and the *corrected* ones, as shown in Eq. 1. Essentially, this is a measure of the accuracy of the label correction method in obtaining the original labels. However, to avoid confusion, we will refer to it as *reconstruction score*, r.

$$r = \frac{\sum_{i=1}^{N} \hat{y}_i = y_i}{N} \tag{1}$$

For each training set (D_o^{train}, D_n^{train}, and D_c^{train}), we then apply the chosen ML algorithm to it, obtaining the classifiers M_o, M_n, and M_c, respectively. These models are then evaluated under different scenarios.

Firstly, we consider the testing scenario where we only have access to corrupted data both for training and testing, evaluating the *corrected* (M_c) and *noisy* (M_n) models on the *noisy* test set, D_n^{test}. The aim is to understand the effect of correcting biased training data when discrimination still exists at testing time. An adequate noise correction would result in less discriminatory predictions without significant loss in predictive performance.

Our next objective is to understand the effect of correcting biased training data when the discrimination has been eliminated in the meantime and the testing data is unbiased. To achieve this, the models (M_o, M_n, and M_c) are evaluated on the *original* test set D_o^{test}.

Finally, we extend the previous scenario to remove the assumption that the original data is unbiased. In other words, we analyze the effect of correcting training data when the discrimination has been eliminated in the meantime but the original data was already biased and, thus, its labels are noisy. To achieve this, the *corrected* model, M_c, is evaluated on a test set with labels without noise. However, since we do not have access to the clean labels, we use a label noise correction method to correct the test data as well. We employ the same method that is being analyzed, but a more extensive empirical validation could use different methods or a combination of them. In any case, the results should be interpreted carefully, as the unbiased labels cannot be determined.

When working with fairness benchmark datasets, the same methodology can be applied, but the noise injection step is skipped, and we only have two datasets: an *originally biased* and a *corrected* one.

The diagram presented in Fig. 1 illustrates the explained methodology.

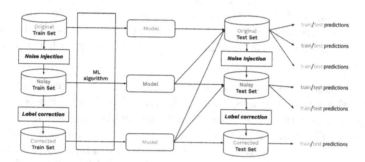

Fig. 1. Diagram of the proposed methodology for the systematic analysis of the impact of label noise correction on ML Fairness.

4   Experiments

To illustrate the use of the proposed methodology, we perform an empirical evaluation of six label noise correction methods. We focused on methods that are not based on deep learning approaches (e.g. [15]) for computational efficiency reasons, given the large number of experiments carried out. However, the methodology is independent of the label noise correction method.

In this section, we present the considered label noise correction methods, describe the key aspects of the experimental setup, and analyze the obtained results. The code implementing the proposed methodology is available at https://github.com/reluzita/fair-lnc-evaluation.

Label Noise Correction Methods. The **Bayesian Entropy Noise Correction** (BE) [16] method utilizes multiple Bayesian classifiers trained on different subsets of the data to derive a probability distribution for each sample, indicating the likelihood of it belonging to each considered class. This probability distribution is then used to calculate the instance's information entropy. Mislabeled samples are corrected if their entropy falls below the defined threshold.

Polishing Labels. (PL) [14] is a method that replaces the label of each instance with the most frequent label predicted by a set of models obtained with different training samples.

The **Self-Training Correction** (STC) [14] algorithm first divides the data into a noisy and a clean set using a noise-filtering algorithm. These methods identify and remove noisy instances from data; in this case, the Classification Filter [17] is used. A model is obtained from the clean set and is used to estimate the confidence that each instance in the noisy set is correctly labeled. The most likely mislabeled instance is relabeled and added to the clean set. These steps are repeated until the desired proportion of labels is corrected.

Clustering-Based Correction. (CC) [14] involves clustering the data multiple times with a varying number of clusters, calculating a set of weights for each cluster based on its label distribution and size. These weights are meant to benefit the most frequent class in the cluster. The weights obtained from each clustering are added up for each instance, and the label with the maximum weight is chosen.

Ordering-Based Label Noise Correction. (OBNC) [8] leverages an ensemble classifier trained from the noisy data. The ensemble decides labels through a voting scheme, and these votes are used to calculate an ensemble margin for each instance. The samples that were misclassified by the ensemble are ordered based on their margin, and the most likely mislabeled samples are relabeled.

Finally, the **Hybrid Label Noise Correction** (HLNC) [20] method starts by separating high and low-confidence samples based on their clustering. The high-confidence samples are used to simultaneously train two very different models, using the SSK-means [2] and Co-training [3] algorithms. These are iteratively applied to each low-confidence sample, and if both models agree on its label, it is relabeled and set as high-confidence.

Datasets and Algorithm. The datasets used in the noise injection experiments are available on OpenML[2] and are summarized in Table 1.

The learning algorithm used was Logistic Regression due to its simplicity and popularity for tabular data.

Evaluation Measures. To evaluate the obtained models, we tested the predictive performance of the predictions by calculating the Area Under the ROC Curve (AUC) metric [10]. In terms of fairness, we consider the Predictive Equality [4] metric. This metric requires both protected and unprotected groups to have the same false positive rate (FPR), which is related to the fraction of subjects in the negative class that were incorrectly predicted to have a positive value. We obtain the Predictive Equality difference:

$$PE_{dif} = |P(\hat{y} = 1|y = 0, g = 0) - P(\hat{y} = 1|y = 0, g = 1)| \tag{2}$$

---

[2] https://www.openml.org/.

Table 1. Characterization of the used datasets. Abbreviations: $(+,\cdot)$ – positive class instances; $(\cdot,p)$ – protected group instances; $(+,p)$ – positive class, protected group instances; $(+,u)$ – positive class, unprotected group instances.

dataset	OpenML id	# instances	# features	$(+,\cdot)$	$(\cdot,p)$	$(+,p)$	$(+,u)$
ads	40978	1377	1558	33 %	76 %	34 %	33 %
bank	1461	15111	30	33 %	51 %	24 %	43 %
biodeg	1494	1055	41	34 %	15 %	5 %	39 %
churn	40701	2121	22	33 %	23 %	21 %	37 %
credit	29	653	43	45 %	31 %	47 %	45 %
monks1	333	556	6	50 %	49 %	49 %	51 %
phishing	4534	11055	30	56 %	66 %	59 %	49 %
sick	38	636	26	33 %	39 %	12 %	47 %
vote	56	312	14	58 %	52 %	54 %	63 %

We further considered other fairness measures, namely Demographic Parity [6], Equalized odds [11], and Equal opportunity [4]. However, we only present the results in terms of the Predictive Equality difference since the same conclusions can be derived from the results that were obtained using any of the aforementioned fairness metrics.

5   Results

In this section, we examine the results obtained from the conducted experiments, analyzing the robustness of label correction methods in terms of predictive accuracy as well as fairness.

Similarity to Original Labels After Correction. Firstly, we compare the *corrected* labels to the *original* ones at different levels of injected noise. Figure 2 shows, on average, how similar each method's correction was to the original labels, considering both types of bias. Regardless of the type of bias, the OBNC method was able to achieve the highest similarity to the original labels

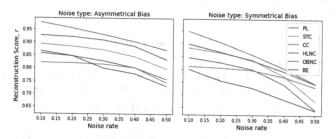

Fig. 2. Reconstruction score (r), representing the similarity between *original* and *corrected* labels at different noise rates.

Performance Evaluation on the Noisy Test Set. This scenario represents cases where only biased data is available for training and testing, and as such, we evaluate the models on the *noisy* test set. The trade-off between the AUC metric and the Predictive Equality difference metric for different noise rates is shown in Fig. 3.

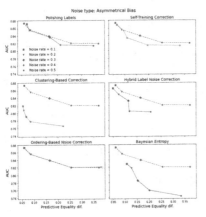
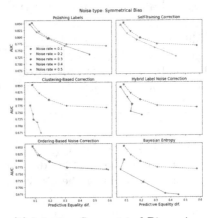

(a) Injecting Asymmetrical Bias noise.   (b) Injecting Symmetrical Bias noise.

Fig. 3. Trade-Off between AUC and Predictive Equality difference obtained on the *noisy* test set when using each of the label correction methods. The red dashed line shows the performance of the model obtained from the *noisy* train set at each noise rate. (Color figure online)

The OBNC method achieved performance similar to using the *noisy* data, while PL and STC show small improvements, mainly in terms of fairness. The CC method performs the best in achieving fairness, being able to keep discrimination at a minimum even at higher noise rates, as shown in Fig. 3b, but losing significant predictive performance to do so. The HLNC method maintained its ability to improve fairness at minimum expense to the predictive performance of the resulting models.

Performance Evaluation on the Original Test Set We further evaluate the obtained models on the *original* test set to investigate how the methods would perform in a testing scenario where the biases that were present in the training data have been corrected in more recent data. The trade-off between AUC and Predictive Equality difference for each method at different noise rates is shown in Fig. 4.

In this testing scenario, the methods still behave in a similar way to the previous one in relation to each other. The OBNC method was shown to correct the labels in a way that is the most similar to the *original* train set. Still, the performance of the resulting model is comparable to using the *noisy* train set.

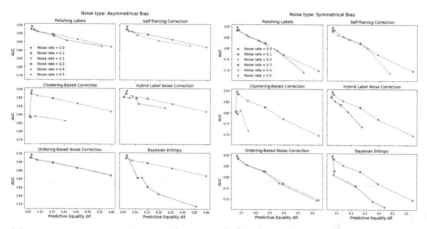

(a) Injecting Asymmetrical Bias noise.   (b) Injecting Symmetrical Bias noise.

Fig. 4. Trade-Off between AUC and Predictive Equality difference obtained on the *original* test set using each of the label correction methods. The red dashed line shows the performance of the model obtained from the *noisy* train set at each noise rate. (Color figure online)

The PL and STC methods achieve a slightly better trade-off between predictive performance and fairness. On the other hand, the CC method shows significant improvements in terms of fairness, but at the expense of a lower AUC score. The BE method achieves a low score in both metrics. Finally, the HLNC method was found to be the best at simultaneously improving both predictive performance and fairness.

Performance Evaluation on the Corrected Test Set. Finally, we analyze the effect of correcting training data when the discrimination has been eliminated in the meantime, but the original data was already biased. We evaluate the performance of the models obtained using *corrected* train data on the test set corrected using the same method, comparing it to the performance obtained when testing the same models on the *original* test set. The results for the AUC metric are presented in Fig. 5 and for the Predictive Equality difference metric in Fig. 6.

In terms of AUC, the PL, STC, and BE methods tend to result in an overestimation of the predictive performance of the resulting model. At the same time, the OBNC method appears to slightly underestimate it. The HLNC method shows similar performance to testing on the *original* test set in the presence of both types of noise, while the CC method only achieves this when dealing with Asymmetrical Bias noise. Regarding the Predictive Equality difference metric, all methods show a performance very similar to using the *original* test set. A slight underestimation of discrimination can be seen for the PL, STC, and BE methods for the higher noise rates.

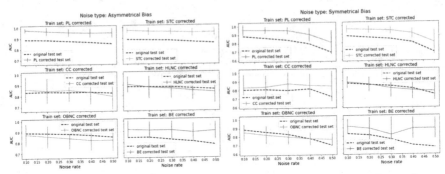

(a) Injecting Asymmetrical Bias noise.   (b) Injecting Symmetrical Bias noise.

Fig. 5. Comparison in AUC between testing the model obtained from the data corrected by each method on the original test set and on the test set corrected by the same method.

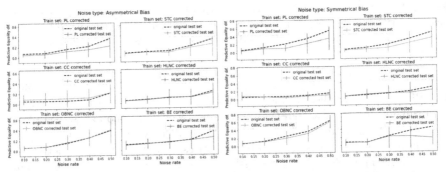

(a) Injecting Asymmetrical Bias noise.   (b) Injecting Symmetrical Bias noise.

Fig. 6. Comparison in Predictive Equality difference between testing the model obtained from the data corrected by each method on the original test set and on the test set corrected by the same method.

6   Discussion

The ability to correct the labels does not necessarily guarantee a good compromise between accuracy and fairness. For instance, the OBNC method obtained the highest similarity with the original labels. However, when assessing the compromise between predictive performance and fairness, the OBNC method had a much less satisfactory performance, showing barely any difference from training with the noisy training set.

On the other hand, the CC method, which did not show a high reconstruction score, kept discrimination at minimum values, even at the highest noise rates. However, this was achieved at the cost of lower predictive performance. The nature of the fairness metrics can explain this: e.g., the Predictive Equality metric calculates the difference between the FPR of each group, meaning that if both groups have a high but similar FPR, the predictions are technically fair but not accurate.

We must acknowledge some limitations of this study, as they can impact the generalizability of our findings.

The first is related to an essential feature of the proposed methodology: it may use standard benchmark datasets to assess the robustness of label correction methods. While these datasets are much larger than typical fairness studies, the sensitive attribute and the positive class are arbitrarily chosen, which may result in differences from cases where label noise is caused by discrimination. Nevertheless, the methodology can also be applied to benchmark fairness datasets to assess the generalizability of the results. Additionally, the predicted classes were based on a decision threshold of 0.5, which is unrealistic in many problems where discrimination might be an issue. A systematic evaluation of different threshold points would be desirable, as the choice of the decision boundary impacts fairness metrics. Regarding benchmark fairness datasets, problem-specific decision thresholds can also be used. Therefore, future research would involve applying this methodology to benchmark fairness datasets.

7 Conclusions

This work tackles the problem of learning fair ML classifiers from biased data. In such a scenario, we look at the inherent discrimination in datasets as label noise that can be eliminated using label noise correction techniques. This way, the corrected data could be used to train fair classifiers using standard ML algorithms without further application of fairness-enhancing techniques. We propose a methodology to empirically evaluate the effect of different label noise correction techniques in improving the fairness and predictive performance of models trained on previously biased data. Our framework manipulates the amount and type of label noise and can be used on fairness benchmarks and standard ML datasets. In the conducted experiments, we analyzed six label noise correction methods. The Hybrid Label Noise Correction method achieved the best trade-off between fairness and predictive performance.

References

1. Barocas, S., Hardt, M., Narayanan, A.: Fairness and machine learning: limitations and opportunities. fairmlbook.org (2019). https://www.fairmlbook.org
2. Basu, S.: Semi-supervised clustering by seeding. In: Proceedings of the ICML-2002 (2002)
3. Blum, A., Mitchell, T.: Combining labeled and unlabeled data with co-training. In: Proceedings of the Eleventh Annual Conference on Computational Learning Theory, pp. 92–100 (1998)
4. Chouldechova, A.: Fair prediction with disparate impact: a study of bias in recidivism prediction instruments. Big Data 5(2), 153–163 (2017)

5. Datta, A., Tschantz, M.C., Datta, A.: Automated experiments on ad privacy settings: a tale of opacity, choice, and discrimination. arXiv preprint arXiv:1408.6491 (2014)
6. Dwork, C., Hardt, M., Pitassi, T., Reingold, O., Zemel, R.: Fairness through awareness. In: Proceedings of the 3rd Innovations in Theoretical Computer Science Conference, pp. 214–226 (2012)
7. Favier, M., Calders, T., Pinxteren, S., Meyer, J.: How to be fair? a study of label and selection bias. Mach. Learn. (2023). https://doi.org/10.1007/s10994-023-06401-1
8. Feng, W., Boukir, S.: Class noise removal and correction for image classification using ensemble margin. In: 2015 IEEE International Conference on Image Processing (ICIP), pp. 4698–4702. IEEE (2015)
9. Fogliato, R., Chouldechova, A., G'Sell, M.: Fairness evaluation in presence of biased noisy labels. In: International Conference on Artificial Intelligence and Statistics, pp. 2325–2336. PMLR (2020)
10. Hanley, J.A., McNeil, B.J.: The meaning and use of the area under a receiver operating characteristic (ROC) curve. Radiology **143**(1), 29–36 (1982)
11. Hardt, M., Price, E., Srebro, N.: Equality of opportunity in supervised learning. In: Advances in Neural Information Processing Systems, vol. 29 (2016)
12. Jiang, H., Nachum, O.: Identifying and correcting label bias in machine learning. In: International Conference on Artificial Intelligence and Statistics, pp. 702–712. PMLR (2020)
13. Mehrabi, N., Morstatter, F., Saxena, N., Lerman, K., Galstyan, A.: A survey on bias and fairness in machine learning. ACM Comput. Surv. (CSUR) **54**(6), 1–35 (2021)
14. Nicholson, B., Zhang, J., Sheng, V.S., Wang, Z.: Label noise correction methods. In: 2015 IEEE International Conference on Data Science and Advanced Analytics (DSAA), pp. 1–9. IEEE (2015)
15. Patrini, G., Rozza, A., Krishna Menon, A., Nock, R., Qu, L.: Making deep neural networks robust to label noise: a loss correction approach. In: Proceedings of the IEEE Conference on Computer Vision and Pattern Recognition, pp. 1944–1952 (2017)
16. Sun, J.W., Zhao, F.Y., Wang, C.J., Chen, S.F.: Identifying and correcting mislabeled training instances. In: Future Generation Communication and Networking (FGCN 2007), vol. 1, pp. 244–250. IEEE (2007)
17. Triguero, I., Sáez, J.A., Luengo, J., García, S., Herrera, F.: On the characterization of noise filters for self-training semi-supervised in nearest neighbor classification. Neurocomputing **132**, 30–41 (2014)
18. Wang, J., Liu, Y., Levy, C.: Fair classification with group-dependent label noise. In: Proceedings of the 2021 ACM Conference on Fairness, Accountability, and Transparency, pp. 526–536 (2021)
19. Wu, S., Gong, M., Han, B., Liu, Y., Liu, T.: Fair classification with instance-dependent label noise. In: Conference on Causal Learning and Reasoning, pp. 927–943. PMLR (2022)
20. Xu, J., Yang, Y., Yang, P.: Hybrid label noise correction algorithm for medical auxiliary diagnosis. In: 2020 IEEE 18th International Conference on Industrial Informatics (INDIN), vol. 1, pp. 567–572. IEEE (2020)

Part-Aware Prototype-Aligned Interpretable Image Classification with Basic Feature Domain

Liangping Li, Xun Gong[✉], Chenzhong Wang, and Weiji Kong

School of Computing and Artificial Intelligence, Southwest Jiaotong University, Chengdu 610031, People's Republic of China
{liangpingli,chenzhongwang,kongweiji}@my.swjtu.edu.cn,
xgong@home.swjtu.edu.cn

Abstract. In recent years, the interpretive *this looks like that* structure has gained significant attention. It refers to the human tendency to break down images into key parts and make classification decisions by comparing them to pre-existing concepts in their minds. However, most existing prototypical-based models assign prototypes directly to each category without considering that key parts with the same meaning may appear in images from different categories. To address this issue, we propose dividing prototypes with the same meaning into the same latent space (referred to as Basic Feature Domain) since different category parts only slightly affect the corresponding prototype vectors. This process of integrating prototypes based on the feature domain is referred to as prototype alignment. Additionally, we introduce the concept of part-aware optimization, which prioritizes prototypical parts of images over simple category labels during optimizing prototypes. Moreover, we present two feature aggregation methods, *by row* and *by cluster*, for the basic feature domain. We demonstrate competitive results compared to other state-of-the-art prototypical part methods on the CUB-2011-200 dataset and Stanford Cars dataset using our proposed self-explanatory part-aware proto-aligned network (PaProtoPNet).

Keywords: Interpretable image classification · Reasoning process · Prototype alignment

1 Introduction

In recent years, there has been a growing interest in Explainable Artificial Intelligence (XAI) research due to the urgent requirements for model interpretability in high-risk domains such as autonomous driving [16] and medical diagnosis [1]. As a result, numerous effective post-hoc explanation and self-explanation methods have been proposed successively. To make black-box models more interpretable, researchers use concept vectors to identify features that humans can understand and use to explain the model's decision-making process [4,6,17]. One such self-explanation method is the prototype learning-based network model, which uses concept embeddings referred to as prototypes [3]. According to ProtoPNet and related studies, bird images contain distinctive features like the head, wings, and feet that vary across different categories. These features are captured during network training and stored as prototypes. The *this looks like that* framework is widely adopted in this type of network, whereby an image that shares some similarity with a class prototype is likely to belong to that class. Specifically, the similarity between prototypes representing a certain class and the feature map is calculated, and the class with the highest similarity is considered as the recognition result of the network. Importantly, by mapping the prototype similarity back to the original image pixels, the model can identify which part of the image played a critical role in the decision-making process, thus making the classification process fully transparent.

The ProtoPNet variants [2, 3, 9, 12, 15] utilize a unique set of prototype vectors for each category to represent their specific features. During classification, the image's the highest similarity with the prototypes from each category determines the classification category. However, they overlook the possibility of shared similar features among different images. Consequently, the prototypes exhibit a scattered distribution in the feature space, lacking clear interrelationships. We argue that prototypes representing similar parts across different categories should exhibit similarity. In Fig. 1(c), despite slight dif-

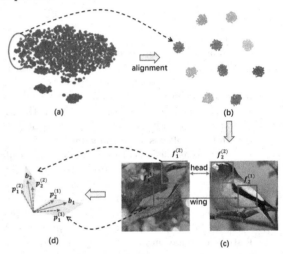

Fig. 1. The essential idea of PaProtoPNet: (a) traditional prototypes are typically scattered; (b) prototypes with similar meanings are clustered together; (c) different birds still exhibit similar features (such as head and wings) on the image; (d) the relationship between basic feature domains and prototypes. $p_c^{(m)}$, $f^{(m)}$ and b_m are defined in Sect. 2.1.

ferences in form and color, the heads of both the first and second birds belong to the *head* category. Deformable ProtoPNet [2] introduces deformable prototypes and ProtoTree [8] incorporates a tree structure to improve interpretability. While TesNet [15] and ProtoPool [9] address variations within and between classes, respectively, neither they nor others have fully resolved this issue. We introduce basic feature domains to capture features within similar part, aligning prototypes of the same part within one domain. In other words, prototypes representing *head* cluster together in Fig. 1(b). Therefore, our proposed PaProtoPNet takes a part-aware approach, deviating from the category-aware framework in

TesNet. Given the constraint that prototypes representing the same part are limited to the confines of a specific basic feature domain, we term this restriction as *prototype alignment*.

With no doubt, there are three critical issues that need to be addressed in the construction of basic feature domains and their corresponding concept vectors: (1) ensuring that the basic feature domains do not overlap and represent distinct meanings, i.e., *head* and *wing* are not the same; (2) ensuring that the feature domains and their corresponding prototype vectors are similar, which guarantees that the prototype is assigned to the correct feature rather than others; and (3) enabling feature maps to be effectively divided into multiple parts corresponding to multiple feature domains during training, which is the foundation for the model to function properly. It is evident that the conventional clustering and separation losses used in ProtoPNets are inadequate for addressing these challenges. Drawing inspiration from the concept of concept vector orthogonalization, we have devised several losses to address the first two issues. Moreover, we have introduced a feature separation module to allocate suitable features to each basic feature domain and employed a discriminative loss to enhance feature distinctiveness.

The main contributions of this paper are as follows:

(1) By introducing the concept of basic feature domains, we have designed the first part-aware prototypical case-based interpretable neural network, which enables the alignment of prototypes within their respective domains and allows for the clear identification of prototypes with the same meaning, without the need for additional computations.
(2) We have designed a feature separation module that allocates features to basic feature domains. This module provides two feature aggregation methods, ensuring effective grouping of features within each domain.
(3) Compared to state-of-the-art ProtoPNets, our proposed method has demonstrated competitive accuracy on the CUB-2011-200 dataset [14] and Stanford Cars dataset [7].

2   Method

2.1   The Overview of PaProtoPNet Architecture

The class label of the image x_i in the dataset is represented as y_i. For convenience, the subscript i is omitted for the computation values and matrices related to x_i. The PaProtoPNet comprises four critical parts shown in Figture 2: the feature extraction layer that includes a convolutional neural network, the Feature Separation Module with a fixed number M of basic feature domains, the prototype layer with $M \times C$ prototype vectors, and the classification layer with a weight matrix W.

The convolutional neural network we employed consists of the backbone and the add layer. The former comprises pre-trained classic networks such as VGGs [11] and ResNets [10]. Unlike the previous ProtoNets method, the add

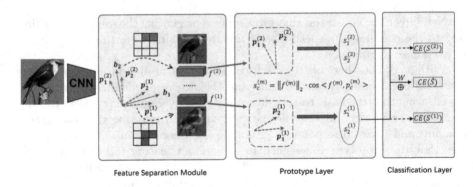

Fig. 2. The overall architecture of our method.

layer uses *BatchNorm2d, Conv2d(1×1), Conv2d(1×1), BatchNorm2d* instead of *Conv2d(1×1), ReLU, Conv2d(1×1), Sigmoid* since the sigmoid operation transforms every element of the tensor to a value between 0 and 1, which is not conducive to fully partitioning the feature map. $Z \in R^{W \times H \times D}$ denotes the features extracted by the convolutional layer, $W \times H$ represents the spatial resolution, and D indicates the number of channels.

In our network model, M basic feature domains are represented as a matrix $B = \{b_m : b_m \in R^D, m = 1, ..., M\}$, which aims to capture feature vectors that are close to themselves in the feature map. In other words, ignoring the missing part of basic features in the image, we need to partition the WH vectors in Z into M parts, assign them to each basic feature domain, and form M integrated feature vectors, which we call fusion vectors represented by $F = \{f^{(m)} : f^{(m)} \in R^D, m = 1, ..., M\}$. To implement this allocation process, we propose a Feature Separation Module (FSM) in this paper. In each feature domain, the network can distinguish among C categories, corresponding to C prototypes. So all prototypes are equally divided and correspond to M feature domains, where $P^{(m)} = \{p_c^{(m)} : p_c^{(m)} \in R^D, c = 1, ..., C\}$ represents the prototype set under the m-th feature domain. Through these operations, both the feature map and prototype are represented in a feature-aware form.

In each basic feature domain, a classification score matrix $S^{(m)} = \{s_c^{(m)} : s_c^{(m)} \in R, c = 1, ..., C\}$ is generated between the integrated vector $f^{(m)}$ and the corresponding prototype vector set $P^{(m)}$. The purpose of the classification layer is to use a learnable parameters W to weight the classification results in each domain and generate the final model result $\hat{S}$.

2.2 Basic Feature Domain and Prototype Alignment

We believe that features in different parts of an image often have significant differences in high-dimensional space. However, as the category changes, they only exhibit minor differences. Therefore, this section aims to construct a certain number of basic feature domains to correspond to these different parts of features and to shield the changes brought by the category. Obviously, there are two necessary requirements posed for the basic feature domains to address.

(1) The domains should not overlap with each other, meaning that the domain vectors should be far apart from each other. To better illustrate this point, we use the example of a bird. A bird is mainly composed of several parts such as *head, back (feathers), wings* and *feet* which are referred to as *basic feature domains* in this article and have clear semantic distinctions. Inspired by TesNet, orthogonality can make a cluster of vectors spread out in high-dimensional space. Therefore, we introduce the orthogonality loss for feature domain vectors by

$$\mathcal{L}_{orth\_d} = \frac{1}{M \times (M-1)} \sum_{i \neq j}^{M} \frac{|b_i \cdot b_j^\top|}{\|b_i\|_2 \|b_j\|_2} \tag{1}$$

where b_i and b_j refer to the i-th and j-th elements whose meanings are distinct from each other in the domain vector list B, $\|\cdot\|_2$ denotes the calculation of the 2-norm of a vector, $|\cdot|$ means to compute a absolute value. Furthermore, solely ensuring orthogonality among domain vectors cannot guarantee optimal discriminability of prototype vectors across different domains. Therefore, the orthogonalization loss for prototype vectors is represented as:

$$\mathcal{L}_{orth\_p} = \frac{\sum_{m_1 \neq m_2}^{M} \left\| P^{(m_1)} \cdot P^{(m_2)^\top} \right\|_F}{M \times (M-1) \times C} \tag{2}$$

where $\|\cdot\|_F$ represents the matrix Frobenius norm, and the reason for dividing by C is because each $P^{(m)}$ contains C prototypes. So the overall orthogonalization loss $\mathcal{L}_{orth}$ is represented as $\mathcal{L}_{orth} = \mathcal{L}_{orth\_d} + \beta_1 \cdot \mathcal{L}_{orth\_p}$ where β_1 is a balance factor.

(2) In a comprehensive way, domain vectors should represent prototypes that imply the feature for all classes at the feature level. In our structure, the *head* domain vector is supposed to serve as the cluster center for these prototypes assigned to all classes to stand for *head*. We directly calculate the mean vector of the prototypes in the domain as the domain vector:

$$b_m = \frac{1}{C} \sum_c^C p_c^{(m)} \tag{3}$$

where b_m is the m-th vector in domain tensor B and $p_c^{(m)}$ denotes the c-th prototype in the equivalent domain. In contrast to orthogonal loss, a clustering constraint draws prototypes closer together to form a cluster of vectors centered around its corresponding domain vector b_m. The proximity cost $\mathcal{L}_{closer}$ is computed as

$$\mathcal{L}_{close} = -\frac{1}{M \times C} \sum_m^M \sum_c^C \frac{|p_c^{(m)} \cdot b_m^\top|}{\left\| p_c^{(m)} \right\|_2 \|b_m\|_2} \tag{4}$$

calculating the cosine similarity between each feature domain vector and its respective prototypes. During the training process, the loss $\mathcal{L}_{close}$ that is optimized to a reasonable value will force the prototypes to cluster together, with the center of the cluster being B, ultimately achieving the overall goal of the relationship between the prototypes and domains.

Therefore, the loss used to construct the basic feature domain and align prototypes is defined as $\mathcal{L}_{BFD} = \mathcal{L}_{orth} + \beta_2 \mathcal{L}_{close}$.

2.3   Feature Separation Module

This subsection will primarily focus on detailing how the Feature Separation Module (FSM) allocates and integrates suitable per-pixel features for each basic feature domain from the feature map Z_i. This step presents a notable departure from other ProtoPNets. FSM facilitates the learning of crucial image regions by basic feature domains and serves as a concentrated manifestation of the feature-aware concept. Essentially, it involves reducing the allocation of the $W \times H$ points in the feature map to M feature vectors.

The primary objective is to establish the correlation matrix, denoted as G by cosine similarity, between the pixel points in feature map Z_i and all domain vectors. The size of the similarity map G should be $(WH \times M)$, and the similarity in G between feature point $z_{a,b}$ in the a-th row and b-th column of Z and the m-th domain vector can be defined as $g(z_{a,b}, b_m) = cos < z_{a,b}, b_m >$ where $cos < \cdot, \cdot >$ representing the similarity between two vectors. Following the computation of similarity map G, two approaches proposed in this article can be employed to identify the feature map points that are associated with domain vectors.

Aggregate Feature by Cluster. The first method involves selecting feature point vectors that are closest to the basic feature domain b_m while being distant from others. This can be mathematically expressed using the following formula:

$$Mask_{a,b}^{(m)} = \begin{cases} 1, & g(z_{a,b}, b_m) > g(z_{a,b}, b_k), \\ & \forall b_k \in B \ \& \ k \neq m \\ 0, & others \end{cases} \tag{5}$$

where $Mask \in R^{M \times W \times H}$ is a binary tensor whose elements are constrained to 0 or 1. When $Mask_{a,b}^{(m)} = 1$, it implies that the feature point located at the position (a, b) in the map should be assigned to the m-th feature domain. Conversely, if $Mask_{a,b}^{(m)} = 0$, it can not be assigned .

Aggregate Feature by Row. The second approach is as follows: with the feature map having a resolution of $W \times H$, we can select the feature point with the maximum similarity g from each row, so that the $Mask$ is represented as

$$Mask_{a,b}^{(m)} = \begin{cases} 1, & g(z_{a,b}, b_m) = \max_{1 \leq j \leq H} g(z_{a,j}, b_m) \\ 0, & others \end{cases} \tag{6}$$

Evidently, it is possible that more than one feature vector can be obtained by each feature domain. Therefore, the mean operator of vectors is employed as the way of aggregating multiple feature points. The specific calculation steps are as follows:

$$N^{(m)} = \sum_a^W \sum_b^H Mask_{a,b}^{(m)}$$
$$f^{(m)} = \frac{\sum_a^W \sum_b^H (Mask_{a,b}^{(m)} \cdot z_{a,b})}{N^{(m)}} \tag{7}$$

where the numerator component is designed to combine all qualifying feature vectors while $N^{(m)}$ indicates the calculated number of vaild feature points and $f^{(m)} \in R^{1 \times D}$ denotes the final aggregated feature vectors for computation together with the prototypes.

To facilitate the differentiation of $z_{a,b}$ across diverse fundamental feature domains, a meticulously crafted cost $\mathcal{L}_{dis}$ is optimized to make the feature points attracted by a certain domain far away from other domains. This discriminative loss function $\mathcal{L}_{dis}$ is written as

$$\mathcal{L}_{dis} = \mathcal{L}_{negative} - \mathcal{L}_{positive}$$
$$\mathcal{L}_{negative} = \frac{1}{M} \sum_m^M \frac{\sum_a^W \sum_b^H (1 - Mask_{a,b}^{(m)}) g(z_{a,b}, b_m)}{(WH - N^{(m)})} \tag{8}$$
$$\mathcal{L}_{positive} = \frac{1}{M} \sum_m^M \frac{\sum_a^W \sum_b^H Mask_{a,b}^{(m)} g(z_{a,b}, b_m)}{N^{(m)}}$$

where $\mathcal{L}_{negative}$ denotes the average similarity between feature points assigned to a specific domain and other domains' vectors. Typically, a smaller $\mathcal{L}_{negative}$ value indicates greater ease in distinguishing the feature map. On the other hand, $\mathcal{L}_{positive}$ denotes the average similarity between feature points and their own domain vectors, and a larger value is generally considered better.

2.4   Computing Scores for Classification

Once the integrated features F are extracted from the feature map Z, we can leverage them along with domain-specific prototypes to compute scores for image classification. The introduction of the concept of feature domain distinguishes the utilization of such feature-aware vectors F in conjunction with the prototypes from the operations similar to other ProtoPNets. As same as TesNet, we still use the projection of $f^{(m)}$ onto $p_c^{(m)}$ as the value of the score denoted as $s_c^{(m)} = \|f^{(m)}\|_2 \cdot cos < f^{(m)}, p_c^{(m)} >$. In each domain, there are C scores available for classification. Therefore, we can consider each domain as a separate classifier, with a total of M classifiers. To obtain the final classification result, we use a learnable weight $W \in R^M$ to synthesize scores that belong to the same class but come from different domains that are followed as

$$\hat{s}_c = \frac{\sum_m^M e^{w_m} \cdot s_c^{(m)}}{\sum_m^M e^{w_m}} \tag{9}$$

where w_m is the adaptive parameter in weight W of the m-th feature domain.

To ensure optimal performance of each basic feature domain, it is necessary to compute the cross-entropy loss not only for the final result but also for the scores of each domain, as shown in Fig. 1. Then, the classification loss is defined as

$$CE(S^{(m)}) = -log\frac{exp((1-\alpha)s_{y_i}^{(m)})}{exp((1-\alpha)s_{y_i}^{(m)}) + \sum_{k\neq y_i}^{C} exp(s_k^{(m)})} \tag{10}$$

$$\mathcal{L}_{ce} = \frac{1}{N}\sum_i^N (CE(\hat{S}) + \beta_3 \sum_m^M CE(S^{(m)}))$$

where the subscript i is omitted from both $S^{(m)}$ and $\hat{S}$ to indicate scores computed from the i-th sample. It should be emphasized that the model's classification result depends only on the weighted average $\hat{S}$.

2.5   Overall Loss Function

We have presented a comprehensive elucidation of three loss functions employed in training our model. Hence, the collective loss amalgamating these functions can be expressed as

$$\mathcal{L} = \lambda_1 \cdot \mathcal{L}_{BFD} + \lambda_2 \cdot \mathcal{L}_{dis} + \lambda_3 \cdot \mathcal{L}_{ce} \tag{11}$$

wherein $\mathcal{L}_{BFD}$ pertains to the construction of the fundamental feature domain and alignment of prototypes within the domain, $\mathcal{L}_{dis}$ is directed towards optimizing the discriminative capability of features extracted from images, and $\mathcal{L}_{ce}$ serves to guarantee the precision of the classification task.

3   Experiments

Dataset: We conducted experiments on two widely-used fine-grained datasets, CUB and CAR. The CUB dataset comprises 200 bird species, with 5,994 images for training and 5,794 for testing. We applied the same data augmentation methods as prior studies, including random rotation, skewing, shearing, and horizontal flipping, to the training set. The CAR dataset contains 16,185 images of 196 car models, split into 8,144 training images and 8,041 testing images. We augmented the CAR training set using the same techniques as CUB.

Experimental Details: We employed various backbones, namely VGG16, VGG19, ResNet34, ResNet50, ResNet152, DenseNet121, and DenseNet161 [5], followed by two 1×1 convolution layers, to test the efficacy of our proposed method. To ensure a fair comparison with ProtoTree, Deformable ProtoPNet, and ProtoPool, we first pre-trained ResNet50 on iNaturalist2017 [13] before fine-tuning it on the CUB dataset, while all other backbones were pre-trained on ImageNet. Like other ProtoPNets, we utilized 10 prototypes with 64 dimensions for each category, resulting in 2,000 and 1,960 prototypes for CUB and CAR, respectively. Prior to inputting the images into the network, each image in the dataset was resized to 224×224. In this study, we set β_1, β_2, β_3, and λ_2 to 0.5, while alpha was set to 0.8. Additionally, λ_1 was set to 10, and λ_3 was set to 1.

Table 1. Comparison of PaProtoPNet with other prototypical methods trained on the Stanford Car(CAR), cropped CUB-200-2011 (CUB-b) and full CUB-200-2011 (CUB-f).

dataset	model	V16	V19	R34	R50	R152	D121	D161
CAR	ProtoPNet [3]	88.3	89.4	88.8	-	88.5	87.7	89.5
	TesNet [15]	90.3	90.6	90.9	-	92.0	91.9	92.6
	ProtoPool [9]	-	-	89.3	88.9	-	-	-
	Ours(by row)	90.6	90.9	90.9	91.0	91.5	91.2	91.9
	Ours(by cluster)	**92.9**	**93.0**	**91.9**	**92.9**	**93.0**	**92.0**	**92.7**
CUB-b	ProtoPNet [3]	77.2	77.6	78.6	-	79.02	79.0	80.8
	TesNet [15]	81.3	81.4	82.8	-	82.7	84.8	84.6
	ProtoPool [9]	-	-	80.3	85.5	81.5	-	-
	Ours(by row)	81.2	82.6	**83.1**	**87.7**	**84.5**	83.3	84.0
	Ours(by cluster)	**81.5**	**82.8**	82.3	86.7	83.9	81.7	83.7
CUB-f	ProtoPNet [3]	70.3	72.6	72.4	81.1	74.3	74.0	75.4
	Def. ProtoPNet [2]	75.7	76.0	76.8	86.4	79.6	79.0	81.2
	ProtoKNN [12]	77.5	77.6	**77.6**	87.2	80.6	79.9	81.4
	Ours(by row)	**78.8**	**79.1**	76.8	**87.4**	80.4	79.8	80.8
	Ours(by cluster)	75.3	78.9	**77.6**	86.5	79.5	79.5	79.2

3.1 Performance Comparison

In this paper, we compare the classification accuracy of our approach with other prototypical parts-based methods on CAR images, CUB images cropped by bounding box (CUB-b), and full CUB images (CUB-f). Similar to other works, we have trained models on various backbones and VGG, ResNet and DenseNet are abbreviated as V, R and D, respectively. The original papers of the corresponding methods did not conduct relevant experiments using the same backbone, resulting in the absence of certain data in Table 1. Despite not achieving

Table 2. Comparison of optimal results between PaProtoPNet and other prototypical methods on the CUB-200-2011

model	acc on CUB
ProtoPNet* [3]	84.8(b), 81.1(f)
ProtoTree** [8]	87.2(f)
TesNet* [15]	86.2(b), 83.5(f)
Def. ProtoPNet** [2]	87.8(f)
ProtoPool** [9]	87.6(f)
ProtoKNN [12]	87.5(b), 87.2(f)
PaProtoPNet(ours)	**87.7(b), 87.4(f)**
PaProtoPNet*(ours)	**88.6(b), 88.2(f)**

higher results on certain backbones, our PaProtoPNet demonstrated superior performance overall, outperforming other methods on both the CAR and CUB datasets.

On the CUB dataset, there are two different comparison methods for the ProtoPNets variant in Table 1: cropped images using bounding boxes or full images. We also followed the convention of the ProtoPNets variant and compared our model integration method with other similar explainable models on

CUB. In Table 2, the * and ** symbols represent the results of using 3-model and 5-model integration, respectively, while b and f indicate that the model was trained and tested on images cropping by bounding box cropping and full images, respectively. As shown in Table 2, our approach has achieved results of considerable competitiveness, with our single model even outperforming the accuracy of some ensemble models.

3.2 Model Analysis

In Fig. 1(b), we visualize the two-dimensional distribution of prototypes, demonstrating the capability of PaProtoPNet to achieve prototype alignment. Furthermore, the visualizations of the upsampling images of M asks in Eqs. 5 and 6 can be examined to observe the function of the Feature Separation Module. In addition, ablation experiments were conducted on the margin-based cross-entropy loss, as shown in Table 3. We varied the value of alpha as 0, 0.3, 0.5, 0.8, and 0.9. Keep the other parameters in the *Experimental Details* unchanged. From the results, it can be con-

Table 3. Ablation studies on margin-based cross-entropy loss

margin(α)	method			
	by row		by cluster	
	V16	R34	V16	R34
0	72.1	80.6	72.2	78.8
0.3	75.3	80.7	78.2	81.2
0.5	78.0	81.3	80.8	81.8
0.8	81.2	83.1	81.6	82.3
0.9	79.6	82.7	79.2	82.0

cluded that the margin-based cross-entropy loss improves the performance of PaProtoPNet, with 0.8 generally yielding the best results. It is noteworthy that our model must employ the add_layer in the form of batchnorm2d-conv-conv-batchnorm2d to fulfill the orthogonality requirements of points within the feature map. We empirically demonstrate that the PaProtoPNet with the original add_layer could not converge effectively.

3.3 Reasoning Process

Although prototypes are aligned at the level of basic feature domains, the main reasoning logic of our PaProtoPNet still follows that of ProtoPNet, with the only difference being that a trained prototype may correspond to several points in an image. Figure 3 illustrates the decision-making process of our proposed network for the Yellow-headed Blackbird test image. Each classification score in this process can be obtained through the network, making our model a classification parameter transparent and self-explanatory model like other variants of ProtoPNet. After being given a test image, the model extracts feature points from the feature map based on different basic feature domains, which is reflected in the activate map column. We observe that different and effective domains may activate different parts of the image.

Taking PaProtoP-Net by cluster (the upper part of Fig. 3) as an example, the first feature domain tends to activate the head of the bird, while the second one focuses on the chest and feathers. During the training phase, the model finds prototypical parts of the Yellow-headed Blackbird for each basic feature domain in the training set images (the third column of the Fig. 3). The model finds

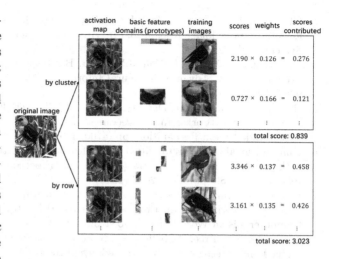

Fig. 3. The reasoning process of PaProtoPNet.

that the test image is sufficiently similar to the prototype in the *head* with a similarity score of 2.190. Finally, all similarity scores from different feature domains are weighted and summed to obtain the final classification score. Although it is difficult to intuitively understand what parts of the bird are represented by the basic feature domains in PaProtoPNet by row, its reasoning process is similar.

4   Discussion and Future Work

In summary, we employed the basic feature domain to achieve prototype alignment, proposed our PaProtoPNet model, and provided a preliminary attempt at feature point selection. Our method has been validated through experiments on the CUB and CAR datasets. However, there are still some limitations in our method. We assumed that every image contains mappings for all basic feature domains during the design phase, but in reality, some images may be missing certain features. For example, the feet of some birds may not be visible in the image due to the angle of the shot. Additionally, although our two proposed FSM have achieved our initial expectations, this simple design has limited the diversity of the feature domain. To address these challenges effectively requires an adaptive and flexible approach which will be explored in future work. Finally, applying prototypical parts and the basic feature domain to other visual tasks is also our next work.

References

1. Afnan, M.A.M., et al.: Interpretable, not black-box, artificial intelligence should be used for embryo selection. Hum. Reprod. Open **2021**(4), hoab040 (2021)
2. Donnelly, J., Barnett, A.J., Chen, C.: Deformable ProtoPNet: an interpretable image classifier using deformable prototypes. In: Proceedings of the IEEE/CVF Conference on Computer Vision and Pattern Recognition, pp. 10265–10275 (2022)

3. Ghorbani, A., Wexler, J., Zou, J.Y., Kim, B.: This looks like that: deep learning for interpretable image recognition. In: Advances in Neural Information Processing Systems, vol. 32 (2019)

4. Ghorbani, A., Wexler, J., Zou, J.Y., Kim, B.: Towards automatic concept-based explanations. In: Advances in Neural Information Processing Systems, pp. 9273–9282 (2019)

5. Huang, G., Liu, Z., Van Der Maaten, L., Weinberger, K.Q.: Densely connected convolutional networks. In: Proceedings of the IEEE Conference on Computer Vision and Pattern Recognition, pp. 4700–4708 (2017)

6. Kim, B., et al.: Interpretability beyond feature attribution: quantitative testing with concept activation vectors (TCAV). In: Proceedings of Machine Learning Research, vol. 80, pp. 2668–2677 (2018)

7. Krause, J., Stark, M., Deng, J., Fei-Fei, L.: 3d object representations for fine-grained categorization. In: Proceedings of the IEEE International Conference on Computer Vision Workshops, pp. 554–561 (2013)

8. Nauta, M., Van Bree, R., Seifert, C.: Neural prototype trees for interpretable fine-grained image recognition. In: Proceedings of the IEEE/CVF Conference on Computer Vision and Pattern Recognition, pp. 14933–14943 (2021)

9. Rymarczyk, D., Struski, Ł, Górszczak, M., Lewandowska, K., Tabor, J., Zieliński, B.: Interpretable Image Classification with Differentiable Prototypes Assignment. In: Avidan, S., Brostow, G., Cissé, M., Farinella, G.M., Hassner, T. (eds.) ECCV 2022. LNCS, vol. 13672, pp. 351–368. Springer, Cham (2022). https://doi.org/10.1007/978-3-031-19775-8_21

10. Shafiq, M., Gu, Z.: Deep residual learning for image recognition: a survey. Appl. Sci. **12**(18), 1–43 (2022)

11. Simonyan, K., Zisserman, A.: Very deep convolutional networks for large-scale image recognition. arXiv preprint arXiv:1409.1556 (2014)

12. Ukai, Y., Hirakawa, T., Yamashita, T., Fujiyoshi, H.: This looks like it rather than that: ProtoKNN for similarity-based classifiers. In: The Eleventh International Conference on Learning Representations, pp. 1–17 (2023)

13. Van Horn, G., et al.: The inaturalist species classification and detection dataset. In: Proceedings of the IEEE Conference on Computer Vision and Pattern Recognition, pp. 8769–8778 (2018)

14. Wah, C., Branson, S., Welinder, P., Perona, P., Belongie, S.: The Caltech-UCSD Birds-200-2011 Dataset (2011)

15. Wang, J., Liu, H., Wang, X., Jing, L.: Interpretable image recognition by constructing transparent embedding space. In: Proceedings of the IEEE/CVF International Conference on Computer Vision, pp. 895–904 (2021)

16. Wiegand, G., Schmidmaier, M., Weber, T., Liu, Y., Hussmann, H.: I drive-you trust: explaining driving behavior of autonomous cars. In: Extended Abstracts of the 2019 CHI Conference on Human Factors in Computing Systems, pp. 1–6 (2019)

17. Zhang, R., Madumal, P., Miller, T., Ehinger, K.A., Rubinstein, B.I.: Invertible concept-based explanations for CNN models with non-negative concept activation vectors. In: Proceedings of the AAAI Conference on Artificial Intelligence, vol. 35, pp. 11682–11690 (2021)

Hybrid CNN-Interpreter: Interprete Local and Global Contexts for CNN-Based Models

Wenli Yang[✉], Guan Huang, Renjie Li, Jiahao Yu, Yanyu Chen, and Quan Bai

School of ICT, College of Sciences and Engineering, University of Tasmania, Hobart, Australia
yang.wenli@utas.edu.au

Abstract. Convolutional neural network (CNN) models have seen advanced improvements in performance in various domains, but lack of interpretability is a major barrier to assurance and regulation during operation for acceptance and deployment of AI-assisted applications. There have been many works on input interpretability focusing on analyzing the input-output relations, but the internal logic of models has not been clarified in the current mainstream interpretability methods. In this study, we propose a novel hybrid CNN-interpreter through: (1) An original forward propagation mechanism to examine the layer-specific prediction results for local interpretability. (2) A new global interpretability that indicates the feature correlation and filter importance effects. By combining the local and global interpretabilities, hybrid CNN-interpreter enables us to have a solid understanding and monitoring of model context during the whole learning process with detailed and consistent representations. Finally, the proposed interpretabilities have been demonstrated to adapt to various CNN-based model structures.

Keywords: Local context · global context · Hybrid interpreter · filter importance · correlation

1 Introduction

Although the performance of convolutional neural network (CNN) models has significantly increased over the past decade, yet without reliable interpretabilities that effectively represent the learning processes, humans still consider untrust¬worthy when continue to view CNN models to be used as unreliable when utilized for real-world decision making [1]. The lack of trust undermines the deployment of CNN-based technologies into many domains, such as medical diagnosis [2], healthcare [3], autonomous driving [4]. This motivates the need for building trust in these CNN-assisted decision making. Building trust by adding interpretabilities has significant theoretical and practical value for both improving model performance and enhancing transparency. Interpretabilities can

T. Liu et al. (Eds.): AI 2023, LNAI 14472, pp. 197–208, 2024.
https://doi.org/10.1007/978-981-99-8391-9_16

be represented by using a variety of expressions, such as visual interpretability and semantic interpretability [5]. Among different expressions, the visual interpretability of CNN-based models is the most fundamental and direct way to explain the network representations.

In terms of visual interpretability methods, the interpretability of CNN-based models emphasizes either visualization of training data rules [6] or the visualization inside the model [5]. Presently, most visual interpretability methods focused more on the understanding of input features on the model performance ensured rather than on the context of CNN-based models, which can reveal the learning process through each layer and the generic features learned from any black box models. Additionally, the internal logic of models has not been clarified in the current mainstream visual interpretability methods. Most existing internal logical discussion about models is used for tree-based models [7]. For CNN-based models, to the best of our knowledge, correlation interpretabilities have not been discussed so far.

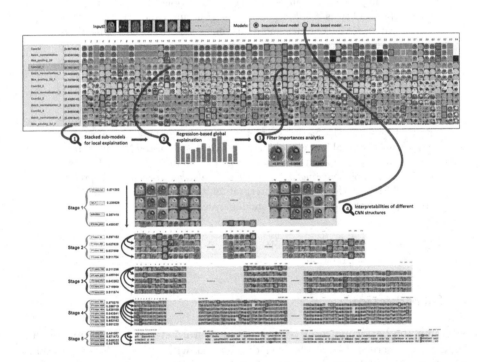

Fig. 1. Hybrid CNN-interpreter: understand how different convolutional neural networks learn through layers and filters: 1) Local interpretability to represent each layer's output and learning ability; **2)** Global interpretability to explore the representations ability of feature maps (learned by different filters) by using regression models; **3)** feature importance analytic to indicate the contributions of convolution filters when making predictions. Overall, it enables for both local and global interpretability for different CNN structures.

We present a novel hybrid CNN interpreter, which builds stacking ensemble models of each layer for local interpretabilities and extends local interpretabilities to compute global correlation to assess the importance of convolution filters. The results are illustrated by binary classification of brain tumor data in the different model structures as shown in Fig. 1. It makes three innovative improvements:

1. Build an original stacking forward propagation algorithm by computing the contributions of each layer to the final prediction. In stacking, each time takes the outputs of the specific layer as input and connects to the final probability mapping layer directly, which can use the set of predictions as a local context and examine feature maps learned by different layers' ability for the final prediction contributions.
2. Extend local interpretabilities to a global interpretation by layer-based regression models, which are constructed by using all the local interpretabilities as dependent variables and each feature map representation as an independent variable across the entire dataset. This enables the examination of the representation ability of feature maps (learned by different filters) for interpreting a model's global behavior.
3. Extract filter importance by assigning scores to each filter in each layer of a model that indicates the correlation of convolution filters when making a prediction, which can help people understand the relationship between the filters and the target variable and can conditionally improve the performance of the model, potentially making the model lighter and speeding up the model's working by removing the unimportant filters.
4. Applicable for a broad range of CNN-based models, which can be utilized to interpret both sequence and nonsequence-focused models. For the sequence models, the interpretabilities can be through each layer to provide the learning context, whereas for the nonsequence models, the interpretabilities can be customized for different stages, blocks, or specific layers. This reveals how different learning is in different models and in different training processes.

2   Related Work

2.1   Convolutional Neural Network Structures

CNNs with strong representation ability of deep structures have ever-increasing popularity in many applications. AlexNet [8] is a leading structure of convolutional neural networks and has huge applications for classification tasks. The evolution after the AlexNet can be mainly summarized in two ways: sequence-focused models by increasing the depth of networks and nonsequence-focused models by adding units or modules.

Sequence focused models: by leveraging the sequence information through layers to improve the model performance, such as VGG16 and VGG19 [9], MSRANet [10]. All of these models are focused on using multiple convolutional layers and activation layers to increase the depth of network, which can extract deeper and better features than the simple structure.

Non-sequence-focused models: by using modular structures to add functional units or components to extend the width of network. Typical examples such as GoogLeNet [11], Inception [12]. GoogLeNet utilizes multiple branches, which allows the network to choose between multiple convolutional filter sizes in each branch. An inception network comprises repeating components referred to as inception modules to increase the representational power of a neural network.

By combining them together, ResNet [13] uses four stages made up of residual blocks, each of which uses several residual blocks with convolutional blocks and identity blocks. Finally, residual blocks are stacked together on top of each other to form the whole network, which can gain better performance from considerably increased depth.

In this paper, we will pick up AlexNet and ResNet as two representative models to interpret the learning process of different model structures.

2.2   Visual Interpretabilities of CNN-Based Models

According to different implementation methods, the visual interpretability methods can be divided into three different categories: input visualization method, model visualization method, and mixed visualization method.

The input visualization explains processes simultaneously from the initial input stage to the final output result. For example, Jeyaraj and Nadar [14] stated a regression-based partitioned method for oral cancer diagnosis. Another example was the importance estimation network produced by Gu et al. [15], which was diagnosed by the classification network by investigating the irrelevant information. The input visualization merely provides a user-friendly interface for seeing and understanding input data, but no information on how the relevant features contribute to the prediction is supplied.

Generally, model visualization is based on the level of layers and finds out the prediction results of the model between different layers. For example, Graziani et al. [16] employed "The concept activation vector (TCAV)" to transform medical pictures into quantitative characteristics by using radiomics. Additionally, Villain et al. stated a novel GradCAM method on brain MRI image datasets [17]. The model visualization could find hidden problems inside the model, but the currently selected features used for model interpretability may not be repeatable in each input.

Hybrid visualization model refers to a combination of both input visualization and model visualization. This model tries to focus on not only explaining the local prediction but also exploring global knowledge inside the model. Hybrid visualization model concentrates on finding the connections or correlations between the input aspect and model layer aspect. Unfortunately, to my knowledge, there are no obvious research on mixed visualization method.

In this paper, we discuss how to interpret CNN-based models using hybrid visualization to make the interpretability appropriate for different model structures. We describe the local interpretability by building a set of stacked ensemble models to provide various input visualization. Then, we extend local interpretability to model visualization by using regression-based analytics. Finally, we identify the importance of each filter in each layer to represent the global context of the models.

3   Method

Hybrid CNN-interpreter aims to explain the deep learning model's different layers and the filter's feature representation ability for the final prediction contribution. It provides global insight into not only the model's layer level, but also the filter level along that layer. The output of the hybrid CNN-interpreter is a layer and filter-based importance distribution matrix, representing the importance level of feature maps learned from different layers and filters by the model. The hybrid CNN-interpreter consists of the original CNN-model forward propagation module, linear regression module and filter importance analysis module.

3.1   Stacking Forward Propagation

In the first original forward propagation module, each image is sent to the original model. We skip and connect each layer $(l_i(X_{ij}))$ directly to the final probability mapping layer $(f(\cdot))$, and the output probability (Y_i) is recorded. This is to examine feature maps learned by different layers' ability for the final classification contribution (Eq. 1).

$$f[l_i(X_{ij})] = Y_i \tag{1}$$

where X_{ij} represents the jth feature map at the ith layer.

3.2   Linear Regression Module

In the second step, a linear regression model is constructed by using the output probability as a dependent variable and each feature map's mean value as independent variables. This is to examine feature maps' (learned by different filters) representation ability for the classification (Eq. 2).

$$Y_i = b_{i0} + \sum b_{ij} \overline{X}_{ij} \tag{2}$$

where $\overline{X}_{ij}$ represents the mean value of feature map X_{ij}. We applied L2 regularization and the objective function is represented in Eq. 3.

$$L = \sum (\hat{Y}_i - Y_i)^2 + \lambda \sum b_{ij}^2 \tag{3}$$

where $\hat{Y}_i$ is the predicted value and Y_i is the true value.

Algorithm 1 indicates the process of how $\overline{X}_{ij}$ are extracted from the model outputs.

In our experiment, we use Ridge regression model to calculate the variance of our feature maps. The Ridge model solves regression problems with an $L2$ regularization loss function, the equation of the loss is shown in Eq. 3. In the

---

Algorithm 1. Mean values of feature maps extraction

1: **procedure** EXPORT-FEATURES($img, lnames$)   ▷ The input of this algorithm are images and the layer names of the model

2:   $model.predict(img) \leftarrow features$ ▷ The feature maps are obtained by visualisation the model prediction.

3:   **for** $lname, features$ **do**

4:     $mean\text{-}list = [\,]$

5:     **for** i in range($features\text{-}number$) **do**

6:       $x = features[0, :, :, i]$

7:       $x = x - x.mean()$ ▷ $x.mean()$ the mathematical mean value of x

8:       $x = x\ /\ x.std()$ ▷ $x.std()$ the mathematical standard deviation value of x

9:       $x = x * 32$

10:       $x = x + 64$

11:       $x = numpy.clip(x, 0, 255)$ ▷ Use numpy.clip() function to limit values outside the interval are clipped to the interval edges

12:       $mean\text{-}list.append(x)$ ▷ Store the mean value of the feature maps into $mean\text{-}list$

13:     **end for**

14:   **end for**

15:   **return** $mean\text{-}list$ ▷ The output of the algorithm is a list with the mean value of the feature maps

16: **end procedure**

Ridge model, the parameter α ($\alpha = 1$ in our experiment) is used to control the regularization strength. The input of our regression model is the mathematical mean value of our feature maps and the output of the CNN models after prediction (confidence score in our experiment).

Then, the regression model will return four parameters to represent the importance of our feature map, i.e. the coefficient of determination (R^2), standard-errors (SEs), t-values and p-values. The equation of R^2 is demonstrated in 4. The meaning of R^2 is the proportion of the variance of the variable that is explained by this predicted model. In our experiment, we use R^2 to evaluate the reliability of the prediction, if the value of R^2 close to 1, which means the model predictions are reliable and effective. Conversely, if the value of R^2 is close to 0, the predictions of the model are not reliable and ineffective.

$$R^2 = \frac{\sum_i e_i^2}{\sum_i (y_i - \overline{y})^2} \tag{4}$$

The equation of SE is shown in Eq. 5. The SE is another statistical tool for us to evaluate the degree of deviation of the sample mean from the overall true mean. Normally, a higher SE value indicates that the mean value of the feature maps is more distinct, whereas a smaller SE value means the features extracted are similar.

$$SE = \frac{\sigma}{\sqrt{n}} \tag{5}$$

where SE is the standard error of the sample, σ is the sample standard deviation and n is the number of samples.

3.3 Filter Importance Analysis Module

In the third filter importance analysis module, a matrix of different filters' representation scores has been calculated, the matrix helps people understand the contribution of the feature map learned by different filters in each layer of the network to the final classification result.

In our settings, the t-value (Eq. 6) is used to indicate the level of difference between features. The t-value is calculated as the difference expressed in standard error units. It means that the standard deviation estimated is far away from 0. A large t-value indicates the evidence against the null hypothesis, in other words, we could declare a relationship between the CNN-predictions and the feature maps.

$$t = \frac{\overline{x} - \mu_0}{\frac{s}{\sqrt{n}}} \tag{6}$$

where $\overline{x}$ is the sample mean, μ_0 is the population mean, s is the sample standard deviation and n is sample size.

We also used p-value (Eq. 7) to evaluate two variables, namely, they are the importance of the features and the CNN model prediction accuracy. We want to examine the relationship between the predictor variables and the response variable to find out if higher accuracy in model prediction will result in higher importance values in the feature maps' mean values.

$$p = \frac{\widehat{p} - p0}{\frac{\sqrt{p_0 (1 - p_0)}}{n}} \tag{7}$$

where $\widehat{p}$ is sample proportion, p_0 is assumed population proportion in the null hypothesis and n is the sample size.

4   Experiment and Discussion

A total of 253 public brain MRI images dataset (155 tumorous images and 98 nontumorous images) are used to build the classification models. Since CNNs can be independent of translation, view position, size, and lighting [16], the data augmentation was applied by manually flipping and rotating the image sets. Moreover, considering the dimension of the image may be changed after rotation, the image will be only rotated arbitrarily between 0 and 10° in our experiments. Finally, after the above data augmentation, the dataset has 1,085 positive and 980 negative examples. We use 70% of the images for training and

15% of the images for validating to generate the AlexNet and ResNet classifier models. The remaining of 15% of the images are for model interpretability in our proposed hybrid CNN-interpreter.

4.1 Local Interpretability for CNN-Based Models

The hybrid CNN-interpreter enables local interpretability with individual repeatable capacity by building a set of stacked ensemble submodels. Each submodel is developed by connecting the output of each layer in the original model and the final probability mapping layer to output prediction results, representing the model's learning ability. For the AlexNet model, the local interpretability will focus on each layer. For the ResNet model, the local interpretability will concentrate on each stage and component in each stage. The hybrid CNN interpreter will collect this set of prediction results for individual predictions.

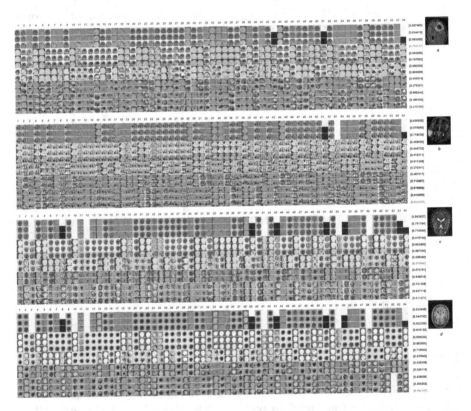

Fig. 2. Local interpretability results of AlextNet model. a, Wrong prediction results of brain tumour. **b,** Correct prediction results of brain tumour. **c,** Wrong prediction results of non-brain tumour. **d,** Correct prediction results of non-brain tumour

In our experiments, by using AlexNet as example, four different samples are input into the AlexNet model to explore the local interpretability, where we pick up both tumour and nontumour images as well as both the final correct prediction and wrong prediction. Figure 2 represents the prediction results of each layer in the AlexNet model, which is guaranteed consistent for repeatable running. From the local interpretability, we can see that each layer of different samples makes different individual predictions. Moreover, it is not an actual rule "more layers = better performance". For example, the $Conv2d\_1$ layer demonstrates the best performance in Fig. 2 (a) and the $Batch\_normalization\_2$ layer achieved the better performance in Fig. 2(c).

4.2 Global Interpretability for CNN-Based Models

By combining the local interpretability of each layer across the entire training data, we present the regression-based methods to provide global insight into each filter in each layer, which will capture the global pattern of filters and the relationship between feature representations and prediction results. The experiments from the hybrid CNN-interpreter for global interpretability cover: 1) summary plots of the linear coefficient to represent the strength and direction of the linear relationship between each filter and prediction results in each layer, stage or internal block; 2) correlation analysis through layers of the AlexNet model and stages/blocks of the ResNet model respectively.

We summarize the linear coefficients of all layers in the AlexNet model as shown in Fig. 3, and the distribution of each filter's coefficient in each layer can also be displayed by highlighting the positive as blue and the negative as red. Based on the results, we can see that filters 20 and 31 show the majority of positive effects among all the layers, whereas filters 28 and 35 show negative effects in most layers. We also get the linear coefficients of all stages and every internal block in the ResNet model.

Based on the linear coefficient results above, we analyze the correlation between layers in the AlexNet model and the correlation between stages and internal blocks in each stage. Figure 4 (a), we shows that the pairwise layers such as $Conv2d$ and $Batch\_normalization(Bn)$, $Conv2d\_1$ and $Batch\_normalization\_1(Bn\_1)$ always have strong correlation with each other, which can reflect the local interpretability that these pairwise layers get the similar prediction results as well. Moreover, we can clearly see the $Con2d\_4$ and Bn_4 layers have global negative effects for the final prediction results, compared with the local interpretability in Fig. 2, the prediction results in the $Max\_pooling\_2d\_2$ layer almost get noticeably different results compared with $Conv2d\_4$ layer and Bn_4 layers, which can approve the consistency between local and global interpretability.

For the ResNet model, Fig. 4 (b) shows that $stage4$ has negative effects from $stage1$ to $stage3$. This reflects the consistency with the local interpretability that the performance increased from $stage1$ to $stage3$ but dropped starting from the $stage4$. Moreover, the correlations among the internal blocks in $stage4$ indicate a generally positive correlation with each residual block. However, the correlations get less and less along with the blocks getting deeper, such as the correlation between $Conv_block$ and following identity blocks, is getting lower as shown in Fig. 4 (c).

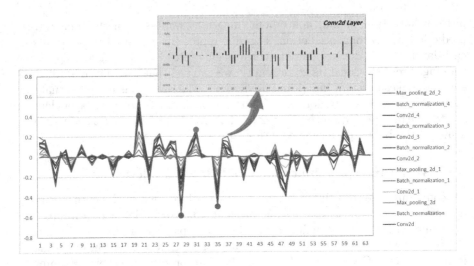

Fig. 3. Summary plot of linear coefficients in AlextNet Model

a	Conv2d	Bn	Mp_2d	Conv2d_1	Bn_1	Mp_2d_1	Conv2d_2	Bn_2	Conv2d_3	Bn_3	Conv2d_4	Bn_4	Mp_2d_2
Conv2d	1												
Bn	0.779536	1											
Mp_2d	0.30161	0.364406	1										
Conv2d_1	0.571508	0.518985	0.485607	1									
Bn_1	0.328443	0.321118	0.595418	0.764426	1								
Mp_2d_1	0.51741	0.433018	0.419729	0.537876	0.365535	1							
Conv2d_2	0.27546	0.256676	0.352218	0.277827	0.199354	0.488807	1						
Bn_2	0.21541	0.163118	-0.38656	0.000208	-0.18027	-0.27187	-0.14924	1					
Conv2d_3	0.276401	0.439407	0.118666	0.097576	0.022728	0.335027	0.469468	-0.12299	1				
Bn_3	0.775794	0.667391	0.328202	0.559419	0.348241	0.62236	0.460947	0.15056	0.429298	1			
Conv2d_4	-0.22818	-0.0805	-0.1846	-0.3098	-0.18986	0.009627	0.074398	-0.15418	0.131848	-0.05294	1		
Bn_4	-0.16237	-0.3659	-0.4977	-0.38621	-0.36683	-0.27815	-0.19903	0.084495	-0.07138	-0.28235	0.73575	1	
Mp_2d_2	0.390962	0.380329	0.429784	0.264844	0.215349	0.427755	0.491625	0.095726	0.289114	0.680224	-0.22435	-0.54445	1

b	Stage 1	Stage 2	Stage 3	Stage 4	Stage 5
Stage 1	1				
Stage 2	0.220311	1			
Stage 3	0.095078	-0.05877	1		
Stage 4	-0.04158	-0.00036	-0.05279	1	
Stage 5	0.037952	0.00956	-0.00903	0.018162	1

c	Conv_block	identity_block_1	identity_block_2	identity_block_3	identity_block_4	identity_block_5
Conv_block	1					
identity_block_1	0.178773861	1				
identity_block_2	0.122210974	0.31108795	1			
identity_block_3	0.11480204	0.245819442	0.328700444	1		
identity_block_4	0.091460254	0.09441372	0.21198078	0.273242666	1	
identity_block_5	0.048173515	0.11637556	0.144813479	0.237425768	0.367817927	1

Fig. 4. Correlation analysis among layers, stages or blocks for global interpretability. a, Correlation coefficients between layers in AlexNet Model. **b,** Correlation coefficients between stages in ResNet Model. **c,** Correlation coefficients between internal blocks in ResNet Model (by using stage 4 as reference).

By combining local interpretability and global interpretability, the experiments show that when the forward probability of the submodel is high, the features learned by the selected top five positive filters are consistent with human cognition with more significant semantic information, such as both the five filters (20,26,62,45,and 31) of $Conv\_2d\_1$ layer in AlexNet model and the filters (151, 24, 118, 190, and 48) of $identity\_block\_2$ output layer in ResNet model showing the part of important features in the tumor area. Moreover, when the forward probability of the submodel is low, filters marked as negative effects can learn valuable semantic information. For example, the feature representation of filter 15 of the Mp_2d_2 layer in the AlexNet model actually covers part of the tumor area, and the filter 45 of the input layer in *stage*3 in the ResNet model can represent parts of features in the tumor area as well. In contrast, filters considered positive by the model cannot learn the critical parts of the image (e.g., 42 and 59 filters in Bn_3 layer in the AlexNet model as well as 63 and 107 filters of the input layer in *stage*3) can be understood that wrong cognition of the model leads to a low forward probability result. These could also verify the validity and consistency of our hybrid CNN-interpreter between the local and global interpretabilities.

5   Conclusion and Future Work

The hybrid CNN interpreter helps users to have a deep conceptual understanding of CNN-based models by providing both local and global interpretabilites. The local interpretability by using original forward propagation can reveal how image data progresses through the layers of the CNN-based models, whereas the global filter importance based on the linear regression module can indicate how much each filter contributes to the model prediction. The hybrid interpreter can be flexibly adapted to various CNNs (layer-based, stage-based, or internal-block-based). The correlations between layers, stages, or blocks can also be generalized to understand the context of complex CNN structures. By demonstrating how to apply the hybrid CNN interpreter to explain different types of CNN-based models, we take brain tumor classification tasks to provide an interpretation of the AlexNet and ResNet models. The experiment results showed the efficiency and consistency between local and global interpretabilities, as well as among different models.

The proposed interpreter can be used to improve and debug models and to make CNN-based models more accurate and reliable in the future. The local interpreter can reveal prediction results for specific samples of different learning processes, and based on these interpretabilities, we can set the gating and memory mechanisms to debug models that can help us identify where to access that memory and where to ignore it during the whole learning process. For example, the model can skip some layers with bad performances and tell where to access and make connections again. The filter importance and feature correlations identified by the global interpreter enable developers to determine which filter can be eliminated to get better performance.

References

1. Liu, H., et al.: Trustworthy AI: a computational perspective. arXiv preprint arXiv:2107.06641 (2021)
2. Zhang, Y., Weng, Y., Lund, J.: Applications of explainable artificial intelligence in diagnosis and surgery. Diagnostics **12**(2), 237 (2022)
3. Pawar, U., O'Shea, D., Rea, S., O'Reilly, R.: Explainable AI in healthcare. In: 2020 International Conference on Cyber Situational Awareness, Data Analytics and Assessment (CyberSA), pp. 1–2 (2020)

4. Atakishiyev, S., Salameh, M., Yao, H., Goebel, R.: Explainable artificial intelligence for autonomous driving: a comprehensive overview and field guide for future research directions. arXiv preprint arXiv:2112.11561 (2021)
5. Zhang, Q.-S., Zhu, S.-C.: Visual interpretability for deep learning: a survey. Front. Inf. Technol. Electron. Eng. **19**(1), 27–39 (2018)
6. Bińkowski, M., Sutherland, D.J., Arbel, M., Gretton, A.: Demystifying mmd GANs. arXiv preprint arXiv:1801.01401 (2018)
7. Waghen, K., Ouali, M.-S.: Multi-level interpretable logic tree analysis: a data-driven approach for hierarchical causality analysis. Expert Syst. Appl. **178**, 115035 (2021)
8. Krizhevsky, A., Sutskever, I., Hinton, G.E.: Imagenet classification with deep convolutional neural networks. Commun. ACM **60**(6), 84–90 (2017)
9. Simonyan, K., Zisserman, A.: Very deep convolutional networks for large-scale image recognition. arXiv preprint arXiv:1409.1556 (2014)
10. Zheng, Q., Chen, Z., Liu, H., Lu, Y., Li, J.: Msranet: learning discriminative embeddings for speaker verification via channel and spatial attention mechanism in alterable scenarios. Available at SSRN 4178119
11. Szegedy, C., et al.: Going deeper with convolutions. In: Proceedings of the IEEE Conference on Computer Vision and Pattern Recognition, pp. 1–9 (2015)
12. Szegedy, C., Vanhoucke, V., Ioffe, S., Shlens, J., Wojna, Z.: Rethinking the inception architecture for computer vision. In: Proceedings of the IEEE Conference on Computer Vision and Pattern Recognition, pp. 2818–2826 (2016)
13. He, K., Zhang, X., Ren, S., Sun, J.: Deep residual learning for image recognition. In: Proceedings of the IEEE Conference on Computer Vision and Pattern Recognition, pp. 770–778 (2016)
14. Jeyaraj, P.R., Samuel Nadar, E.R.: Computer-assisted medical image classification for early diagnosis of oral cancer employing deep learning algorithm. J. Cancer Res. Clin. Oncol. **145**(4), 829–837 (2019)
15. Gu, D., et al.: VINet: a visually interpretable image diagnosis network. IEEE Trans. Multimedia **22**(7), 1720–1729 (2020)
16. Graziani, M., Andrearczyk, V., Marchand-Maillet, S., Müller, H.: Concept attribution: explaining CNN decisions to physicians. Comput. Biol. Med. **123**, 103865 (2020)
17. Villain, E., Mattia, G. M., Nemmi, F., Péran, P., Franceries, X., le Lann, M. V.: Visual interpretation of CNN decision-making process using simulated brain MRI. In: 2021 IEEE 34th International Symposium on Computer-Based Medical Systems (CBMS), pp. 515–520 (2021)

Impact of Fidelity and Robustness of Machine Learning Explanations on User Trust

Bo Wang[✉], Jianlong Zhou, Yiqiao Li, and Fang Chen

University of Technology Sydney, Sydney, Australia
{Bo.Wang-11,Yiqiao.Li-1}@student.uts.edu.au,
{Jianlong.Zhou,Fang.Chen}@uts.edu.au

Abstract. EXplainable machine learning (XML) has recently emerged as a promising approach to address the inherent opacity of machine learning (ML) systems by providing insights into their reasoning processes. This paper explores the relationships among user trust, fidelity, and robustness within the context of ML explanations. To investigate these relationships, a user study is implemented within the context of predicting students' performance. The study is designed to focus on two scenarios: (1) *fidelity-based scenario*—exploring dynamics of user trust across different explanations at varying fidelity levels and (2) *robustness-based scenario*—examining dynamics of in user trust concerning robustness. For each scenario, we conduct experiments based on two different metrics, including self-reported trust and behaviour-based trust metrics. For the fidelity-based scenario, we find that users trust both high and low-fidelity explanations compared to without-fidelity explanations (no explanations) based on the behaviour-based trust results, rather than relying on the self-reported trust results. We also obtain consistent findings based on different metrics, indicating no significant differences in user trust when comparing different explanations across fidelity levels. Additionally, for the robustness-based scenario, we get contrasting results from the two metrics. The self-reported trust metric does not demonstrate any variations in user trust concerning robustness levels, whereas the behaviour-based trust metric suggests that user trust tends to be higher when robustness levels are higher.

Keywords: Human computer interaction · Machine learning explanation · User trust · Fidelity · Robustness

1 Introduction

Machine learning (ML) finds widespread applications in various domains, playing a pivotal role in numerous contexts. However, the lack of interpretability poses a significant challenge in understanding the inner workings of ML models. Hence, the explanation of machine learning holds the utmost importance. Explaining

ML involves elucidating the intricate connections between input and outcome within ML models, facilitating user comprehension of the underlying reasoning. By elucidating the mechanisms of ML models, users can enhance their trust in the model's decisions, gain interpretability of the results, and gain insights into the decision-making process [20]. Moreover, explaining machine learning provides researchers, developers, and decision-makers with opportunities to gain deeper insights and improve the models. Recently, the field of ML explanation has obtained considerable attention from researchers. For instance, in the domain of recommender system [18], image classifier [12], and medicine [7], the researchers demonstrate that users express deeper insights when provided with explanations than systems lacking explanatory capabilities.

Furthermore, the selection of appropriate ML explanation methods with superior performance hinges upon the quality of the explanations. The quality of ML explanations encompasses three crucial aspects: user-related factors (e.g. user trust and satisfaction), explanation-related factors (e.g. fidelity), and model-related factors (e.g. robustness and fairness) [8]. User trust, as a critical aspect in ML explanations, represents one of the primary objectives in the explanatory process. It serves as a measurable criterion for quantifying subjective evaluation and enables assessing the quality of ML explanation methods. Further, fidelity holds significant importance in eXplainable machine learning (XML) as it ensures the provision of reliable explanations that align with the internal mechanisms of the underlying ML model. Recent studies confirm the correlation between explanation fidelity and user trust, emphasizing the need for high-fidelity explanations [11]. However, a fundamental question arises: **Does the user exclusively trust high-fidelity explanations?** To comprehensively address this inquiry, we devise a *fidelity-based scenario*, which builds upon the work of Papenmeier et al. [11] and further explore user trust variation at different levels of fidelity by visualizing explanations from two distinct methods. Alternatively, robustness in XML methods refers to their inherent capability to consistently provide reliable and consistent explanations, even when subjected to diverse perturbations. It is natural to raise the question: **Does robustness affect user trust?** In response to this research question, we design a *robustness-based scenario*, which explores user trust variation at different levels of robustness through visualization of explanations from a single method.

To thoroughly evaluate the impact of explanations on human trust within ML systems, researchers can adopt a combined approach utilizing a self-reported trust scale and behavioural metrics [13]. Thus, our paper assesses user trust from subjective and objective components by developing self-reported and behaviour-based trust metrics. To facilitate this evaluation, we employ a user survey that predicts student performance levels and measures variations in user trust across different levels of fidelity and robustness.

In this study, we make the contributions as follows:

- We evaluate how user trust varies on different explanation methods over different levels of fidelity. Specifically, we conduct a comparative analysis of user trust within the explanations generated by LIME and SHAP, incorporating three distinct levels of fidelity: high, low, and without-fidelity.

- We investigate how user trust fluctuates across different levels of robustness. To evaluate user trust, we employ visualizations of the explanations from LIME, encompassing both high and low levels of robustness.
- We employ a developed self-reported trust questionnaire and behaviour-based trust metrics to measure user trust. These measurement approaches allow us to capture subjective perceptions of trust reported by users, along with objective indicators derived from user behaviours and interactions.

2   Related Work

2.1   User Trust

User trust in XML has been identified as a pivotal factor influencing human behaviour in human-machine interactions [21]. Users tend to base their behaviours on the guidance provided by well-performing XML systems when they trust the system. Conversely, if an XML system makes noticeable mistakes, it can lead to mistrust or even complete distrust of the system, causing users to deviate from following its recommendations. Several researchers emphasize that user trust would affect the adoption of ML. On specific, Asan et al. [2] argue that user trust is recognized as one mediator that influences clinicians' use and adoption of ML. Similarly, Shin and Park [15] highlight that user trust plays a crucial role in shaping potential adopters' willingness to undertake the inherent risks involved in adopting algorithm services. Likewise, the results in this paper [16] show that user trust acts as a liaison and interface between heuristic and systematic processing, facilitating ML service adoption. While user trust is inherently a subjective experience, Schmidt and Biessmann [14] build a function that establishes a link between the quality of ML explanation and user trust. This metric aids in identifying whether individuals are more biased toward the predictions made by ML systems.

2.2   Fidelity

Fidelity stands out as a vital property that impacts the quality of ML explanation. Indeed, its importance lies in the fact that a high-fidelity explanation can provide accurate and valuable information, encompassing the identification of important features and their significance to users. On the contrary, a low-fidelity explanation may result in the provision of meaningless insights. In essence, fidelity serves as a metric to measure how well an explanation method can mimic the behaviours of the underlying model [8]. Specially, Moradi and Samwald [9] employ fidelity as a metric to evaluate their proposed explanation method called Confident Itemsets Explanation, demonstrating its superior simulation of the underlying model compared to other ML explanation methods. Numerous researchers have devoted substantial efforts to exploring fidelity in

ML explanations. For instance, Dai et al. [6] develop a quantitative metric for measuring fidelity to evaluate the precision of the explanations. Besides, the paper [11] incorporates a user study focusing on textual explanations, which reveals that explanations with low fidelity significantly decrease user trust levels. Also, we recognize the significance of the relationship between fidelity and user trust in this study. However, our primary focus lies in exploring the associations between different explanation methods at varying levels of fidelity and user trust.

2.3 Robustness

The robustness of an explanation method refers to its sensitivity towards minor changes in the input, resulting in both prediction variations and corresponding adjustments in the explanation [3]. Alvarez-Melis and Jaakkola [1] utilize the metrics quantifying the robustness of explanation to evaluate the performance of current explanation methods (e.g. LIME and SHAP). Their findings highlight that while these methods provide explanations, they exhibit notable sensitivity to even slight input variations, thereby affecting their reliability. Moreover, the significance of robustness in XML has been extensively discussed in the literature [4,19]. Chan and Darwiche [4] introduce their algorithms to maintain the robustness of the Most Probable Explanation, thus facilitating the design and debugging of Bayesian networks in the presence of parameter changes. Furthermore, Tocchetti et al. [19] emphasize the importance of robustness in Graph Neural Networks (GNN) due to their vulnerability to adversarial attacks, where minor input alterations can lead to substantial output impacts. Considering the critical role of robustness, the expectation for the explanation method goes beyond mere reasonability, demanding even greater robustness. Our work focuses on exploring the connection between robustness in explanations and user trust through a user study.

3   Hypotheses

To elucidate our research questions, we formulate three hypotheses: H1 and H2 for the fidelity-based scenario, and H3 for the robustness-based scenario.

- H1: The level of user trust is impacted by the fidelity of explanations, wherein high-fidelity explanations uniquely contribute to higher user trust;
- H2: The level of user trust is influenced by distinct explanation methods when the level of fidelity remains constant;
- H3: The level of user trust is contingent upon the robustness of the XML. We posit that the higher robustness of the XML results in higher user trust.

4   Methodology

To investigate our three hypotheses, we conduct a case study focused on predicting student performance levels. We design two rounds, each comprising eight tasks, with consideration for both fidelity (six tasks) and robustness (two tasks).

4.1   Fidelity-Based Scenario Study Design

Fidelity denotes the extent to which an explanation accurately reflects the underlying model. To assess hypotheses H1 and H2, we incorporate three defined conditions: high-fidelity, low-fidelity, and without-fidelity. Under these conditions, we present visualizations employing two distinct explanation methods, LIME and SHAP (as delineated in Table 1). We establish the importance of features using the permutation importance method, as illustrated in Fig. 1, which serves as a ground truth for generating varying levels of fidelity in explanations.

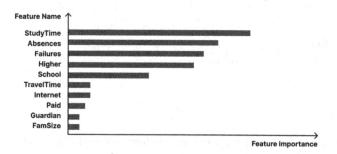

Fig. 1. Ground truth of feature importance.

- High-Fidelity: Under this condition, explanations effectively identify significant features that influence the predictions made by machine learning models. In this study, obtaining high-fidelity explanations involves generating explanations from LIME and SHAP, followed by manual adjustments to align them with the ground truth of feature importance.
- Low-Fidelity: In this condition, the explanations insufficiently identify significant features that impact the predictions made by machine learning models. To create low-fidelity explanations, substantial modifications are applied to the important features, deviating significantly from the ground truth of feature importance.
- Without-Fidelity: In this condition, both explanation methods exhibit none fidelity, effectively equating to a complete absence of any explanations.

4.2   Robustness-Based Scenario Study Design

Robustness in XML reflects that similar input should lead to similar explanations, characterized as insensitivity. To validate H3, two conditions are designed, high-robustness and low-robustness, wherein the LIME explanation method is exclusively employed (as illustrated in Table 1).

- High-robustness: In this condition, it is expected that the explanations remain relatively stable under minor feature modifications. In our implementation, achieving high-robustness explanations involves deliberately introducing slight variations in the weight and value of both important and unimportant features within subsequent explanations derived from high-fidelity explanations.

– Low-robustness: In this condition, explanations undergo significant changes when minor feature modifications are made. To deliver participants the low-robustness explanations, the condition is distinguished from the high-robustness condition by visualizing and simulating the explanations with considerable fluctuations of feature weights.

Table 1. Task set up in the experiments.

Methods	Level of Fidelity			Level of Robustness	
	High	Low	None	High	Low
LIME	Task 1.1	Task 2.1	Task 3.1	Task 4.1	Task 4.2
SHAP	Task 1.2	Task 2.2	Task 3.2	——	——

4.3 Metrics

To measure user trust, self-reported and behavior-based trust metrics are used.

Self-reported Trust Metric. The self-reported user trust (subjective user trust) level is assessed through a series of five subjective questions, with the first three questions derived from the measurements [17] and the last two questions formulated based on the metrics [10]. The self-reported user trust level is assessed on a 5-point Likert scale ranging from 1 (strongly disagree) to 5 (strongly agree).

– I trust the results from the ML explanation system.
– The ML explanation system is trustworthy.
– I believe that the results from the ML explanation system are reliable.
– I believe that the ML explanation system can explain well the reasons behind students' performance.
– I believe that the ML explanation system can provide a detailed explanation for each student's performance.

Behaviour-Based Trust Metric. We introduce this approach that takes into account the participants' behaviours in fidelity and robustness-based scenario experiments. It quantifies behaviour-based trust (objective user trust) as the frequency of appropriate decisions made by participants out of the total decisions undertaken. A higher average frequency indicates that participants exhibit higher user trust level in completing the tasks.

5 Experiment

5.1 Dataset

The student performance dataset in secondary education is employed as the foundational data source for this study. The original dataset, sourced from the UCI machine learning repository [5], contains 649 instances and comprises 30 nominal attributes which determine student numeric grades from 0 to 20. To streamline the complexity of this study, the dataset is reduced to 10 attributes, namely: school, guardian, paid, higher, famsize, traveltime, studytime, failures, absences, and internet. Furthermore, students' grades are categorized into distinct performance levels (A^+, A, B, C, D, and F). XGBoost is employed for performance level prediction, utilizing 70% of data for training and the rest for testing.

5.2 Participants

30 participants (15 males, 14 females, and 1 participant who prefers not to disclose their gender) were invited to participate in this study through our social networks. The participants' ages were distributed as follows: 18–24 years (6 participants), 25–34 years (24 participants), and 34–50 years (1 participant). Among the participants, 15 had completed their master's degree, followed by 12 participants who held bachelor's degrees, 2 participants were current Ph.D. students, and 1 who had completed an honours degree.

5.3 Experimental Procedure

The experiment is deployed on the Qualtrics platform. Participants begin with a welcome page that introduces the researchers and outlines the study's objectives. Following this, to proceed with the study, participants are required to provide their explicit consent from the consent form page. Prior to commencing the study tasks, participants are afforded the opportunity to familiarize themselves with the process through example and feature information pages.

Subsequently, the main study commences, where participants are presented with a random task. Each task involves one student instance with ten features, accompanied by its corresponding performance level and the explanations generated by a specific explainability method. Depending on the task types (fidelity or robustness), the requirements to complete the tasks are different. For fidelity-based tasks, participants are asked to make predictions with the supply of new student instances based on the ground truth of feature importance and provided explanations. In contrast, robustness-based tasks require participants to make predictions based on the new student instance, considering two successive minor modifications in the student instance and updated visualizations of the explanations. Upon the completion of each task, participants rate their level of user trust concerning the explanations using a 5-point Likert scale questionnaire.

Following the completion of the 16 tasks across the two rounds, participants are then required to fill out a demographic questionnaire, providing information on their gender, age, education level, and familiarity with machine learning explanations. The distribution of 16 tasks is randomized to ensure the prevention of bias in the collected results. The entire study is estimated to take participants between 15 to 25 min to complete.

6   Results

This section presents the results obtained from the experiments conducted in both the fidelity and robustness scenarios. To acquire these results, we perform a series of statistical tests, including one-way ANOVA tests, Tukey's HSD post-hoc tests, and paired t-tests to analyse the variations in user trust within these two scenarios.

6.1   Correlations Between User Trust and Fidelity

In order to examine the validity of H1 and H2, respectively, we initially implement one-way ANOVA tests and Tukey's HSD post-hoc tests to assess user trust variations in the different fidelity levels for each explanation. Subsequently, we conduct paired t-tests to further evaluate user trust differences when comparing two explanation methods. The implementation of Tukey's HSD post-hoc tests is more effective in reducing Type I errors when conducting multiple comparisons within a condition compared to paired t-tests.

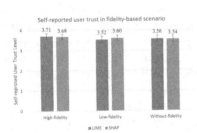

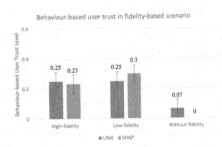

(a) Variations in user trust across three fidelity levels based on the self-reported trust metric for both LIME and SHAP

(b) Variations in user trust across three fidelity levels based on the behaviour-based trust metric for both LIME and SHAP

Fig. 2. Fidelity-based scenario: self-reported and behaviour-based trust metrics.

In the fidelity-based scenario experiment, variations in user trust level are first investigated among three fidelity levels when explanations are derived from the same methods. Self-reported user trust results indicate that participants trust high, low, and without-fidelity explanations. However, behaviour-based user trust results show trust in both high and low-fidelity explanations, while no trust in without-fidelity explanations. Figure 2(a) depicts the average subjective user trust levels across two different explanations at high, low, and without-fidelity levels (error bars correspond to standard errors and it is the same in other figures). One-way ANOVA tests at a 5% significance level (it is the same in other tests) indicate no significant user trust differences for both of the explanation methods based on the self-reported trust metric, thereby rejecting H1.

Nonetheless, one-way ANOVA tests find that there are significant user trust differences in the LIME ($F(2, 87) = 3.839$, $p = .025$) and SHAP ($F(2, 87) = 10.500$, $p < .000$) respectively when examining the behaviour-based user trust (see Fig. 2(b)). Then Tukey's HSD post-hoc tests are performed to explore the pair-wise user trust differences between the fidelity conditions for each explanation method. In the explanations generated by LIME and SHAP, Tukey's HSD post-hoc tests find that participants have higher user trust when the explanations have high-fidelity than those without-fidelity (LIME: $p = .048$; SHAP: $p = .003$). Participants also show higher user trust when the explanations have low-fidelity compared to those without-fidelity (LIME: $p = .048$; SHAP: $p < .000$). However, participants show the same user trust when the explanations are with high-fidelity compared to those low-fidelity (LIME: $p > .05$; SHAP: $p > .05$). These findings reject H1 by implying that users trust high and low-fidelity explanations while untrusting ones without fidelity.

On the other hand, variations in user trust level are then analysed between explanations from two different methods when keeping fidelity consistent. No statistically significant differences in user trust among the consistent fidelity levels have been observed when comparing the two explanation methods, as evidenced by both self-reported and behaviour-based trust metrics (see Fig. 2(a) and Fig. 2(b), respectively). The paired t-tests do not reveal any statistically significant differences in user trust across different explanation methods at each fidelity level. These outcomes, based on both self-reported and behaviour-based trust metrics, lead to the rejection of H2.

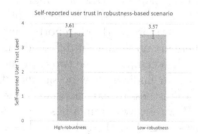

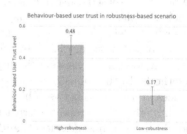

(a) Variations in user trust in the high and low-robustness based on the self-reported trust metric for LIME.

(b) Variations in user trust in the high and low-robustness based on the behaviour trust metric for LIME.

Fig. 3. Robustness-based scenario: self-reported and behaviour-based trust metrics.

6.2   Correlations Between User Trust and Robustness

To evaluate the validity of H3, we perform paired t-tests to quantify user trust differences with respect to the robustness level based on both the self-reported and behaviour-based user trust levels.

In the robustness-based scenario, variations in user trust level are examined between the high and low levels of robustness when a single explanation method is employed. The results of the scenario demonstrate no significant differences in user trust between the high and low robustness levels through the analysis of the self-reported user trust levels. Conversely, analysing the behaviour-based user trust levels shows a positive correlation between user trust and robustness level. Figure 3(a) illustrates the average subjective user trust level for the high and low-robustness levels, based on the self-reported trust metric. The results of the paired t-tests, suggesting no statistically significant differences in user trust levels between the high and low robustness levels, reject our H3 based on the self-reported trust metric.

However, Fig. 3(b) depicts the mean objective user trust levels for the high and low-robustness levels, respectively. Another paired t-test is conducted to analyse the objective user trust levels, revealing statistically significant differences in user trust levels between the high and low robustness conditions ($t = 3.842$, $p < .000$). These results suggest a positive correlation between user trust and robustness, indicating that higher robustness levels lead to higher user trust. As a result, our H3 is confirmed based on the behaviour-based trust metric.

7   Discussion

This study investigates the user trust differences in the fidelity and robustness of ML explanation under a specific condition, respectively. In the fidelity-based scenario, the analysis of the results of the self-reported trust metric indicates that participants do not exhibit significant differences in trust toward explanations with different levels of fidelity. On the contrary, the analysis of the results of the behaviour-based trust metric shows that participants trust both high and low-fidelity explanations, while a majority of them do not trust the absent explanations when fidelity levels are comparable. Furthermore, both self-reported and behaviour-based trust metrics indicate that there are no significant differences in user trust among the consistent fidelity levels when comparing the two explanation methods. In the robustness-based scenario, our findings show no significant differences in user trust between high and low-robustness explanations, as determined through the self-reported trust metric. However, we obtain a positive correlation between robustness levels and user trust based on the behaviour-based trust metric.

Our findings carry significant implications for the evaluation of the quality of ML explanations. For example, if a unified method for measuring the quality of ML explanations is established, fidelity and robustness could not only serve as contributing factors to this method but also complement the evaluation of user trust. Moreover, participants significantly trust both high and low-fidelity explanations compared to without-fidelity explanations. This suggests that explanations incorporating fidelity can effectively build user trust in ML applications. However, trust in high and low fidelity may pose a challenge for users as they attempt to discern the reliability of the provided explanations. Additionally, the observed positive correlation between user trust and robustness, as assessed through the behaviour-based trust metric, can be applied in the implementation of algorithms to enhance user trust in ML systems.

8   Conclusion and Future Work

The quality of ML explanations significantly influences the selection of effective explanation methods, encompassing user-related, explanation-related, and model-related aspects. Among these aspects, user trust holds particular importance and serves as a key objective in the explanatory process. Both fidelity and robustness also play essential roles in shaping the quality of explanations and ultimately impacting user trust. Through an investigation of the relationships between user trust, fidelity, and robustness, we contribute to a comprehensive understanding of these factors in the context of ML explanations. Our findings indicate that 1) participants trust both high and low-fidelity explanations than without-fidelity explanations in the behaviour-based user trust, contrary to the results from self-reported user trust; 2) no significant differences in user trust exist when comparing explanations with consistent fidelity level; 3) a higher robustness level positively influences user trust in the behaviour-based user trust analysis rather than in the self-reported user trust analysis. Moving forward, our future work aims to establish a comprehensive metric that combines subjective perceptions and objective properties to effectively evaluate the quality of ML explanations. Additionally, we recognize the need for improvement in the visualization of explanations in our user study.

References

1. Alvarez-Melis, D., Jaakkola, T.S.: On the robustness of interpretability methods (2018). https://arxiv.org/abs/1806.08049. arXiv:1806.08049
2. Asan, O., Bayrak, A.E., Choudhury, A.: Artificial intelligence and human trust in healthcare: focus on clinicians. J. Med. Internet Res. **22**(6), e15154 (2020). https://doi.org/10.2196/15154. Company: Journal of Medical Internet Research Distributor: Journal of Medical Internet Research Institution: Journal of Medical Internet Research Label: Journal of Medical Internet Research Publisher: JMIR Publications Inc., Toronto, Canada
3. Carvalho, D.V., Pereira, E.M., Cardoso, J.S.: Machine learning interpretability: a survey on methods and metrics. Electronics **8**(8), 832 (2019). https://doi.org/10.3390/electronics8080832
4. Chan, H., Darwiche, A.: On the robustness of most probable explanations (2012). https://arxiv.org/abs/1206.6819. arXiv:1206.6819
5. Cortez, P.: Student performance. UCI Machine Learning Repository (2014). https://doi.org/10.24432/C5TG7T
6. Dai, J., Upadhyay, S., Aivodji, U., Bach, S.H., Lakkaraju, H.: Fairness via explanation quality: evaluating disparities in the quality of post hoc explanations. In: Proceedings of the 2022 AAAI/ACM Conference on AI, Ethics, and Society, pp. 203–214 (2022). https://doi.org/10.1145/3514094.3534159. arXiv:2205.07277
7. Holzinger, A., Biemann, C., Pattichis, C.S., Kell, D.B.: What do we need to build explainable AI systems for the medical domain? (2017). https://arxiv.org/abs/1712.09923. arXiv:1712.09923
8. Löfström, H., Hammar, K., Johansson, U.: A meta survey of quality evaluation criteria in explanation methods (2022). https://arxiv.org/abs/2203.13929. arXiv:2203.13929

9. Moradi, M., Samwald, M.: Post-hoc explanation of black-box classifiers using confident itemsets. Exp. Syst. Appl. **165**, 113941 (2021). https://doi.org/10.1016/j.eswa.2020.113941arXiv:2005.01992

10. Pan, Y., Froese, F., Liu, N., Hu, Y., Ye, M.: The adoption of artificial intelligence in employee recruitment: the influence of contextual factors. Int. J. Hum. Res. Manage. **33**(6), 1125–1147 (2022). https://doi.org/10.1080/09585192.2021.1879206

11. Papenmeier, A., Englebienne, G., Seifert, C.: How model accuracy and explanation fidelity influence user trust (2019). https://arxiv.org/abs/1907.12652. arXiv:1907.12652

12. Ribeiro, M.T., Singh, S., Guestrin, C.: Why should i trust you?: explaining the predictions of any classifier. In: Proceedings of the 22nd ACM SIGKDD International Conference on Knowledge Discovery and Data Mining, pp. 1135–1144. ACM, San Francisco California USA (2016). https://doi.org/10.1145/2939672.2939778

13. Sanneman, L., Shah, J.A.: The situation awareness framework for explainable AI (SAFE-AI) and human factors considerations for XAI systems. Int. J. Hum.-Comput. Interact. **38**(18–20), 1772–1788 (2022). https://doi.org/10.1080/10447318.2022.2081282

14. Schmidt, P., Biessmann, F.: Quantifying interpretability and trust in machine learning systems (2019). https://arxiv.org/abs/1901.08558. arXiv:1901.08558

15. Shin, D.: Role of fairness, accountability, and transparency in algorithmic affordance. Comput. Hum. Behav. **98**, 277–284 (2019)

16. Shin, D.: How do users interact with algorithm recommender systems? The interaction of users, algorithms, and performance. Comput. Hum. Behav. **109**, 106344 (2020)

17. Shin, D.: The effects of explainability and causability on perception, trust, and acceptance: implications for explainable AI. Int. J. Hum Comput Stud. **146**, 102551 (2021). https://doi.org/10.1016/j.ijhcs.2020.102551

18. Tintarev, N.: Explaining recommendations. Ph.D. thesis, University of Aberdeen, UK (2009)

19. Tocchetti, A., et al.: A.I. robustness: a human-centered perspective on technological challenges and opportunities (2022). https://arxiv.org/abs/2210.08906. arXiv:2210.08906

20. Zhou, J., Gandomi, A.H., Chen, F., Holzinger, A.: Evaluating the quality of machine learning explanations: a survey on methods and metrics. Electronics **10**(5), 593 (2021). https://doi.org/10.3390/electronics10050593

21. Zhou, J., Verma, S., Mittal, M., Chen, F.: Understanding relations between perception of fairness and trust in algorithmic decision making (2021). https://arxiv.org/abs/2109.14345. arXiv:2109.14345

Interpretable Drawing Psychoanalysis via House-Tree-Person Test

Yaowu Xie, Ting Pan, Baodi Liu, Honglong Chen, and Weifeng Liu[✉]

China University of Petroleum (East China), Qingdao, China
liuwf@upc.edu.cn

Abstract. As the number of people with psychological disorders continues to increase in recent years, it is particularly important to identify patients at an early stage. As one of the widely recognized methods of drawing psychoanalysis methods, the House-Tree-Person (HTP) test is commonly used for psychological assessment. Among the standards for HTP drawing analysis, the analysis of holistic features such as size and position is the most important evaluation standard. Traditional manual analysis for test results is time-consuming and highly subjective. Although image classification models provide fast and accurate predictions, they lack reliable decision-making interpretation. Existing methods for interpreting classification models can offer a certain degree of interpretation, but they are unable to provide specific interpretation related to psychology. In this paper, we propose a method to quantitatively analyze the interpretable results of decisions made by image classification models in relation to expert knowledge. Specifically, we initially utilize the interpretable method to identify the important features that influence the model's decision-making process. Then, we use self-annotated files containing important symbols relevant to psychology to segment important features, obtaining psychological features associated with the model's decision-making process. Finally, we quantitatively analyze the size and position through the psychological features respectively, comparing them with the widely used expert knowledge and rules. The experimental results reveal that the decisions of image classification model on the HTP dataset are consistent with the relevant universal rules of psychological features. The proposed method improves the reliability of image classification algorithm in psychotherapy and psychodiagnosis.

Keywords: Drawing psychoanalysis · HTP test · Feature quantification · Explainable artificial intelligence

1 Introduction

With the development of society and people's aspiration for a better life, the incidence of psychological disorders continues to rise. Early detection and timely treatment of mental illnesses are crucial. However, psychological disorders often have latent and covert characteristics, making them difficult to detect. Drawing

T. Liu et al. (Eds.): AI 2023, LNAI 14472, pp. 221–233, 2024.
https://doi.org/10.1007/978-981-99-8391-9_18

psychological analysis is a projection test that allows patients to express their true emotions through drawings. Its characteristics lie in its flexibility and ease of administration [1]. Among these methods, the House-Tree-Person (HTP) test is the most classical technique in psychological drawing analysis [2]. It primarily involves analyzing the drawings of houses, trees, and human figures by the subjects to assess their psychological states. American psychologist Buck pointed out that houses, trees, and human figures can project psychological meanings and personality traits in the American Journal of Clinical Psychology [3].

The analysis of the HTP test by psychological experts encompasses at least three levels. The overall analysis of the drawings is the most significant evaluation criterion, which involves assessing the position and size of the drawings. For instance, an excessively large or small drawing or deviations in position are considered negative features, indicating possible negative emotions in the subject. However, this analysis method is subject to the therapist's subjective intentions and experiential judgment, thus limiting its large-scale application.

In recent years, artificial intelligence has been extensively employed in various complex tasks, and computer vision has made remarkable progress, especially in image recognition tasks. The drawings in the HTP test are a form of images. As a result, some researchers have attempted to use deep learning methods to detect and analyze these drawings, replacing the role of therapists and achieving objective and reliable assessments. Liu et al. [4] proposed a method that employs convolutional neural networks to classify the content of HTP drawings. Meanwhile, Kim et al. [5] utilized deep learning to identify and classify drawings to assist in art therapy processes. However, as the depth of neural networks increases, their complexity also grows, and the internal decision mechanisms become harder to comprehend, resembling a black box.

Due to the need for explaining the reasons behind the model to users and experts, Explainable Artificial Intelligence (XAI) has emerged and garnered increasing attention in the field of artificial intelligence. Researchers have proposed various methods aimed at interpreting the decisions of deep learning models in a human-understandable manner, particularly concerning image classification tasks. Depending on the scope of interpretation, interpretable methods can be categorized into global interpretability and local interpretability. Global XAI methods seek to comprehend the behavior of models across the entire dataset, while local XAI methods aim to understand models' predictions for individual samples. Furthermore, they strive to identify the influence of each input feature on specific outputs [6]. Among local interpretability methods, Class Activation Mapping (CAM) stands out as a popular category, often employed to enhance the interpretability of Convolutional Neural Networks. CAM, initially proposed by Zhou et al. [7], generates heatmap visualizations that reveal the most salient regions within task-specific images of CNN models. These visualizations are typically obtained by linearly weighting feature maps produced by the final convolutional layer of the network. Researchers have introduced various variations to enhance the original CAM formulation [8–13], aiming to render the interpretable results more trustworthy.

The class activation mapping can provide interpretations for the feature regions influencing the decisions of classification models. However, it falls short in offering human-understandable interpretations that are directly related to psychological factors. In this paper, we propose a quantification method for interpreting the results of image classification networks based on relevant expert knowledge. Specifically, we first use image classification model ResNet50 to learn from our constructed HTP dataset and utilize GradCAM++ to interpret the network's decisions. Subsequently, we combine the interpretable results with our self-annotated files containing crucial psychological symbols to extract the psychological features relevant to the decisions. Finally, we perform an overall analysis of the psychological features, including their positions and sizes. Experimental results demonstrate that our proposed quantification method aligns well with widely accepted expert knowledge, providing a more reliable interpretation for image classification networks, so as to provide a more reasonable decision basis for psychological experts and provide better service for the patients. The main contributions of our research are as follows:

1. We propose a method for quantifying the interpretable results of image classification networks based on relevant expert knowledge, which can categorize the interpretable results of image classification models into psychological features, providing psychologists with trustworthy decision bases.
2. We construct an HTP dataset containing important psychological symbols, with the drawings sourced from a university's psychological testing center, and provide annotations for it.
3. We conduct extensive experiments, and the overall analysis of psychological features reveals that our proposed method is consistent with widely applied expert knowledge and rules.

2   Related Works

2.1   Drawing Psychoanalysis

In the field of psychology and clinical application research, the subjects are given pencils, erasers and white papers, requiring them to draw on the white paper. According to certain standards, these pictures are analyzed and assessed, aiming to understand the psychological phenomenon, intelligence state or personality characteristics of the subjects, for clinical psychological assessment, diagnosis and treatment services. All of this form and types of psychological tests are collectively referred to as drawing tests. Drawing psychoanalysis can reflect the personality characteristics of people in psychology and is widely used in individual and group tests [14]. Carter et al. [15] applied drawing psychoanalysis to find differences between normal and autistic children.

HTP drawing projection test, referred to as HTP test, belongs to the psychological projection test method. The test requires subjects to draw pictures of houses, trees and people on white paper with a pen, and then professionals analyze and evaluate the pictures according to the psychological projection principles, so as to visualize the psychological characteristics of subjects. Quickly understand the general intelligence level, personality integration, interpersonal relationship and the current emotional state, so as to facilitate the follow-up clinical psychological interview and evaluation. HTP test is one of the widely used methods in drawing psychoanalysis, which can be used to evaluate and treat patients with schizophrenia, anxiety and depression. Some researchers have used qualitative methods to study the HTP drawing characteristics of computers and drug users and children who have experienced earthquakes [16]. The HTP test is a psychodynamically based projection test that allows the subject to express unlimited mental states and activities hidden in the subconscious [17].

2.2   Class Activation Mapping

CAM (class activation mapping) algorithm [7] interprets the decision of the network by locating the region related to network decision in the input image. Zhou et al. [7] proposed the class activation mapping algorithm in 2016. It is illustrated that convolutional neural networks used for classification tasks have extraordinary positioning capabilities, and it shows the features concerned by the high-level neurons of the network. CAM algorithm needs to use the fully connected layer of the convolutional network with global average pooling layer [18]. Instead, the output feature graph of the last convolution layer is up-sampled to obtain the feature graph of the same size as the input image, and finally the class activation graph is obtained by the calculated linear combination.

In order to apply CAM algorithm to the general network, Guided Grad-CAM algorithm is proposed in literature [19]. By combining GBP algorithm with Grad-CAM algorithm, the features obtained from the input of each layer of neurons can be located. There is no longer any restriction on the network. By calculating the gradient of all feature maps of a certain layer by the output class score, and averaging the gradient on the channel granularity, the linear combination coefficient of the feature maps of the layer can be obtained. The final class activation graph is also obtained by linear combination of feature maps, but the difference is that ReLU function required for the combined results is activated to extract positive gradient information.

In order to solve the problem that GradCAM can not locate multiple same objects in the same graph clearly, Chattopadhay et al. [20] proposed Grad-CAM++ which by measuring the importance of each pixel in the feature map to the overall decision of CNN, it is used to visually interpret the CNN-based architecture. A weighted combination of the positive partial derivatives of a particular class score is used as a weight by the last convolutional layer feature plot to generate a visual interpretation of the corresponding class label. In addition, the closed-form solution is derived, and the high-order accurate expression is obtained, and only one backpropagation is required. The computational amount is consistent with that of previous gradient-based methods, which is an effective generalization of gradient-based visual interpretable methods.

3   Method

Our method mainly includes two modules, quantization of size and quantization of position, as shown in the Fig. 1. In this section, we will introduce these two quantization methods in detail.

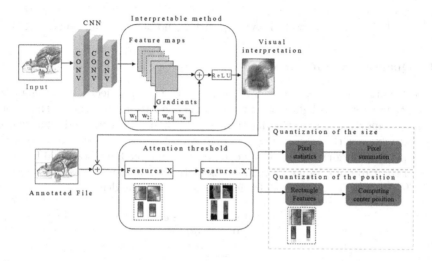

Fig. 1. The framework of our method. The interpretable algorithm interpretable classification model, interpretable results fuse and segment annotated files containing mental symbols to obtain mental features, and quantified the position and size of the mental feature region concerned by the network.

Quantification of the size. This module quantifies the size of various psychological features concerned by the network in the sample to reflect the impact of some psychological features on network decision-making.

Quantification of the position. This module quantifies the position of various psychological features concerned by the network in the samples, and reflects the influence of some psychological features on network decision-making.

Using pre-trained models in deep learning is called transfer learning. The core idea is that the convolutional neural network learns general low-level features in the shallow layer, and the last layer learns specific categories of information. Pre-trained models such as VGGNet [21], ResNet [22] and other networks are trained with the ImageNet dataset [23]. In order to use the pre-trained model on the dataset, it is usually to remove the last layer and adds a new connection layer. So we use ResNet50 as the pre-training model, removing the last layer and adding a two-class fully connected layer.

After the HTP dataset is trained by the ResNet50. The GradCAM++ is used to obtain visual interpretations that the network makes the decision. HTP test involves some important psychological symbols, so we mark the symbols and fuse them with the interpretable results to obtain the psychological features that the network decision depends on. Each psychological feature region is composed of two parts, the region related to network decision making and the region unrelated to network decision making. The analysis of the whole feature is mainly aimed at the position and size of the psychological feature region related to the network decision. In the evaluation criteria, the size is too large means that the size is greater than two-thirds of the drawing, and the size is too small means that the size is less than one-ninth of the drawing. Position deviation refers to the deviation between the overall drawing position and the drawing center [3,24].

3.1   Quantization of the Size

The psychological feature regions depended on network decision making are mainly the regions highlighted by interpretable algorithms. The redder the color, the more important it is to the decisions made by the network. And what we've done is to extract this size.

We set a attention threshold range, compare each pixel in the mental feature, leave in the threshold range, and then calculate the number of pixels left in the area, that is, the size of the mental feature related to the decision.

$$\alpha_{ij} = \begin{cases} 1, & p_{ij} \in \phi \\ 0, & others \end{cases} \tag{1}$$

$$S_{gc} = \sum_i \sum_j \alpha_{ij} \tag{2}$$

Because there are multiple psychological features in each sample, we sum the sizes of psychological features in the same class according to the features corresponding to each psychological feature.

$$S_l = \sum_{g=1}^{m} S_{gc} \tag{3}$$

Finally, the total sizes of all categories are summed to obtain the total size of mental features related to network decision making in the whole drawing.

$$S_{total} = \sum_{l=1}^{\lambda} S_l \tag{4}$$

3.2   Quantization of the Position

By setting the attention threshold, the psychological feature regions concerned by the network are extracted, and the average coordinates of the center points of all the psychological feature regions related to the network decision are obtained.

The four coordinates of the coordinates (x, y, w, h) of the minimum external rectangle are calculated by fitting the extracted mental feature region into one, where (x, y) is the center point coordinate.

$$P_{gc}(x,y) = Rect(p_{ij} \in \phi) = (x,y) \tag{5}$$

Because there are multiple psychological features in each sample, we sum and average the central point coordinates of the same category according to the corresponding features of each psychological feature, and obtain the central point coordinates of each psychological feature.

$$P_l(x,y) = \left(\frac{\sum_{g=1}^{m} P_{gc}(x)}{m}, \frac{\sum_{g=1}^{m} P_{gc}(y)}{m} \right) \tag{6}$$

Finally, the center coordinates of each psychological feature are summed and averaged to get the total center coordinates of the psychological features related to network decision making in the sample.

$$P_{total}(m,n) = \left(\frac{\sum_{l=1}^{\lambda} p_l(x)}{\lambda}, \frac{\sum_{l=1}^{\lambda} p_l(y)}{\lambda} \right) \tag{7}$$

4 Experiments

4.1 The HTP Dataset

We creat an HTP dataset to support relevant HTP test research. We collect over 3000 images from the university's psychological testing center. In terms of the range of participants, the tested individuals cover 8 academic disciplines, aiding in reflecting the psychological states of university students and enhancing the diversity of the dataset. The testing period spans from 2010 to 2021, enabling capturing the changes in participants' psychological states over time.

To fully utilize test images and minimize the impact of objective disturbances, we conduct the following operations on all images in the dataset:

1. Image scanning: We use two resolutions for scanning, 3504×2480 (horizontal) and 2480×3504 (vertical). After scanning, images are transformed into color images with three channels for studying in the deep learning models.
2. Image denoising: We apply denoising processing to the images to avoid unnecessary impact from objective factors.
3. Image annotation: We annotate objects related to psychological states in the images, comprising 9 categories, including house, tree, person, sun, cloud, dog, bird, chimney, and fruit.

Evaluation standards for drawing results are mainly based on the application of HTP test evaluation criteria and the causes, characteristics, prevention, and intervention strategies for psychological crises among university students [25]. Due to the subjective nature of psychologists' understanding of evaluation criteria, we invite three psychologists to evaluate the drawings. We grouped all test drawings, with each group containing 50 images. Each psychologist rests for at least half an hour after completing an evaluation group to avoid errors. Finally, by analyzing each psychologist's evaluations of each image, we retained images with consistent evaluation results. If there were differences in evaluations, we re-evaluated the images. Images with significantly divergent evaluation opinions were deleted. In the end, we collected a total of 3046 accurate drawing results. In our experiment, we randomly selected 250 images with positive connotations to ensure a 1:1 ratio of positive to negative images. Finally, the proportion of training set to validation set was 7:3, with 70% of the training set used for model training and 30% for model evaluation as shown in the Table 1.

Table 1. The number of drawings in the HTP datatset.

The HTP dataset	Positive samples	Negative samples	Total samples
Training set	179	171	350
validation set	71	79	150

4.2   Implementation Detials

This experiment uses ResNet50 for the image classification task of HTP test. ResNet50 uses training sets for training and validation sets for validation. The hyperparameters are as follows: The classification loss uses the cross entropy loss. The optimizer is stochastic gradient descent, the learning rate is 0.001, the momentum is 0.8, and the weight-decay is 5.00E−05. The lower bound of the threshold for extracting the area of interest is [140, 60, 60] and the upper bound is [255, 190, 204].

4.3   Experiment Results

The ResNet50 model is trained in the HTP dataset, and the training set accuracy reached 94.6%. Note ResNet50 can perform high-performance classification of samples in HTP datasets. The model has such high accuracy, but not knowing the basis of its decision, whether the basis is reasonable, does not provide any useful information for the therapist. To provide trusted visual explanations to the therapist. We apply GradCAM++ to interpret how ResNet50 classifies the samples. Because GradCAM++ is a gradient-based CANM method, which do not change model architecture and has a better explanatory ability in convolutional neural networks.

GradCAM++ highlights the features used by the network. From the Fig. 2, it can be observed that the network has focused on some features related to psychology, indicating that the network's decisions are related to the content of psychological drawings. However, it does not provide interpretations for the finer and more meaningful features, making it difficult to understand the specific psychological characteristics the network is focusing on. To further refine the interpretable results, we integrate the annotated files containing 9 classes of psychologically relevant symbols with the interpretable method's results, aiming to achieve a more detailed interpretation of the psychological features.

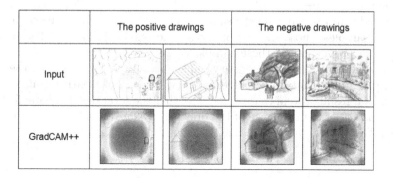

Fig. 2. Interpretable renderings of the interpretable algorithm GradCAM++ for the image classification network

We conducted a quantity analysis of the 9 classes of psychological symbols in the annotated files. Among the Fig. 3, the symbols representing houses, trees and person appear most frequently, while symbols of other classes appear relatively less frequently in both positive and negative samples. Through the fusion and segmentation of annotated files and interpretable results, we can obtain fine-grained psychological characteristics on which models rely for decision making.

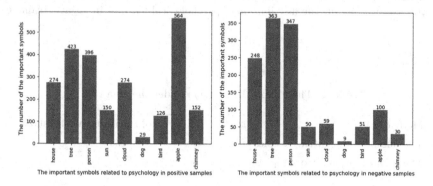

Fig. 3. The statistics of 9 classes important symbols related to psychology in positive and negative samples. Left: the mathematical statistics of mental symbols in positive samples. Right: the mathematical statistics of mental symbols in negative samples.

We quantify the total sizes of the three main psychological features (house, tree, person), and present the statistical results in the Fig. 4. The distribution of negative samples is more divergent than that of positive samples. The rule of dividing samples by size reveals that the size of psychological features between 1/9 and 2/3 of the drawing size is classified as a positive sample, and vice versa. Our sample size is 224 × 224, and the quantified size of each sample is associated with the rule. The result is listed in the Table 2. There are 243 positive samples that meet the above rule, while 81 samples in the negative sample meet the rule. For those samples whose size is not within the regular interval of the positive sample, the proportion of those sample is 2.8%, while the proportion of the negative sample is 32.4%. Negative samples accounted for 92.0% of the total number of samples that do not fit the regular interval of positive samples. The experimental results reveal that the samples that the total size of mental features is less than 1/9 of the total size of the drawing or more than 2/3 is highly likely the negative samples, and there may be mental illness.

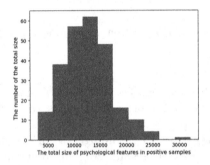 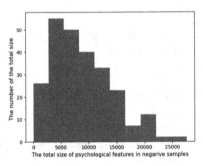

Fig. 4. The distribution of total size of psychological features concerned by the network in positive and negative samples. Left: the distribution of psychological features in positive samples. Right: the distribution of psychological features in negative samples.

Table 2. The sample distribution under the area rule

Methods	outside the threshold	within the threshold
The number of negative samples	81	169
The number of positive samples	7	243

We quantitatively and analyze the positions of the psychological features that the network focuses on across all samples. According to criteria related to location, the central positions of psychological features for positive samples are situated within the middle region, while for negative samples, they are located at the edges. Consequently, we define a square region with a side length equal to half of the sample's length, centered at the midpoint of the sample. Within this defined square area, the central position of psychological features that the network attends to is considered a positive sample; otherwise, it is deemed a negative sample. As shown in the Table 3, there are 249 positive samples with their central positions falling within this area. Outside this square area, there are 21 negative samples, accounting for 8.1%, and only 1 positive sample, representing 0.4%. The experimental results disclose that when the central position of the psychological features that the network attends to is situated outside this square area, the sample is highly likely to be a negative one. In other words, samples with central coordinates of psychological features located along the edges are more likely to be negative samples.

Table 3. The sample distribution under the location threshold

	outside the threshold	within the threshold
The number of negative samples	21	229
The number of positive samples	1	249

5   Conclusion

Image classification networks demonstrate strong performance but lack interpretability. Local interpretable methods can only reveal the features on which network decisions rely, falling short of comprehensible psychological features. This study proposes a quantification method that correlates interpretable results of network decisions with expert knowledge to achieve credible interpretations of psychologically relevant features in network decisions. Experimental results on the HTP dataset validate the alignment of interpretive outcomes with widely applied expert knowledge of psychological features. Analyzing classification model decisions with psychological relevance still faces numerous challenges. This paper requires a lot of manual work in the early stage. Future research will continue to explore more automatic interpretable analysis methods, providing real-time and reliable interpretations for patients and psychological experts.

Acknowledgements. The paper has been supported by the National Natural Science Foundation of China (Grant No. 62372468, 61671480), the Shandong Natural Science Foundation (Grant No. ZR2023MF008), the Qingdao Natural Science Foundation(Grant No. 23-2-1-161-zyyd-jch), and the Major Scientific and Technological Projects of CNPC (Grant No. ZD2019-183-008).

References

1. Leibowitz, M.: Interpreting Projective Drawings: A Self-psychological Approach. Routledge, London (2016)

2. Zhou, A., Xie, P., Pan, C.: Performance of patients with different schizophrenia subtypes on the synthetic house-tree-person test. Soc. Behav. Personal. Int. J. **47**(11), 1–8 (2019)
3. Buck, J.N.: The H-T-P test. J. Clin. Psychol. **6**, 78503–78512 (1948)
4. Liu, L.: Image classification in HTP test based on convolutional neural network model. Comput. Intell. Neurosci. (2021)
5. Kim, T., Yoon, Y., Lee, K.: Application of deep learning in art therapy. Int. J. Mach. Learn. Comput. **11**(6) (2021)
6. Du, M., Liu, N., Hu, X.: Techniques for interpretable machine learning. Commun. ACM **63**(1), 68–77 (2019)
7. B, Zhou., Khosla, A., Lapedriza, A.: Learning deep features for discriminative localization. In: Proceedings of the IEEE Conference on Computer Vision and Pattern Recognition, pp. 2921–2929 (2016)
8. Ramaswamy, H G.: Ablation-CAM: visual explanations for deep convolutional network via gradient-free localization. In: proceedings of the IEEE/CVF Winter Conference on Applications of Computer Vision, pp. 983–991. IEEE (2020)
9. Draelos, R.L., Carin, L.: Use HiResCAM instead of Grad-CAM for faithful explanations of convolutional neural networks. arXiv preprint arXiv:2011.08891 (2020)
10. Fu, R., Hu, Q., Dong, X.: Axiom-based grad-CAM: towards accurate visualization and explanation of CNNs. arXiv preprint arXiv:2008.02312 (2020)
11. Jiang, P.T., Zhang, C.B., Hou, Q.: LayerCAM: exploring hierarchical class activation maps for localization. IEEE Trans. Image Process. **30**, 5875–5888 (2021)
12. Muhammad, MB., Yeasin, M.: Eigen-CAM: class activation map using principal components. In: 2020 International Joint Conference on Neural Networks (IJCNN), pp. 1–7 (2020)
13. Wang, H., Wang, Z., Du, M.: Score-CAM: score-weighted visual explanations for convolutional neural networks. In: Proceedings of the IEEE/CVF Conference on Computer Vision and Pattern Recognition Workshops, pp. 24–25 (2020)
14. Lan, Y., Nagai, Y.: Research about children's painting education method based on house tree person test. West East J. Soc. Sci. **8** (2019)
15. Carter, C.K., Hartley, C.: Are children with autism more likely to retain object names when learning from colour photographs or black-and-white cartoons. J. Autism Dev. Disord. **51**(9), 3050–062 (2021)
16. Afolayan, A.: Haitian children's house-tree-person drawings: global similarities and cultural differences. Ph.D. dissertation, Antioch University (2015)
17. Di Leo, J.H.: Interpreting Children's Drawings. Routledge, London (2013)
18. Lin, M., Chen, Q., Yan, S.: Network in network. arXiv preprint arXiv:1312.4400 (2013)
19. Selvaraju, RR., Cogswell, M., Das, A.: Grad-CAM: visual explanations from deep networks via gradient-based localization. In: Proceedings of the IEEE International Conference on Computer Vision, pp. 618–626 (2017)
20. Chattopadhay, A., Sarka, rA., Howlader, P.: Grad-CAM++: generalized gradient-based visual explanations for deep convolutional networks. In: 2018 IEEE Winter Conference on Applications of Computer Vision (WACV), pp. 839–847. IEEE (2018)
21. Simonyan, K., Zisserman, A.: Very deep convolutional networks for large-scale image recognition. arXiv preprint arXiv:1409.1556 (2014)
22. He, K., Zhang, X., Ren, S.: Deep residual learning for image recognition. In: Proceedings of the IEEE Conference on Computer Vision and Pattern Recognition, pp. 770–778 (2016)

23. Deng, J., Dong, W., Socher, R.: ImageNet: a large-scale hierarchical image database. In: 2009 IEEE Conference on Computer Vision and Pattern Recognition, pp. 248–255 (2009)
24. Burns, R.C.: Kinetic-House-Tree-Person Drawings (KHTP): An Interpretative Manual. Brunner/Mazel (1987)
25. Hongtao, J.: Usage of painting art therapy in mental health education of Chinese college students. High. Educ. Oriental Stud. 1(2) (2021)

A Non-asymptotic Risk Bound for Model Selection in a High-Dimensional Mixture of Experts via Joint Rank and Variable Selection

TrungTin Nguyen[1] , Dung Ngoc Nguyen[2]([✉]) , Hien Duy Nguyen[3] , and Faicel Chamroukhi[4]

[1] Univ. Grenoble Alpes, Inria, CNRS, Grenoble INP, LJK, Inria Grenoble Rhone-Alpes, 655 av. de l'Europe, 38335 Montbonnot, France
trung-tin.nguyen@inria.fr
[2] Department of Statistical Sciences, University of Padova, Padua, Italy
ngocdung.nguyen@unipd.it
[3] School of Mathematics and Physics, University of Queensland, Brisbane, Australia
h.nguyen7@uq.edu.au
[4] IRT SystemX, Palaiseau, France
faicel.chamroukhi@irt-systemx.fr

Abstract. We are motivated by the problem of identifying potentially nonlinear regression relationships between high-dimensional outputs and high-dimensional inputs of heterogeneous data. This requires regression, clustering, and model selection, simultaneously. In this framework, we apply the mixture of experts models which are among the most popular ensemble learning techniques developed in the field of neural networks. In particular, we consider a more general case of mixture of experts models characterized by multiple Gaussian experts whose means are polynomials of the input variables and whose covariance matrices have block-diagonal structures. More especially, each expert is weighted by a gating network that is a softmax function of a polynomial of the input variables. These models require several hyper-parameters, including the number of mixture components, the complexity of the softmax gating networks and Gaussian mean experts, and the hidden block-diagonal structures of the covariance matrices. We provide a non-asymptotic theory for model selection of such complex hyper-parameters using the slope heuristic approach in a penalized maximum likelihood estimation framework. Specifically, we establish a non-asymptotic risk bound on the penalized maximum likelihood estimation, which takes the form of an oracle inequality, given lower bound assumptions on the penalty function.

Keywords: Dimensionality reduction · Low rank estimation · Mixture of experts · Finite mixture regression · Non-asymptotic model selection · Oracle inequality · Variable selection

Supplementary Information The online version contains supplementary material available at https://doi.org/10.1007/978-981-99-8391-9_19.

1   Introduction

Mixture of experts (MoE) models, introduced by Jacobs et al. [16] are widely applied to decompose the prediction model through a combination of gating models and expert models, both of which depend on the input variables. These flexible models are specific instances of conditional computation [3], where different model experts are responsible for different regions of the input space. Thus, by applying only a subset of parameters to each example, MoE can increase model capacity while keeping training and inference costs roughly constant. For reviews on this topic, we refer to [19,25]. Furthermore, they have gained popularity due to universal approximation properties in various special cases, including mixture models [28,30], mixture of regression models [15], and fully-parameterized mixture of experts models [26,27]. In high-dimensional multivariate multiple regression for heterogeneous data, we refer to outputs $\mathbf{Y} \in \mathcal{Y} \subset \mathbb{R}^Q$ as target or response variables, and inputs $\mathbf{X} \in \mathcal{X} \subset \mathbb{R}^P$ as explanatory or predictor variables, where Q and P are both much larger than the sample size. Additionally, hidden interactions may exist in the graphical structure between response variables. In such cases, regression, clustering, and model selection need to be performed simultaneously. Consequently, we employ MoE models to identify potential non-linear relationships between output and input variables in the high-dimensional heterogeneous data. We assume that $\mathbf{Y}$, conditional on $\mathbf{X}$, follows a distribution with the true but unknown probability density function $s_0(\cdot \mid \mathbf{X} = \mathbf{x})$. Motivated by universal approximation theorems for MoE models, s_0 can be estimated by

$$s_{\psi_K}(\mathbf{y} \mid \mathbf{x}) = \sum_{k=1}^{K} g_k(\mathbf{w}(\mathbf{x}))\phi(\mathbf{y}, \mathbf{v}_k(\mathbf{x}), \boldsymbol{\Sigma}_k(B_k)), \quad \text{with}$$

$$g_k(\mathbf{w}(\mathbf{x})) = \frac{\exp(w_k(\mathbf{x}))}{\sum_{l=1}^{K} \exp(w_l(\mathbf{x}))}, \quad \text{for } k = 1, \dots, K. \qquad (1)$$

Here, on each cluster $k \in \{1, \dots, K\}$, g_k is called a softmax gating network corresponding to the weight functions, $\mathbf{w}(\mathbf{x}) = (w_1(\mathbf{x}), \dots, w_K(\mathbf{x}))$, of $\mathbf{x}$, and $\phi(\cdot, \mathbf{v}_k(\mathbf{x}), \boldsymbol{\Sigma}_k(B_k))$ is a Gaussian expert with the mean function $\mathbf{v}_k(\mathbf{x})$ and covariance matrix $\boldsymbol{\Sigma}_k(B_k)$ depending on the block-diagonal structure B_k. We call $s_{\psi_K}(\mathbf{y} \mid \mathbf{x})$, defined as in (1), the *softmax-gated block-diagonal MoE* (SGaBloME) models with unknown functional parameters $\boldsymbol{\psi}_K = (w_k, \mathbf{v}_k, \boldsymbol{\Sigma}_k(B_k))_{k \in \{1, \dots, K\}}$. Furthermore, when the weights and the means of the SGaBloME model s_{ψ_K} are the functions depending on polynomials of the input variables $\mathbf{x}$ which are specified, for $k \in \{1, \dots, K\}$, respectively, as

$$w_k(\mathbf{x}) = \omega_{k0} + \sum_{d=1}^{D_W} \boldsymbol{\omega}_{kd}^T \mathbf{x}^d, \quad \text{with } \omega_{k0} \in \mathbb{R}, \ \boldsymbol{\omega}_{kd} \in \mathbb{R}^P, \qquad (2)$$

$$\mathbf{v}_k(\mathbf{x}) = \boldsymbol{v}_{k0} + \sum_{d=1}^{D_V} \boldsymbol{\Upsilon}_{kd} \mathbf{x}^d, \quad \text{with } \boldsymbol{v}_{k0} \in \mathbb{R}^Q, \ \boldsymbol{\Upsilon}_{kd} \in \mathbb{R}^{Q \times P}. \qquad (3)$$

Here, $\omega_0 = (\omega_{k0})_{k\in\{1,...,K\}}$, $\omega = (\omega_{k1},\ldots,\omega_{kD_W})_{k\in\{1,...,K\}}$ and $v_0 = (v_{k0})_{k\in\{1,...,K\}}$, $\Upsilon = (\Upsilon_{k1},\ldots,\Upsilon_{kD_V})_{k\in\{1,...,K\}}$ are K-tuples of unknown coefficients with the maximum degrees D_W and D_V of polynomials for the weight and mean functions, respectively, and $\mathbf{x}^d = (x_1^d,\ldots,x_P^d)$ is a vector of all components of $\mathbf{x}$ with power d. Then we call an SGaBloME model s_{ψ_K} defined by (1) with the weights and mean experts specified as in (2)–(3) the polynomial SGaBloME model.

Motivation for Block-Diagonal Covariance Matrices. It is worth mentioning that the block-diagonal covariance matrices $\boldsymbol{\Sigma}(\mathbf{B}) = \left(\boldsymbol{\Sigma}_k(B_k)\right)_{k\in\{1,...,K\}}$ depend on the block structures $\mathbf{B} = (B_k)_{k\in\{1,...,K\}}$ that are the partitions of the outputs' index set $\{1,\ldots,Q\}$ for each cluster. This structure is not only a trade-off between the model complexity and sparsity but is also motivated by some real-world applications, where one wishes to perform prediction on data sets with heterogeneous observations and graph-structured hidden interactions between the outputs. A relevant example is the gene expression data where, subject to phenotypic response, genes interact with only a few other genes, there are small modules of correlated genes, see e.g. [14] for more details.

Motivation for Polynomial Regression. To solve the high-dimensional regression problem, some authors applied SGaBloME models with certain simplifying assumptions. More specifically, Devijver [13] focused on a mixture of Gaussian linear regression models where the gating networks do not depend on the input variables. On the other hand, Chamroukhi et al. [7] considered MoE for multiple regression models with the univariate output variable, however, the weights and means are linear functions of the inputs and thus the capacity of MoE models is limited. In fact, in the context of convolutional neural networks, Chen et al. [8] have empirically found that the mixture of linear experts performs better than a single expert, but is still significantly worse than the mixture of non-linear experts. Within this framework, we are motivated to integrate nonlinearities into SGaBloME models by defining the weights and mean experts as *linear combinations of bounded functions* (LinBo) whose coefficients belong to a compact set. Such a general setting may include the polynomial basis with a bounded input domain, the suitable re-normalized wavelet dictionaries, or the Fourier basis on an interval. If the dimensions of the inputs and outputs are not too large, it is not necessary to select relevant variables and/or use rank sparse models. Then we can work on the softmax-gated MoE models with the linear combinations of bounded functions for weight and mean functions as in [24]. However, to deal with high-dimensional data and simplify the interpretation of sparsity, we consider a special case of LinBo-SGaBloME models to explore the presence of nonlinearities that is the class of polynomial SGaBloME models defined by (1)–(3). On the convergence rates of polynomial SGaBloME models, we refer to [23] for a discussion of the optimal convergence rate of an MoE model where each expert is associated with a polynomial regression model.

Model Selection for Polynomial SGaBloME Models. The estimation of SGaBloME models can be performed by using a well-known expectation-maximization (EM) algorithm [11], which obtains the global convergence in regression mixture models [18]. However, it crucially requires data-driven hyper-parameter choices, including the number of mixture components, the degree of complexity of each softmax gating network and each Gaussian expert mean function, and the hidden block-diagonal structures of the covariance matrices. Hyper-parameter choices from the data-driven learning algorithms belong to the class of model selection problems that select the model with the lowest risk from the data. Typically, penalization is one of the main strategies proposed for model selection that minimizes the sum of the empirical risk with a term of penalty so that the model can be fitted to data while avoiding the overfitting problem.

Related Works. Typically, model selection for MoE models is performed using the asymptotic criteria [4,31], whose uses in small samples are limited. Birgé et al. [5] proposed a novel approach, called *slope heuristic*, supported by a non-asymptotic oracle inequality via a general model selection theorem, see [2] and the references therein for recent reviews. This method leads to an optimal data-driven choice of multiplicative constants for penalties. In fact, oracle inequalities for the least absolute shrinkage and selection operator (Lasso) [32] and general penalized maximum likelihood estimators were established in the spirit of the methods based on concentration inequalities developed by [20]. These results include work on the simplified assumptions of MoE models such as high-dimensional Gaussian graphical models [14], Gaussian mixture model selection [21], and finite mixture regression models [12,13], or softmax-gated MoE models with linear combinations of bounded functions for weight and mean functions without consideration of variable selection in the high-dimensional setting [24].

1.1 Main Contributions

In this work, we established an oracle inequality for model selection, as shown in Theorem 1, under lower bound assumptions on penalty terms. This allows us to obtain non-asymptotic risk bounds in the form of weak oracle inequalities allowing the numbers of predictor and response variables that grow or are even much larger than the sample size. More concretely, the constructed oracle inequality shows that the performance of our penalized maximum likelihood estimations is comparable to that of oracle models with sufficiently large constant multiples of the penalties. The forms of these constants are only known up to multiplicative constants and are proportional to the dimensions of the models. Moreover, the flexibility of polynomial SGaBloME models requires the hyper-parameters comprising the number of mixture components, the degree of polynomial mean functions, and the potential hidden block-diagonal structures of the covariance matrices of the multivariate output. Therefore, the aforementioned theoretical justifications for the penalty shapes motivate the use of the heuristic slope criterion to select these hyper-parameters of the models under consideration.

Notations. For any matrix $\mathbf{A}$ with the elements A_{ij}, we denote $|\!|\!|\mathbf{A}|\!|\!|_\infty =$ $\max_{i,j} |A_{ij}|$ the max-norm, and $\mathbf{A}_{.j}$ the jth column, of $\mathbf{A}$. Furthermore, the smallest and largest eigenvalues of $\mathbf{A}$ are denoted by $\mathrm{eig}(\mathbf{A})$ and $\mathrm{Eig}(\mathbf{A})$, respectively. The notation $\mathbf{A} \succ 0$ indicates that $\mathbf{A}$ is positive definite. For any vector $\mathbf{a}$, we denote $\|\mathbf{a}\|_p$ the l_p-norm of $\mathbf{a}$ for $0 < p \leq \infty$. We call $\dim(\mathcal{S})$ the total number of parameters to be estimated or the dimension of a parametric model $\mathcal{S}$. If S is a finite set, we denote $\mathrm{card}(S)$ the cardinality, $\mathcal{P}(S)$ the set of all subsets, and $\mathcal{B}(S)$ the set of all partitions, of S. The set of all natural numbers without zero is denoted by $\mathbb{N}^*$. For any $K \in \mathbb{N}^*$, the notation $[K]$ corresponds to the set $\{1, \ldots, K\}$. Finally, we refer to $a \wedge b$ as $\min\{a, b\}$ for $a, b \in \mathbb{R}$.

Paper Organization. The rest of the paper is organized as follows. Section 2 is devoted to the construction of a collection of polynomial SGaBloME models for high-dimensional heterogeneous data. In Sect. 3, we state the main theoretical results of oracle inequality for the penalized maximum likelihood estimations under some conditions on the parameter space and input domain of the models. Finally, Sect. 4 contains concluding remarks and future directions.

2   Collection of Polynomial SGaBloME Models

For high-dimensional data, it is necessary to work with parsimonious models by combining two well-known approaches: *selection of relevant variables* and *rank sparse models*. Within this framework, the collection of polynomial SGaBloME models is then constructed.

2.1   Variable Selection via Selecting Relevant Variables

In this section, we introduce the index sets for the input and output variables so that they are related to each other. This facilitates the variable selection of the models in a highly dimensional framework. In particular, for every $p \in [P], q \in [Q]$, we call a couple (X_p, Y_q) *irrelevant* if the elements $(\boldsymbol{\Upsilon}_{kd})_{qp} = 0$ and $(\boldsymbol{\omega}_{kl})_p = 0$ for all cluster $k \in [K]$ and degrees $d \in [D_V], l \in [D_W]$. Therefore, the variables (X_p, Y_q) are *relevant* if they are not irrelevant. Formally, we denote $I = \{(p, q) \in [P] \times [Q] : (X_p, Y_q)\text{is irrelevant}\}$ the set of indices of irrelevant couples, and the complement of I, called $J = ([P] \times [Q]) \setminus I$, is thus the set of indices of relevant couples with $J \in \mathcal{P}([P] \times [Q])$. In addition, we also denote $J_{in} = \{p \in [P] : \exists q \in [Q], (p, q) \in J\}$ the set of indices of input variables that are relevant to the outputs so that $J_{in} \subseteq [P]$.

     We notice that, for every cluster $k \in [K]$ and degree $d \in [D_V]$, all entries of $\boldsymbol{\Upsilon}_{kd}$ belonging to columns indexed by $[P] \setminus J_{in}$ equal to 0, in other words, $\boldsymbol{\Upsilon}_{kd}$ has the relevant columns indexed by J_{in}. Hence, the matrix $\boldsymbol{\Upsilon}_{kd}$ will have $Q \times \mathrm{card}(J_{in})$ coefficients to be estimated, which are smaller than $Q \times P$ when all variables are considered. The number of parameters in the regression matrices is therefore considerably reduced when the cardinality of J_{in} is much smaller than the number of input variables P. The subsets J or J_{in} can be constructed by the

Lasso [32] and has been extended to deal with multiple multivariate regression models for column sparsity using the Group-Lasso [33].

2.2 Variable Selection via Rank Sparse Models

Anderson et al. [1] introduced rank sparse models in the regression framework which is if regression matrices have low rank or at least can be well approximated by low-rank matrices, then the corresponding regression models are said to be rank sparse. In the polynomial SGaBloME models, we assume that for every cluster $k \in [K]$ and degree $d \in [D_V]$, the matrix Υ_{kd} has the associated rank R_{kd} and therefore it is completely determined by $R_{kd} \times (P - (Q - R_{kd}))$ coefficients, which can be less than $Q \times P$. Combined with the selection of relevant variables method, we denote a rank matrix by $\mathbf{R} = (R_{kd})_{k \in [K], d \in [D_V]}$ with the element $R_{kd} \in [\text{card}\,(J_{in}) \wedge Q]$ for each $k \in [K], d \in [D_V]$.

2.3 Collection of Polynomial SGaBloME Models

So far, each polynomial SGaBloME model defined by (1)–(3) can be characterized by the set $\mathbf{m} = (K, D_W, D_V, \mathbf{B}, J, \mathbf{R})$ where K is the number of clusters, D_W and D_V are the maximum degrees of polynomials of the weight and mean functions, respectively, $\mathbf{B}$ is the set of the block-diagonal structures of the covariance matrices, J is the set of relevant variables, and $\mathbf{R}$ is the rank matrix of coefficient matrices. Let $\mathcal{S}_{\mathbf{m}}$ be a class of (conditional) densities of polynomial SGaBloME models with respect to $\mathbf{m}$, which is specified as

$$
\begin{aligned}
\mathcal{S}_{\mathbf{m}} = \Big\{ & s_{\psi\,\mathbf{m}} \equiv s_{\psi\,K} \text{ defined by } (1)-(3) \text{ with} \\
& \psi_{\mathbf{m}} = (\omega_0, \omega, v_0, \Upsilon, \Sigma(\mathbf{B})) \in \Psi_{\mathbf{m}}, \\
& \Psi_{\mathbf{m}} = \mathbb{R}^K \times \mathbf{W}_J^{K \times D_W} \times \mathbb{R}^{K \times Q} \times \mathbf{V}_{J,\mathbf{R}}^{K \times D_V} \times \Omega_{\mathbf{B}}^K, \\
& \mathbf{W}_J = \{\alpha = (\alpha_1, \ldots, \alpha_P) \in \mathbb{R}^P : \alpha_j = 0, \ \forall j \in ([P] \setminus J_{in})\}, \\
& \mathbf{V}_{J,\mathbf{R}} = \{\mathbf{A} \in \mathbb{R}^{Q \times P} : \mathbf{A}_{\cdot j} = \mathbf{0}, \ \forall j \in [P] \setminus J_{in} \text{ and } \text{rank}(\mathbf{A}) = R \in \mathbf{R}\} \\
& \Omega_{\mathbf{B}} = \{\Sigma(B) \in \mathbb{R}^{Q \times Q} : \Sigma(B) \succ 0 \text{ and } \Sigma(B) \text{ depends on block } B \in \mathbf{B}\}\Big\},
\end{aligned}
\tag{4}
$$

for every $\mathbf{m} \in \mathbb{N}^* \times \mathbb{N}^* \times \mathbb{N}^* \times \mathcal{B}([Q])^K \times \mathcal{P}([P] \times [Q]) \times [\text{card}\,(J_{in}) \wedge Q]^{K \times D_V}$. The collection of polynomial SGaBloME models defined in (4) is generally large and therefore not feasible in practice. Therefore, we restrict the set of (K, D_W, D_V) to a finite set of $(\mathcal{K}, \mathcal{D}_W, \mathcal{D}_V)$ where, for $K^*, D_W^*, D_V^* \in \mathbb{N}^*$, $\mathcal{K} = [K^*]$, $\mathcal{D}_W = [D_W^*]$, $\mathcal{D}_V = [D_V^*]$. Accordingly, the collection of polynomial SGaBloME models on the deterministic set of hyper-parameters can be defined as

$$
\begin{aligned}
\mathcal{S} = \Big\{ & \mathcal{S}_{\mathbf{m}} \text{ defined by } (4) \text{ such that } \mathbf{m} \in \mathcal{M}, \\
& \mathcal{M} = \mathcal{K} \times \mathcal{D}_W \times \mathcal{D}_V \times \mathcal{B}([Q])^K \times \mathcal{P}([P] \times [Q]) \times [\text{card}\,(J_{in}) \wedge Q]^{K \times D_V} \Big\}.
\end{aligned}
\tag{5}
$$

Furthermore, because the block structures are specified by the partitions of the index set $\{1, \ldots, Q\}$, the number of such structures follows the so-called Bell number, which grows exponentially even for a moderate number of variables Q and clusters K. Therefore, it is infeasible to consider an exhaustive exploration of the combination of all the partitions to detect the block structures for covariance matrices. Motivated by the recent work of [14], for a set of thresholds $\mathcal{E}$ and on each cluster $k \in [K]$, we restrict our attention to the sub-collection $\mathcal{B}_{k,\mathcal{E}} = (\mathcal{B}_{k,\epsilon})_{\epsilon \in \mathcal{E}}$ of $\mathcal{B}([Q])$, where $\mathcal{B}_{k,\epsilon}$ is the partition of the output variables corresponding to the block-diagonal structure of the adjacency matrix $\mathbf{E}_{k,\epsilon}$ based on the thresholded absolute values of the sample covariance matrix $\mathbf{S}_k$. More formally, for each $\epsilon \in \mathcal{E}$, $(\mathbf{E}_{k,\epsilon})_{qq'} = 1$ if $|(\mathbf{S}_k)_{qq'}| > \epsilon$, otherwise it is equal to 0 for $q, q' \in [Q]$. In fact, Mazumder et al. [22] have shown that the class of block-diagonal structures detected by the graphical Lasso algorithm is identical to the block-diagonal structures detected by the thresholding of the sample covariance matrices, which supports our motivation for this restriction.

For the set of relevant variables, we focus on a random subset $\mathcal{J}$ of $\mathcal{P}([P] \times [Q])$ with the controlled size of $\mathcal{J}$ required in the high-dimension case. Accordingly, the number of possible vectors of ranks is reduced by working on a random subset of $[\text{card}(J_{in}) \wedge Q]^{K \times D_V}$, which is denoted by $\mathcal{R}_{(K,J,D_V)}$ depending on $J \in \mathcal{J}$ with the dimension of $K \in \mathcal{K}$ and $D_V \in \mathcal{D}_V$. As a result, the collection of polynomial SGaBloME models based on a random sub-collection of hyper-parameters can be specified as

$$\widetilde{\mathcal{S}} = \Big\{ \mathcal{S}_{\mathbf{m}} \text{ defined by (4) such that } \mathbf{m} \in \widetilde{\mathcal{M}}, \tag{6}$$
$$\widetilde{\mathcal{M}} = \mathcal{K} \times \mathcal{D}_W \times \mathcal{D}_V \times (\mathcal{B}_{k,\mathcal{E}})_{k \in [K]} \times \mathcal{J} \times \mathcal{R}_{(K,J,D_V)} \Big\}.$$

3   Main Theoretical Results

In this section, we begin by introducing conditions on the parameter space of the models and give an overview of loss functions that are useful for comparing two (conditional) probability density functions. A general principle of penalized maximum likelihood estimation is also derived. Next, we show a finite sample oracle inequality used to ensure that if we penalize the log-likelihood in an approximate approach, we are able to select a model that is as good as the oracle.

3.1   Boundedness Conditions on the Parameter Space

By motivation of integrating the nonlinearities into SGaBloME models discussed in Sect. 1, we consider the class of linear combinations of bounded functions for the weights and mean experts whose coefficients belong to compact sets with a bounded input domain. More specifically, we let $(\mathbf{X}_{[N]}, \mathbf{Y}_{[N]}) = ((\mathbf{X}_1, \mathbf{Y}_1), \dots, (\mathbf{X}_N, \mathbf{Y}_N))$ be N pairs of real-valued random variables $(\mathbf{X}, \mathbf{Y})$ where the covariates $\mathbf{X}$ are assumed to belong to a hypercube, that is $\mathcal{X} = [0, 1]^P$. Then, there exist the constants $C_\omega, C_\Upsilon, c_\Sigma, C_\Sigma > 0$ such that, for every $k \in [K]$,

$$\|\boldsymbol{\omega}_{kd}\|_\infty \leq C_\omega, \quad \|\boldsymbol{\Upsilon}_{kl}\|_\infty \leq C_\Upsilon, \quad \text{for every } d \in [D_W], l \in [D_V], \tag{7}$$

moreover, the eigenvalues of the block-diagonal covariances of the Gaussian experts lie on a positive interval, that is

$$0 < c_{\boldsymbol{\Sigma}} \leq \mathrm{eig}\left(\boldsymbol{\Sigma}_k\left(B_k\right)\right) \leq \mathrm{Eig}\left(\boldsymbol{\Sigma}_k\left(B_k\right)\right) \leq C_{\boldsymbol{\Sigma}}. \tag{8}$$

This setting can be applied to the case of polynomial functions for the weights of the softmax gates and the means of the Gaussian experts as we described in (2)–(3). More generally, the oracle inequality provided by Theorem 1 still holds for monomials of weights, allowing for the interaction between different inputs.

3.2 Loss Function

To evaluate the maximum likelihood estimate, the Kullback-Leibler (KL) divergence is the most natural loss function, which is generally defined by

$$KL(s,t) = \begin{cases} \int_{\mathbb{R}^D} \ln\left(\frac{s(x)}{t(x)}\right) s(x) dyx & \text{if } sdx \text{ is absolutely continuous w.r.t. } tdx, \\ +\infty & \text{otherwise,} \end{cases}$$

where $s(\cdot)$ and $t(\cdot)$ are two density functions. In our work, we will apply the *tensorized KL divergence* to capture the structure of the density functions conditional on the random variables $\mathbf{X}$, that is

$$\mathrm{KL}^{\otimes \mathrm{N}}(s,t) = \mathbb{E}_{\mathbf{X}_{[N]}}\left[\frac{1}{N}\sum_{n=1}^{N} KL\left(s\left(\cdot \mid \mathbf{X}_n\right), t\left(\cdot \mid \mathbf{X}_n\right)\right)\right].$$

Another case of the tensorized KL divergence is the *tensorized Jensen-KL divergence* [9], which is given, for any $\rho \in (0,1)$, by

$$\mathrm{JKL}_{\rho}^{\otimes \mathrm{N}}(s,t) = \mathbb{E}_{\mathbf{X}_{[N]}}\left[\frac{1}{N}\sum_{n=1}^{N}\frac{1}{\rho} KL\left(s\left(\cdot \mid \mathbf{X}_n\right), (1-\rho) s\left(\cdot \mid \mathbf{X}_n\right) + \rho t\left(\cdot \mid \mathbf{X}_n\right)\right)\right].$$

A relationship between the tensorized KL and the tensorized Jensen-KL divergence can be found in [10, Proposition 1].

3.3 Penalized Maximum Likelihood Estimation (PMLE)

In the context of maximum likelihood estimation, given a collection $S_{\mathbf{m}}$, we aim to estimate s_0 by the conditional density $\widehat{s}_{\mathbf{m}}$ that minimizes the negative log-likelihood (NLL) as

$$\widehat{s}_{\mathbf{m}} = \arg\min_{s_{\mathbf{m}} \in S_{\mathbf{m}}} \sum_{n=1}^{N} -\ln\left[s_{\mathbf{m}}\left(\mathbf{Y}_n \mid \mathbf{X}_n\right)\right].$$

It is important to us to look for almost minimizer of this quantity and thereby define an η-log-likelihood minimizer (LLM) that satisfies

$$\sum_{n=1}^{N} -\ln\left[\widehat{s}_{\mathbf{m}}\left(\mathbf{Y}_n \mid \mathbf{X}_n\right)\right] \leq \inf_{s_{\mathbf{m}} \in \mathcal{S}_{\mathbf{m}}} \sum_{n=1}^{N} -\ln\left[s_{\mathbf{m}}\left(\mathbf{Y}_n \mid \mathbf{X}_n\right)\right] + \eta, \tag{9}$$

where the error term $\eta > 0$ is added to avoid any existence issue such as the infimum may not be reached. See [6, Chapter 2], [9,10,24] for more details of this literature. However, this approach underestimates the risk of the estimation and leads to the selection of overly complex models. Therefore, a trade-off between good data fit and model complexity can be found by adding an appropriate penalty term pen(m). More concretely, for a given choice of pen(m), the *selected model* $S_{\widehat{m}}$ is chosen as the one whose index $\widehat{m}$ is a η'-minimizer of the sum of the NLL and penalty function, that is

$$\sum_{n=1}^{N} -\ln\left[\widehat{s}_{\widehat{m}}\left(\mathbf{Y}_n|\mathbf{X}_n\right)\right] + \operatorname{pen}\left(\widehat{m}\right) \le \inf_{m \in \mathcal{M}}\left\{\sum_{n=1}^{N} -\ln\left[\widehat{s}_m\left(\mathbf{Y}_n|\mathbf{X}_n\right)\right] + \operatorname{pen}(m)\right\} + \eta',$$

$$(10)$$

for $\eta' > 0$. We then call $\widehat{s}_{\widehat{m}}$ the η'-PMLE that depends on both error terms η and η'. From now on, the term *selected model* or *best data-driven model* is used to indicate the model that satisfies (10).

3.4   Oracle Inequality

In this section, we provide the construction of an oracle inequality that guarantees a non-asymptotic theory for model selection in high-dimensional polynomial SGaBloME models.

Theorem 1. *Let* $\left(\mathbf{X}_{[N]}, \mathbf{Y}_{[N]}\right)$ *be a random sample of* $(\mathbf{X}, \mathbf{Y})$ *where* $\mathbf{Y}|\mathbf{X}$ *arises from the unknown conditional density* s_0. *For every* $\mathbf{m} = (K, D_W, D_V, \mathbf{B}, J, \mathbf{R}) \in \mathcal{M}$, *the model* $S_{\mathbf{m}}$ *can be specified by (4). Assume that there exists* $\tau > 0$ *and* $\epsilon_{KL} > 0$ *such that, for all* $\mathbf{m} \in \mathcal{M}$, *one can find* $\bar{s}_{\mathbf{m}} \in S_{\mathbf{m}}$ *such that*

$$\operatorname{KL}^{\otimes N}(s_0, \bar{s}_{\mathbf{m}}) \le \inf_{s \in S_{\mathbf{m}}} \operatorname{KL}^{\otimes N}(s_0, s) + \frac{\epsilon_{KL}}{N}, \quad \text{and} \quad \bar{s}_{\mathbf{m}} \ge e^{-\tau} s_0. \quad (11)$$

Furthermore, we construct a random sub-collection $\widetilde{S}$ *of* S *such that every model of* $\widetilde{S}$ *depends on the sets of* $\widetilde{\mathcal{M}} \subset \mathcal{M}$ *as in (5)–(6). Then, there is a constant* C *such that, for any* $\rho \in (0,1)$ *and* $C_1 > 1$, *there are two constants* κ *and* C_2 *depending only on* ρ *and* C_1 *such that, for every* $\mathbf{m} \in \mathcal{M}$, $\xi_{\mathbf{m}} \in \mathbb{R}^+$, $\Xi = \sum_{\mathbf{m} \in \mathcal{M}} e^{-\xi_{\mathbf{m}}} < \infty$ *and*

$$\operatorname{pen}(\mathbf{m}) \ge \kappa \left[(C + \ln N) \dim(S_{\mathbf{m}}) + (1 \vee \tau)\xi_{\mathbf{m}}\right],$$

and the η'-PMLE $\widehat{s}_{\widehat{m}}$ *defined in (10) on the subset* $\widetilde{\mathcal{M}} \subset \mathcal{M}$ *satisfies*

$$\mathbb{E}_{\mathbf{X}_{[N]}, \mathbf{Y}_{[N]}}\left[\operatorname{JKL}_{\rho}^{\otimes N}(s_0, \widehat{s}_{\widehat{m}})\right] \le C_1 \mathbb{E}_{\mathbf{X}_{[N]}, \mathbf{Y}_{[N]}}\left[\inf_{\mathbf{m} \in \widetilde{\mathcal{M}}}\left(\inf_{s \in S_{\mathbf{m}}} \operatorname{KL}^{\otimes N}(s_0, s) + 2\frac{\operatorname{pen}(\mathbf{m})}{N}\right)\right]$$

$$+ C_2(1 \vee \tau)\frac{\Xi^2}{N} + \frac{\eta' + \eta}{N}. \quad (12)$$

Remarks. Theorem 1 guarantees that a penalized criterion leads to a good model selection and that the penalty is only known up to multiplicative constants κ, and is proportional to the dimensions of the models $\dim(\mathcal{S}_\mathbf{m})$. In particular, in the small and finite sample setting, these multiplicative constants can be calibrated using the slope heuristic approach. We notice that (11) is not a strong assumption and is satisfied in the case s_0 is bounded with compact support. This oracle inequality compares the performance of our PMLE with the best model in the collection. However, Theorem 1 allows us to approximate well a rich class of conditional PDFs if we use polynomials of weights and Gaussian expert means of sufficient degrees, or enough clusters due to the universal approximation of MoE models. This results in the term on the right of (12) being small, for $\mathcal{D}_W, \mathcal{D}_V$ and $\mathcal{K}$ well chosen. It should be emphasized that Theorem 1 extends the main result of [24], which is only valid for a full collection of LinBoSGaME models in the low-dimensional setting. Furthermore, in the context of MoE models, our non-asymptotic oracle inequality for SGaME models in Theorem 1 can be seen as a complementary result to a classical asymptotic theory [17, Theorems 1, 2, and 3], and an l_1 oracle inequality that focuses on the properties of the Lasso estimator rather than the model selection procedure [29].

Main Challenges on the Proof of Theorem 1. To prove Theorem 1 it is inspired by [24] for handling LinBo-SGaME models, however, our method and most of the technical details differ. This is because their approach is not directly applicable to our high-dimensional SGaBloME models, due to restrictions on relevant predictor variables and rank reduction, and Gaussian experts with block-diagonal covariance matrices. In particular, the main difficulty in proving our oracle inequality lies in bounding the bracketing entropy for the collections of SGaBloME models. This requires several regularity assumptions, which are not easy to verify due to the complexity of SGaBloME models and technical reasons. Therefore, our proofs require the development of several new ideas. Furthermore, unlike [24], which uses a model selection theorem for a deterministic collection of models from [9,10], we need to find a way to use the model selection theorem for MLE among a random sub-collection (cf. [12, Theorem 5.1] and [14, Theorem 7.3]). We refer readers to Sections S-1 and S-2 in the supplementary materials for a sketch of proof and detailed proof of Theorem 1.

4 Conclusion and Perspectives

We have studied PMLEs for polynomial SGaBloME models in high-dimensional heterogeneous data. Our main contribution is to establish a non-asymptotic risk bound in the form of an oracle inequality, provided that lower bounds of the penalty hold. The future direction is to empirically evaluate our oracle inequality and to extend the current oracle inequality to more general settings where Gaussian experts are replaced by elliptic distributions.

References

1. Anderson, C.W., Stolz, E.A., Shamsunder, S.: Multivariate autoregressive models for classification of spontaneous electroencephalographic signals during mental tasks. IEEE Trans. Biomed. Eng. **45**(3), 277–286 (1998)
2. Arlot, S.: Minimal penalties and the slope heuristics: a survey. J. Soc. Française Stat. **160**(3), 1–106 (2019)
3. Bengio, Y.: Deep learning of representations: looking forward. In: Dediu, A.-H., Martín-Vide, C., Mitkov, R., Truthe, B. (eds.) SLSP 2013. LNCS (LNAI), vol. 7978, pp. 1–37. Springer, Heidelberg (2013). https://doi.org/10.1007/978-3-642-39593-2_1
4. Biernacki, C., Celeux, G., Govaert, G.: Assessing a mixture model for clustering with the integrated completed likelihood. IEEE Trans. Pattern Anal. Mach. Intell. **22**(7), 719–725 (2000)
5. Birgé, L., Massart, P.: Minimal penalties for Gaussian model selection. Probab. Theory Relat. Fields **138**(1), 33–73 (2007)
6. Borwein, J.M., Zhu, Q.J.: Techniques of Variational Analysis. Springer, Heidelberg (2004). https://doi.org/10.1007/0-387-28271-8
7. Chamroukhi, F., Huynh, B.T.: Regularized maximum-likelihood estimation of mixture-of-experts for regression and clustering. In: 2018 International Joint Conference on Neural Networks (IJCNN), pp. 1–8 (2018)
8. Chen, Z., Deng, Y., Wu, Y., Gu, Q., Li, Y.: Towards understanding the mixture-of-experts layer in deep learning. In: NeurIPS (2022)
9. Cohen, S.X., Le Pennec, E.: Partition-based conditional density estimation. ESAIM: Probab. Stat. **17**, 672–697 (2013)
10. Cohen, S., Le Pennec, E.: Conditional density estimation by penalized likelihood model selection and applications. Technical report, INRIA (2011)
11. Dempster, A.P., Laird, N.M., Rubin, D.B.: Maximum likelihood from incomplete data via the EM algorithm. J. Roy. Stat. Soc. Ser. B **39**(1), 1–38 (1977)
12. Devijver, E.: Finite mixture regression: a sparse variable selection by model selection for clustering. Electron. J. Stat. **9**(2), 2642–2674 (2015)
13. Devijver, E.: Joint rank and variable selection for parsimonious estimation in a high-dimensional finite mixture regression model. J. Multivar. Anal. **157**, 1–13 (2017)
14. Devijver, E., Gallopin, M.: Block-diagonal covariance selection for high-dimensional Gaussian graphical models. J. Am. Stat. Assoc. **113**(521), 306–314 (2018)
15. Ho, N., Yang, C.Y., Jordan, M.I.: Convergence rates for Gaussian mixtures of experts. J. Mach. Learn. Res. **23**(323), 1–81 (2022)
16. Jacobs, R.A., Jordan, M.I., Nowlan, S.J., Hinton, G.E.: Adaptive mixtures of local experts. Neural Comput. **3**(1), 79–87 (1991)
17. Khalili, A.: New estimation and feature selection methods in mixture-of-experts models. Can. J. Stat. **38**(4), 519–539 (2010)
18. Kwon, J., Qian, W., Caramanis, C., Chen, Y., Davis, D.: Global convergence of the EM algorithm for mixtures of two component linear regression. In: COLT, vol. 99, pp. 2055–2110. PMLR (2019)
19. Masoudnia, S., Ebrahimpour, R.: Mixture of experts: a literature survey. Artif. Intell. Rev. **42**(2), 275–293 (2014)
20. Massart, P.: Concentration Inequalities and Model Selection: Ecole d'Eté de Probabilités de Saint-Flour XXXIII-2003. Springer, Heidelberg (2007). https://doi.org/10.1007/978-3-540-48503-2_7

21. Maugis, C., Michel, B.: A non asymptotic penalized criterion for Gaussian mixture model selection. ESAIM: Probab. Stat. **15**, 41–68 (2011)
22. Mazumder, R., Hastie, T.: Exact covariance thresholding into connected components for large-scale graphical lasso. J. Mach. Learn. Res. **13**(1), 781–794 (2012)
23. Mendes, E.F., Jiang, W.: On convergence rates of mixtures of polynomial experts. Neural Comput. **24**(11), 3025–3051 (2012)
24. Montuelle, L., Le Pennec, E., et al.: Mixture of Gaussian regressions model with logistic weights, a penalized maximum likelihood approach. Electron. J. Stat. **8**(1), 1661–1695 (2014)
25. Nguyen, H.D., Chamroukhi, F.: Practical and theoretical aspects of mixture-of-experts modeling: an overview. Wiley Interdisc. Rev.: Data Min. Knowl. Discov. **8**(4), e1246 (2018)
26. Nguyen, H.D., Chamroukhi, F., Forbes, F.: Approximation results regarding the multiple-output Gaussian gated mixture of linear experts model. Neurocomputing **366**, 208–214 (2019)
27. Nguyen, H.D., Nguyen, T., Chamroukhi, F., McLachlan, G.J.: Approximations of conditional probability density functions in Lebesgue spaces via mixture of experts models. J. Stat. Distrib. Appl. **8**(1), 13 (2021)
28. Nguyen, T., Chamroukhi, F., Nguyen, H.D., McLachlan, G.J.: Approximation of probability density functions via location-scale finite mixtures in Lebesgue spaces. Commun. Stat. - Theory Methods **52**, 1–12 (2022)
29. Nguyen, T., Nguyen, H.D., Chamroukhi, F., McLachlan, G.J.: An l_1-oracle inequality for the Lasso in mixture-of-experts regression models. arXiv preprint arXiv:2009.10622 (2020)
30. Nguyen, T., Nguyen, H.D., Chamroukhi, F., McLachlan, G.J.: Approximation by finite mixtures of continuous density functions that vanish at infinity. Cogent Math. Stat. **7**(1), 1750861 (2020)
31. Schwarz, G., et al.: Estimating the dimension of a model. Ann. Stat. **6**(2), 461–464 (1978)
32. Tibshirani, R.: Regression shrinkage and selection via the lasso. J. Roy. Stat. Soc. B **58**(1), 267–288 (1996)
33. Yuan, M., Lin, Y.: Model selection and estimation in regression with grouped variables. J. Roy. Stat. Soc. B **68**(1), 49–67 (2006)

Reinforcement Learning

Auction-Based Allocation
of Location-Specific Tasks

Fahimeh Ramezani[1] , Brendan Sims[2](✉) , and Haris Aziz[1]

[1] School of Computer Science, UNSW, Sydney, Australia
{f.ramezani,haris.aziz}@unsw.edu.au
[2] Defence Science and Technology Group, Adelaide, Australia
brendan.sims2@defence.gov.au

Abstract. We consider a task allocation problem in which agents and tasks have locations, and the goal is to allocate tasks among agents so as to minimise the distance travelled. We analyse two important algorithms under a generalised setting that puts additional feasibility constraints on allocations. We provide matching lower and upper bounds on the approximation guarantees achieved by the algorithms. We then conduct an experimental analysis of the relative performance of the algorithms. Our results indicate the relative performance of the algorithms as well as the effect of feasibility constraints on the guarantees of the algorithms.

Keywords: Task allocation · Routing problems · Multi-agent systems

1 Introduction

We consider the fundamental, ubiquitous, and widely-applicable problem of allocating tasks to agents. Both the tasks and agents have specific locations. The goal is to allocate sequences of tasks to agents so as to minimise the total distance travelled by the agents. The problem is encountered in scenarios when robots or unmanned vehicles need to coordinate in a distributed manner to undertake tasks. The centralised version of the problem is equivalent to the travelling salesman problem (TSP) when there is exactly one agent. Hence the problem is NP-hard which motivates the need for efficient heuristic approaches as well as approximation algorithms.

Lagoudakis et al. in [10] presented two elegant distributed polynomial-time algorithms that provide a guarantee 2-approximation. In other words, the total distance incurred by any solution of the algorithm is not more than twice the optimal distance required to undertake all the tasks. One is based on making bids while considering underlying paths. The other takes a tree-based approach for bidding. The purpose of this paper is to theoretically and experimentally analyse these algorithms under more general feasibility constraints.

T. Liu et al. (Eds.): AI 2023, LNAI 14472, pp. 249–260, 2024.
https://doi.org/10.1007/978-981-99-8391-9_20

Contributions. In this paper, we revisit the paper of Lagoudakis et al. [10] and generalise their model and algorithms by handling realistic feasibility constraints such as a maximum number of tasks per agent. Such constraints are widely encountered in practice. Our first insight is that even simple feasibility constraints significantly affect the theoretical approximation guarantees of the algorithms. We provide matching lower and upper bounds on the approximation guarantee achieved by the algorithms. In particular, the algorithms achieve an approximation of $O(m)$ where m is the number of tasks.

After conducting a theoretical analysis of the algorithms, we undertake a thorough experimental comparison of the algorithms. Our experimental findings show that the path-based approach performs consistently better than the tree-based approach when the objective is to minimise the total travel distance. The introduction of feasibility constraints results in performance degradation in terms of the total distance travelled. The effect on both approaches is similar. When compared to optimal allocations for small problem sizes with randomised locations, both approaches perform quite well with the gap to optimal slightly growing as the problem size increases.

Related Work. Our problem can be viewed as a multi-agent version of the TSP, which is well-known to be NP-hard [1]. There are several papers in robotics, multi-agent systems, operations research, and theoretical computer science that have examined variants of the problem we consider (see e.g., [4,6,7,13]). Several of the papers consider scenarios where agents or robots have valuations over subsets of tasks and the goal is to maximise the sum of the agents' valuations [5]. A comprehensive taxonomy of robot-task allocation problems is presented by Korsah, Stentz and Dias in [9]. According to this taxonomy, our problem is classified as a time-extended assignment problem for single-task robots and single-robot tasks with in-schedule dependencies.

Our setting can be categorised as a Multiple Depot Vehicle Routing Problem (MDVRP) where a set of agents (vehicles) are initially located at one or more different sites and they are required to visit all other sites whilst the total travel distance by all agents is minimised. However, we do not require agents to return to a depot. Carlson et al. [3] consider a min-max objective function for MDVRPs where the maximum travel distance by any vehicle must be minimised. They presented an LP-based heuristic to solve large problems and performed a theoretical analysis relating to the optimal problem solution. Bae et al. [2] also consider min-max MDVRPs for problems involving two robots and each robot has its own cost function. They refer to the problem as a min-max Multiple Depot Heterogeneous Travelling Salesman Problem and present heuristics for solving the problem. Clearly the problem generalises the TSP and is NP-hard.

The capacity-constrained vehicle routing problem is a general version of the vehicle routing problem where vehicles have limited carrying capacity and must pick up and deliver items between locations. The objective is that all delivery demands are served while minimising the total distance travelled by vehicles [11]. Sarkar et al. [12] present a heuristic algorithm that solves the problem, which is scalable in terms of the number of tasks.

2   Setup

We consider a variant of task allocation in a metric space where a set of m tasks, $T = \{t_1, \ldots, t_m\}$, is allocated to a set of n agents, $A = \{a_1, \ldots, a_n\}$, and every task and agent is characterised by a location. We define a distance function $d : A \cup T \times A \cup T \to \mathbb{R}^+$, that specifies the distance between any two locations. Moreover, we assume that the distance function is symmetric and satisfies the triangle inequality, which means for every location $x, y, z \in A \cup T$, we have $d(x, y) = d(y, x)$, and $d(x, y) + d(y, z) \geq d(x, z)$.

An *allocation* $X = (X_1, \ldots, X_n)$ is a partitioning of set T into n bundles where X_i is the bundle allocated to agent i. An *ordered allocation*, $X^O = (X_1^O, \ldots, X_n^O)$ specifies an ordering of tasks in each ordered bundle X_i^O for $1 \leq i \leq n$. We denote the unordered set of tasks pertaining to X_i^O by X_i. In this paper, we seek to assign an ordered bundle X_i^O to every agent i, $1 \leq i \leq n$, such that the cost incurred by agent i, denoted by $d_i(X_i^O)$ is the travel cost to visit all tasks in X_i^O according to its ordering. We assume that all agents are the same, so in this case, without loss of generality we can assume the travel cost is the distance traversed. Let $X_i^O = (t_1, t_2, ..., t_k)$, then $d_i(X_i^O) = d(a_1, t_1) + \sum_{j=1}^{k-1} d(t_j, t_{j+1})$. Our goal is to find an ordered allocation of tasks to agents with minimum total distance, which is $\min_{X^O} \sum_{i=1}^{n} d_i(X_i^O)$. Lagoudakis et al. in [10] presented an auction-based approximation algorithm to solve this problem, the so-called *multi-robot routing problem* with *MiniSum* objective function. Their algorithm finds an ordered allocation X^O with cost (i.e., $\sum_{i=1}^{n} d_i(X_i^O)$) at most twice the optimal cost incurred by agents to visit all tasks.

In what follows, we demonstrate that our main problem, that is, finding an ordered allocation with minimum total distance, has an alternative interpretation as a *social welfare maximisation problem*. Let $\Pi(T)$ be the set of all ordered bundles in T and every agent i, has a valuation function $v_i : \Pi(T) \to \mathbb{R}^+$, which gives a value for every ordered bundle. For an ordered allocation X^O, the social welfare $SW(X^O)$ is the sum of individual valuations for the assigned ordered bundles, $\sum_{i=1}^{n} v_i(X_i^O)$. Our goal is to find an ordered allocation of tasks to agents which maximises the social welfare $SW(X^O)$. We denote it by $\text{SWM} = \max_{X^O} \sum_{i=1}^{n} v_i(X_i^O)$, where the maximum is taken over all ordered allocations. In this paper, we consider a specific class of *reward-cost-based* valuation functions. Note that a valuation function $v_i : \Pi(T) \to R^+$ is *reward-cost-based* if there exists a corresponding cost function $d_i : \Pi(T) \to R^+$ and a reward function $r_i : T \to R^+$ such that $v_i(X_i^O) = \sum_{t \in X_i^O} r_i(t) - d_i(X_i^O)$, where $d_i(X_i^O)$ denotes the travel distance incurred by agent i to visit all tasks in X_i^O according to its ordering. Moreover, we assume that the reward function is the same for all agents and we replace r_i by $r : T \to R^+$.

Notice that for any given ordered allocation $X^O = (X_1^O, \ldots, X_n^O)$

$$\sum_{i=1}^{n} v_i(X_i^O) = \sum_{i=1}^{n} \left\{ \sum_{t \in X_i} r(t) - d_i(X_i^O) \right\} = \sum_{i=1}^{n} \sum_{t \in X_i} r(t) - \sum_{i=1}^{n} d_i(X_i^O) = \sum_{j=1}^{m} r(t_j) - \sum_{i=1}^{n} d_i(X_i^O).$$

$$(1)$$

The third equality follows from the fact that X^O is a partition of tasks and $\cup_{i=1}^{n} X_i = T$. Therefore, by (1), we get

$$SWM = \max_{X^O} \sum_{i=1}^{n} v_i(X_i^O) = \sum_{j=1}^{m} r(t_j) + \max_{X^O}(-\sum_{j=1}^{n} d_i(X_i^O)) = \sum_{j=1}^{m} r(t_j) - \min_{X^O} \sum_{i=1}^{n} d_i(X_i^O).$$

This implies that finding an allocation that maximises social welfare in our setting is equivalent to find an allocation that minimises the total distance travelled by agents to visit all tasks.

3   Auction-Based Algorithms

Lagoudakis et al. [10] showed that finding an ordered allocation of tasks to agents which has minimum total distance is an NP-complete problem. More specifically, they showed that there is no polynomial time algorithm for solving multi-robot routing problems optimally with the *MiniSum* objective function unless $P = NP$. They also presented an auction-based approximation algorithm that proceeds in sequential rounds where agents are considered as bidders and tasks as goods. Initially, no tasks are allocated to any agent. In each round of the algorithm, every agent bids on a single unallocated task. For the *MiniSum* objective, this corresponds to the task with the minimum increase to an agent's total travel distance. Every agent computes its bid in parallel and then broadcasts it to the other agents. Once an agent has received bids from the other agents, it locally determines the winning bid. The agent with lowest bid wins the current round and is assigned to corresponding task. This procedure is repeated until all tasks are allocated. Upon allocation of all tasks, each agent determines a path that specifies the order in which it will visit its allocated tasks. The algorithm is decentralised and proceeds in m sequential rounds with each agent receiving $n - 1$ bids per round. Therefore, $O(nm)$ messages need to be communicated during this algorithm. This auction-based approach has two advantages. First, it is simple and second, it can be applied as a decentralised algorithm, where agents make task allocation decisions without a central auctioneer. In the following sections, we explain some approaches by which agents may construct bids for tasks.

3.1   Bidding Rule

In every round of the auction algorithm described above, agents apply the same bidding rule to compute bids. The general bidding rule in this algorithm is analogous to hill climbing in that it seeks to find a good, but not necessarily optimal allocation. Each agent bids on unallocated task t, the difference between the minimum cost to visit its current allocation and the minimum cost to visit its current allocation with additional task t.

Suppose that $(X_1, \ldots, X_n)$ is a current partial allocation of tasks to agents in some round. $MinVisitCost(a_i, X_i)$ denotes the minimum cost for agent a_i to visit all the tasks in X_i. Then agent a_i's bid for unallocated task t is

$$MinVisitCost(a_i, X_i \cup \{t\}) - MinVisitCost(a_i, X_i).$$

We present the auction algorithm with $MinVisitCost$ bidding rule in Algorithm 1.

Algorithm 1. Auction algorithm with $MinVisitCost$ bidding rule

---

Require: (A, T, d)
Ensure: allocation $X^O = (X_1^O, \ldots, X_n^O)$
 1: **while** there exists some unallocated task $t \in T$ **do**
 2:    Among all i, $1 \le i \le n$ and $t \in T$, choose the lowest bid:
       $MinVisitCost(a_i, X_i \cup \{t\}) - MinVisitCost(a_i, X_i)$.
       Suppose that the winner is agent is a_i and the task is t.
 3:    Insert t in X_i, update X_i.
 4:    $T \longleftarrow T \setminus \{t\}$
 5: **end while**
 6: **return** X^O

3.2 BidSumPath Bidding Rule

In order for each agent a_i to compute their lowest bid, $MinVisitCost(a_i, X_i \cup \{t\})$, they should consider all permutations of $X_i \cup \{t\}$, and choose the permutation with minimum total cost. This problem is equivalent to the TSP and is NP-hard. Heuristic algorithms can be used by agents to find an approximate solution for $MinVisitCost(a_i, X_i \cup \{t\})$. Suppose that we give an ordered path X_i^O to agent a_i. To determine $MinVisitCost(a_i, X_i \cup \{t\})$ approximately, agent a_i inserts task t into all possible extensions of the existing path X_i, which means it can be inserted between any two adjacent tasks or added at the end of the existing path. It then chooses the position in the path that minimises the total cost of the new path. We denote the cost of this new path by $PathCost(a_i, X_i^O \oplus \{t\})$. Therefore, agent a_i's bid for unallocated task t is:

$$PathCost(a_i, X_i^O \oplus \{t\}) - PathCost(a_i, X_i^O).$$

This bidding rule is referred to by Lagoudakis et al. in [10] as *BidSumPath*. Applying this bidding rule, they obtained the following theorem: the performance ratio of the *BidSumPath* bidding rule for the *MiniSum* objective function is at most 2.

3.3 BidSumTree Bidding Rule

Another approach to calculate $MinVisitCost(a_i, X_i \cup \{t\})$ in order for an agent to compute it's bid on a task is by constructing trees. For any agent a_i and any set of tasks X_i, we can estimate $MinVisitCost(a_i, X_i \cup \{t\})$ by $MinTreeCost(a_i, X_i \cup \{t\})$.

Let $MinTreeCost(a_i, X_i \cup \{t\})$ denote the minimum spanning tree (MST) with $X_i \cup \{t\} \cup \{a_i\}$ nodes and root a_i. Then agent a_i's bid on unallocated task t is

$$MinTreeCost(a_i, X_i \cup \{t\}) - MinTreeCost(a_i, X_i).$$

This bidding rule is referred to by Lagoudakis et al. in [10] as *BidSumTree*. Unlike *BidSumPath*, *BidSumTree* can be computed exactly since finding an MST can be achieved in polynomial time. In addition, a tree over X_i remains unchanged within a tree over $X_i \cup \{t\}$. As task t was not allocated in earlier rounds to X_i, it does not offer a better way to connect the nodes in X_i, so t can be added as a leaf to X_i. When the auction algorithm concludes and all tasks are allocated, a depth first search algorithm can be used to find an ordered set of tasks for every agent. Applying this bidding rule, Lagoudakis et al. in [10] obtained the following theorem: the performance ratio of the *BidSumTree* bidding rule for the *MiniSum* objective function is at most 2.

4 Theoretical Analysis Under Task Capacities

We can incorporate any constraints that specify which ordered task bundles are feasible for an agent. We encode all such constraints by a general feasibility function $f : \{1, 2, ..., n\} \times \Pi(T) \rightarrow \{0, 1\}$ where $f(i, X_i^O) = 1$ if $X_i^O \in \Pi(T)$ is feasible for agent a_i and $f(i, X_i^O) = 0$ otherwise. For any sequence of tasks X_i^O and $X_i'^O$, we will denote by $X_i^O \leq_{sub} X_i'^O$ the relation that X_i^O is a subsequence of $X_i'^O$. We assume that if X_i^O is infeasible for agent a_i, then any $Y^O \geq_{sub} X_i^O$ is infeasible for agent a_i. Furthermore, if X_i^O is feasible for agent i, then any $Y^O \leq_{sub} X_i^O$ is feasible for agent a_i.

A standard constraint is to put an upper limit on the number of tasks an agent can be allocated, i.e. task capacities. The constraint is used for example by Choi et al. [5]. Let α_i indicate the maximum number of tasks that agent a_i can be allocated and we want to find an allocation for the *MiniSum* problem. We first note that the algorithms that we have considered can naturally adapt to the case of complex feasibility constraints: we only give an additional task to an agent if no feasibility constraint is violated.

We prove that if we add task capacity constraints for agents, the auction algorithm does not give a constant approximation ratio. We assume that the sum of task capacities for all agents is at least the number of tasks. Otherwise, the agents cannot complete all tasks and this raises the issue of which tasks should be completed. Assume we have m tasks $\{t_1, t_2, ...t_m\}$ and n agents $\{a_1, a_2, ..., a_n\}$ with capacities $\{\alpha_1, \alpha_2, ..., \alpha_n\}$ such that $\sum_{i=1}^{n} \alpha_i \geq m$. In this case we show that the performance ratio of the auction algorithm described in Sect. 3 is $O(m)$.

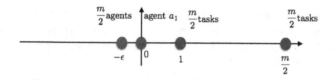

Fig. 1. A configuration for m tasks and $\frac{m}{2} + 1$ agents.

Proposition 1. *The performance ratio of the* BidSumPath *bidding rule when agents have variable capacities and are in one place initially for the* MiniSum *objective function is* $O(m)$. *In particular, there exists an instance where the approximation ratio is at least* $\frac{m}{6}$.

Proof. Consider instance $\mathcal{I}$ consisting of m tasks, and $\frac{m}{2} + 1$ agents, placed on a 2-dimensional plane. Let m be an even number. One agent, say a_1, is located at $(0,0)$ and the remaining $\frac{m}{2}$ agents are located at $(-\epsilon, 0)$. Also, we have $\frac{m}{2}$ tasks at $(1,0)$ and $\frac{m}{2}$ tasks at $(\frac{m}{2}, 0)$. Assume agent a_1 has capacity $\frac{m}{2}$ and each remaining agent has capacity of one (see Fig. 1).

Considering the distance between tasks and agents, the auction algorithm with the *BidSumPath* bidding rule gives the first $\frac{m}{2}$ tasks located at $(1,0)$ to agent a_1 and all other tasks $\{\frac{m}{2} + 1, ..., m\}$ to agents $\{2, ..., \frac{m}{2} + 1\}$ respectively. In this case, the total cost is $1 + \frac{m}{2} \times (\frac{m}{2} + \epsilon)$ while the optimal solution allocates the $\frac{m}{2}$ agents at $(-\epsilon, 0)$ to the $\frac{m}{2}$ tasks located at $(1,0)$ and a_1 to the tasks at $(\frac{m}{2}, 0)$. In this case, the total optimal cost is $\frac{m}{2}(1 + \epsilon) + \frac{m}{2}$. By the following calculation for every $0 \le \epsilon \le 1$, we have:

$$\frac{\text{Auction Alg with capacity}(\mathcal{I})}{\text{OPT with capacity}(\mathcal{I})} = \frac{1 + \frac{m^2}{4} + \frac{m}{2}\epsilon}{\frac{m}{2} + \frac{m}{2}\epsilon + \frac{m}{2}} \ge \frac{\frac{m^2}{4}}{m + \frac{m}{2}\epsilon} = \frac{\frac{m}{4}}{1 + \frac{1}{2}\epsilon} \ge \frac{m}{6}.$$

The last inequality is correct since $0 \le \epsilon \le 1$, therefore $\frac{1}{2} \le 1 + \frac{1}{2}\epsilon \le \frac{3}{2}$ and $\frac{1}{1 + \frac{1}{2}\epsilon} \ge \frac{2}{3}$. Now assume we have m tasks $\{t_1, t_2, ...t_m\}$, n agents $\{a_1, a_2, ..., a_n\}$ with capacities $\{\alpha_1, \alpha_2, ..., \alpha_n\}$ and all agents are initially in one place. Consider the optimal allocation where agents do not have any capacity restriction, then the total distance travelled by the agents is not more than the optimal allocation with capacity constraints. We have

$$\text{OPT with capacity} \ge \text{OPT without capacity.}$$

When we assume all agents are in one location, then the feasible solution consists of the union of paths rooted at this location. Each path indicates the tasks to be completed by an agent. Since these paths have a common root, they form a spanning tree and the optimal cost is not less than the minimum spanning tree, that is

$$\text{OPT without capacity} \ge \text{MST.}$$

Hence we have,

$$\frac{1}{\text{OPT with capacity}} \le \frac{1}{\text{MST}}.$$

Now consider the case when agents have capacity constraints and are co-located. In this case, for example, agent a_1 wins tasks $\{t_1, t_2, t_3\}$ but its capacity is exhausted and it cannot travel to task t_4. Agent a_2 then repeats the path of agent a_1 in order to travel to task t_4. Since there are m tasks, this scenario may be repeated at most m times. Therefore a path may be repeated at most m times and

$$\text{Auction Alg with capacity} \leq m \times \text{MST}.$$

Finally, we have

$$\frac{\text{Auction Alg with capacity}}{\text{OPT with capacity}} \leq \frac{m \times \text{MST}}{\text{MST}} = m.$$

The same example works for the *BidSumTree* bidding rule, since the tree solution is the same as path solution.

Proposition 2. *The performance ratio of the* BidSumTree *bidding rule with variable agent capacities for the* MiniSum *objective function is* $O(m)$.

5   Experimental Comparison of Algorithms

In this section, we analyse the performance of the path-based auction approach against the tree-based auction approach as presented in the paper by Lagoudakis et al. in [10] and summarised in Sect. 3.

The focus of our experiments is a comparison of both approaches when subject to capacity-based feasibility constraints on the allocation of tasks to agents. Such feasibility constraints are motivated by realistic settings where agents are likely to have a limited capacity of tasks that they can be allocated due to limited fuel or battery capacity. We investigate the impact of feasibility constraints on performance in terms of minimising the total distance travelled by all agents to their allocated tasks. Finally, we compare solutions produced by both approaches with an optimal solver for small problem sizes to situate the performance of the considered algorithms including the impact of feasibility constraints.

The results show that for both approaches, the introduction of feasibility constraints was slightly detrimental to the overall performance in terms of distance travelled. In addition, the path-based approach consistently outperforms the tree-based approach.

5.1   Experimental Setup and Design

The performance of the path-based and tree-based auction algorithms is analysed through the use of Monte Carlo experimentation. In our analysis, the total distance travelled by all agents is considered as the main performance metric. This is equivalent to the *MiniSum* objective function. For the path-based approach, agents bid according to the *BidSumPath* bidding rule and for the tree-based approach, agents bid according to the *BidSumTree* bidding rule. Additionally, we also report on the runtime of both approaches in Sect. 5.2.

The analysis in Sect. 5.2 considers variable numbers of agents and tasks, to align with the analysis performed by Johnson et al. in [8]. In scenarios with increasing numbers of tasks, the number of agents is fixed at two and in scenarios with increasing numbers of agents, the number of tasks is fixed at 200. We consider smaller problem sizes in Sect. 5.3 given the large time required to solve the problem optimally. Each data point in the figures and tables presented in Sects. 5.2 and 5.3 and represents the average of 500 simulation runs.

The simulation environment consists of a two-dimensional 100 by 100 area. Agents' initial locations and tasks are distributed randomly within the area. For all scenarios, we have generated our own location data for agents and tasks. Scenarios have been specified such that all tasks are guaranteed to be allocated (i.e. we ignore scenarios where the sum of agent task capacities does not meet or exceed the number of tasks), each agent is guaranteed to be allocated at least one task and each task is required to only be completed once. Given that we compare performance in terms of total distance travelled, it is important that all tasks are allocated to ensure a consistent comparison.

A Dell Latitude computer with an Intel Core i7 processor and 8 GB of RAM was used to run the simulations. In terms of communication, a simplified model was used which assumed agents communicated across a fully-connected network with perfect communications. As both algorithms have identical convergence characteristics, it was unnecessary to introduce complexity here.

5.2 Impact of Feasibility Constraints

In this section, we analyse the impact of feasibility constraints on allocations for the path-based and tree-based auction approaches. The results compare the case where no feasibility constraints are applied and all agents can be allocated an unlimited number of tasks with the case where feasibility constraints are applied and each agent has a limited number of tasks that it can be allocated. This introduces an element of heterogeneity to the agents, but they are otherwise identical. In this analysis, when feasibility constraints are applied, each agent's task capacity is variable and set randomly. The intention of this analysis was to identify the general impact of feasibility constraints on the performance of the path and tree-based auction algorithms as opposed to how performance changes for different task capacities.

Figure 2a presents performance in the two-agent case as the number of tasks increases. Firstly, it can be seen that without feasibility constraints the path-based approach outperforms the tree-based approach in terms of total distance with the performance gap widening as the number of tasks increases. The trade-off is that the tree-based approach is computationally more efficient [10], but this is not evident in these results. When constraints are applied, the total travel distance increases for both approaches, as seen in Fig. 2a. This can be attributed to better-positioned agents being unable to be allocated tasks when their allocation is full with the effect slightly more significant for the tree-based approach. For both approaches, the performance gap between the case with feasibility constraints and the case without also widens as the number of tasks increases.

The impact of feasibility constraints on the 200 task case is shown in Fig. 2b. Again, the path-based approach outperforms the tree-based approach. For both approaches, it can be seen that as the number of agents increases, the effect of feasibility constraints becomes negligible. In these cases, with an increased density of agents, it is less likely that the agents will be fully allocated and the feasibility constraints will have less effect.

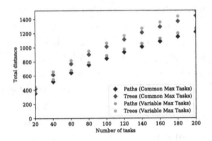

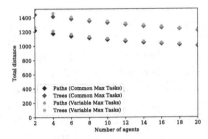

(a) Total distance for two agents vs tasks, unlimited vs limited/variable maximum tasks.

(b) Total distance for 200 tasks vs agents, unlimited vs limited/variable maximum tasks.

Fig. 2. Comparison of path-based and tree-based auction approaches with feasibility constraints applied for increasing numbers of (a) tasks and (b) agents.

Finally, Fig. 3 presents the runtime of the path-based approach and the tree-based approach for increasing numbers of tasks. Although both approaches converge in an equal number of steps, the tree-based approach is computationally more efficient, as previously stated. This is demonstrated in Fig. 3 where the path-based approach has a longer runtime that grows at a greater rate as the number of tasks increases. Note that the introduction of feasibility constraints reduces the overall runtime slightly.

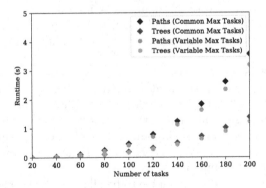

Fig. 3. Runtime for two agents vs tasks, unlimited vs limited/variable max. tasks.

5.3   Performance Against Optimal

In this section, we compare the performance of the path-based and tree-based auction approaches against an optimal brute force solver in terms of total travel distance. We report on cases subject to no feasibility constraints and cases where feasibility constraints limit the number of tasks an agent is allocated. The task capacity of each agent is variable and set randomly, hence this information is not shown in Tables 1a and 1b. The analysis in this section considers much smaller numbers of agents (two to three) and tasks (three to eight) than in Sect. 5.2 due to the large amount of time required to solve the problem optimally.

Table 1a shows the path-based approach is close to optimal without feasibility constraints for small problem sizes. With feasibility constraints applied, the gap is slightly larger as in Table 1b, but it can be seen that feasibility constraints negatively affect the performance of all approaches. Furthermore, it can be seen that the gap to optimal grows for both approaches as the problem size increases whether or not feasibility constraints are applied. As proven by Lagoudakis et al. in [10], the case without feasibility constraints has lower and upper bounds on the performance ratio of 1.5 and 2 respectively whilst we have proven that the performance ratio when feasibility constraints are applied is $O(m)$.

Table 1. Total distance travelled for path and tree-based approaches vs optimal.

Agents	Tasks	Optimal Total Dist	Path Total Distance	Tree Total Distance
2	3	97.7	98.7 (1.011)	106.1 (1.086)
2	5	140.5	143.7 (1.023)	159.0 (1.132)
2	8	191.5	198.3 (1.035)	225.3 (1.177)
3	4	106.1	107.7 (1.016)	116.0 (1.093)
3	6	143.4	146.7 (1.023)	163.4 (1.139)
3	8	173.4	178.5 (1.029)	200.5 (1.157)

(a) No feasibility constraints.

Agents	Tasks	Optimal Total Dist	Path Total Distance	Tree Total Distance
2	3	101.6	106.7 (1.051)	112.8 (1.111)
2	5	151.0	161.1 (1.068)	177.5 (1.175)
2	8	202.3	221.0 (1.092)	248.0 (1.226)
3	4	111.8	117.9 (1.055)	127.0 (1.136)
3	6	147.4	158.6 (1.076)	174.7 (1.185)
3	8	181.1	196.5 (1.085)	220.8 (1.219)

(b) Feasibility constraints applied.

6   Conclusions

In this paper, we examined a widely applicable problem encountered in logistics where the goal is to coordinate the routes of agents to undertake tasks and to do so whilst minimising the total distance travelled. We generalised two important algorithms for the problem and theoretically and experimentally analysed them. When there are constraints on the number of tasks per agent, the performance guarantee of the generalised algorithms is $O(m)$. Under the same constraints, it will be interesting to understand the best achievable approximation for optimal total cost when agents have task capacities. Analysing other feasibility constraints (such as distance capacities per agent) is another direction for future work. More generally, we show that while additional feasibility constraints can be imposed on various allocation algorithms, these feasibility constraints can affect the performance and the theoretical guarantees of the algorithms. Obtaining general insights on the interplay between feasibility constraints and performance guarantees is an important research direction in this line of work.

Acknowledgements. The research is supported by DST Group through the Centre for Advanced Defence Research in Robotics and Autonomous Systems under the project "Task Allocation for Multi-Vehicle Coordination" (UA227119).

References

1. Applegate, D.L., Bixby, R.E., Chvatal, V., Cook, W.J.: The Traveling Salesman Problem: A Computational Study. Princeton Series in Applied Mathematics. Princeton University Press, USA (2007)
2. Bae, J., Lee, J., Chung, W.: A heuristic for task allocation and routing of heterogeneous robots while minimizing maximum travel cost. In: International Conference on Robotics and Automation, ICRA 2019, Montreal, QC, Canada, 20–24 May 2019, pp. 4531–4537 (2019)
3. Carlsson, J., Ge, D., Subramaniam, A., Wu, A., Ye, Y.: Solving min-max multi-depot vehicle routing problem. Lect. Glob. Optim. **55**, 31–46 (2009)
4. Chien, S.A., Barrett, A., Estlin, T.A., Rabideau, G.: A comparison of coordinated planning methods for cooperating rovers. In: Proceedings of the Fourth International Conference on Autonomous Agents, AGENTS 2000, Barcelona, Catalonia, Spain, 3–7 June 2000, pp. 100–101 (2000)
5. Choi, H., Brunet, L., How, J.P.: Consensus-based decentralized auctions for robust task allocation. IEEE Trans. Rob. **25**(4), 912–926 (2009)
6. Dias, M., Stentz, A.: A free market architecture for distributed control of a multirobot system. In: Proceedings of the International Conference on Intelligent Autonomous Systems, pp. 115–122 (2000)
7. Gerkey, B.P., Mataric, M.J.: Sold!: auction methods for multirobot coordination. IEEE Trans. Robot. Autom. **18**(5), 758–768 (2002)
8. Johnson, L.B., Choi, H.L., Ponda, S.S., How, J.P.: Decentralized task allocation using local information consistency assumptions. J. Aerosp. Inf. Syst. **14**(2), 103–122 (2017)
9. Korsah, G.A., Stentz, A., Dias, M.B.: A comprehensive taxonomy for multi-robot task allocation. Int. J. Robot. Res. **32**(12), 1495–1512 (2013)
10. Lagoudakis, M.G., et al.: Auction-based multi-robot routing. In: Robotics: Science and Systems I, 8–11 June 2005, pp. 343–350. Massachusetts Institute of Technology, Cambridge (2005)
11. Lenstra, J.K., Kan, A.H.G.R.: Complexity of vehicle routing and scheduling problems. Networks **11**(2), 221–227 (1981). https://doi.org/10.1002/net.3230110211
12. Sarkar, C., Paul, H.S., Pal, A.: A scalable multi-robot task allocation algorithm. In: 2018 IEEE International Conference on Robotics and Automation, ICRA 2018, Brisbane, Australia, 21–25 May 2018, pp. 1–9 (2018)
13. Zlot, R., Stentz, A., Dias, M.B., Thayer, S.: Multi-robot exploration controlled by a market economy. In: Proceedings 2002 IEEE International Conference on Robotics and Automation (Cat. No. 02CH37292), vol. 3, pp. 3016–3023 (2002)

Generalized Bargaining Protocols

Yasser Mohammad[1,2,3]([envelope]) [iD]

[1] NEC CORPORATION, Minato City, Japan
y.mohammad@nec.com
[2] National Institute of Advanced Industrial Science and Technology (AIST), Tokyo, Japan
[3] Assiut University, Asyut, Egypt

Abstract. Automated Negotiation (AN) is a research field with roots extending back to the mid-twentieth century. There are two dominant AN research directions in recent years: (1) designing new heuristic strategies for the simplest bargaining protocol called the Alternating Offers Protocol (AOP) and (2) defining new mediated protocol that require a trusted third party. Intelligence lies in the strategy in the first direction and the protocol in the latter. This paper argues for a third way that aims at designing unmediated AN protocols with desired properties. We introduce a generalization of AOP to a wide class of unmediated protocols that keep its main advantages while providing the designer with the freedom to design protocols with desired properties. We also introduce the first fruits of this research direction in the form of an unmediated protocol and a corresponding simple strategy that can be shown theoretically to be exactly rational, optimal, and complete for bilateral negotiations with no information about partner's preferences.

Keywords: Automated Negotiation · Multiagent Systems · Mechanism Design

1 Introduction

Negotiation is used frequently in modern societies and its most common form is bargaining in which agents exchange offers until an agreement is reached or one of the agents walks away. Automated negotiation is receiving more interest from the research community recently due to the accelerated pace at which businesses employ intelligent agents to manage different production and business processes and the need for these agents to work together as self-interested agents representing their institutions. Automated negotiation has a long history as a research discipline dating back to Nash's solution of the bargaining game [23]. Nash's Bargaining game was a single round protocol. It was later extended by Rubinstein in his classical work that described a perfect equilibrium for a more realistic bargaining protocol with time-pressure but no extrinsic limit on the number of exchanged offers [24]. The preferences of all players were assumed to be common knowledge in that work.

T. Liu et al. (Eds.): AI 2023, LNAI 14472, pp. 261–273, 2024.
https://doi.org/10.1007/978-981-99-8391-9_21

One of the most commonly used automated negotiation protocols that assume private preferences is the Alternating Offers Protocol (AOP) with its multilateral extension (the Stacked Alternating Offers Protocol—SAOP byagent [1]). Several strategies have been proposed for negotiation under this protocol [5]. These methods can be classified into exact solutions of simplified problems [4] and heuristic solutions to the most general case [12]. Most recently, MiCRO was proposed for AOP guaranteeing that agreements – if found – are Pareto-optimal [10].

Alternatives to the AOP protocol have been proposed that are mostly mediated. A commonly used example is the Single Text (ST) negotiation protocol in which a third-party mediates the interaction between agents by offering a tentative agreement for them and updating it based on their responses [16].

The main issue with mediated protocols for business applications is that the third-party has some control over the final agreement and must be trusted. For example, in the ST protocol, the final agreement will be one that was proposed by the mediator. This makes the mediator a player in the negotiation game and as it will be a self-interested player – in real-world applications – it may be biased in its proposal generation (e.g. larger businesses).

Research in unmediated automated negotiation has focused on designing intelligent heuristic strategies for small variations of the simple alternating offers protocol in which agents exchange offers till agreement. Auction theory went the other direction by focusing on designing auction mechanisms for which specific simple strategies are both dominant and easy to execute (DSIC mechanisms). This work is a first step toward a similar research program in automated negotiation that aims at developing *intelligent* unmediated negotiation protocols that have simple dominant strategies and equilibria.

In this paper, we present a new formulation of negotiation in which AOP is generalized to a large set of generalized bargaining protocols (GBPs) that are clearly defined allowing designers to reason more effectively about them. Generalized Bargaining Protocols provide a *common language* to speak and reason about negotiation protocols and can describe both unmediated and some mediated protocols. We also – briefly – present the first result of this research program in the form of a specific GBP called the Tentative Agreement Unique Offers Protocol (TAU) and propose a specific strategy that renders it rational, optimal and complete. TAU is then evaluated against AOP utilizing baseline and SOTA negotiation strategies and is shown to provide superior results.

2   Automated Negotiation

A Negotiation Scenario s is defined as a tuple $(\mathcal{A}^s, \mathcal{D}^s, \mathcal{P}, \mathcal{I}^s)$ where $\mathcal{A}^s$ is the set of agents numbered from 1 to n_A, $\mathcal{D}^s$ is the negotiation domain, $\mathcal{P}$ is the negotiation protocol, and $\mathcal{I}^s$ is a tuple defining the information available to each agent.

The negotiation domain $\mathcal{D}^s \equiv (\Omega^s, \mathcal{F}^s)$ defines (1) The outcome space (Ω^s of size n_o) comprising all possible agreements. A special outcome $\Phi \notin \Omega^s$ is assumed to always exist to represent disagreement and we define the *extended outcome space* Ω^+ as the $\Omega^s \bigcup \{\Phi\}$. (2) Agent preferences $\mathcal{F}^s$ which can either be ordinal[1] ($\succeq$) defining an ordering[2] over Ω^+ – per agent – or cardinal (u) defining a mapping per agent from Ω^+ to $\Re$ with higher values indicating better outcomes[3]. We use $\succeq_a$ and u_a to refer to the preferences for agent a.

The *Information Set Tuple* $\mathcal{I}^s$ is a tuple of n_A information sets $(\{I_a^s(\mathcal{F}^s) : a \in \mathcal{A}^s\})$ where $I_a^s(\mathcal{F}^s)$ represents all the information available to agent a about the preferences $\mathcal{F}^s$ of all agents including itself[4]. In this work, we assume that the agent has no information about the strategy used by its partner(s). If such information is available, they can be included in I_a^s.

We define the following sets for a negotiation scenario s:

Rational Outcome Set $\overline{\Omega}^s$: The subset of the outcome space Ω^s that is not worse than disagreement for all agents: $\{\omega : \omega \in \Omega \wedge \omega \succeq_a \Phi \forall a \in \mathcal{A}^s\}$.

Win-Win deals $\widehat{\Omega}^s$: Rational outcomes strictly better than disagreement for someone: $\left\{\omega \in \overline{\Omega}^s : \exists a \in \mathcal{A}^s \to \omega \succ_a \Phi\right\}$.

Pareto Outcome Set $\mathcal{P}$: The rational outcomes that cannot be improved for one agent without making at least one other agent worst off:

$$\left\{\psi : \psi \in \overline{\Omega}^s \wedge \left(\neg\exists\left(\omega \in \Omega^s \setminus \{\psi\}, b \in \mathcal{A}^s\right) : \omega \succ_b \psi \wedge (\omega \succeq_a \psi \forall a \in \mathcal{A}^s)\right)\right\}.$$

A negotiation strategy describes the behavior of an agent in accordance with a protocol $\mathcal{P}$ for the set of scenarios it was designed to handle. Its input is the tuple (n_A, a, Ω, I_a), where a is the agent index controlled by the strategy.

In order to compare different negotiation protocols, we need a way to assign strategies to negotiation scenarios (a *Strategy Assignment Rule* (SAR) hereafter). A *deterministic* SAR assigns strategies deterministically. A *pure* SAR π assigns the strategy π to all agents in every scenario.

Given a negotiation scenario s and a SAR Υ, the result of all agents $\mathcal{A}^s$ following their assigned strategies selects a member of the extended outcome space called the *negotiation outcome* $\omega_*^s \in \Omega^+$. We indicate executing the negotiation according to protocol $\mathcal{P}$ on s using the SAR π resulting on ω_*^s as a negotiation outcome by $s(\Upsilon, \mathcal{P}) = \omega_*^s$.

Because the strategy used by agent a has access only to (Ω, I_a^s), while the full game induced by the scenario (G^s) is defined by $(\Omega, \mathcal{F}^s)$, the agent faces a Game with Incomplete Information (GII) in which nature moves first assigning agent and partner(s) types (defined as Ω, and preferences $\mathcal{F}^s$ sampled according to $\mathcal{I}^s$). Nature never plays again. The size of this GII is huge (and in most cases infinite). No wonder we only have heuristic approaches except for the case with complete information ($I_a^s = \mathcal{F}^s$) [7,23,24]. Most recent research in automated negotiation—including this paper—assumes that I_a^s is simply the agent's own

---

[1] We assume standard transitivity on preferences.

[2] $a \succeq b$ means that a is not worse than b. Symbols $\succ, \approx$ are defined accordingly.

[3] If the codomain of u within the range $[0, 1]$, the utility function is called normalized. Time-pressure can be modeled here by a discounting factor as inagent [24].

[4] We will drop the scenario superscript when known from the context.

preferences ($\mathcal{F}_a^s$) but in most realistic situations, agents have some partial information about partner preferences. Sometimes, information about the partner's preferences is assumed implicitly to be available. For example, several works assume no agreement is worse than disagreement for any agent [10] or that the utility function has a specific functional form [8].

The Alternating Offers Protocol and its Stacked multilateral version (SAOP) [1] are the most widely used negotiation protocols in recent years. SAOP is a multilateral negotiation protocol in which agents are arranged in a circle (defining an ordering). The first agent sends an initial offer to the next agent in the circle. When receiving an offer, an agent can either accept it, end the negotiation immediately or respond with a counter offer which is sent to the next agent in the circle. This process continues until an agent ends the negotiation (on its turn) or an offer is accepted by all agents (except the one offering it which is implicitly assumed to accept the offer). AOP is the bilateral version of SAOP.

Different authors use slightly different variations of AOP for different negotiation contexts. For example,agent [25] used a version in which a final confirmation message is needed to finalize an agreement. A similar variation was recently used inagent [2]. Another variation involves requiring that agents never repeat offers was used by [3] and others. agent [20] proposed another variation in which an agent can only repeat its last offer (not any other offer it sent earlier).

3   Proposed Framework

Our main motivation for introducing Generalized Bargaining Protocols (GBPs) is to provide a class of negotiation protocols that keep the essence of bargaining while providing a clear set of knobs that can be adjusted by the designer to generate new protocols with desirable features in general or for specific applications.

Two common features of all GBPs are 1) All offers originate from the agents. 2) agents can leave the negotiation anytime preventing agreement (ensuring exact rationality). An extra desired feature of a GBP protocol is to be implementable distributedly in the agent machines with no need for third-parties. This subclass is called Distributed Generalized Bargaining Protocols DGBP hereafter.

Figure 1 shows a timing diagram for all the components involved in a bilateral negotiation using a GBP. A GBP negotiation between n_A agents is carried out in n_T negotiation threads ($\mathbb{T}$). Each negotiation thread is assigned to exactly one agent called its owner. One or more agents are assigned as responders on each thread to respond to offers from the thread owner. A negotiation trace T is defined as an ordered tuple of all offers sent in every thread with the associated responses. The domain of all possible traces is called $\mathcal{T}$ hereafter. A GBP is defined as a tuple $(\alpha, \rho, \chi, f, \sigma)$ where:

Assignment Rule $\alpha : \mathbb{T} \to \mathcal{A}$ Defines the negotiation threads ($\mathbb{T}$) and selects exactly one agents to own each thread.

Responders Rule $\rho : \mathbb{T} \to \mathbb{P}(\mathcal{A})$ Selects one or more agents to respond to offers in every thread.

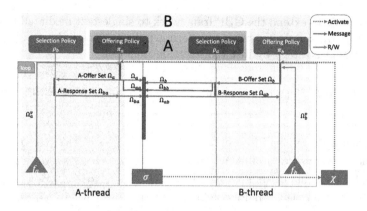

Fig. 1. Timing diagram of a bilateral GBP.

Activation Rule $\chi : \mathcal{T} \to \mathbb{T} \to \mathbb{P}(\mathbb{T})$ Selects one or more threads to activate based on the negotiation trace (i.e. runs the owner's offering policy followed by responders' selection policies).

Filtering Rule $f : \mathcal{T} \to \mathbb{P}(\mathbb{P}(\Omega))^{n_T}$ Filters out invalid outcome-sets for the current state of the protocol from the outcome space leading to the Valid Offering Space for each agent $\Omega_a^v \subset \mathbb{P}(\Omega)$ for thread a.

Evaluation Rule $\sigma : \mathcal{T} \to \mathbb{P}(\Omega) \cup \{\triangleright\}$ Reads the negotiation trace after every agent decision and returns one of two decisions: 1) an outcome-set ending the negotiation or 2) The value $\triangleright$ which activates the Activation Rule to select the next set of Offering Policies to activate.

Despite its apparent complexity, in most cases the mechanism designer can focus almost completely on the filtering, and evaluation rules because the remaining three rules and parameters are mostly fixed by the negotiation context. Nevertheless, having the expressiveness of the full GBP framework is important in modeling different negotiation contexts (e.g. negotiations designed to reduce the search space rather than reaching an agreement, concurrent negotiations).

A GBP agent strategy is thus defined by two components (very similar to AOP strategies):

Offering policy $\pi_a : \mathbb{P}(\mathbb{P}(\Omega)) \to \mathbb{P}(\Omega)$ Receives a set of outcomes (the Valid Offering Space) $\Omega_a^v \subseteq \mathbb{P}(\Omega)$ from which it selects an element (i.e. set of outcomes) Ω_a (called an offer) to send through a thread it owns. Offering the empty set $\emptyset$ is interpreted as ending the negotiation leading to disagreement.

Selection policy $\rho_a : \mathbb{P}(\Omega) \to \mathbb{P}(\Omega)$ Receives an offer (Ω_b) on a given thread b and returns a subset of it as a response $\Omega_{ba} \subseteq \Omega_b$. The empty set $\emptyset$ is interpreted as rejection.

An attractive feature of the GBP formulation is that all rules and policies are defined as simple set operations. This simplifies the process of analysis.

It is possible to extend the GBP framework to single-text mediated protocols (e.g. [15,16]) by simply moving the offering policies *inside* the mediator and define the agent strategy by just the selection policy.

Table 1. Mapping some common bargaining protocols to GBP.

Protocol	α	ρ	χ	f	σ
Rubinstein's Bargaining Protocol [24]	α_1	ρ_o	χ_{rr}	f_a	σ_{aot}
Alternating Offers Protocol (AOP)	α_1	ρ_o	χ_{rr}	f_a	σ_{aot}
Stacked Alternating Offers Protocol (SAOP) [1]	α_1	ρ_n	χ_{rr}	f_a	σ_{sao}
Alternating Multiple Offers Protocol (AMOP) [1]	α_1	ρ_o	χ_{rr}	f_a	σ_{aot}
William's Concurrent Negotiation Protocol (Single) [25]	α_1	ρ_a	χ_{rr}	f_a	σ_{aot}
Concurrent Negotiations with Global Preferences [20]	α_n	ρ_o	χ_{rr}	f_a	σ_{aot}
Bargaining Chips [2]	α_n	ρ_c	χ_{rr}	f_a	σ_{aot}
Negotiation-Based Branch & Bound (NB3) [11]	α_n	ρ_n	χ_{at}	f_a	σ_{aot}
Decomposable Alternating Offers Protocol [18]	α_n	ρ_o	χ_{at}	f_d	σ_{aot}

Assignment and Responders Rules α, ρ. The assginment and responders rules define the message exchange graph for the protocol. They determine which agents responds to what offers.

The most commonly used assignment rule is the One2One rule (α_1) where each agent is assigned exactly one thread. We will assume this assignment rule for the result of this paper. Some protocols (e.g. NB3 [11]) can be modeled by assuming n_A threads per agent with one responder for each (α_n).

The most commonly used responders rules is the Others rule (ρ_o) used by AOP and its multilateral AMOP extension [1] which sets all agents except the owner as responders.

SAOP uses another rule we call the Next rule (ρ_n) which defines an ordering of agents in a circle and assigns the next agent in this ordering as a sole responder for the thread owned by every agent. Another rule that is used mostly in the context of concurrent negotiation is the All rule (ρ_a) which assigns all agents as responders in all threads and can be used to model each negotiation individually (e.g. [2,25]). The last two rules can be combined (NextConfirm: ρ_c) to model cases of concurrent negotiations with a confirmation message.

All assginment and responders rules mentioned so far define a stationary graph for message exchange. More complicated rules that change the responders in each thread during the negotiation can also be defined by allowing the graph structure to depend on negotiation history. We are not aware of any protocols that require this extension.

Activation Rule χ. The activation rule controls the timing of offers and responses. An offering policy can send offers ONLY when it is activated by the activation rule. Negotiation starts when the activation rule selects a subset of the agents to send their first offers. Once an offering policy sends an offer, it becomes inactive until activated again by the activation rule.

The most flexible activation rule is the Any-Time rule (χ_{at}) which activates all agents at the first step, then immediately activates each offering policy once it becomes inactive until a response is received from all agents in the corresponding negotiation thread. All offering policies are then activated again if the evaluation rule returns the value "$\triangleright$" [2,11].

Probably the most widely used activation rule is the Round-Robin rule (χ_{rr}) in which a single offering policy is activated at any point of time based on a predefine ordering. Moreover, once an offering policy sends an offer, it is not re-activated until all other offering policies and selection policies are run. This models the timing of AOP and SAOP [1].

Filtering Rule f. The filtering rule is used to limit the freedom of offering policies in choosing the offer to send at any time by setting Ω_a^v. The simplest filtering rule (All f_a) is to simply set $\Omega_a^v = \{\{\omega\} : \omega \in \Omega\}$. This is the most widely used filtering rule.

Another filtering rule rule that is used sometimes is the No-Repetition rule (f_{nr}) which sets $\Omega_a^v = \{\{\omega\} : \omega \in \Omega \setminus \mathbb{P}_a\}$ (where $\mathbb{P}_a$ is the union of all offers sent by the agent so far) which simply means that agents are not allowed to repeat any offers. This was used byagent [3].

f_{nr} and f_a, both limit agents to single-outcome offers. Some bargaining mechanisms allow (or dictate) offers with higher cardinality (i.e. partial agreements). For example, in the Decomposable Alternating Offers Protocol (DAOP) [18], agents negotiate over subsets of negotiation issues sequentially until some agreement is reached for each issue which is then fixed for the rest of the negotiation. This is equivalent to selecting subsets of the outcome space with progressively smaller sizes (fixed issue values). This filtering rule is called f_d hereafter.

Evaluation Rule σ. The evaluation rule decides when a negotiation should end and the final output of the negotiation. The simplest evaluation rule is the Accept-Or-Timeout rule (σ_{aot}) which ends the negotiation withe last selected outcome-set if it is non-empty, ends the negotiation with disagreement if a time-out condition is met (n. rounds or wall time) and otherwise, continues the negotiation. This is used by all AOP variants.

Table 1 shows examples of unmediated negotiation protocols variations as GBPs. Both single negotiation and concurrent negotiations can be modeled.

3.1 Evaluating Negotiation Protocols

Different problems call for different features in the negotiation protocol. In this section we define some of the most important measures for comparing of a negotiation protocol $\mathcal{P}$ and SAR Υ for a predefined set scenarios $\mathbb{S}$.

Rationality is defined as the fraction of $\mathbb{S}$ that lead to a *rational* outcome under $\mathcal{P}$, Υ: $\left|\left\{s : s \in \mathbb{S} \wedge s(\Upsilon, \mathcal{P}) \in \overline{\Omega}^s\right\}\right| / |\mathbb{S}|$. **Completeness** is is defined as the fraction of $\mathbb{S}$ with win-win deals that lead to a *rational* win-win outcome under $\mathcal{P}$, Υ: $\left|\left\{s : s \in \mathbb{S} \wedge \widehat{\Omega}^s \neq \emptyset \wedge s(\Upsilon, \mathcal{P}) \in \widehat{\Omega}^s\right\}\right| / |\mathbb{S}|$.

Optimality is defined as the expected value of one minus the distance of the negotiation outcomes to the Pareto-Outcome-Set under $\mathcal{P}$, Υ: $\mathbb{E}_\mathbb{S}\left[1 - \min_{\omega \in \mathcal{P}} \sqrt{\sum_{a \in \mathcal{A}}(u_a(s(\Upsilon, \mathcal{P})) - u_a(\omega))^2}\right]$. **Fairness** is defined as the expected value of one minus the distance of the negotiation outcomes to any bargaining solution[5] for each scenario under $\mathcal{P}$, Υ: $\mathbb{E}_\mathbb{S}$ $\left[1 - \min_{\omega \in \{Nash, Kalai, OrdinalNash, OrdinalKalai\}} \sqrt{\sum_{a \in \mathcal{A}}(u_a(s(\Upsilon, \mathcal{P})) - u_a(\omega))^2}\right]$. **Welfare** is defined as the expected sum of the achieved utility for all agents under $\mathcal{P}$, Υ: $\mathbb{E}_\mathbb{S}\left[\mathbb{E}_{a \in \mathcal{A}^s}[u_a(s(\Upsilon, \mathcal{P}))]\right]$. Finally, the **Advantage** is defined as the expected agent advantage (i.e. difference between achieved utility and reserved value) for all agents under $\mathcal{P}$, Υ: $\mathbb{E}_{\mathbb{S}, \mathcal{A}^s}[u_a(s(\Upsilon, \mathcal{P})) - u_a(\Phi)]$.

An Outcome-Perfect Negotiation Protocol (OPNP) for scenarios $\mathbb{S}$ is one that is *exactly* rational, complete, optimal, and fair (i.e. achieving 100% in all of these criteria) with a single strategy assignment rule for all members of $\mathbb{S}$. Intuitively, if a protocol is OPNP, it is guaranteed to always find the best win-win deals possible or lead to disagreement if none are available. An exactly fair OPNP protocol with a Bayes-Nash Equilibrium strategy assignment rule for scenarios $\mathbb{S}$ is called a Perfect Negotiation Protocol (PNP) for $\mathbb{S}$.

3.2 Tentative Agreements Unique Offers (TAU)

Now that we described GBPs, we introduce a novel specific instantiation of it called Tentative Agreements Unique Offers Protocol (TAU) that can be analyzed using simple set-theoretic tools and shown to be an Outcome-Perfect Negotiation Protocol.

The main contribution of this paper is the GBP framework itself and the details of TAU will be the subject of a future publication. The main reason for introducing it here is to show the effectiveness of GBP in modeling new protocols and the simplicity it provides for analyzing them.

Agents are allowed to repeat offers but once an agent starts repeating an offer, it is forced to continue repeating the same offer forever. Moreover, an agreement is reached only if the same offer was offered and selected by every agent. We need one more definition to describe this protocol: the Recently Proposed Tuple $\mathbb{P}_a$ is the set of a's past offers and the last of them is P_a^{last}. Let, i be the current step and n_o be the size of the outcome space, and Ω_a^i be the offer from agent a at step i, the Tentative Agreement Unique Offers Protocol (TAU) is defined as GBP($\alpha_{11}, \rho_o, \chi_{at}|\chi_{rr}, f_\tau, \sigma_\tau$): The filtering rule simply removes all previous offers of the agent except its last one:

$$f_\tau(a) = \left\{\{\omega\} : \omega \in P_a^{last} \bigcup (\Omega^+ \setminus \mathbb{P}_a)\right\} \tag{1}$$

The evaluation rule starts by adding the offer to a thread-specific tentative agreement list (Ω_a^*) iff it was accepted by every responder:

$$\Omega_a^* \leftarrow \Omega_a^* \bigcup \bigcap_b \Omega_{ba}^i \tag{2}$$

---

[5] We use the Nash [23] and Kalai [13] Bargaining solutions (and ordinal extensions).

The evaluation rule then generates an agreement if an outcome is in the tentative agreement list of all threads, ends the negotiation if an agent explicitly ends it or all agents are repeating offers, and continues the negotiation otherwise:

$$\sigma_\tau = \begin{cases} \Phi & \exists a \in \mathcal{A} : \Omega_a^i = \emptyset \vee \Omega_a^i \subseteq \mathbb{P}_a^i \forall a \in \mathcal{A} \\ \{\omega\} & \omega \in \Omega_a^* \forall a \in \mathcal{A} \\ \triangleright & otherwise. \end{cases} \tag{3}$$

The Slow Concession Strategy (SCS) is a simple strategy for TAU with discrete outcome-spaces. It offers outcomes from best to worst (breaking ties in a prede-fined common full lexical ordering $\succ_l$)[6] repeating the worst outcome not worse than disagreement once it is reached. The offering policy of SCS (playing as agent a) can then be expressed as follows:

$$\pi_{SCS}(i) : \Omega_a^i = \left\{ \sup_{\succ_l}(\overline{\Omega}_a^v(i)) \right\} \tag{4}$$

where $\overline{\Omega}_a^v(i) \equiv \left\{ \omega : \{\omega\} \in \Omega_a^v(i) \cap \overline{\Omega}_a \wedge \omega \succ_a \psi \forall \{\psi\} \in \Omega_a^v(i) \right\}$.

SCS's selection policy is even simpler: select the offered outcome if and only if the current offer is better than the best outcome it received from this partner so far:

$$\rho_{SCS}(\omega, i) : \Omega_{ba}^i = \begin{cases} \left\{ \omega : \omega \in \Omega_b^i \wedge \omega \succ_a \overline{\Omega}_{ba}^i \right\} & \overline{\Omega}_{ba}^i \neq \emptyset \\ \Omega_b^i \cap \overline{\Omega}_a & \overline{\Omega}_{ba}^i = \emptyset \end{cases} \tag{5}$$

where $\overline{\Omega}_{ba}^i \equiv \sup_{\succ_a} \overline{\Omega}_a \cap \left(\cup_{j<i} \Omega_b^j \right)$ is the best outcome received from b before the current step with. It can be shown that TAU is an OPNP (using SCS) for bilateral negotiations with discrete outcome-spaces, no knowledge of partner's preferences in which all outcomes are rational. Moreover, TAU(SCS) is exactly rational, optimal and complete without the assumption that all outcomes are rational and empirically to be fairer than SOTA strategies for AOP.

4   Empirical Evaluation

The main contribution of this paper is the GBP framework for defining nego-tiation protocols which describes the whole process as set-theoretic operations simplifying proofs of theoretical properties. In this section we evaluate our pro-posed GBP (i.e. TAU) against AOP empirically using baseline and top strate-gies for AOP. All experiments were conducted using a fork of NegMAS 0.9 [21] platform with its NegMAS-GENIUS bridge for the official implementations of state-of-the-art (SOTA) AOP strategies running on GENIUS [19].

We use six baselines for our experiments: Time-based Conceder, Linear and Boulware strategies, behavior based Nice Tit for Tat (NTfT) strategy [6], and

---

[6] It can be shown that no agent can benefit from deviating from this tie-breaking strategy.

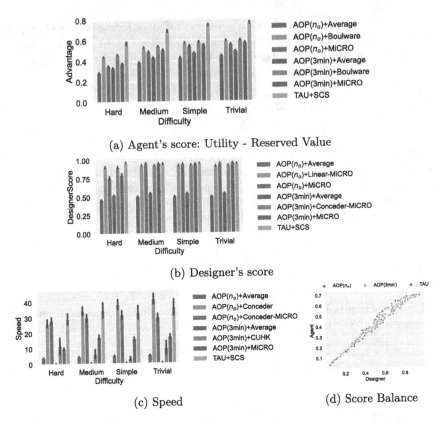

(a) Agent's score: Utility - Reserved Value

(b) Designer's score

(c) Speed

(d) Score Balance

Fig. 2. Evaluation against AOP + top strategy combinations on ANAC domains.

the recently proposed MiCRO [10]. To compare the proposed approach against using SOTA strategies with AOP, we also used the following set of strategies for AOP: Caduceus, Atlas3 [22](most recent winners of non-repeated tracks with no uncertainty in 2015 and 2016), AgentK by [14], Hardheaded by [17], CUHK by [9] (most recent winners of non-repeated bilateral negotiation tracks at ANAC in 2010, 2011, 2012). All combinations of these strategies (121) were used.

We used all publicly available ANAC's domains from its inception in 2010 to 2022 except 2014 (nonlinear, continuous), leading to 184 different domains. For each domain, we varied the reserved value of the second agent so that 100%, 90%, 50%, and 10% of the outcome space is rational to simulate outside options leading to 736 different scenarios. For AOP we ran negotiations with either a wall time limit of 3 min (usually used in ANAC competitions) or $n_o + 1$ rounds (intrinsic limit for TAU) (190,026 negotiations in total). All differences were checked using factorial t-test (and Wilcoxon's nonparametric test) with Bonferroni's multiple-comparisons correction and differences reported hereafter are all statistically significant if not otherwise indicated.

Figure 2a shows the average and *top-2* agent advantage of all the 121 AOP strategy combinations compared with the advantage of using the proposed approach. All AOP strategy combinaions achieve lower advantage in all difficulty levels compared with TAU (with SCS) The harder the negotiation, the smaller the advantage of TAU over the *top* AOP strategy combination. Figure 2b shows the *top-2* overall designer score[7]. There is no statistically significant difference between SCS and the top AOP strategy (MiCRO) except for *hard* scenarios for which TAU (with SCS) clearly outperforms all AOP strategy combinations including MiCRO. Figure 2d shows that the proposed approach dominates all of the 121 AOP strategy combinations in terms of agent and designer scores. Figure 2c shows that the proposed approach is faster than all AOP strategy combinations except Conceder for all difficulty levels despite the fact that it has no inherent time-limit. Taken together, these results show that TAU does not only have nice theoretical properties but is practically effective specially for difficult negotiations with few rational agreements.

5 Conclusion

Our main goal in this work is to advocate for a new research program in automated negotiation mechanism design with the aim of finding new protocols with desired properties for different negotiation conditions. We believe that Generalized Bargaining Protocols provides a useful framework in this direction while keeping most of the advantages of AOP. By providing a large yet clearly defined search space, it will be possible to design new negotiation protocols that move the intelligence from the strategy to the protocol providing theoretical guarantees that are much harder with AOP and opening new avenues of research.

References

1. Aydoğan, R., Festen, D., Hindriks, K.V., Jonker, C.M.: Alternating offers protocols for multilateral negotiation. In: Fujita, K., et al. (eds.) Modern Approaches to Agent-based Complex Automated Negotiation. SCI, vol. 674, pp. 153–167. Springer, Cham (2017). https://doi.org/10.1007/978-3-319-51563-2_10
2. Baarslag, T., Elfrink, T., Nassiri Mofakham, F., Koça, T., Kaisers, M., Aydogan, R.: Bargaining chips: coordinating one-to-many concurrent composite negotiations. In: IEEE/WIC/ACM International Conference on Web Intelligence and Intelligent Agent Technology, pp. 390–397 (2021)
3. Baarslag, T., Gerding, E.H.: Optimal incremental preference elicitation during negotiation. In: IJCAI, pp. 3–9 (2015)
4. Baarslag, T., Gerding, E.H., Aydogan, R., Schraefel, M.: Optimal negotiation decision functions in time-sensitive domains. In: 2015 IEEE/WIC/ACM International Conference on Web Intelligence and Intelligent Agent Technology (WI-IAT), vol. 2, pp. 190–197. IEEE (2015)

---

[7] Defined as the product of optimality, completeness, fairness, and welfare.

5. Baarslag, T., Hindriks, K., Hendrikx, M., Dirkzwager, A., Jonker, C.: Decoupling negotiating agents to explore the space of negotiation strategies. In: Marsa-Maestre, I., Lopez-Carmona, M.A., Ito, T., Zhang, M., Bai, Q., Fujita, K. (eds.) Novel Insights in Agent-based Complex Automated Negotiation. SCI, vol. 535, pp. 61–83. Springer, Tokyo (2014). https://doi.org/10.1007/978-4-431-54758-7_4

6. Baarslag, T., Hindriks, K., Jonker, C.: A tit for tat negotiation strategy for real-time bilateral negotiations. In: Ito, T., Zhang, M., Robu, V., Matsuo, T. (eds.) Complex Automated Negotiations: Theories, Models, and Software Competitions. Studies in Computational Intelligence, vol. 435, pp. 229–233. Springer, Heidelberg (2013). https://doi.org/10.1007/978-3-642-30737-9_18

7. Di Giunta, F., Gatti, N.: Bargaining over multiple issues in finite horizon alternating-offers protocol. Ann. Math. Artif. Intell. **47**(3), 251–271 (2006)

8. Fujita, K., Aydogan, R., Baarslag, T., Hindriks, K., Ito, T., Jonker, C.: ANAC 2016 (2016). https://web.tuat.ac.jp/~katfuji/ANAC2016/

9. Hao, J., Leung, H.: CUHKAgent: an adaptive negotiation strategy for bilateral negotiations over multiple items. In: Marsa-Maestre, I., Lopez-Carmona, M.A., Ito, T., Zhang, M., Bai, Q., Fujita, K. (eds.) Novel Insights in Agent-based Complex Automated Negotiation. SCI, vol. 535, pp. 171–179. Springer, Tokyo (2014). https://doi.org/10.1007/978-4-431-54758-7_11

10. de Jonge, D.: An analysis of the linear bilateral ANAC domains using the MiCRO benchmark strategy. In: Proceedings of the 31st International Joint Conference on Artificial Intelligence (IJCAI) (2022)

11. Jonge, D., Sierra, C.: NB3: a multilateral negotiation algorithm for large, non-linear agreement spaces with limited time. Auton. Agent. Multi-Agent Syst. **29**(5), 896–942 (2015)

12. Jonker, C.M., Aydogan, R., Baarslag, T., Fujita, K., Ito, T., Hindriks, K.V.: Automated negotiating agents competition (ANAC). In: AAAI, pp. 5070–5072 (2017)

13. Kalai, E.: Proportional solutions to bargaining situations: interpersonal utility comparisons. Economet. J. Econom. Soc. 1623–1630 (1977)

14. Kawaguchi, S., Fujita, K., Ito, T.: AgentK2: Compromising Strategy Based on Estimated Maximum Utility for Automated Negotiating Agents. In: Ito, T., Zhang, M., Robu, V., Matsuo, T. (eds.) Complex Automated Negotiations: Theories, Models, and Software Competitions. SCI, vol. 435, pp. 235–241. Springer, Heidelberg (2013). https://doi.org/10.1007/978-3-642-30737-9_19

15. Klein, M., Faratin, P., Sayama, H., Bar-Yam, Y.: Negotiating complex contracts. Group Decis. Negot. **12**(2), 111–125 (2003)

16. Klein, M., Faratin, P., Sayama, H., Bar-Yam, Y.: Protocols for negotiating complex contracts. IEEE Intell. Syst. **18**(6), 32–38 (2003)

17. van Krimpen, T., Looije, D., Hajizadeh, S.: HardHeaded. In: Ito, T., Zhang, M., Robu, V., Matsuo, T. (eds.) Complex Automated Negotiations: Theories, Models, and Software Competitions. SCI, vol. 435, pp. 223–227. Springer, Heidelberg (2013). https://doi.org/10.1007/978-3-642-30737-9_17

18. Li, Z., Hadfi, R., Ito, T.: Divide-and-conquer in automated negotiations through utility decomposition. IIAI Lett. Inform. Interdisc. Res. **3** (2023)

19. Lin, R., Kraus, S., Baarslag, T., Tykhonov, D., Hindriks, K., Jonker, C.M.: Genius: an integrated environment for supporting the design of generic automated negotiators. Comput. Intell. **30**(1), 48–70 (2014)

20. Mohammad, Y.: Concurrent local negotiations with a global utility function: a greedy approach. Auton. Agent. Multi-Agent Syst. **35**(2), 1–31 (2021)

21. Mohammad, Y., Nakadai, S., Greenwald, A.: NegMAS: a platform for situated negotiations. In: Aydoğan, R., Ito, T., Moustafa, A., Otsuka, T., Zhang, M. (eds.) ACAN 2019. SCI, vol. 958, pp. 57–75. Springer, Singapore (2021). https://doi.org/10.1007/978-981-16-0471-3_4

22. Mori, A., Ito, T.: Atlas3: a negotiating agent based on expecting lower limit of concession function. In: Fujita, K., et al. (eds.) Modern Approaches to Agent-based Complex Automated Negotiation. SCI, vol. 674, pp. 169–173. Springer, Cham (2017). https://doi.org/10.1007/978-3-319-51563-2_11

23. Nash Jr., J.F.: The bargaining problem. Econom. J. Econom. Soc. 155–162 (1950)

24. Rubinstein, A.: Perfect equilibrium in a bargaining model. Econom. J. Econom. Soc. 97–109 (1982)

25. Williams, C.R., Robu, V., Gerding, E.H., Jennings, N.R.: Negotiating concurrently with unknown opponents in complex, real-time domains. In: Proceedings of the Twentieth European Conference on Artificial Intelligence, pp. 834–839 (2012)

SAGE: Generating Symbolic Goals for Myopic Models in Deep Reinforcement Learning

Andrew Chester$^{(\boxtimes)}$ⓘ, Michael Dannⓘ, Fabio Zambettaⓘ,
and John Thangarajahⓘ

School of Computing Technologies, RMIT University, Melbourne, Australia
{andrew.chester,michael.dann,fabio.zambetta,john.thangarajah}@rmit.edu.au

Abstract. Model-based reinforcement learning algorithms are typically more sample efficient than their model-free counterparts, especially in sparse reward problems. Unfortunately, many interesting domains are too complex to specify complete models, and learning a model takes a large number of environment samples. If we could specify an incomplete model and allow the agent to learn how best to use it, we could take advantage of our partial understanding of many domains. In this work we propose SAGE, an algorithm combining learning and planning to exploit a previously unusable class of incomplete models. This combines the strengths of symbolic planning and neural learning approaches in a novel way that outperforms competing methods on variations of taxi world and Minecraft.

Keywords: Reinforcement Learning · Deep Learning

1 Introduction

Deep reinforcement learning (DRL) agents have shown remarkable progress in a range of application domains, from video games [19] to robotic control [9], but still lag far behind humans in their ability to learn quickly from limited experience. Some of the best performing DRL agents are in board games, largely due to the fact that these agents are provided with a full deterministic model of their environment [28]. Model-based reinforcement learning allows agents to plan ahead, which can guide exploration efficiently. In most real world domains it is impractical (if not impossible) to provide a complete model of the environmental dynamics, and despite some promising recent work [26], learning a model is often highly sample inefficient. However, in many cases it is feasible to specify an incomplete model which captures only some aspects of the domain.

T. Liu et al. (Eds.): AI 2023, LNAI 14472, pp. 274–285, 2024.
https://doi.org/10.1007/978-981-99-8391-9_22

The key contribution of this paper is a novel integrated planning and learning algorithm which enables the use of incomplete symbolic domain models that cannot be exploited by prior work, providing more efficient learning in challenging environments. More specifically, it allows agents to use models that are simultaneously *abstract* – meaning the model cannot be used for short term planning – and *myopic* – meaning the model cannot be used for long term planning.

To illustrate the concepts of myopic and abstract models, consider a single taxi operating in a known road network. Passengers appear at random times and locations according to an unknown distribution, and have random destinations drawn from a separate unknown distribution. The taxi driver has access to a map, which shows the location and destination of any currently available passengers. The taxi can only pick up one passenger at a time, and the goal of the taxi driver is to maximise the number of passengers delivered in a fixed time period.

A human faced with this task is likely to approach it hierarchically, deciding on long term goals before filling in the details. First, they decide what they want to accomplish at a high level: "should I pick up passenger A, B, or refuel"? Then, they would plan out a route to navigate to their intended destination: "To get to that passenger I will need to take the next right and then go straight for three kilometres". Finally, in the short term they control the car by steering, braking, etc. In this example it is relatively easy to define an abstract and myopic model that is appropriate for mid-range planning, i.e. how to navigate through the streets. However, defining a model that would allow us to maximise daily income (long term) would require, at minimum, an understanding of where passengers are likely to be at different times of day, and controlling the vehicle at the low-level (short term) would need a full physics simulator.

Abstract planning models often do not even contain the objects that would be needed to perform short term planning; for example, a navigation model has no concept of a steering wheel. Myopic models suffer similarly, for example a navigation model with no concept of money cannot plan to earn the most cash by the end of the day. They may also be deficient in other ways, such as not capturing stochasticity, or partial observability. These inaccuracies in the model accumulate over the lifetime of the plan, making long term predictions futile, but they can still be used effectively for mid-range planning.

A hierarchical system could exploit such incomplete models by using model-free RL to learn control policies for the short and/or long term. To this end, we propose SAGE (Symbolic Ancillary Goal gEneration), a three-tier architecture consisting of: a meta-controller that sets symbolic goals, a planner which produces a sequence of symbolic steps to attain the chosen goal, and a low-level controller which uses these steps as guidance for atomic actions. SAGE's top-level meta-controller uses RL to set *ancillary* symbolic goals. Ancillary goals are not defined in terms of the primary objective for the agent (maximising reward), as the myopic model may not capture expected reward. Instead, they are used to guide the lower layers of the hierarchy, allowing the meta-controller to focus on *what* should be done, as opposed to *how* it should be done. Compared to other recent methods which blend RL and symbolic planning [7,17,31] SAGE allows the meta-controller to set arbitrary goals in the symbolic space. It can also make use of models that are both myopic and abstract, unlike prior work. We demonstrate that SAGE learns more efficiently than competing approaches in a small taxi domain, and continues to learn near-optimal policies as the environment scales up in size. We show that planning is crucial to its success by comparing against an ablated hierarchical method on a CraftWorld task.

2 Background

RL. We now briefly review the notation we will be using in this paper. A Markov Decision Process (MDP) $\mathcal{M}$ consists of a tuple $\mathcal{M} = (\mathcal{S}, \mathcal{A}, T, R)$, where $s \in \mathcal{S}$ are the possible environment states, $a \in \mathcal{A}$ are the allowable actions, $T : (\mathcal{S} \times \mathcal{A} \times \mathcal{S}) \rightarrow [0, 1]$ is the transition matrix specifying the environment transition probabilities, and $R : (\mathcal{S} \times \mathcal{A} \times \mathcal{R}) \rightarrow [0, 1]$ specifies the reward probabilities ($\mathcal{R} \subset \mathbb{R}$ is the range of allowable rewards). The objective of the agent is to maximise the total *return* U, which is exponentially discounted by a *discount rate* $\gamma \in [0, 1]$, that is $U_t = \sum_{k=0}^{T} \gamma^k r_{t+k+1}$, where T denotes the length of the episode remaining. The agent takes actions according to some *policy* $\pi(a|s) : (\mathcal{S} \times \mathcal{A}) \rightarrow [0, 1]$, which defines the probability of taking each action in each state. Sampling of an action according to the policy is given by $a_t \sim \pi(\cdot|s_t)$.

Symbolic Planning. In this work we use the planning domain definition language (PDDL) to describe symbolic planning problems [18]. Formally, a planning task T is a pair: (D, N), where D is the general planning *domain*, and N is the particular *instance* to be solved. The domain defines the tuple $(\mathcal{T}, \mathcal{P}, \mathcal{F}, \mathcal{O})$, where $\mathcal{T}$ is a set of types for objects, $\mathcal{P}$ is a set of Boolean predicates of the form $P(b_1, \ldots, b_k)$, and $\mathcal{F}$ is a set of numeric functions of the form $f(b_1, \ldots, b_k)$, where $b_1, \ldots, b_k$ are object variables. $\mathcal{O}$ is the set of planning operators (referred to throughout as *symbolic steps*), which have *preconditions* (Boolean formulas over predicates and functions which must hold for the operator to be valid) and *postconditions* (assignments to predicates or functions which are changed by the operator). The instance contains (B, I, G), where B are definitions of the objects, I is a conjunction of predicates (or their negations) and functions that describe the initial state of the world, and G is the goal which likewise describes the desired end-state of the world. The goal may only partially constrain the state of the world by omitting predicates and functions that may take any value.

3 Related Work

Hierarchical reinforcement learning (HRL) agents consist of a number of layers, each acting at a higher level of abstraction than the last. Given suitable abstractions, this approach improves sample efficiency in sparse reward learning tasks, primarily due to temporally extended and semantically consistent exploration [21]. The most common formalism for HRL is the options framework [29], in which the top layer chooses from a set of *options*, rather than the base actions in the original MDP. Each option is implemented as a different policy by the lower layer of the agent, and executes until a termination condition is reached. There are a variety of different ways to extend HRL with deep learning [2,15,16,20]. Hierarchical DQN (hDQN) [13] is most relevant to our work as it allows manually specified options, similar to our provided planning model. Unlike our work, these hierarchical approaches cannot use a provided planning model to reduce sample complexity.

Relevant recent model-based work uses provided [6] or learned [30] hierarchical planning models to further increase efficiency and allow transfer of components to related tasks in a tabular environment. Similarly Roderick et al. [23] define an abstract symbolic representation which is used to communicate goals to a low-level controller that uses deep RL to learn options. Unfortunately these methods cannot scale to environments with large abstract state spaces as the meta-controller still uses tabular count-based models. With the rise of deep RL, most recent model-based work instead uses models in a learned latent space. These latent models can be applied in a variety of ways, such as input to: a model-free controller [8], a hybrid policy network with planning lookahead [5,22,26], or a model predictive controller [10].

Prior work integrating *symbolic* planning and reinforcement learning is similar to HRL except the meta-controller has been replaced by a planner using an abstract model. The controller learns options for each of the planning actions, as well as expected values for taking these options, and these are chained together at runtime by a plan constructed over the abstract domain [31]. This allows the planner to take into account rewards that are unknown in the original planning model, and avoid actions which are unable to be executed reliably by the controller. This work was subsequently extended to remove the need for the designer to specify goals, and instead a meta-controller provides reward thresholds for the planner to achieve [17]. These thresholds do not provide guidance for the states to be achieved by the planner, but merely ask it to piece together a sequence of sufficiently rewarding actions.

An alternate approach is to use the output from a planner to modify the MDP given to the learning agent. DARLING uses a model (without rewards) of the state space to generate a set of plans that are close to optimal in the number of planning steps [14]. Any action that does not appear in these plans is occluded from the agent, lowering sample complexity by reducing the search space. This presupposes that short plans are close to optimal in terms of environmental reward, which may not be the case. Recently Illanes et al. [12] combine HRL with symbolic planning, and similarly prune unhelpful options from consideration during execution.

None of these approaches are directly applicable to our problem, as they require a single planning goal to be specified at the start of the episode. While this can be mitigated by replanning periodically throughout the episode, they are all inherently reactive. The agent is restricted to respond to what is currently visible; there is no mechanism for the planner to predict and prepare for future events that have not occurred. By contrast, our technique allows online ancillary goal generation during episodes to allow the agent to anticipate events which are not captured by its myopic model.

4 SAGE

SAGE has a three tier hierarchical architecture (Fig. 1):

1. The meta-controller - an RL component which takes state observations and outputs ancillary goals.
2. The symbolic planner - an off-the-shelf planner which takes ancillary goals and outputs a plan composed of symbolic steps.
3. The low-level controller - an RL component which takes an individual symbolic step and state observation, and outputs atomic environment actions.

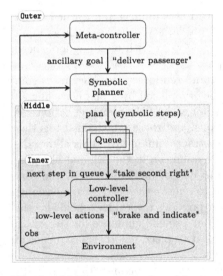

Fig. 1. SAGE conceptual architecture showing data flow.

```
 1: for k = 1, ..., total_episodes do
 2:     while episode not finished do Outer
 3:         G ~ πG(·|s; θG)
 4:         P ← plan(s, G)
 5:         for each step o in P do Middle
 6:             τ ← τmax
 7:             while step o not completed
      do                                Inner
 8:                 if τ = 0 go to 3
 9:                 a ~ πL(·|s, o; θL)
10:                 s, r ← act(a)
11:                 θG ← train(B)
12:                 τ ← τ − 1
13:             end while
14:         end for
15:     end while
16: end for
```

Algorithm 1: Conceptual sketch of main control loop

To understand how these components interact, we return to the taxi example. The meta-controller sees the current map, and needs to make a decision about what to do next. In this scenario, its options are to pick up a new passenger, move to a particular location, deliver a passenger already in the taxi, or some combination of these. The chosen ancillary goal is then passed to the planner, which uses a symbolic representation of the map to construct an abstract symbolic plan to achieve that goal. This plan consists of high-level steps such as "head to the end of the road, pick up the passenger there, then make two right turns". Each step of the plan is passed in sequence to the low-level controller, which steers and accelerates the car to achieve the current step. Both the low-level and meta-controllers are trained by interacting with the environment.

In line with prior work [17] we assume access to a symbolic planning domain model and a high-level (PDDL) state representation. Unlike in most prior work, this is *not an onerous burden* as our approach allows the use of myopic and abstract models. This allows the designer to omit dynamics from the model which are unknown or difficult to capture formally, and focus solely on the aspects of the problem which are amenable to symbolic reasoning. In the remainder of this section we formalise each component before presenting the unified planning and learning algorithm.

4.1 Meta-controller

The meta-controller is responsible for the strategic direction of the agent. It acts on an abstract semi-MDP with modified action space, transition and reward functions. Since the actions correspond to goals in the symbolic planning language, for each predicate in the planning domain we include a binary output which determines if it is included in the goal, as well as a variable number of discrete or continuous outputs which determine the arguments for that predicate (see Sect. 5 for examples).

The meta-controller acts on an extended time scale; it receives a single transition when the entire symbolic plan has been completed (or abandoned due to time out). That is, an experience step is of the form $(s_t, G_t, U_{t:t+n}, s_{t+n})$, where s_t, s_{t+n} are the state observations at the start and end of the plan respectively, G_t is the action (i.e. ancillary goal) chosen, and $U_{t:t+n} = \sum_t \gamma^t r_t$ is the accumulated environmental reward over the plan of duration n. This results in a slow rate of experience generation, we outline some techniques to accelerate learning in the supplemental material. SAGE is not restricted to a single learning algorithm, and we show results for both on- and off-policy approaches; for generality we assume experience is stored in a replay buffer, $\mathcal{B}$ to train on.

4.2 Symbolic Planner

The planner decomposes the ancillary goals (G) set by the meta-controller into a series of achievable symbolic steps ($o \in \mathcal{O}$). The planning domain is fixed and provided up front, so each new plan only requires a description of the initial and goal states. The planning goal is determined by the meta-controller as described above, while the initial state is the high-level PDDL description of the current state observation. Given the planning instance, the symbolic planner outputs a sequence of symbolic steps required to achieve the goal. These steps are stored in a queue, and each is passed one by one to the low-level controller for use in goal-directed RL. Once the final step in the queue has been completed, we deem the plan to be successful; otherwise if any individual step of the plan timed out, we say the plan failed.

4.3 Low-Level Controller

The low-level controller operates on a universal MDP [25] $\mathcal{M}_L = (\mathcal{S}, \mathcal{O}, \mathcal{A}, T, R_L)$, where $\mathcal{O}$ is the set of possible goals which correspond to the symbolic steps given by the planning model. The reward function R_L : $(\mathcal{S}, \mathcal{O}, \mathcal{A}) \rightarrow \mathbb{R}$ is equal to the environment reward plus an intrinsic bonus of $r_{complete}$ when the step has been completed. This encourages the controller to achieve the intended goal while still allowing it to be sensitive to the underlying reward signal. The state space $\mathcal{S}$, action space $\mathcal{A}$, and transition function T are unchanged from the base environment.

The low-level controller must be able to determine when it has successfully completed the symbolic step so it can request the next one from the queue. This is done by checking whether the postconditions of the current step are met in the high-level PDDL state description of the current observation. In addition, in order to prevent the agent from getting stuck trying to complete a step which is beyond the controller's capabilities, we implement a timeout limit of τ_{max} atomic actions per symbolic step. If the step has not been completed by then, the entire plan fails and control passes back to the meta-controller to choose a new goal.

We use PPO as our learning algorithm for the low-level controller [27], and we stress that this is independent of the choice of algorithm for the meta-controller, as they use different underlying experience to train on. In our experiments we have pre-trained the low-level controller to improve the convergence of the meta-controller.

4.4 Learning Algorithm

For each episode, the learning algorithm takes the form of three nested loops. At the innermost level, the low-level controller acts to complete the current symbolic step. In the middle level, the planner provides a new symbolic step when the previous one is complete. At the outermost level, the meta-controller chooses a new ancillary goal after the entire plan is finished. The pseudocode is presented in Algorithm 1.

5 Experiments

We test the following hypotheses. H1: SAGE can achieve higher scores than a scripted planning approach by learning regularities in the environment which are not part of the provided model. H2: SAGE can scale to larger domains than other learning approaches by exploiting its provided myopic model. H3: SAGE requires the planner to learn effectively, not just the hierarchical decomposition of the problem. By providing evidence for these hypotheses, we show that SAGE can effectively solve environments that are out of reach of prior approaches.

Our first experimental domain, an extension of the classic taxi gridworld setting [3], uses a myopic world model that is not abstract. This allows us to focus on ancillary goal generation without the complications of low-level control, and addresses H1 and H2. Second is a 2D Minecraft inspired world similar to those used in other work [1,12]. We have extended the environment from a grid world to a simple physics based motion system and provided a model which is both abstract and myopic. This showcases the full three tier architecture of SAGE, addressing H2 and H3. Experimental details, hyperparameters, and code, are provided in the supplemental material.

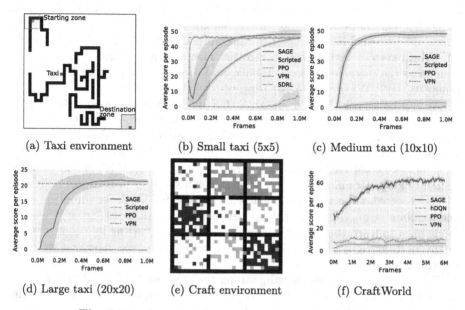

(a) Taxi environment   (b) Small taxi (5x5)   (c) Medium taxi (10x10)

(d) Large taxi (20x20)   (e) Craft environment   (f) CraftWorld

Fig. 2. Results achieved in Taxi and CraftWorld domains

5.1 Taxi

A $n \times n$ grid world (Fig. 2a) is randomly populated with a sparse maze at the start of each episode, and the taxi starts in a random unoccupied space. At each timestep if there is no passenger on the map or in the taxi, another passenger is spawned with probability p. The passenger's location is sampled uniformly from the starting zone, a $k \times k$ square at the top left of the map, with their destination sampled from a corresponding destination zone in the bottom right. A reward of 1 is given for each passenger delivered to their destination, and each episode is terminated after t timesteps. To showcase the scalability of our method we perform experiments on small, medium and large versions of this environment, with values $n = (5, 10, 20)$, $p = (0.12, 0.08, 0.05)$, $k = (2, 2, 3)$, $t = (1000, 2000, 2000)$ respectively.

The provided myopic model captures the road network and the requirements for picking up and dropping off passengers, but crucially does not know the location of the starting or destination zones. The meta-controller acts in a factored hybrid action space with Boolean variables indicating the presence of PDDL predicates, and discrete and continuous variables indicating the corresponding predicate arguments. For example, the goal empty ∧ in(4,6) ∧ delivered(p0) indicates that passenger 0 should be delivered to their destination, the taxi should be empty and in square (4,6). To accommodate this, we have used PPO

[27] as the algorithm for the meta-controller in these experiments as it can handle hybrid discrete and continuous action spaces [4]. For this domain, we benchmarked against four methods. Scripted: A planning approach which uses the same symbolic model as SAGE, but which always chooses the goal of delivering the closest passenger (H1). PPO: A standard model-free PPO algorithm using the same network architecture as our meta-controller (H2), VPN [22]: a strong model-based RL precursor to MuZero [26] (H2). SDRL [17]: a symbolic-RL hybrid system with a meta-controller. This is only applicable to the small domain as the meta-controller is tabular and so cannot run on our larger environments. (H2) We used the Fast Downward planner [11].

As can be seen in Figs. 2b to 2d, SAGE quickly learns a near-optimal policy across all environment sizes. This is unsurprising as the ancillary goals the meta-controller needs to learn are equivalent in each case: deliver any existing passenger and then return to the top left to wait for the next one. The visual input is more complex and varied in larger environments, which can be seen by a slightly longer training time and some minor instability in the large policy.

As expected, the scripted planner is also unaffected by the environment size, and exhibits strong but sub-optimal performance in all cases. Since it has no mechanism to learn the regularities in the passengers' starting locations it waits in the bottom right of the map after delivering each passenger, rather than returning to the starting zone.

For the other methods though, scaling the environment up introduces substantial challenges. SDRL learns very quickly, but cannot reliably outperform the scripted agent. While it can learn how good certain actions are, the planning component is still inherently reactive and cannot foresee future passengers. The learning speed can be partially attributed to a different observation space; the meta controller requires tabular input, as opposed to the image based input of the other methods. This helps it to learn rapidly in small environments, but prevents it from being applied to the larger domains.

PPO learns relatively quickly in the small environment, as random policies often produce rewards, whereas in the large environment it fails to ever get sufficient rewards under the random policy to kickstart learning. The medium environment is an interesting middle ground; the agent learns to pick up passengers from the top left and deliver them to the bottom right, but cannot reliably navigate around the randomised walls and so gets stuck in many episodes.

Similarly, while VPN can eventually learn reasonable policies in the small and medium environments (see supplemental material for experiments with more experience), it requires far more experience to learn the model. Furthermore, the planning horizon of VPN was set at 10 for these experiments (increased from a maximum of 5 in the original paper), and as the large environment requires plans of up to 80 steps before seeing a reward, even with an accurate model it is challenging for VPN to operate effectively.

5.2   CraftWorld

CraftWorld is a 2D domain where the agent is trying to navigate through its surroundings to collect coins (Fig. 2e). The randomly generated environment is filled with a variety of blocks of different materials, some of which can be cleared with the right equipment crafted from other blocks. For example, the agent can clear tree blocks to collect wood to make a pickaxe, which will then allow it to remove a stone block.

The key difference between the CraftWorld and taxi domains is that the agent moves continuously rather than in discrete steps. Like in Minecraft, the agent turns left and right and accelerates to move in the direction it is facing. This is designed to mimic many real world tasks in that the low-level dynamics are unable to be captured accurately by a symbolic model, i.e. the model is abstract, as well as myopic.

The planner receives knowledge of the recipes for crafting equipment and the equipment required for clearing different block types. In contrast to the taxi domain where the planner uses a detailed map, here it only knows about the macro-level room map. Since this model is abstract, we pre-train a goal-directed policy using PPO to act as the low-level controller. For example policies are trained to "collect coins" or "craft a pickaxe". Because the planner needs to reason about quantities of materials, we use ENHSP [24].

As before we benchmark against PPO and VPN (H2), but since the model is abstract we can no longer meaningfully compare against a scripted planning approach. We instead use hDQN [13]; a two layer hierarchical model which has the same pre-trained skills as SAGE and receives the same representation as input. This is equivalent to an ablated version of our full architecture without the planning layer (H3).

SAGE significantly outperforms the benchmark methods, learning to collect approximately 65 of the 75 total coins in the domain after 6M frames of experience (Fig. 2f). Neither PPO nor hDQN meaningfully improve after a short initial period; the agents can collect the coins in the first room but frequently get stuck there and cannot reliably navigate between rooms. It should be noted that hDQN and SAGE have the benefit of 40M frames of pretrained low-level controllers, so a direct comparison to PPO and VPN in this graph is not possible. We ran these methods up to 50M frames for a fair comparison, but neither meaningfully improves after the first 6M, averaging only 5 for PPO and 0.1 for VPN at the end of training (see supplemental material for full training graphs).

To understand why, consider what is needed to perform well. If the agent is in a room without any coins in it, the next possible reward may be hundreds of atomic actions away, even with an optimal policy. This could correspond to a dozen sequential symbolic steps, or only two sequential ancillary goals. As a result, it is a challenging environment for a model-based method like VPN to operate in, even if it could learn a sufficiently accurate model in the first place. By contrast, SAGE's meta-controller does not need to reason over many steps to perform well. We hypothesise that hDQN's comparatively poor performance is due to the fact that constructing essential pieces of equipment such as pickaxes requires precise sequencing of symbolic steps, and a single incorrect action can result in the irreversible waste of the raw components.

Another advantage of SAGE's planning model is that it enables temporally extended and semantic exploration [21]; it effectively explores in the state space, not the action space. Consider the difference in the actions chosen by the two methods: for SAGE the meta-controller's actions (ancillary goals) correspond to states of the world it wants to be in: "I should have a pickaxe". On the other hand, hDQN is exploring by taking random (albeit high-level) actions: "face a tree". This is evidenced by SAGE outperforming hDQN at the start of training, even though both methods are acting randomly with the same low-level controllers.

6   Conclusion

In this paper we have introduced SAGE, a hierarchical agent combining deep reinforcement learning and symbolic planning approaches. This neuro-symbolic architecture allows us to exploit easily defined explicit knowledge about the environment while maintaining the flexibility of learning unknown dynamics. SAGE outperforms both planning and learning based approaches on complex problems, particularly as the environments are scaled up. We believe this is the first work in which an RL agent sets symbolic goals for a planner independently, rather than selecting between a handful of predetermined templates.

A limitation of our method is that all actions must go through the planner, which requires the provided model to address all aspects of the domain. In Real Time Strategy games like StarCraft II, building and unit production is nicely captured by a myopic and abstract model, but combat is not. An interesting line of future work would be to allow the meta-controller to selectively bypass the planner, directly setting ancillary goals for the low-level controller when appropriate.

References

1. Andreas, J., Klein, D., Levine, S.: Modular multitask reinforcement learning with policy sketches. In: ICML (2017)
2. Bagaria, A., Konidaris, G.: Option discovery using deep skill chaining. In: ICLR (2020)
3. Dietterich, T.G.: Hierarchical reinforcement learning with the MAXQ value function decomposition. JAIR 13, 227–303 (2000)
4. Fan, Z., Su, R., Zhang, W., Yu, Y.: Hybrid actor-critic reinforcement learning in parameterized action space. In: IJCAI (2019)
5. François-Lavet, V., Bengio, Y., Precup, D., Pineau, J.: Combined reinforcement learning via abstract representations. In: AAAI (2019)
6. Gopalan, N., et al.: Planning with abstract Markov decision processes. In: ICAPS (2017)
7. Gordon, D., Fox, D., Farhadi, A.: What should i do now? Marrying reinforcement learning and symbolic planning. arXiv preprint arXiv:1901.01492 (2019)
8. Ha, D., Schmidhuber, J.: Recurrent world models facilitate policy evolution. In: NeurIPS (2018)

9. Haarnoja, T., Zhou, A., Abbeel, P., Levine, S.: Soft actor-critic: off-policy maximum entropy deep reinforcement learning with a stochastic actor. In: ICML (2018)

10. Hafner, D., et al.: Learning latent dynamics for planning from pixels. In: ICML (2019)

11. Helmert, M.: The Fast Downward planning system. JAIR **26**, 191–246 (2006)

12. Illanes, L., Yan, X., Icarte, R.T., McIlraith, S.A.: Symbolic plans as high-level instructions for reinforcement learning. In: ICAPS (2020)

13. Kulkarni, T.D., Narasimhan, K., Saeedi, A., Tenenbaum, J.: Hierarchical deep reinforcement learning: integrating temporal abstraction and intrinsic motivation. In: NeurIPS (2016)

14. Leonetti, M., Iocchi, L., Stone, P.: A synthesis of automated planning and reinforcement learning for efficient, robust decision-making. AIJ **241**, 103–130 (2016)

15. Levy, A., Konidaris, G., Platt, R., Saenko, K.: Learning multi-level hierarchies with hindsight. In: ICLR (2019)

16. Li, A.C., Florensa, C., Clavera, I., Abbeel, P.: Sub-policy adaptation for hierarchical reinforcement learning. In: ICLR (2020)

17. Lyu, D., Yang, F., Liu, B., Gustafson, S.: SDRL: interpretable and data-efficient deep reinforcement learning leveraging symbolic planning. In: AAAI (2019)

18. McDermott, D., et al.: PDDL - the planning domain definition language. Technical report, Yale Center for Computational Vision and Control (1998)

19. Mnih, V., et al.: Human-level control through deep reinforcement learning. Nature **518**(7540), 529 (2015)

20. Nachum, O., Gu, S., Lee, H., Levine, S.: Data-efficient hierarchical reinforcement learning. In: NeurIPS (2018)

21. Nachum, O., Tang, H., Lu, X., Gu, S., Lee, H., Levine, S.: Why does hierarchy (sometimes) work so well in reinforcement learning? arXiv preprint arXiv:1909.10618 (2019)

22. Oh, J., Singh, S., Lee, H.: Value prediction network. In: NeurIPS (2017)

23. Roderick, M., Grimm, C., Tellex, S.: Deep abstract Q-networks. In: AAMAS (2018)

24. Scala, E., Haslum, P., Thiébaux, S.: Heuristics for numeric planning via subgoaling. In: IJCAI (2016)

25. Schaul, T., Horgan, D., Gregor, K., Silver, D.: Universal value function approximators. In: ICML (2015)

26. Schrittwieser, J., et al.: Mastering Atari, Go, chess and shogi by planning with a learned model. Nature **588**(7839), 604–609 (2020)

27. Schulman, J., Wolski, F., Dhariwal, P., Radford, A., Klimov, O.: Proximal policy optimization algorithms. arXiv preprint arXiv:1707.06347 (2017)

28. Silver, D., et al.: A general reinforcement learning algorithm that masters chess, shogi, and Go through self-play. Science **362**(6419), 1140–1144 (2018)

29. Sutton, R.S., Precup, D., Singh, S.: Between MDPs and semi-MDPs: a framework for temporal abstraction in reinforcement learning. AIJ **112**(1–2), 181–211 (1999)

30. Winder, J., et al.: Planning with abstract learned models while learning transferable subtasks. In: AAAI (2020)

31. Yang, F., Lyu, D., Liu, B., Gustafson, S.: PEORL: Integrating symbolic planning and hierarchical reinforcement learning for robust decision-making. In: IJCAI (2018)

Leaving the NavMesh: An Ablative Analysis of Deep Reinforcement Learning for Complex Navigation in 3D Virtual Environments

Dale Grant[1]([✉]) [iD], Jaime Garcia[1] [iD], and William Raffe[2] [iD]

[1] University of Technology Sydney, Sydney, NSW 2000, Australia
dale.t.grant@student.uts.edu.au, jaime.garcia@uts.edu.au
[2] Deakin University, Melbourne, VIC 3125, Australia
william.raffe@deakin.edu.au

Abstract. Expanding non-player character (NPC) navigation behavior in video games has the potential to induce novel player experiences. Current industry standards represent traversable world geometry by utilizing a Navigation Mesh (NavMesh); however NavMesh complexity scales poorly with additional navigation abilities (e.g. jumping, wall-running, jet-packs, etc.) and increasing world scale. Deep Reinforcement Learning (DRL) allows for an NPC agent to learn how to navigate environmental obstacles with any navigation ability without NavMesh dependence. Despite the promise of DRL navigation, adoption in industry remains low due to the required expert knowledge in agent design and the poor training efficiency of DRL algorithms. In this work, we utilize the off-policy Soft-Actor Critic (SAC) DRL algorithm to investigate the importance of different local observation types and agent scalar information to agent performance across three topologically distinct environments. We implement a truncated n-step returns method for minibatch sampling which improves early training efficiency by up to 75% by reducing inaccurate off-policy bias. We empirically evaluate environment partial observability with observation stacking where we find that 4–8 observation stacks renders the environments sufficiently Markovian.

Keywords: Deep Reinforcement Learning · Virtual Environment Navigation · Video Games · Depth Vision · Occupancy Grid Maps

1 Introduction

For a product to be successful in the competitive video games industry, it must both distinguish itself from the competition whilst minimizing development costs. Despite the ubiquity of Artificial Intelligence (AI) in modern video games, further development in AI is seldom explored in AAA games; resulting in developers not considering the limitations imposed by the current industry standards on game design and player experiences.

T. Liu et al. (Eds.): AI 2023, LNAI 14472, pp. 286–297, 2024.
https://doi.org/10.1007/978-981-99-8391-9_23

The primary utilization for AI in video game development is controlling the behavior of non-playable characters (NPC), which fulfil many roles in the game world (e.g. vendors, enemies, allies, or bystanders) to improve player experience and immersion. A core aspect of NPC behavior is how they navigate the environment, typically in either static routes or in reaction to player actions.

The industry standard for NPC navigation stores a graph representation of the world geometry with a Navigation Mesh (NavMesh). During run-time, a path between two locations is found using a path-finding algorithm (such as A* or Djikestra's [14]). The NPC moves between the planned path locations using a character controller. Despite providing a compact world representation, the NavMesh is often comprised of disconnected regions that require additional edges to traverse between. As manual placement of these link edges is time-consuming for large and/or complex environments [1], automatic link edge generation is often employed. However, tuning the number of link edges is a non-trivial task as too few can result in predictable NPC behavior whereas an overabundance bloats graph complexity and run-time cost [5].

The standard NavMesh approach does not effectively accommodate for additional complex navigation abilities (e.g. jumping, teleportation, grapple hooks, wall running, etc.) as they not only increase graph complexity, but also tend to require specialized tests to ensure trajectory validity [5]. To reduce development time and cost, game developers opt to avoid such navigation abilities; resulting in simple, exploitable NPC behavior. Similarly, developers seldom implement dynamic environmental features, such as moving platforms, to reduce the likelihood of NPCs moving off-mesh; resulting in limited level design. By removing the reliance of NPC navigation on the NavMesh, the resulting constraints on NPC behavior and level design could be lifted, expanding potential player experiences and game design.

Deep Reinforcement Learning (DRL) has successfully demonstrated itself across multiple domains such as continuous control, robotic navigation, board games, and Atari video games. Recent works in video game navigation have successfully utilized DRL agents to navigate 3D environments for applications in quality assurance testing [4,11,24], Turing test discrimination [8], adversarial level generation [10], and large-scale environment navigation [1,2]. Despite these successes, adoption of DRL navigation solutions by video game developers remains low due to high expert knowledge requirements (e.g. observation type selection, hyper-parameter tuning, reward shaping) and extensive training times from poor data efficiency.

In this paper we implement an end-to-end DRL agent with continuous player-like inputs that performs point-to-point navigation in 3D virtual environments without the limitations of a NavMesh-based approach. Our DRL agent utilizes the off-policy learning algorithm Soft Actor-Critic (SAC) [12] as it offers greater data efficiency and lower hyper-parameter sensitivity than its on-policy competitors. Using this DRL agent we explore the influence of different observation types and algorithmic alterations on navigation success rates in multiple, topologically distinct, 3D testbed environments. From these results, we determine which different observation types are effective for resolving particular environment topologies, propose observation stacking to evaluate and reduce the partial observability of the environment state, and demonstrate how n-step returns can improve training efficiency by reducing the impact of inaccurate value estimates during early training. By providing transparency of common observation types and algorithmic techniques which improve training efficacy; we seek to improve the appeal of 3D DRL navigation solutions in the video games industry.

Our key contributions are: 1) Implementation details for 2D depth map and 3D occupancy grid map observations for reproducibility; 2) Assessment of the importance of observation types for different environments through ablation studies; 3) Utilizing truncated n-step returns to reduce the negative impact of inaccurate critic value estimates during early training; and 4) Empirical evaluation of the partial observability of the environment using observation stacking to determine the necessity of hidden state memory methods.

2   Background and Related Work

2.1   DRL Navigation in Video Games

Deep Reinforcement Learning (DRL) has been successfully utilized across many different video games, such as Doom [17] and ATARI games [3]. Recent works have utilized DRL navigation agents for point-to-point navigation [1,2], level solving [10], and automatic gameplay testing [4,11,24]. Agents often observe their environment with up to three observation types: agent state information [4,11], 2D depth vision [2,8,21], and 3D topological information [1,24]. From these observations, the agent policy decides on a value for each available action. Basic agent movement involves walking forward/backwards, left/right strafing, and yaw rotation [19] with additional movement options such as jumping [1,2] and climbing [4,11]. Navigated environments range from smaller (100 m × 100 m) testbed environments [1,4,21] to large (1 km × 1 km) environments [2]. Despite successful navigation, little research has been performed to determine the relative importance of each observation type, the environment topology they best represent, and their influence on agent performance.

2.2   Popular DRL Algorithms for Navigation

Video game DRL navigation approaches typically utilize either the on-policy Proximal Policy Optimization [23] (PPO) or the off-policy Soft-Actor Critic (SAC) algorithms. Both are model-free algorithms that learn through trial-and-error with the environment without modelling environment transitions. On-policy algorithms are easier to implement than their off-policy counterparts but have lower data efficiency as they cannot recycle experiences generated from past policies. As AAA game engines are considered slow simulators [1], off-policy algorithms are deemed more desirable for video game navigation tasks.

2.3   Environment Partial Observability

As local observations by navigating agents do not encompass the entire environment state, navigation can be considered as a Partially Observable Markov Decision Process (POMDP), where part of the environment state is hidden from the agent observation. Partial observability is often handled with either observation stacking [20] or unstructured memory such as Long Short-Term Memory (LSTM) [19] to learn a hidden state representation. The benefits of LSTM for 3D navigation vary with environmental scale, with small environments demonstrating little benefit and larger environments showing improved training efficiency [1,2] with minor success rate improvement [2]. To our knowledge, a method to empirically assess how partial the observations of a navigation environment are has yet to be demonstrated in the literature.

2.4   Enhancing Training Efficiency

Modern off-policy algorithms utilize a combination of bootstrapping to learn value function estimation, experience replay buffers (ERB) to break temporal correlation, and function approximation for state generalization; resulting in a learning divergence called *the deadly triad* [25]. Modern mitigation methods for the deadly triad often focus on ERB sampling strategies [22,27] and n-step returns [7,26]. While sampling prioritization methods have proven successful in some environments, they have been noted to increase learning instability [26]. N-step returns reduce the impact of inaccurate critic estimates at early timesteps at the cost of increased training variance [7]; however this variance can be mitigated when combined with large ERB capacities [9] due to the larger state-action coverage of the ERB samples. Despite this potential benefit, n-step returns have yet to be explored for navigation environments.

3   Method

Environments are built using the Unity ML-Agents [16] framework as it provides appropriate tools for DRL navigation and a built-in inference engine for industry implementation. To support 3D topological observations and efficient 2D depth observations, we extend the ML-Agents sensor suite with custom sensors components. Environment observations are output to a custom SAC implementation, to accommodate for 3D convolutional networks and other algorithmic alterations, which supplies actions to the agents in the environment. The received actions are utilized by the agent character controller to move a 1.8 m tall character. Further implementation details and hyper-parameters of this work can be found at https://code.research.uts.edu.au/DaleGrant/LeavingTheNavmesh.

3.1   Environments

We investigate three small hand-made testbed environments (up to 60 m × 60 m) as large-scale environments can be broken down into a sequence of smaller point-to-point traversals [2]. We employ three different environments (shown in Fig. 1):

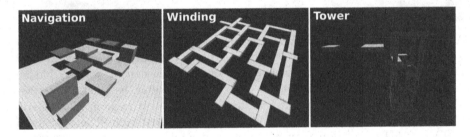

Fig. 1. The Navigation (60 m × 60 m × 10 m), Winding (60 m × 60 m × 1 m), and Tower (7 m × 7 m × 21 m) testbed environments. Each environment is designed to challenge the agent with distinct topology to assess the importance of different observations.

1) *Navigation* - A low-risk environment where the agent must utilize jumps to traverse platforms of varying heights and avoid obstacles; 2) *Winding* - A flat, narrow path with a high risk of falling; and 3) *Tower* - A dense, vertical tower which requires double jumps to traverse between levels. Each environment is designed to test the agent on a particular topology common in video game environments in isolation from other topology. Available 3D navigation testbed environments [2,21] we not utilized for this work as their procedurally generated environments would obscure topological understanding without significant modification.

3.2   Task

The agent performs point-to-point navigation on a 3D environment from start position p_{start} to goal position p_{goal} utilizing all available navigation abilities. All potential start and goal positions are sampled and cached during environment initialization using Poisson Disk Sampling [6] as it provides a distribution of locations separated by a minimum distance r_{min}. At episode start, the start and goal positions are randomly sampled from all cached positions.

3.3   Agent Observations

Agents observations are composed of agent state information and local perception, where the latter is comprised of 2D Depth Map (DM) and 3D Occupancy Grid Map (OGM) observations (shown in Fig. 2). The agent state information is comprised of the agent's: relative vector to the goal ($\mathbb{R}^3$), distance to the goal ($\mathbb{R}$), velocity ($\mathbb{R}^3$), acceleration ($\mathbb{R}^3$), yaw and pitch rotations ($\mathbb{R}^2$), grounded state ($\mathbb{B}$), and jump availability state ($\mathbb{B}$). Each agent information observation is normalized between $[-1, 1]$ in the environment to improve learning stability with the Boolean observations being converted to continuous values $[0, 1]$ for use in the DRL algorithm.

As rendering a 2D depth map with a virtual camera is prohibitively expensive for the many concurrent agents typically utilized in a production setting, we construct the depth map by projecting a 9 × 9 cone of rays in the agent looking direction with 20 m ray length and maximum cone angle of 45 degrees. The depth value is normalized and decreases over the length of the ray.

The 3D occupancy map is generated using boxcasts at environment initialization and cached for efficient sampling at run-time. The occupancy map (or Semantic Map [24]) is a categorical discretization of the environment where each voxel is represented by a semantic integer value. Our voxels are either empty or solid with values 0 and 1, respectively. During run-time the agent samples a 9 × 9 × 9 region of cells around its position from the cached occupancy map.

Fig. 2. An agent's depth observation (white rays) and occupancy grid map observation (voxels). Blue voxels represent empty cells and red voxels contain solid objects. The inset shows the normalized 2D depth map. (Color figure online)

Each input observation is stacked four times as it has been shown to improve partial observability of the environment for some visual tasks [20].

3.4   Actions

The agent performs continuous actions equivalent to common player controller inputs: forward/backward, left/right strafe, turn left/right (yaw), look up/down (pitch), and jump. The jump action can be performed when grounded and once in mid-air. The jump action is treated as continuous for the DRL policy and binarized for the environment.

To reduce inference cost and promote longer time scale decisions [2], we utilize frame skipping/action repetition [20] to repeat a supplied policy action for multiple environment steps. A frame-skipping of four was chosen for our environments as the loss of fine-grained control at higher repetition values proved detrimental for the higher risk environments in this work.

3.5   Reward Function

The reward signal the agent receives from the environment contains: 1) A sparse terminal reward of +1 for success, −1 for failure, and 0 for episode time out; 2) A time-step penalty of –0.001 every environmental step to incentivize shorter trajectories; and 3) A dense euclidean reward for each environment step in which the agent reduces its current shortest euclidean distance from the goal. The euclidean reward signal is normalized by the initial goal distance and multiplied by a euclidean scaling factor of 0.5 to balance reducing the goal distance with successfully reaching the goal. To maintain a consistent peak reward signal during frame skipping, the dense reward signals are accumulated during repeated environment actions and obtained when either a new action is requested or the episode terminates. Episode termination occurs when: 1) the agent is within 1 m of the goal (Success); 2) the agent falls off the map (Failure); or 3) the agent takes 1000 environment steps (Timeout).

3.6   Algorithm Architecture

As game engines are considered slow simulators, we implement the off-policy DRL algorithm Soft Actor-Critic [12] with a learned entropy coefficient and no state-value network [13]. Each input observation type is encoded accordingly before entering the policy and critic networks to improve state generalization. The Agent state information is normalized with a Batch Normalization [15] layer to resolve the internal covariant shift of observations over time. The DM and OGM local observations are encoded using 2D and 3D Convolutional Neural Networks (CNNs) [18], respectively. As our OGM utilizes binary values, we do not implement an embedding layer as in existing semantic OGMs [24].

The action-value estimate $Q(s_t, a_t)$ for SAC is trained using an entropy regularized expected return:

$$G_t = r_t + \gamma(Q_{\bar{\theta}}(s_{t+1}, a) - \alpha \log(\pi_\phi(a|s_{t+1}))) \tag{1}$$

for a transition t, log probability $\log(\pi_\phi(a|s_{t+1}))$, actor parameters ϕ, and critic and target network parameters θ and $\bar{\theta}$, respectively. To reduce the influence of inaccurate Q-value estimates, we augment the single-step expected return with an entropy-regularized truncated n-step return:

$$G_t^{(n)} = \sum_{k=0}^{n-1} \gamma^k r_{t+k} + \gamma^n((Q_{\bar{\theta}}(s_{t+n+1}, a) - \alpha \log(\pi_\phi(a|s_{t+n+1}))) \tag{2}$$

The discounted n-step return is calculated for each batch sampled from the replay buffer. The n-step for a sample i is given by $\min((T_i - t_i), n)$ for sample step t_i and sample episode terminal step T_i, truncating n_i between $[1, n]$. We default to an n-step of three as recommended in the literature [9,26].

Off-policy algorithms implement experience replay buffers to improve data efficiency by performing updates with past transitions many times; however as the average policy age of a sample increases, training efficiency tends to decrease [26]. To reduce the maximum sample policy age [26], we perform gradient updates once every four decision steps.

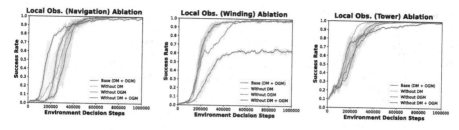

Fig. 3. Local observation ablation study for the Navigation, Winding, and Tower environments. All agents were trained over 3 seeds. The shaded regions represent 95% confidence intervals. Depth Map observations effective for Navigation environment whereas occupancy grid map observation more effective in topologically complex Winding and Tower environments.

4   Results and Discussion

To understand the influence of observation types, algorithmic architecture, and environmental features on navigation performance, we perform four experiments for a base agent (uses both local observation types) across three distinct environments: 1) local observation ablation study to evaluate the impact of removing Depth Map (DM) and/or Occupancy Grid Map (OGM) observations on agent performance; 2) agent information ablation study to assess relative importance of agent state observations and its encoding architecture; 3) observation stack parameter study to empirically investigate environment partial observability; and 4) n-step return parameter study to investigate the influence of our n-step return implementation on training efficiency. Agent performance is evaluated with two metrics: 1) the *stable success rate* value; and 2) *training efficiency*, the number of environment decision steps to achieve a stable success rate.

4.1   Local Observation Ablation Study

From the results of the local observation ablation shown in Fig. 3, we find that the inclusion of at least one local observation enables similar peak stable success rates between observation types for each environment; however training efficiency can vary significantly between local observation types with the DM and OGM observations proving most effective in the Winding/Tower and Navigation environments, respectively.

These findings indicate that in environments with dense, complex topology, e.g. many winding edges or layered vertical topology, OGM observations prove more beneficial as the agent can resolve topology in all directions simultaneously whereas DM observations are more useful for environments where coarser navigation around simple geometry is required. Contrary to findings in the literature [1], we found use of both local observation types to have lower training efficiency than the best ablated model in all studied environments. We attribute this trend to either learning being distributed over more training parameters

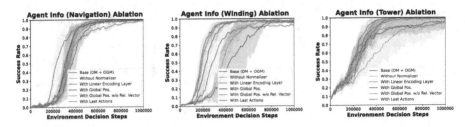

Fig. 4. Agent Information ablation study for the Base agent in the Navigation, Winding, and Tower environments. All agents were trained over 3 seeds. The shaded regions represent 95% confidence intervals. Batch normalization improvements indicate presence of covariant input shift.

thereby diluting the effects of the more beneficial observation; or that no studied environment contains topology (or a combination of) that required both local observation types to resolve. Finally, we note that the agent with no local observations demonstrated a high success rate in the Navigation and Tower environments, indicating that the available actions of the agent trivialize the risk and/or difficulty of the environment.

4.2   Agent Information Ablation Study

The results of the agent information ablation study of common agent information alterations for the base agent in all three environments is shown in Fig. 4. The studied agent information variations include: 1) the observation of the global agent and goal positions (with and without the existing relative agent to goal vector observation), the observation of the agent's last actions, the implementation of a linear encoding layer with ReLU activation post batch normalization, and the removal of the batch normalization layer.

We find that the inclusion of the global position observations always negatively impacts training efficiency for all environments as the relative vector provides similar information with fewer weights to optimize. We observe that the inclusion of the linear encoding layer negatively impacts training efficiency and stable success rate from the base agent which indicates that encoding agent state information is unnecessary for sufficient feature representation. We note that the removal of batch normalization reduced training efficiency and variance for all environments which indicates notable internal covariant shift of observations over the course of training. Lastly, we find that the addition of the last action observation negatively impacts training efficiency for all environments due to increased number trainable weights introduced.

Fig. 5. Varying number of stacked observations for the base agent in the Navigation, Winding, and Tower environments. All agents were trained over 3 seeds. The shaded regions represent 95% confidence intervals. Improvements to training efficiency are attributed to improved state observability.

4.3 Investigating Partial Observability

We investigate the partial observability of the base agent in all environments by varying the number of stacked observations. From the results shown in Fig. 5, we find that both training efficiency and stable success rate increase with observation stacks until 3–4 observation stacks depending on environment. After which a range of observation stacks tend to result in similar performance (4–8 obs), with further stacks (10+) reducing training efficiency and stable success rate in all environments. As such, we deem all studied environments to be sufficiently observable at four observation stacks with additional stacking proving either ineffective or harmful to agent performance. These results combined with the marginal improvement observed in the navigation literature [1,2] when utilizing LSTM for similarly sized environments allows us to infer that the additional complication and computation cost of LSTM to be unnecessary for our environments.

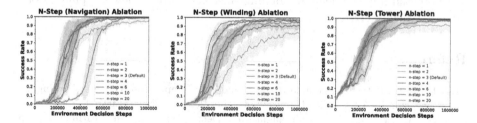

Fig. 6. Varying n-step returns for the base agent in the Navigation, Winding, and Tower environments. All agents were trained over 3 seeds. The shaded regions represent 95% confidence intervals. Improvements to training efficiency are attributed to a reduction in inaccurate estimated Q-value bias to critic updates in early training.

4.4 Improving Data Efficiency

We investigate effectiveness of our truncated n-step return in improving training efficiency for the base agent in each environment. From the results in Fig. 6, we find that increasing the n-step return improves training efficiency across all environments until around 4–6 steps, where a decrease in both stable success rate and training efficiency are observed. The improvement to training efficiency can be attributed to a reduction in the impact of inaccurate Q-value estimates to Q-value updates during early training by biasing towards the episodic return; however too many n-step returns limits bootstrapping effectiveness and increases training variance.

5 Conclusion

In this paper, we implement an off-policy DRL agent with continuous actions to navigate 3D virtual environments to avoid the limitations introduced by traditional NavMesh-based approaches. Using this agent, we demonstrate how the importance of different local observations varies with the topological nature of the navigable environment. Through investigation of agent information we demonstrate the impact of covariant input shift on training efficiency. We demonstrate that observation stacking can sufficiently reduce environment partial observability without the use of recurrent memory. To improve the data efficiency of our approach, we implement a truncated n-step return, which by reducing inaccurate bootstrapping in early training, can improve training efficiency by up to 75%. By providing insight into the environment topology that benefits from particular observation types along with methods to improve data efficiency for our DRL agent, we aim to improve both reproducibility and industry appeal for 3D DRL navigation solutions. Potential future works could include implementation of the DRL agent to dynamic environments to investigate any changes to environment partial observability, the influence of additional navigation abilities such as grapple hooks and wall-running on local observation type importance, and potential improvements to training efficiency through new sampling prioritization schemes [22].

References

1. Alonso, E., Peter, M., Goumard, D., Romoff, J.: Deep Reinforcement Learning for Navigation in AAA Video Games, November 2020. arXiv:2011.04764 [cs]
2. Beeching, E., et al.: Graph augmented Deep Reinforcement Learning in the GameRLand3D environment, December 2021. arXiv:2112.11731 [cs]
3. Bellemare, M.G., Naddaf, Y., Veness, J., Bowling, M.: The arcade learning environment: an evaluation platform for general agents. J. Artif. Intell. Res. **47**, 253–279 (2013). https://doi.org/10.1613/jair.3912
4. Bergdahl, J., Gordillo, C., Tollmar, K., Gisslén, L.: Augmenting Automated Game Testing with Deep Reinforcement Learning, March 2021. arXiv:2103.15819 [cs]
5. Budde, S.: Automatic Generation of Jump Links in Arbitrary 3d Environments. Master's thesis, Humboldt-Universität zu Berlin (2013)
6. Cook, R.L.: Stochastic sampling in computer graphics. ACM Trans. Graph. **5**(1), 51–72 (1986). https://doi.org/10.1145/7529.8927

7. Daley, B., Amato, C.: Reconciling λ-returns with experience replay. In: Advances in Neural Information Processing Systems, vol. 32. Curran Associates, Inc. (2019)

8. Devlin, S., et al.: Navigation Turing Test (NTT): Learning to Evaluate Human-Like Navigation, July 2021. arXiv:2105.09637 [cs]

9. Fedus, W., et al.: Revisiting Fundamentals of Experience Replay, July 2020. arXiv:2007.06700 [cs]

10. Gisslén, L., Eakins, A., Gordillo, C., Bergdahl, J., Tollmar, K.: Adversarial Reinforcement Learning for Procedural Content Generation, June 2021. arXiv:2103.04847 [cs]

11. Gordillo, C., Bergdahl, J., Tollmar, K., Gisslén, L.: Improving Playtesting Coverage via Curiosity Driven Reinforcement Learning Agents, June 2021. arXiv:2103.13798 [cs]

12. Haarnoja, T., Zhou, A., Abbeel, P., Levine, S.: Soft Actor-Critic: Off-Policy Maximum Entropy Deep Reinforcement Learning with a Stochastic Actor, August 2018. arXiv:1801.01290 [cs, stat]

13. Haarnoja, T., et al.: Soft Actor-Critic Algorithms and Applications, January 2019. arXiv:1812.05905 [cs, stat]

14. Hart, P.E., Nilsson, N.J., Raphael, B.: A formal basis for the heuristic determination of minimum cost paths. IEEE Trans. Syst. Sci. Cybern. 4(2), 100–107 (1968). https://doi.org/10.1109/TSSC.1968.300136

15. Ioffe, S., Szegedy, C.: Batch Normalization: Accelerating Deep Network Training by Reducing Internal Covariate Shift, March 2015. arXiv:1502.03167 [cs]

16. Juliani, A., et al.: Unity: A General Platform for Intelligent Agents, May 2020. arXiv:1809.02627 [cs, stat]

17. Kempka, M., Wydmuch, M., Runc, G., Toczek, J., Jaśkowski, W.: ViZDoom: A Doom-based AI Research Platform for Visual Reinforcement Learning, September 2016. arXiv:1605.02097 [cs]

18. Lecun, Y., Bottou, L., Bengio, Y., Haffner, P.: Gradient-based learning applied to document recognition. Proc. IEEE 86, 2278–2324 (1998). https://doi.org/10.1109/5.726791

19. Mirowski, P., et al.: Learning to Navigate in Complex Environments, January 2017. arXiv:1611.03673 [cs]

20. Mnih, V., et al.: Human-level control through deep reinforcement learning. Nature 518(7540), 529–533 (2015). https://doi.org/10.1038/nature14236

21. Roy, J., Girgis, R., Romoff, J., Bacon, P.L., Pal, C.: Direct Behavior Specification via Constrained Reinforcement Learning, January 2022. arXiv:2112.12228 [cs]

22. Schaul, T., Quan, J., Antonoglou, I., Silver, D.: Prioritized Experience Replay, February 2016. arXiv:1511.05952

23. Schulman, J., Wolski, F., Dhariwal, P., Radford, A., Klimov, O.: Proximal Policy Optimization Algorithms, August 2017. arXiv:1707.06347 [cs]

24. Sestini, A., Gisslén, L., Bergdahl, J., Tollmar, K., Bagdanov, A.D.: CCPT: Automatic Gameplay Testing and Validation with Curiosity-Conditioned Proximal Trajectories, February 2022. arXiv:2202.10057 [cs]. https://doi.org/10.48550/arXiv.2202.10057

25. Sutton, R.S., Barto, A.G.: Reinforcement Learning: An Introduction. MIT Press, Cambridge (1998)

26. van Hasselt, H., Doron, Y., Strub, F., Hessel, M., Sonnerat, N., Modayil, J.: Deep Reinforcement Learning and the Deadly Triad, December 2018. arXiv:1812.02648 [cs]. https://doi.org/10.48550/arXiv.1812.02648

27. Zhang, S., Sutton, R.S.: A Deeper Look at Experience Replay, April 2018. arXiv:1712.01275 [cs]. https://doi.org/10.48550/arXiv.1712.01275

Transformed Successor Features for Transfer Reinforcement Learning

Kiyoshige Garces[✉][ID], Junyu Xuan[ID], and Hua Zuo[ID]

Australian Artificial Intelligence Institute (AAII), University of Technology Sydney,
15 Broadway, Sydney, NSW 2007, Australia
Oscar.K.GarcesAparicio@student.uts.edu.au,
{Junyu.Xuan,Hua.Zuo}@uts.edu.au

Abstract. Reinforcement learning algorithms require an extensive number of samples to perform a specific task. To achieve the same performance on a new task, the agent must learn from scratch. Transfer reinforcement learning is an emerging solution that aims to improve sample efficiency by reusing previously learnt knowledge in new tasks. Successor feature is a technique aiming to reuse representations to leverage that knowledge in unseen tasks. Successor feature has achieved outstanding results on the assumption that the transition dynamics must remain across tasks. Initial Successor feature approach omits settings with different environment dynamics, common among real-life tasks in reinforcement learning problems. Our approach transformed successor feature projects a set of diverse dynamics into a common dynamic distribution. Hence, it is an initial solution to relax the restriction of transference across fixed environment dynamics. Experimental results indicate that the transformed successor feature improves the transfer of knowledge in environments with fixed and diverse dynamics under the control of a simulated robotic arm, a robotic leg, and the cartpole environment.

Keywords: reinforcement learning · successor features · transfer reinforcement learning

1 Introduction

Reinforcement learning (RL) has demonstrated remarkable performance in various complex games and in robotic control tasks [6,19]. However, the classical RL has low sample efficiency and the solutions are task-specific. An example of this is an agent controlling a robot arm. The agent needs millions of interactions with the environment to solve a simple task, such as reaching objects. Interestingly, any modification to the robot or a change in the task to perform [19], requires a new start of the learning process anew.

Intuitively, one can expect the agent to reuse previously learnt knowledge in the new conditions. Transfer RL (TRL) [20] aims to increase sample efficiency

---

Supplementary Information The online version contains supplementary material available at https://doi.org/10.1007/978-981-99-8391-9_24.

T. Liu et al. (Eds.): AI 2023, LNAI 14472, pp. 298–309, 2024.
https://doi.org/10.1007/978-981-99-8391-9_24

and reduce task specificity. Typically, TRL involves two sets of tasks: a set of *source tasks* to learn *what* knowledge to reuse and a set of *target tasks* applying *how* reuse that knowledge. To answer these questions, many works have proposed reusing knowledge representations [17]. In this vein, Barreto et al. [4] proposed a transferable representation that generalises the successor representation [9] and makes it possible to evaluate a policy across tasks.

Successor features have shown impressive results in transferring knowledge between tasks that differ in reward functions but share transition probability distributions or *dynamics* [4–6,19]. This condition is unusual in real-world problems. For instance, if an agent controls robotic arms with different physical specifications (e.g. degrees of freedom or arm dimensions), this implies handling diverse dynamics. To manage diverse dynamics in successor features, Zhang et al. [19] proposed a linear relationship between the source and target feature functions. Under the assumption that one knows that the source and target dynamics are similar before-hand. Recent work proposed using Gaussian processes to represent target successor features as noised source successor features [1], but the implementation of the Gaussian process is complicated and its cubic complexity makes it difficult to scale in high-dimensional environments [10].

To overcome complexity and the assumption of similarities in the tasks, we propose a novel mechanism based on *successor feature*. Our method assumes the existence of a latent base dynamics and an invertible and differentiable function that can map any dynamics to this base dynamics. In other words, we assume that a given dynamics is a transformation of the base dynamics. We can use the transformed dynamics to build a new successor feature, *transformed successor feature*. This extends the ability of the original successor feature to handle diverse dynamics. To reuse the successor feature in the target tasks, we propose an ensemble method [14].

In the *reacher* environment, our approach shows that it is capable of outperforming the original successor feature sharing dynamics. In a diverse dynamics setting, our approach achieves at least the same performance as the baseline. In *cartpole* with different pole lengths and *hopper* with different reward dynamics, our approach shows that it is capable of transferring previously acquired knowledge outperforming the baseline in diverse dynamics setting.

Finally, our approach makes the following contributions: 1) A novel mechanism to predict state features within a base dynamics. It involves projecting the TSF into the base dynamics using probability density transformation techniques, enabling the capture of both similar and diverse dynamics. 2) A simple but powerful approach using ensemble models to transfer between source and target tasks. This is an expressive and straightforward solution for the successor features to manage diverse dynamics.

2 Background

2.1 Reinforcement Learning (RL)

RL problems are commonly modelled by a Markov decision process (MDP). An MDP is defined by the tuple $\mathcal{M} \equiv (\mathcal{S}, \mathcal{A}, p, r, \gamma)$, where $\mathcal{M}$ is the MDP; $\mathcal{S}$ and $\mathcal{A}$

are the states and action spaces; $p(\cdot|s, a)$ is the transition probability distribution (dynamics) that defines the distribution of the next state given a state, s, taking action, a; r is a reward function that can be defined by $r : \mathcal{S} \times \mathcal{A} \mapsto \mathbb{R}$; and $\gamma \in [0, 1)$ corresponds to the discount factor on future rewards. The main objective of RL is to discover the optimal policy $\pi^* : \mathcal{S} \mapsto \mathcal{A}$, which returns an action given a state to maximise the expected accumulated return $\sum_{t=1}^{\infty} \gamma^{t-1} r(s_t, a_t)$.

There is a category of RL algorithms which finds π^* from the state-action value function (also known as the Q function), which is defined as

$$Q^\pi(s, a) = \mathbb{E}_{\pi, s_{t+1} \sim p} \left[\sum_{t=1}^{\infty} \gamma^{t-1} r(s_t, a_t) \,\middle|\, s_0 = s, a_0 = a \right]. \tag{1}$$

Many dynamic programming-based RL algorithms alternate two processes, *policy evaluation* to compute Q^π and *policy improvement* which obtain policies taking greedy actions upon the Q function, $\pi'(s) \in arg\,max_a Q^\pi(s, a)$. Constant alternation of both processes will converge in π^*. Therefore, getting π^* for one task requires intensive interactions. In this paper, we focus on scenarios where knowledge can be reused in new tasks without restarting the learning process.

2.2   Transfer Reinforcement Learning with Successor Features

TRL is a compelling set of techniques to reuse previously learnt knowledge and commonly defines two sets of tasks. A set of *source tasks*, $\mathcal{M}^s \equiv \{M_i\}_{i=1}^n$, used by the agent to extract knowledge, and a set of target tasks, in which knowledge is transferred to and is defined by $\mathcal{M}^\tau \equiv \{M_j\}_{j=1}^m$. That is, the main objective of TRL is to transfer the knowledge learnt in M^s to M^τ.

Successor feature is a solution to achieve TRL. In the successor feature [4] approach the reward function is decomposed into a simple product of a feature function and a weight as $r_i(s_t, a_t) = \phi(s_t, a_t)^\top \mathbf{w}_i$ where $\phi : \mathcal{S} \times \mathcal{A} \to \mathbb{R}^d$ denotes the feature function of dimension d and $\mathbf{w}_i$ is the i-th task-specific vector.

The basic idea is that the feature function can be shared by all tasks, which facilitates knowledge sharing among them. This new representation of the reward function allows us to rewrite the Q-value function as

$$Q_i^\pi(s, a) = \mathbb{E}_{\pi, p} \left[\sum_{t=1}^{\infty} \gamma^{t-1} \phi(s_t, a_t) \,\middle|\, s_0 = s, a_0 = a \right]^\top \mathbf{w}_i = \psi^\pi(s, a)^\top \mathbf{w}_i \tag{2}$$

where ψ is named *successor feature* and encapsulates the dynamic of the environment. This factorisation of the rewards and Q function grants the agent the possibility of reusing the successor features in the target tasks. Furthermore, this enables fast computation of Q values in tasks with the same dynamics $Q_j^\pi(s, a) = \psi^\pi(s, a)^\top \mathbf{w}_j$ and only requires fitting the preferences over the features for the task-specific $\mathbf{w}_j$. Nevertheless, Barreto et al. [4] approach is limited to scenarios with fixed dynamics and tasks that differ in reward functions, which is unusual in RL settings.

3   Our Method

Our goal is to provide a mechanism to manage diverse dynamics for the successor feature framework. A logical approach is to learn the dynamics of the specific task independently. In successor feature-based solutions, such an idea is a breach of the objective of sharing knowledge on TRL. Instead of assuming a fixed dynamic for all tasks, we assume that there is a latent base dynamics, and all task-specific dynamics are simply different transformations of this base dynamic. Such a transformation idea not only allows for diverse dynamics across different tasks, but also builds a link between them to facilitate knowledge sharing. On top of this, we propose transformed successor features (TSF) and explain how to use them to achieve TRL.

3.1   Transformed Successor Features (TSF)

We first assume and define a base dynamic under all tasks as a state transition probability distribution at timestep t, $p^*(s_{t+1}|s_t, a_t)$, where $s_{t+1}, s_t \in \mathcal{S}$ and $a_t \in \mathcal{A}$. With this base dynamic, we assume that the dynamic of each task $p^i(s_{t+1}|s_t, a_t)$ is a transformation of this base dynamic, $p^i = T^i(p^*)$, where T^i denotes a transformation. To be specific, we assume that there exists an invertible and differentiable function g^i that maps the state from the base dynamic and the task-specific dynamic, $s_t^* = g^{-i}(s_t^i)$ and $s_{t+1}^* = g^{-i}(s_{t+1}^i)$, where g^{-i} is the inverse function of g. Note that we use a linear form $g(x) = b \cdot x + c$ throughout this paper and will consider extending it to a more complex form in the future, such as normalizing flow-based transformation [16]. According to the change of variable theorem, we have *task-specific dynamic*

$$p^i(s_{t+1}^i, s_t^i, a_t) = p^*(s_{t+1}^*, s_t^*, a_t) \begin{vmatrix} \frac{\partial g^{-i}(s_t^i)}{\partial s_t^i} & 0 \\ 0 & \frac{\partial g^{-i}(s_{t+1}^i)}{\partial s_{t+1}^i} \end{vmatrix}, \tag{3}$$

and given Eqs. (2) and (3), we can further derive the following new task-specific Q function,

$$Q_i^\pi(s^i, a) = \mathbb{E}_{\pi, s_{t+1}^* \sim p^*} \left[\sum_{t=1}^\infty \gamma^{t-1} \phi(g^i(s_t^*), a_t) \,\middle|\, s_0^* = g^{-i}(s^i), a_0 = a \right]^\top \mathbf{w}_i. \tag{4}$$

The above derived Q-value function bears an expectation of future state-action features with respect to a base dynamics. That is, these computed Q values can be viewed as generalised values that take advantage of the particularities of the base dynamics and therefore hope to capture different dynamics. Another merit is that it transforms Q functions for all tasks into expectations under the same base dynamic rather than task-specific dynamics, making subsequent knowledge sharing possible. To achieve knowledge sharing, we define

$$\phi_i(g^i(s_t^*), a_t) = \phi(s_t^i, a_t) \odot [h(g^{-i}(s_t^i))] \tag{5}$$

where h is an affine function shared by all tasks. This decouples the base feature function ϕ from the task-specific transformation g^i. h enhances the flexible relationship between the feature function and the dynamics transformation. Similar to the original successor features, our derived Q function expresses the TSF,

$$\hat{\psi}_i^\pi(s^i, a) = \mathbb{E}_{\pi, p^*}\left[\sum_{t=1}^{\infty} \gamma^{t-1} \phi(s_t^i, a_t) \odot \left[h\left(g^{-i}(s_t^i)\right)\right]\right] \tag{6}$$

where i denotes a task. With respect to this TSF, we make the following remarks:

Remark 1. This new representation retains the same capabilities as the original successor feature and decouples the dynamics from the rewards.

Remark 2. The transition dynamics of TSF is the base probability distribution, thus TSF predict the future features visitation in the latent base dynamics.

Remark 3. There is a shared base feature function ϕ across tasks that facilitates knowledge sharing. Furthermore, the task-specific transformations also provides a way to express the similarity or relationship between tasks.

Remark 4. The term $h\left(g^{-i}(s_t^i)\right)$ adds the particularities of the current state of each task to the shared base feature function. We can recover the shared base features when the outputs of h are 1. Different values suggest more dissimilarities among tasks. Therefore, this definition enables the handling of diverse dynamics.

3.2   Transfer RL via Transformed Successor Features

As a TRL setting, our approach consists of two steps. The knowledge extraction in source tasks, *source TSF learning*, and the transference to target tasks, *target TSF learning*.

Source Transformed Successor Feature Learning. To extract knowledge, the agent minimises the total loss function, $L = L_{\hat{\psi}_i^\pi} + \beta L_w$, where β is an hyperparameter, $L_{\hat{\psi}_i^\pi}$ is the TD error to fit source TSF defined by

$$L_{\hat{\psi}_i^\pi} = \left\|\phi_i(s_t^i, a_t) + \gamma\hat{\psi}_i^\pi(s_{t+1}^i, a_{t+1}) - \hat{\psi}_i^\pi(s_t^i, a_t)\right\|. \tag{7}$$

and reward fitting using

$$L_w = \left\|R - \phi_i(s_t^i, a_t)^\top \mathbf{w}_i\right\|. \tag{8}$$

where R is the reward obtained. Furthermore, to minimise total loss, the agent learns all components required for the source TSF, that is, the invertible function g^i, the shared affine function h, and the reward mapper $\mathbf{w}_i$.

Target Transformed Successor Feature Learning. To construct the target TSF to be reused in the target task set, We compute the target TSF as a weighted source TSF using the super-learner ensemble model [14]. The weights express the relevance of each source TSF in the new tasks, and they generate the following definition.

Given weights $\Omega = \{\omega_{ji}|\{i\}_{i=1}^{n}, \{j\}_{j=1}^{m}\}$, a *target TSF*, $\hat{\psi}_j$, is computed as the weighted learnt *source TSF*,

$$\hat{\psi}_j(s^j, a) = \sum_i \omega_{ji}\hat{\psi}_i^{\pi}(s^j, a) = \mathbb{E}_{\pi,p^*}\left[\sum_{t=1}^{\infty}\gamma^{t-1}\underbrace{\phi(s_t^j, a_t) \odot \left[h\left(\sum_i \omega_i g^{-i}(s_t^j)\right)\right]}_{\text{target transformed feature function}}\right]$$

(9)

Moreover, we expand the source TSF and acquire a new target transformed feature function. To learn the weights Ω, we use the same total loss with the new definitions of the target transformed feature function and the target TSF. Thus, our proposed algorithm only needs to fit the reward mappers and Ω for the target task set that can be learnt simultaneously.

4 Related Work

Contrary to other approaches such as [2,3] that focus on techniques to abstract ground states to latent representations and ensure keeping the Markovian properties, our work focusses on dynamics transformations and induced state dynamics instead of ground states. Moreover, such works do not consider successor representation-style transfer knowledge.

Our method is focused on *disentangled representation transfer* specifically successor representations [20]. Successor representation was formally defined by [9] for tabular cases. In this vein, Janner et al. [12] proposed a mixture of methods to capture more information from the dynamics environment. Madarasz et al. [15] proposed a non-parametric solution to the successor representation to improve performance in unsignalling environments. However, these mixture methods do not address the assumption of fixed dynamics.

Authors such as [4,13] proposed generalised successor representations to allow deep reinforcement learning and continuous spaces. Specifically, Barreto et al. [4] stated the reward function could be expressed as a linear combination and derived the *successor feature*, extensively discussed in this paper. However, it is limited to settings that share dynamics and only differ in reward functions.

In relation to successor feature-based approaches, many works have addressed current limitations. Regarding the linearity of the reward, [6] proposed a method to learn flexible successor features, thus relaxing the linearity. To improve policy improvement in the successor feature framework, previous work e.g. [5,7,18] have proposed alternatives to allow the training of multiple policies instead of improving one policy at a time, thus aiming to improve performance in task changes, i.e., agents are more flexible to task changes. Recently, works such as

[8] described approaches to build better mechanisms to represent and predict latent state spaces from reward characteristics.

To the best of our knowledge, two works have attempted to decouple dynamics with successor features. Abolshah et al. [1] used the *Gaussian process* to abstract measurements between tasks, which implies that the agent should have access to the target and source observations to model the successor feature; hence, the successor feature entangles observations from the source and target tasks. Additionally, this work addresses both restrictions, namely environments with diverse dynamics and rewards. However, *Gaussian processes* can have low expressivity in complex and high-dimensional scenarios [10]. In the second approach, Zhang et al. [19] proposed a linear relationship between the source and the target feature functions. This assumption is limited to similar tasks and assumes that we know similarities before-hand.

5   Experiments

5.1   Settings

Metrics. Asymptotic Performance is a common metric to assess TRL. This metric evaluates the impact of TRL on the final performance of the agent [20].

Agents. For our baseline agent, we implemented the successor feature based on [4], SFDQN. For our agent, TSFDQN, we first select the hyperparameters for the functions g and h. To build invertible functions g, we constructed a linear model and proceeded with a grid search for the most desirable dimension. We found that a linear transformation, $g : \mathbb{R}^{|s|} \to \mathbb{R}^{|g|}$, with $|s|$ as the dimension of the state space and $|g| = 100$ as the dimension performs decently in the reacher environment. The affine transformation h has the following hyperparameters $h : \mathbb{R}^{|g|} \to \mathbb{R}^d$, with d as the dimension of the feature function.

Finally, for the code implementation, we extended a base *successor features* framework[1], we kept the project structure and configuration management, but changed the rest to use the PyTorch library. Our code is made public[2].

Dynamics Settings. We designed two settings to test our approach. A *shared dynamics setting* in which the dynamics remain fixed among the different tasks. In the second setting, *diverse dynamics setting*, we introduce specific settings to the environment to simulate diverse dynamics.

5.2   Results and Discussion

To present our results, we organised the discussion into two parts, namely the *shared dynamics setting* and the *diverse dynamics setting*.

---

[1] https://github.com/mike-gimelfarb/deep-successor-features-for-transfer.
[2] https://github.com/okgaces/deep-successor-features-for-transfer.

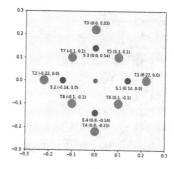

Fig. 1. Robot reacher setup. The blue dots are the coordinates for the source tasks. Orange dots are the coordinates for the target tasks. The green dot is the centre of the reacher arm (Color figure online)

Shared Dynamics Setting. In the first setting, *shared dynamics*, we used the reacher environment. We took the initial configuration of the reacher environment in Barreto et al. [4].

Reacher Environment. The reacher environment is a 2D link reacher. Our agent learns to manipulate robot arms with the objective of reaching different objects in fixed coordinates. The agent learns to achieve different object coordinates and then performs in unseen target objects. The coordinates are shown in Fig. 1

In this section, we describe one of the designed scenarios. This scenario was designed based on Barreto et al. [4]. Each agent interacted with each source task in 10.000 timesteps, where every 1.000 timesteps the agent was moved to perform in target tasks for 500 timesteps. To avoid catastrophic forgetting, agents revisit tasks a total of 50 times.

To build the TSF, we hand-made the shared base feature function. In spite of RL settings lack the privilege of this hand-made function, our definition consists of two components: the shared-based feature function and the transformation functions, g and h, Eq. 6. With this scenario, we can measure the effect of the transformation functions. The output of the feature function depicts the desired tasks, that is, the dimensionality of the feature function depends on the number of tasks, that is, $\phi : \mathcal{S} \times \mathcal{A} \to \mathbb{R}^d$, for d the number of tasks.

The asymptotic performance for the shared dynamics setting is shown in Fig. 2. In the source task, we notice that TSF reuses the optimal policies discovered in the source tasks, generating better performance. The results suggest that the functions h and g, extract more information from the dynamics of the environment, thus generating better representations to transfer. In summary, the performance improvement in source tasks after 200k in which both agents saturate is 20% better than the baseline. The metrics on target tasks bring the effect of the ensemble model into the transfer process. In the target tasks, the agent fits to the most desirable combination of policies for each target task. However, in this experiment, it is noticeable that in target tasks (T.1, T.2, T.3, T.4) the performance is similar to or below the baseline. From Fig. 2b, it can be seen

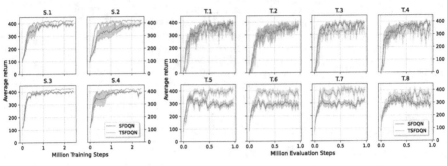

(a) Average return in source tasks

(b) Average return in target tasks

Fig. 2. Scenario 1: Average return in source and target tasks by total training steps. The shading shows one-half of a standard deviation over 5 runs with independent random seeds.

that the TSFDQN algorithm outperforms the target tasks (T.5, T.6, T.7, T.8), which corresponds to the most dissimilar tasks according to their coordinates, as shown in Fig. 1.

From the results, we derived two hypotheses. First, in Scenario 1, Fig. 1, there are four target tasks similar to the source tasks. Second, we wonder about the existence of better mechanisms to transfer apart from the ensemble model, although this hypothesis is outside of the scope of this paper.

Diverse Dynamics Setting. For this setting, we test the agents in the cartpole and hopper environments.

Cartpole Environment. The second environment is the classical cartpole set with *diverse dynamics setting.* We extend the proposed experiment in [1] to balance poles of different lengths. Agents acquire knowledge learning to balance three different poles $\{0.5\,\text{m}, 0.8\,\text{m}, 1.0\,\text{m}\}$, then use the acquired knowledge to balance different lengths $\{0.1\,\text{m}, 3.0\,\text{m}\}$.

In this case, we learn the feature function using the process described in [5]. In a pre-training stage, the agent uses a random policy in the source tasks, to fit the most suitable feature function and weights according to Eq. 8 with a fixed feature dimension, d, as 20. After the pre-training stage, the reward mappers are discarded and the agent proceed with the regular training/evaluation already described. For TSFDQN, we use the same learning rate for the Ω values and the output of the g dimension found in the reacher environment.

The results are shown in Fig. 3. In source tasks, the TSFDQN agent achieves a performance similar to that of the SFDQN agent. In target tasks, our agent performs better than SFDQN with a mean improvement of 49.03%. It is noticeable that the performance on target tasks of the TSFDQN agent is approximately a 40% of the performance on source tasks, while the SFDQN deteriorates about

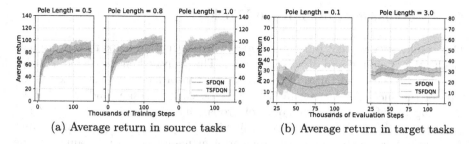

(a) Average return in source tasks   (b) Average return in target tasks

Fig. 3. Average return in source and target tasks by total training steps. Shades show half a standard deviation over 10 runs with independent random seeds.

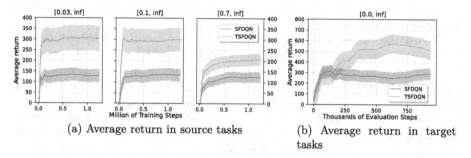

(a) Average return in source tasks   (b) Average return in target tasks

Fig. 4. Average return in source and target tasks by total training steps. Shades show half a standard deviation over 5 runs with independent random seeds.

60%. This suggests that our proposed model acquires similar policies in source tasks but improves those in target tasks.

Hopper Environment. The third environment is based on the hopper environment. The agent aims to control a robot leg to jump forward. The agent receives rewards for every timestep for healthy conditions such as when the leg height is below a threshold and an additional reward for moving forward. Our version has 27 discrete actions. Each action is a 3-tuple with values $\{-0.1, 0, 0.1\}$. To add dissimilarities in the reward function and dynamics, we set different values for healthy states. Despite its simplicity, a robot in a state with a higher value within a healthy z range can take an action that leads to an unhealthy state and a lower reward. In other words, this scenario covers dissimilarities in rewards, and transition dynamics also tests the agent's ability to provide control to different robots and reward conditions. We learn the feature function with the same strategy as in the cartpole environment and the dimension d as 50. With the same learning rate for the Ω values and the g dimension as before.

The results are shown in Fig. 4. Results in source tasks indicate that the transformation model in the transformed feature function provides enhanced features to be predicted by the successor feature, hence induce better policies. Similarly, in the target task, the ensemble model induces a better policy with an average gain of 25% compared to the average return in source tasks.

6   Conclusions and Future Work

Our results suggest that our transformed feature function helps to capture particularities of shared and diverse dynamics. The proposed ensemble model helps to calculate the importance of the source task in order to improve the prediction of the successor feature in the target tasks. Our results indicate that by applying transformation techniques of probability densities, successor features can be extended to manage diverse dynamics. Therefore, our approach is a first step towards resolving the open problem of managing diverse dynamics in the successor feature framework. The results of the study in diverse dynamic settings demonstrate that the proposed approach can learn suboptimal successor features in source tasks, which can then be used to leverage the knowledge of policies in target tasks.

We list the current drawbacks of our work. Our agents which learn feature functions do not generate the optimal policies. To address this, our approach can be combined with methods designed to extract features in successor features context, such as [8]. We hope we can bring ideas from [11] to extend our work to support stochastic policies and continuous action spaces. We believe that utilising more complex invertible and differentiable functions can capture complex dynamics, hence, increase the overall performance of TSF-based agents. To do so, we can use *normalising flows* as a way to apply the change of variable using invertible functions [16], thus capturing complex distributions.

Acknowledgments. This work was supported by the Australian Research Council through Discovery Early Career Researcher Awards DE220101075 and DE200100245, and by the University of Technology Sydney (UTS) and Australian Technology Network (ATN) through UTS ATN-LATAM Research Scholarship Award.

References

1. Abdolshah, M., Le, H., George, T.K., Gupta, S., Rana, S., Venkatesh, S.: A new representation of successor features for transfer across dissimilar environments. In: International Conference on Machine Learning (ICML), vol. 139, pp. 1–9 (2021)
2. Abel, D., Arumugam, D., Lehnert, L., Littman, M.: State abstractions for lifelong reinforcement learning. In: Proceedings of the 35th International Conference on Machine Learning. Proceedings of Machine Learning Research, vol. 80 (2018)
3. Allen, C., Parikh, N., Gottesman, O., Konidaris, G.: Learning Markov state abstractions for deep reinforcement learning. Adv. Neural Inf. Process. Syst. **34**, 8229–8241 (2021)
4. Barreto, A., et al.: Successor features for transfer in reinforcement learning. In: Advances in Neural Information Processing Systems (NIPS), vol. 30. Barcelona, Spain (2017)
5. Barreto, A., Hou, S., Borsa, D., Silver, D., Precup, D.: Fast reinforcement learning with generalized policy updates. Proc. Natl. Acad. Sci. **117**, 30079–30087 (2020)
6. Barreto, A., et al.: Transfer in deep reinforcement learning using successor features and generalised policy improvement. In: International Conference on Machine Learning (ICML), pp. 501–510 (2019)

7. Brantley, K., Mehri, S., Gordon, G.J.: Successor feature sets: generalizing successor representations across policies. In: AAAI Conference on Artificial Intelligence, vol. 35, pp. 11774–11781 (2021)
8. Carvalho, W., Filos, A., Lewis, R.L., Lee, H., Singh, S.: Composing task knowledge with modular successor feature approximators. In: International Conference on Learning Representations (ICLR) (2023)
9. Dayan, P.: Improving generalization for temporal difference learning: the successor representation. Neural Comput. **5**(4), 613–624 (1993)
10. Goumiri, I.R., Priest, B.W., Schneider, M.D.: Reinforcement learning via gaussian processes with neural network dual kernels. In: IEEE Conference on Games (CoG), pp. 1–8 (2020)
11. Hunt, J., Barreto, A., Lillicrap, T., Heess, N.: Composing entropic policies using divergence correction. In: Proceedings of the 36th International Conference on Machine Learning, vol. 97, pp. 2911–2920 (2019)
12. Janner, M., Mordatch, I., Levine, S.: γ-models: generative temporal difference learning for infinite-horizon prediction. Adv. Neural Inf. Process. Syst. (NIPS) **33**, 1724–1735 (2020)
13. Kulkarni, T.D., Saeedi, A., Gautam, S., Gershman, S.J.: Deep successor reinforcement learning (2016). preprint on webpage at https://arxiv.org/abs/1606.02396
14. van der Laan, M.J., Polley, E.C., Hubbard, A.E.: Super learner. Stat. Appl. Genet. Mol. Biol. **6** (2007)
15. Madarasz, T., Behrens, T.: Better transfer learning with inferred successor maps. In: Advances in Neural Information Processing Systems (NIPS), vol. 32. Vancouver, BC, Canada (2019)
16. Rezende, D.J., Mohamed, S.: Variational inference with normalizing flows. In: International Conference on Machine Learning (ICML), vol. 37, pp. 1530–1538 (2015)
17. Schaul, T., Horgan, D., Gregor, K., Silver, D.: Universal value function approximators. In: International Conference on Machine Learning (ICML), vol. 37, pp. 1312–1320. Lille, France (2015)
18. Tasfi, N., Santana, E., Liboni, L., Capretz, M.: Dynamic successor features for transfer learning and guided exploration. Knowl.-Based Syst. **267**, 110401 (2023)
19. Zhang, J., Springenberg, J.T., Boedecker, J., Burgard, W.: Deep reinforcement learning with successor features for navigation across similar environments. In: IEEE/RSJ International Conference on Intelligent Robots and Systems (IROS), pp. 2371–2378 (2017)
20. Zhu, Z., Lin, K., Jain, A.K., Zhou, J.: Transfer learning in deep reinforcement learning: a survey. IEEE Trans. Pattern Anal. Mach. Intell. **45**(11), 13344–13362 (2023)

Cooperative Multi-Agent Reinforcement Learning with Dynamic Target Localization: A Reward Sharing Approach

Helani Wickramaarachchi[1,2]([✉]) [iD], Michael Kirley[1,2] [iD], and Nicholas Geard[1] [iD]

[1] School of Computing and Information Systems, The University of Melbourne, Melbourne, Australia
`hwickramaara@student.unimelb.edu.au,`
`{mkirley,nicholas.geard}@unimelb.edu.au`
[2] ARC Training Centre in Optimisation Technologies, Integrated Methodologies, and Applications (OPTIMA), Melbourne, Australia

Abstract. Cooperation in multi-agent reinforcement learning (MARL) facilitates the acquisition of complex problem-solving skills and promotes more efficient and effective decision-making among agents. Numerous strategies for cooperative learning in MARL exist, including joint action learning, task decomposition, role assignment, and communication protocols. However, deploying these strategies in a complex and dynamic environment remains challenging. To address such challenges, we propose a technique that uses reward sharing to enhance cooperation in partially observable multi-agent environments. As an extension of reward shaping, reward sharing allows agents to work together towards a global objective while still pursuing their local objectives. This approach can foster cooperation and reduce competition between agents without explicit communication, ultimately leading to faster learning and better performance. This study compares three different reward sharing techniques: the Performance Incentive (PI), the Observer's Share (OS), and the Synergy Achievement (SA) in the context of dynamic target localization, focusing on simulation studies. Thereafter, the proposed reward sharing techniques are evaluated under the effects of objective prioritization, various agent counts, and a variety of map sizes. The research reveals that the proposed reward sharing techniques enhance agent performance, scaling the number of agents leads to higher rewards, and demonstrates a negative correlation between map size and average rewards.

Keywords: Multi-Agent Reinforcement Learning · Multiple Objectives · Cooperation

1 Introduction

In Multi-Agent Reinforcement Learning (MARL), agents typically must coordinate their actions to achieve a collaborative (global) objective. This task is

T. Liu et al. (Eds.): AI 2023, LNAI 14472, pp. 310–324, 2024.
https://doi.org/10.1007/978-981-99-8391-9_25

challenging because in the presence of an individual (local) objective, agents might not be willing to collaborate to achieve the global objective. Cooperative learning, a powerful technique in machine learning [1], uses the collective intelligence of multiple agents to solve problems, achieving greater accuracy and efficiency compared to isolated individual agents [2,3].

Implementing cooperative learning is challenging when agents have multiple objectives, including local objective and global objective. Agents must learn to strike a balance between achieving their local and global objectives, which can be complex and often conflicting. In general, cooperative learning strategies are used to encourage agents to effectively collaborate. Joint action learning (JAL) [4,5], task decomposition [6], role assignment [4,7], communication protocols [8], and learning from observation [4,9] are all powerful tools that can be leveraged to enable agents to learn to coordinate their actions, share information and optimize their collective performance.

However the adaptability of such methods remains an issue, as the complexity [5] and dynamicity of the environment increase. For example, task decomposition introduces bottlenecks and coordination difficulties, limiting scalability and generalization capabilities [6]. Learning from observations is hampered by sample inefficiency, lack of contextual information, and handling partial observations [9]. Moreover, the learning process can be hindered by communication overhead caused by message passing and synchronization [8]. These limitations require addressing communication costs, optimizing task decomposition, and developing techniques to enhance sample efficiency and handle partial observation.

Our approach will address limitations related to communication overhead, partial observability, agents' scalability, and map size impact. We propose a Multi-Agent Deep Reinforcement Learning (MDRL) framework [10], while considering the MO nature of the problem. Our overarching aim is to establish an environment of collaboration that empowers agents to effectively work together towards a global objective, simultaneously pursuing an local objective. The key contributions of the paper are:

1. The design of a reward-sharing strategy as a cooperation enhancement technique in the context of multi-agent, multi-objective reinforcement learning (MA-MORL).
2. Adapt a multi-channel observation window to capture extensive contextual information in a partially observable environment.
3. The formulation and modeling of the multi-agent target localization problem in MDRL.
4. Analyze the impact of different decision parameters, such as objective prioritization, various agent counts, and a variety of map sizes on the proposed reward sharing techniques.

The paper is structured as follows: Sect. 1 offers a concise overview, Sect. 2 introduces related work and conducts a comparative analysis of current research, Sect. 3 provides a comprehensive methodology for our proposed approach, Sect. 4 provides extensive information on system parameters, simulation results, and experimental details, and finally, Sect. 5 offers concluding remarks and delineates potential avenues for future research.

2 Related Work

In MARL, potential-based reward shaping (PBRS) is a powerful method to improve the convergence rate of RL agents. Ng et al. [11] originally introduced potential-based reward shaping, which calculates the difference in a potential function $\varphi(s, s')$ between the current state s and the subsequent state s'. Wiewiora et al. [12] further extended this approach to incorporate both states and actions: $\varphi(s, a, s')$ in shaping functions. Though they do not guarantee that the agent will follow the optimal policy in practice, they show that the original reward signal $R(s, a, s')$, where s is the current state, a is the action taken, and s' is the resulting state has the same optimal policy as the dynamically shaped reward function $R(s, a, s') + F(s, a, s')$ where $F(s, a, s')$ is the shaped reward function. Another PBRS based research by Mannion [13,14] discusses the theoretical implications of applying PBRS and difference reward approaches to cooperative MA-MORL situations. The study by Grzes and Kudenko [15,16] shows that knowledge is automatically encoded into a reward signal, so that potential functions can be learned online in parallel with RL. These studies prove that reward shaping with mixed function approximation gives excellent results. In addition Ferreira et al. [17] discusses how socially-inspired rewards can be used to speed up efficient task completion and user adaptation for online learning. For this purpose a potential-based reward shaping method is combined with a sample efficient reinforcement learning algorithm to offer a principled framework to cope with these potentially noisy interim rewards. In a study published in [18], two novel reward functions are compared that combine difference rewards and potential-based reward shaping. This approach has proven to be effective in a single objective MARL environment, and may also be useful when applied to MA-MORL as well.

In MARL, peer communication plays a major role since it can improve overall learning performance [19,20]. By introducing a DRL algorithm, Hostallero et al. [20] introduce peer evaluation based Dual DQN. They gradually reshape their reward function to create more cooperative behavior in response to peer evaluation signals. Despite this extra reward given to peers, it has no proper explanation and violates the budget balance at the same time. Instead of studying algorithmic improvements that address such non-episodic and sparse reward settings, Co-Reyes et al. [21] study the kinds of environment properties that can make acquiring knowledge under such conditions easier.

A recent publication by Huang and Jin [22] shows that reward shaping can be highly effective in guiding agents' learning process. This discusses some factors in detail that might affect successful agent team training. Also, the study on autonomous shaping [23] introduces the use of learned shaping rewards in a sequence of goal-directed reinforcement learning tasks, where an agent uses prior experience on a sequence of tasks to learn a portable predictor that estimates intermediate rewards, resulting in accelerated learning in later tasks that are related but distinct.

3   Methodology

The agents in our study employ the Proximal Policy Optimization (PPO) algorithm to make decisions, and their actions are subsequently rewarded based on these decisions. Our research introduces and evaluates three distinct reward-sharing techniques: Performance Incentive (PI), Observer's Share (OS), and Synergy Achievement (SA). Through a rigorous series of experiments and comprehensive analyses, we assess the effectiveness of these techniques. Additionally, we investigated the effects of objective prioritization, various agent counts, and a variety of map sizes on the dynamic target localization task, as described in [24].

3.1   Problem Specification

Dynamic target localization involves determining a moving target location based on data collected by agents operating in a certain area of interest. We approach this task as a multi-agent system (MAS) [25] problem with N cooperative independent learning agents, each equipped with the same set of actions (UP, DOWN, LEFT, RIGHT, STAY STILL). As the target location is unknown, the problem is formulated as a Partially Observable Markov Game (POMG) [26]. The POMG is characterized by a tuple $(N, S, A, O, P, R, \gamma)$, where:

- N denotes the finite set of agents where $N \geq 2$ and $N \in \mathbf{N}$,
- S denotes the finite set of states,
- $A = A_1 \times A_2 \times \ldots \times A_N$ is the set of joint actions,
- $O = O_1 \times O_2 \times \ldots \times O_N$ set of joint observations,
- $P : S * A * S \rightarrow [0, \infty)$ represents the action-conditioned state transition distribution,
- $R : S * A \rightarrow R^q$ represents reward functions for q different objectives,
- $\gamma \in [0, 1]$ is a discount factor.

The task at hand is episodic in nature, where each episode starts with the sampling of an initial state $p(s_0)$, and at each time-step $t \leq H$, where H is the maximum length of the episode. Actions are sampled based on a parametric stochastic policy $\pi_i(a_t|s_t) : O_i \times A_i \rightarrow [0, 1]$. The successor state at each time-step is given by $p(s_{t+1}|s_t, a_t)$, and a reward r_i is provided by the environment at each time-step. The return for each state-action pair in the episode is a vector defined as the discounted sum of future rewards, $R_t = \Sigma_{t'=t} \gamma^{t'-t} r(s_{t'}, a_{t'})$.

3.2   Observation Space

The grid-based environment in which multiple agents operate is characterized by limited observability and discrete action space. Agents observe cells within a 3-radius distance in every direction excluding the tile they currently occupy. The agents can detect and distinguish between the presence of other agents, the target, or empty tiles, allowing them to make strategic and informed decisions.

The agent's observations serve as input parameters for a Convolutional Neural Network (CNN) [27] which are optimized using the Proximal Policy Optimization (PPO) [28] algorithm to determine the next action and reward. Figure 1 illustrates the channels within the observation space, which store various types of information of obstacles, team members, and the dynamic target. The purpose behind incorporating three windows within the observation space is to provide the CNN with a more comprehensive and elaborate input. This, in turn, enables the agent to make better decisions and select optimal actions [21].

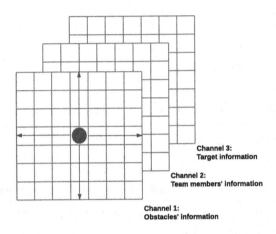

Fig. 1. The 3 channel observation window allows agents to capture extensive contextual information in a partially observable environment. Channel 1 represents information about obstacles within the observation space, channel 2 represents information about team members within the observation space, and channel 3 represents information about the target.

3.3 Agent Configuration

The initial locations of the agents were determined by random seeds. As an agent attempts to move into a wall, it remains in the same location. In spite of this, agents can move into each other and occupy the same cell space. At each step, the order of movement will be randomly selected in order to minimize biases caused by the order of movements. According to this model, agents sequentially see their observations, take actions, and receive rewards from the other agents. The agent is equipped with a sensor that measures the Euclidean distance between the agent's current location and the target's location.

3.4 Target Configuration

We adopted a dynamic target localization scenario in this study, which resulted in the target position changing at each step. The target's initial location was selected using random seeds. Target localization requires a minimum of two agents since the target cannot be located by a single agent alone. The initial value of the target is determined by a function that depends on the number of agents in the environment.

$$r_t \propto N \tag{1}$$

where r_t is the reward provided by the retrieved target and N represents the number of agents within the environment.

3.5   The Policy Optimization Algorithm

The agents update their policies using a stable and efficient optimization algorithm, Proximal Policy Optimization (PPO). In 2017, Schulman et al. introduced PPO [28] as an improved and simplified version of Trust Region Policy Optimization (TRPO) [29].

Algorithm 1 Proximal Policy Optimization (PPO) with Clipped Surrogate Objective

---

1: Initialize policy parameters θ_0
2: **for** $t = 0, 1, 2, \ldots$ **do**
3:     Collect set of trajectories $\mathcal{D}_t$ using policy π_t
4:     Compute rewards-to-go $\hat{R}_t$ for each timestep in $\mathcal{D}_t$
5:     Compute advantage estimates $\hat{A}_t$ using value function V_t
6:     **for** $k = 1, 2, \ldots, K$ **do**
7:         **for** each batch of data $(s_i, a_i, \hat{A}_i, \hat{R}_i)$ in $\mathcal{D}_t$ **do**
8:             Compute ratio $\rho(\theta_k) = \frac{\pi_{\theta_k}(a_i|s_i)}{\pi_{\theta_{k-1}}(a_i|s_i)}$
9:             Compute clipped ratio $\bar{\rho}(\theta_k) = \text{clip}(\rho(\theta_k), 1 - \epsilon, 1 + \epsilon)$
10:             Compute surrogate objective

$$L^{CLIP}(\theta_k) = \min\left(\rho(\theta_k)\hat{A}_i, \bar{\rho}(\theta_k)\hat{A}_i\right)$$

11:             Compute value function loss by regression on mean-squared error:

$$L^{VF}(\theta_k) = \frac{1}{2}\left(V_{\theta_k}(s_i) - \hat{R}_i\right)^2$$

12:             Update policy parameters

$$\theta_{k+1} = \text{argmax}_\theta \left\{ \frac{1}{|\mathcal{D}_t|} \sum_{i=1}^{|\mathcal{D}_t|} \min\left(\rho(\theta)\hat{A}_i, \bar{\rho}(\theta)\hat{A}_i\right) - c_1 L^{VF}(\theta) \right\}$$

13:             Update value function parameters $V_{k+1}(s_i) = V_k(s_i) + \alpha\left(\hat{R}_i - V_k(s_i)\right)$
14:         **end for**
15:     **end for**
16: **end for**

PPO incorporates a Kullback-Leibler (KL) penalty per token from the supervised fine-tuning (SFT) model. The KL divergence measures the similarity of two distribution functions and penalizes extreme distances. By using a KL penalty, the distance between the responses and the outputs of the SFT model trained in step 1 is reduced to avoid over-optimizing the reward model and deviating too much from human intention.

In policy gradient methods, first an estimator of the policy gradient is computed and then plug that estimator into a stochastic gradient ascent algorithm. In general, the gradient estimator has the form of;

$$\hat{g} = \hat{E}_t\left[\nabla_\theta \log \pi_\theta\left(a_t|s_t\right) \hat{A}_t\right] \tag{2}$$

where;

π_θ - a stochastic policy

$\hat{A}_t$ - an estimator of the advantage function at time-step t

$\hat{E}_t[...]$ - the empirical average over a finite batch of samples

In the above Eq. 2 the gradient estimator $\hat{g}$ is obtained by differentiating the objective

$$L^{PG}(\theta) = \hat{E}_t[log\,\pi_\theta\,(a_t|s_t)\,\hat{A}_t] \tag{3}$$

As per the algorithm, multiple steps of optimization will perform on this loss L^{PG} using the same trajectory, which often leads to destructively large policy updates.

With PPO we have a single policy taking care of both the update logic and the trust region. PPO comes up with a clipping mechanism, parameterized by model parameter θ, which clips the probability ratio $\rho(\theta_k)$ between a given range and does not allow it to go further away from the range. i.e. here it modifies the objective, to penalize changes to the policy that move $\rho(\theta_k)$ away from 1. Here ϵ is a hyper parameter controlling the boundary of trust region. So now the main objective is as follows;

$$L^{CLIP}(\theta) = \hat{E}_t\left[min(\rho(\theta_k)\,\hat{A}_t, clip(\rho(\theta_k), 1 - \epsilon, 1 + \epsilon)\hat{A}_t)\right] \tag{4}$$

3.6   Multi-objective Function

Mobile agents in the proposed model have two objectives: local and global. The local (individual) objective of the mobile agent, aims to maximize the precision of target localization. In pursuit of the desired objective, we took two essential parameters into account: the minimization of the Euclidean distance (d) between the agent and the target, and the optimization of resources (E) allocated to each agent.

Let $f(x)$ be the function that maps the input parameters x to the output accuracy of the localization system, where x represents the set of parameters that govern the behavior of the localization system. Then the local objectives can be mathematically formulated as follows:

$$Minimize(Distance) = \min_d [f_1(d)] \tag{5}$$

$$Maximize(Resources) = \max_E [f_2(E)] \tag{6}$$

subject to:

$$E_{min} \leq f_2(E)$$

This constraint ensures that the resource level of the agent does not fall below a minimum threshold E_{min}. The function $f_2(E)$ represents the resource consumption of the agent, and failure to satisfy this constraint would result in the agent being considered non-functional. A hybrid reward derived from above mentioned factors was considered to maximize target localization accuracy. The fitness function is expressed as:

$$F_1(d, E) = \min_d [f_1(d)] + \max_E [f_2(E)] \tag{7}$$

The global(collaborative) objective, aims to foster cooperation and collaboration among agents. To accomplish this objective, our study proposes three reward sharing techniques, namely:

Performance Incentive (PI). This method incentivizes agents that actively engage in the target localization task. Upon successful target detection, the reward (r_t) is distributed exclusively among the participating agents. This approach serves as motivation for agents to actively contribute, resulting in enhanced overall performance and increased team effectiveness.

Observer's Share (OS). This approach is grounded in the careful observation of agents in the environmental context. Given that each agent operates under constrained visibility, it provides rewards for those who actively engage in observation and subsequently share information with the team. Notably, all agents within the observation space receive equitable rewards, irrespective of their direct involvement in achieving the objective. This strategy serves as a strong incentive for agents to foster collaboration, facilitate information sharing, and operate as an effective cohesive unit.

Synergy Achievement (SA). This technique actively cultivates a spirit of teamwork in pursuit of a common objective. The reward allocated for locating the target (r_t) is evenly distributed among all team constituents, irrespective of their discrete contributions. Consequently, it fosters a pervasive atmosphere of mutual ownership and collective responsibility, as each agent recognizes that their accomplishments are inextricably linked to the overall success of the team. This approach effectively stimulates collaboration, cohesion, and the development of a supportive environment.

The fitness function corresponding to the second objective is defined as follows:

$$F_2(m) = r_t/m \tag{8}$$

subject to:

$$r_t > 0$$

$$m \geq 2, \ and \ m \in \mathbf{N}$$

where r_t is the reward provided by the retrieved target and m represents the number of agents who receive the reward. We can further refine the defined objectives by considering a weighted sum of multiple objectives of the dynamic target localization problem. This paper presents a trade-off between ensuring target localization accuracy and encouraging agent collaboration and cooperation. Thus, the multi-objective optimization problem can be formulated as a linear scalarization with importance weights per objective as shown in Eq. 9.

$$R = (1 - \alpha)F_1(d, E) + \alpha F_2(m) \qquad (9)$$

Modifying the priority assigned to each objective yields varying outcomes for the objective function. So the concept of a Pareto front becomes irrelevant as we do not focus on any optimal set of solutions that offer trade-offs. Here, we propose a combination of both reward methods: (i) a non-sparse hybrid reward to guide the agent towards the goal when its resource availability is low, and (ii) a sparse reward to encourage agents to collaborate.

4   Experimental Setup and Analysis

4.1   Experimental Setup

For our experiments, we carried out 200 epochs evaluating each epoch for 5 episodes of maximum length 500, and obtained output measurements by taking the average of 5 episodes. All the agents and the target initially positioned randomly aligned with the given parameter settings in Table 1.

Table 1. Parameters and settings used in the simulator

Parameter Type	Parameter	Symbol	Value
Parameters for the simulator	Map size	map_size	{10, 15, 20, 25}
	Agent count	nagent	{2, 5, 10, 20}
	No of targets	ntarget	1
	Radius of visibility	agent_vision	3
	Learning rate	lr	0.0001
	Gamma	gamma	0.95
	Epsilon	ϵ	0.95
	Epsilon decay	ϵ_{dec}	1e-4
	Epsilon lower bound	ϵ_{end}	0.01
	Importance weights	α	[0, 1]
PPO Parameters	Network	I/P channels	O/P channels
	CNN 1	3	32
	CNN 2	32	64
	Linear 1	147	512
	Linear 2	512	5

4.2   Importance Weight Analysis

In our first experiment, we undertake a comparison of the proposed reward sharing techniques - Performance Incentive (PI), Observer's Share(OS), and Synergy

Achievement (SA) - under diverse α values to gauge their performance. Figure 2 illustrates the average reward observed over 200 epochs for each reward sharing technique, enabling an insightful examination of the impact of importance weight on individual techniques. Our comprehensive analysis reveals a clear negative correlation between the importance weight and the average reward. Furthermore, Pearson's correlation coefficient demonstrates a strong negative correlation, $r(9) = -0.97, p < .01$, indicating a consistent rise in the average reward as the importance weight (α) value decreases.

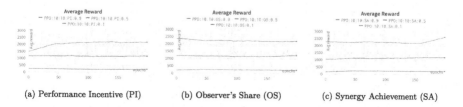

(a) Performance Incentive (PI)   (b) Observer's Share (OS)   (c) Synergy Achievement (SA)

Fig. 2. (a) The average PI reward value, (b) The average OS reward value, (c) The average SA reward value, under varying importance weights {$\alpha = 0.1$ (top), $\alpha = 0.5$ (middle), $\alpha = 0.9$ (bottom)}. The legend description, for example **PPO:10:10:PI:0.5** corresponds to **algorithm:map_size:No_of_agents:reward_sharing_technique: importance_weight**

Figure 3 shows a heat map of the rewards gained by each of the reward sharing techniques at the end of the last epoch. Observation shows that as importance weight increases, reward value decreases. i.e. with the increment of the importance weight value, the global objective may overpower the local objective. This shift in focus might cause agents to sacrifice individual rewards, leading to a decrease in overall rewards.

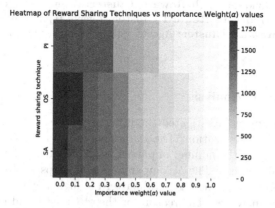

Fig. 3. Heat map indicating the impact of reward sharing techniques (PI, OS, and SA) and importance weights (α values) on reward at the end of final epoch, revealing a negative correlation between importance weight and reward value.

We employed the KMeans clustering algorithm to investigate potential reward distribution patterns within each reward sharing technique. We successfully identified three distinct clusters based on the reward distribution as shown in Fig. 4. We incorporated this information to better target our rewards and optimize our reward system. Table 2 illustrates the corresponding importance weights for each reward cluster.

Table 2. Specification of reward clusters

Reward Cluster	Parameter	α values
High Reward Cluster	HRC	{0.0, 0.1, 0.2}
Moderate Reward Cluster	MRC	{0.3, 0.4, 0.5, 0.6}
Low Reward Cluster	LRC	{0.7, 0.8, 0.9, 1.0}

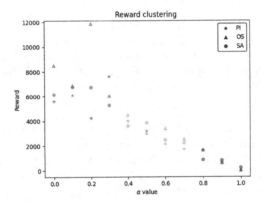

Fig. 4. Cluster analysis of rewards with varying importance weight (α) values for three reward sharing techniques. **High Reward Cluster:** Agent will prioritize their individual objective, **Moderate Reward Cluster:** Agents try to balance between two objectives, **Low Reward Cluster:** Agents will give a higher priority to the collaborate objective

4.3   Agent Count Analysis

Our second experiment investigated the influence of the number of agents participating in target localization. The data presented in Fig. 5 demonstrates a noticeable influence of the number of participating agents on the average reward achieved after 200 epochs. The accompanying heat maps (a, b, and c) visually represent this trend, revealing a general pattern of higher average rewards as the number of agents increases. The reason for the observed trend might be due to enhanced cooperation and coordination among agents. With more agents working together, they can share information, and resources more effectively, leading to better performance in achieving their objectives and, ultimately, higher average rewards. Additionally, having more agents can potentially improve the exploration and exploitation of the environment, leading to better learning and decision-making capabilities.

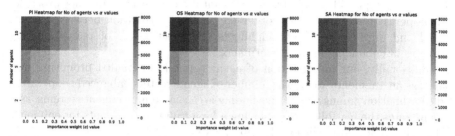

(a) Performance Incentive (PI) based reward.   (b) Observer's Share (OS) based reward.   (c) Synergy Achievement (SA) based reward.

Fig. 5. Heat maps indicating the impact of agent count (2, 5, and 10) and importance weights (α values) on each reward sharing technique.

4.4   Map Size Analysis

The final experiment examined the impact of map size on target localization. Specifically, we investigated a scenario with 10 agents assigning equal priority to both objectives ($\alpha = 0.5$) as well as with the other default parameters mentioned in Table 1. Analyzing Fig. 6, we observed a negative correlation between map size and agents' rewards. The observed decrease in rewards can be attributed to the expanded search space resulting from the map size increase. This expansion results in a more complex environment, making it more difficult for agents to locate the target accurately.

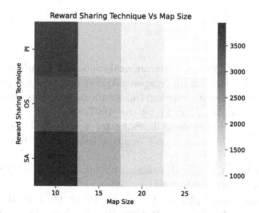

Fig. 6. Heat map indicating the impact of reward sharing techniques (PI, OS, and SA) and map size (10, 15, 20, and 25) on average reward when $\alpha = 0.5$.

5   Conclusions

The comparison between the proposed reward sharing techniques (PI, OS, and SA) helps us understand the behavior of individual agents and the collective group and provides valuable insights into agent decision-making processes and cooperation dynamics. The observed negative correlation between average reward and importance weight highlights the effectiveness of reward sharing strategies in enhancing agent performance.

Additionally, our investigation of agent count has revealed promising outcomes, as an augmentation in the number of agents fosters enhanced cooperation and coordination among them. This finding is crucial in multi-agent systems, as it demonstrates the feasibility of deploying a larger number of agents for enhanced performance and more efficient decision-making.

Furthermore, the exploration into map size has yielded valuable insights into its impact on agent rewards. The identified negative correlation between map size and agents' rewards suggests that the expanded search space resulting from larger maps may challenge agents' performance. This finding underscores the need for careful consideration when scaling up map size and optimizing agent strategies.

Overall, our research highlights a comprehensive investigation and analysis of multiple aspects of multi-agent reinforcement learning applied to dynamic target localization. The combination of time series plots, heat maps, and K-means clustering has proven effective in visualizing and analyzing experimental results.

In conclusion, our study emphasizes the significance of cooperation enhancement techniques, agent count and map size analysis, and effective visualization methods in addressing challenges related to multi-agent target localization. The obtained results provide a solid foundation for future research in this domain and can aid in the design of more efficient and adaptive multi-agent systems for a wide range of applications.

References

1. Yang, J., Borovikov, I., Zha, H.: Hierarchical cooperative multi-agent reinforcement learning with skill discovery. In: Adaptive Agents and Multi-Agent Systems (2019)
2. Multi-agent Reinforcement Learning: Independent vs. Cooperative Agents. Morgan Kaufmann Publishers Inc., San Francisco (1997)
3. Hu, Z., Zhao, D.: Reinforcement learning for multi-agent patrol policy. In: 9th IEEE International Conference on Cognitive Informatics (ICCI'10), pp. 530–535 (2010)
4. Claus, C., Boutilier, C.: The dynamics of reinforcement learning in cooperative multiagent systems. In: AAAI/IAAI (1998)
5. Rashid, T., Samvelyan, M., De Witt, C.S., Farquhar, G., Foerster, J., Whiteson, S.: QMIX: monotonic value function factorisation for deep multi-agent reinforcement learning (2018). ArXiv, abs/1803.11485

6. Marzari, L., Pore, A., Dall'Alba, D., Aragon-Camarasa, G., Farinelli, A., Fiorini, P.: Towards hierarchical task decomposition using deep reinforcement learning for pick and place subtasks. In: 2021 20th International Conference on Advanced Robotics (ICAR), pp. 640–645 (2021)
7. Chaimowicz, L., Campos, M.F., Kumar, V.: Dynamic role assignment for cooperative robots. In: Proceedings 2002 IEEE International Conference on Robotics and Automation (Cat. No.02CH37292), vol. 1, pp. 293–298 (2002)
8. Foerster, J.N., Assael, Y., De Freitas, N., Whiteson, S.: Learning to communicate with deep multi-agent reinforcement learning (2016). ArXiv, abs/1605.06676
9. Gupta, J.K., Egorov, M., Kochenderfer, M.: Cooperative multi-agent control using deep reinforcement learning. In: Sukthankar, G., Rodriguez-Aguilar, J.A. (eds.) AAMAS 2017. LNCS (LNAI), vol. 10642, pp. 66–83. Springer, Cham (2017). https://doi.org/10.1007/978-3-319-71682-4_5
10. Mnih, V., et al.: Playing atari with deep reinforcement learning (2013). ArXiv, abs/1312.5602
11. Ng, A.Y., Harada, D., Russell, S.: Policy invariance under reward transformations: theory and application to reward shaping. In: International Conference on Machine Learning (1999)
12. Wiewiora, E., Cottrell, G.W., Elkan, C.: Principled methods for advising reinforcement learning agents. In: Proceedings of the Twentieth International Conference on International Conference on Machine Learning, ICML'03, pp. 792–799. AAAI Press (2003)
13. Mannion, P., Devlin, S., Mason, K., Duggan, J., Howley, E.: Policy invariance under reward transformations for multi-objective reinforcement learning. Neurocomputing **263**, 60–73 (2017)
14. Mannion, P., Devlin, S., Duggan, J., Howley, E.: Reward shaping for knowledge-based multi-objective multi-agent reinforcement learning. Knowl. Eng. Rev. **33**, e23 (2018). https://doi.org/10.1017/S0269888918000292. Cambridge University Press
15. Grześ, M., Kudenko, D.: Multigrid reinforcement learning with reward shaping. In: Kurková, V., Neruda, R., Koutník, J. (eds.) ICANN 2008. LNCS, vol. 5163, pp. 357–366. Springer, Heidelberg (2008). https://doi.org/10.1007/978-3-540-87536-9_37
16. Grzes, M., Kudenko, D.: Reinforcement learning with reward shaping and mixed resolution function approximation. Int. J. Agent Technol. Syst. **1**, 36–54 (2009)
17. Ferreira, E., Lefèvre, F.: Reinforcement-learning based dialogue system for human-robot interactions with socially-inspired rewards. Comput. Speech Lang. **34**, 256–274 (2015)
18. Devlin, S., Yliniemi, L., Kudenko, D., Tumer, K.: Potential-based difference rewards for multiagent reinforcement learning. In: Adaptive Agents and Multi-Agent Systems (2014)
19. Kim, D., et al.: Learning to schedule communication in multi-agent reinforcement learning (2019). ArXiv, abs/1902.01554
20. Hostallero, D.E., Kim, D., Moon, S., Son, K., Kang, W.J., Yi, Y.: Inducing cooperation through reward reshaping based on peer evaluations in deep multi-agent reinforcement learning. In: AAMAS (2020)
21. Co-Reyes, J.D., Sanjeev, S., Berseth, G., Gupta, A., Levine, S.: Ecological reinforcement learning (2020). ArXiv, abs/2006.12478
22. Huang, B., Jin, Y.: Reward shaping in multiagent reinforcement learning for self-organizing systems in assembly tasks. Adv. Eng. Inform. **54**, 101800 (2022)

23. Konidaris, G.D., Barto, A.G.: Autonomous shaping: knowledge transfer in reinforcement learning. In: Proceedings of the 23rd International Conference on Machine Learning (2006)

24. Rouček, T., et al.: DARPA subterranean challenge: multi-robotic exploration of underground environments. In: Mazal, J., Fagiolini, A., Vasik, P. (eds.) MESAS 2019. LNCS, vol. 11995, pp. 274–290. Springer, Cham (2020). https://doi.org/10.1007/978-3-030-43890-6_22

25. Stone, P., Veloso, M.: Multiagent systems: a survey from a machine learning perspective (2000)

26. Chen, X., Ghadirzadeh, A., Björkman, M., Jensfelt, P.: Meta-learning for multi-objective reinforcement learning (2018)

27. Deep reinforcement learning framework for autonomous driving. Electron. Imaging **2017**(19), 70–76 (2017)

28. Schulman, J., Wolski, F., Dhariwal, P., Radford, A., Klimov, O.: Proximal policy optimization algorithms (2017). ArXiv, abs/1707.06347

29. Schulman, J., Levine, S., Abbeel, P., Jordan, M., Moritz, P.: Trust region policy optimization (2015). ArXiv, abs/1502.05477

Competitive Collaboration for Complex Task Learning in Agent Systems

Dilini Samarasinghe$^{(\boxtimes)}$ [iD], Michael Barlow [iD], and Erandi Lakshika [iD]

University of New South Wales, Canberra, ACT, Australia
{d.samarasinghe,m.barlow,e.henekankanamge}@adfa.edu.au

Abstract. This paper presents a novel competitive collaboration-based reinforcement learning strategy to improve the performance of goal-oriented autonomous agent systems. Competitive characteristics are introduced while agents are required to collaboratively share and update a single knowledge pool. Experimental evaluations are conducted in a reward collection task where the goal is to collect as many reward points as possible across a set space navigating through multiple channels while avoiding costs associated with channel switching. Evaluations across 50 challenge levels of the task in comparison to state-of-the-art reinforcement learning models reveal that introducing a combination of competition and collaboration attributes facilitate statistically significant performance improvements. The compared incremental models carry forward knowledge from previous challenge levels to learn more complex challenges. However, they resort to sub-optimal solutions where cost minimisation is given priority over higher reward collection. Traditional reinforcement learning is generally incapable of identifying a balance between reward collection and cost minimisation leading to poor performance. In contrast, the proposed competitive collaboration-based model is capable of strategically planning the best route while taking appropriate risks with higher costs for better rewards in the long run. Further, investigations with multiple reward schemes illustrate that frequent rewards to guide the agents' strategies as opposed to delayed rewards that are awarded at less frequent intervals are imperative for agent performance. The presented evaluations lend insights into future research for optimising agent behaviours in complex real-world applications with competitive collaboration.

Keywords: Competitive Collaboration · Reinforcement Learning · Agent Systems · Complex Task Learning

1 Introduction

Autonomous agent systems is a widely explored sub-field of Artificial Intelligence (AI) that has received attention in modeling and solving complex real-world problems. Reward collection [4], planning and coordination [16], and pathfinding [1] are a few among the decision making tasks that autonomous agents are

T. Liu et al. (Eds.): AI 2023, LNAI 14472, pp. 325–337, 2024.
https://doi.org/10.1007/978-981-99-8391-9_26

often used for. In addressing such tasks, multi-agent models are used as opposed to single agent systems when interactions between agents can be exploited to improve efficiency and robustness [6]. Such challenges are commonly addressed with collaboration-based [12] or competition-based [21] agent modeling, primarily based on the nature of their goals.

However, for systems with a single goal and no apparent requirement for distributed workload or multi-agent interactions, it is often assumed that single agent systems are invariably the best option [14]. Little attention is paid in such cases to understand if characteristics of competition can be introduced to collaboratively achieve the goal with improved performance. This paper seeks to address this gap by investigating the potential of combining the two approaches, competition and collaboration, for extracting information off competing agents and incorporating the knowledge for collaborative goal achievement. We present a novel competitive collaboration strategy for complex task learning for agents in a reinforcement learning (RL) environment with following contributions:

- A novel competitive collaboration-based RL (CCRL) algorithm is proposed to enhance the performance and learning of agents in complex and dynamic environments.
- Simulations are conducted in a reward collection task environment under increasingly complex conditions to establish the capacity of the model.
- The impact of utilising different reward schemes is investigated with the proposed RL model in terms of performance and computational time.
- The model is evaluated in comparison to a state-of-the-art Flow-based RL model, an incremental RL model, and a traditional RL model.

The rest of the paper is organised as follows. Section 2 summarises the existing literature on competition and collaboration-based learning approaches. The reward collection task used for evaluations and the proposed CCRL model are presented in Sect. 3. The simulation environment, experimental evaluations, and results are illustrated in Sect. 4. Finally, Sect. 5 presents a discussion of the results and concludes the paper with possible future directions.

2   Related Work

Collaboration-based learning is explored in areas where the goals and/or tasks can be sub-divided and distributed across multiple agents. It is often used to achieve coordination across multiple agents and leverage their experience to adapt to individual tasks targeting a common goal [19]. Further, human-robot collaboration (HRC) [10] is a field of study where interactions between humans and robots are utilised for pursuing a shared goal by sharing a workspace and carrying out tasks in parallel. Collaborative search is also used in multi-objective optimisation, where a population of agents is used to tackle heuristics aimed at exploring the search space simultaneously or, where sharing knowledge for optimisation is performed via leader selection methods to select the preferable solution [7].

Conversely, competitive decision making is generally viewed in a non-cooperative game theory perspective [3]. The competing agents try to achieve their individual goals maximising individual payoffs to achieve Nash equilibrium. Generally, all agents either share common knowledge and are aware of the resources and choice set of each individual; or are required to use decision analytic approaches to forecast the actions of competing agents and their outcomes [3].

While each of these approaches have their own benefits, only a limited number of research avenues have considered the union of competition and collaboration to improve system performance with artificial agents. Jiang et al. [5] investigate a novel group consensus protocol designed utilising both cooperative and competitive interactions among agents for heterogeneous agent systems. Consensus is achieved by implementing sub groups where agents within the same sub group cooperate, while different sub groups compete with each other. Zhang et al. [20] presents a model for dynamic collaboration recommendations through a multi-agent reinforcement learning model with a competition function. Multiple agents (authors) simultaneously seek for collaborations introducing competition into the system while also enhancing performance. A multi-agent coordinated deep reinforcement learning (RL) is proposed by Ding et al. [2] to solve the non-convex economic dispatch problem. A distributed coordinated approach is proposed where agents run independent RL algorithms that use a joint reward. A double network structure with competing reward and target networks are used to achieve coordination among agents.

As can be observed from the existing literature, the current research avenues on competitive collaboration are preliminary focused on domains where multiple or divisible goal(s) and/or heterogeneous interactions are anticipated. The potential of introducing competitive collaboration characteristics to systems with a single goal and non-distributive workload has not been adequately investigated. This paper contributes in expanding the understanding across the use of competitive collaboration in domains where multiple goals or heterogeneous interactions are not expected.

3   Proposed Competitive Collaboration-Based RL Model

Reward Collection Task. The reward collection task utilised to investigate the model consists of 52 channels (rows) and each channel consists of 100 cells. Each cell is associated with a reward value and at a given time tick, the agent can move to the adjacent cell of any available channel. A cost will be incurred for switching channels which is calculated based on the distance between the current channel and the new channel being moved to ($\sqrt{|current\ channel - new\ channel|}$). The goal is for the agent to collect as much rewards (rewards from cells - costs of channel switching) as possible by identifying the best possible path to move through the channels across 100 cells.

To investigate how the model behaves with increasing complexity, the evaluations are conducted across multiple task variations. The task commences with only 2 channels at the first complexity level, and then increases in complexity until 52 channels are introduced to the system. To ensure a consistent increase in complexity across the increments, each new channel being added has at least one cell which has a higher reward compared to the rewards at the same cell position of all previous channels. The total reward for all cells in a single channel is maintained within the range of [1000–1500] and the rewards along a single channel are incremented in a sinusoidal stepwise format.

Competitive Collaboration-Based RL Model. The proposed CCRL model implemented in the RL environment is as illustrated in Algorithm 1. Two agents compete in achieving the same goal and the agent with the most rewards at the end of the task wins a tournament. A Q-learning algorithm [18] is used in learning where agents act based on consequences of their actions without an a priori model of the domain. Both agents navigate the environment simultaneously competing for maximum reward collection and reference a single Q-table for optimal policy construction. The competition lies in the vision to collect more rewards compared to the opponent, while both collaboratively contribute towards building the shared knowledge pool through a single Q-table.

The CCRL model starts by initialising the Q-table and the two competing agents (lines 2–3). The exploration versus exploitation tradeoff with action selection is balanced with the use of decaying epsilon-greedy Q-learning [15]. Both agents simultaneously start the learning process and as the learning improves, more chance is given to exploitation of the most rewarding actions by reducing the decay constant ε (lines 6–13). The Q-table is updated successively by each agent based on their learning process and every state, action pair (s, a) of the solution for each agent is recorded separately (lines 14–21).

The termination criterion is designed based on the novelty of the solutions. All state, action pairs of the solution identified by each agent ($\rho_\theta(S', A')$ and $\rho_\vartheta(S', A')$) are fed separately to a novelty calculation function to determine the novelty of the solution compared to that of the previous episode (lines 23–32). Equation 1 [9] is adopted to calculate this value; the more an (s, a) pair is visited over the episodes, the less novel the solution becomes. To determine the descent rate of novelty, the decay constants $0 < \lambda_1, \lambda_2 < 1$ are used. The average novelty across all pairs is used as the novelty of the entire solution derived by each competing agent. The values range within [−1, 0], where 0 corresponds to a solution that is completely novel while a solution with −1 has no novelty. If at least one agent reaches the threshold −1, it suggests that the learning process has reached the end stage as the agent no longer has the capacity to improve the solution, and therefore, the process is terminated (line 4).

$$b_t(s, a) = (\lambda_1^{n_t(s)} + \lambda_2^{n_t(s,a)})/2 - 1 \qquad (1)$$

To further investigate the impact of different reward structures on the proposed CCRL model, three reward schemes were tested: accumulated reward; immediate reward; and delayed reward. With the accumulated reward scheme, the total rewards (rewards from cells - costs of channel switching) up to the

Algorithm 1. Competitive Collaboration-based RL Model

Require: θ : Competitor agent 1

           ϑ : Competitor agent 2

           $\rho_\theta(S', A')$: The state action pairs in the solutions derived for agent θ

           $\rho_\vartheta(S', A')$: The state action pairs in the solutions derived for agent ϑ

           φ : Novelty threshold

           α : Learning rate

           γ : Discount factor

           ε : Decay constant

           $Q(S, A)$: Q table for all state action pairs

           R : Reward for each state

           s : Current state

           a : Current action

           s' : New state

           t : Termination condition

1: **procedure** CCRL
2:      INITIALISE Q(S, A)
3:      INITIALISE θ, ϑ
4:      **while** !($N_\theta <= \varphi \parallel N_\vartheta <= \varphi$) **do**
5:          **for** EACH EPISODE t **do**
6:               **for** EACH AGENT θ AND ϑ **do**
7:                   INITIALISE STATE s
8:                   **while** s is not terminal **do**
9:                       $\tau \leftarrow$ RND(0,1)
10:                      **if** $\tau < \varepsilon$ **then**
11:                         $a \leftarrow$ RANDOM ACTION FROM A
12:                      **else**
13:                         $a \leftarrow$ MAX $Q(s)$
14:                      **while** opponent agent is updating Q **do**
15:                         WAIT
16:                      $Q(s,a) \leftarrow Q(s,a) + \alpha[R + \gamma \max Q(s', A) - Q(s, a)]$
17:                      **if** θ **then**
18:                         ADD (s,a) TO ρ_θ
19:                      **if** ϑ **then**
20:                         ADD (s,a) TO ρ_ϑ
21:                      $s \leftarrow s'$
22:               $\varepsilon \leftarrow$ UPDATE(ε)
23:               **if** θ **then**
24:                   $n_t(\theta) \leftarrow 0$
25:                   **for** ALL $\rho_\theta(S', A')$ **do**
26:                       $n_t \leftarrow n_t(\theta) +$ NOVELTY CALCULATION $\rho_\theta(S', A')$
27:                   $N_\theta \leftarrow n_t(\theta)$ / total pairs in ρ_θ
28:               **if** ϑ **then**
29:                   $n_t(\vartheta) \leftarrow 0$
30:                   **for** ALL $\rho_\vartheta(S', A')$ **do**
31:                       $n_t \leftarrow n_t(\vartheta) +$ NOVELTY CALCULATION $\rho_\vartheta(S', A')$
32:                   $N_\vartheta \leftarrow n_t(\vartheta)$ / total pairs in ρ_ϑ

current time step t from the start is calculated and assigned to the action as illustrated in Eq. 2. A reward is awarded at every action. The immediate reward scheme is as given in Eq. 3, where a reward is awarded at every action, however, only the reward at the current time step t (reward from the current cell - cost of channel switching if a switch occurred) is assigned to the action. The delayed reward scheme rewards the agent only at the end of completing a single run. The total rewards (rewards from cells - costs of channel switching) collected after moving through all the steps as illustrated in Eq. 4 is assigned to the action. No rewards are awarded at intermediate steps.

$$R_{accumulated} = \sum_{i=1}^{t}(cell\_reward_i - switch\_cost_i) \qquad (2)$$

$$R_{immediate} = (cell\_reward_t - switch\_cost_t) \qquad (3)$$

$$R_{delayed} = \sum_{i=1}^{t\_final} (cell\_reward_i - switch\_cost_i) \qquad (4)$$

4   Experimental Evaluations

4.1   Compared Models

Comparisons are conducted with a traditional RL model, where the agent independently learns each challenge to determine the best solution; an incremental learning model, where the knowledge gathered from one challenge level is carried forward when the agent starts learning the subsequent challenge level; and a Flow-based RL model [11], which is a recent and improved learning model where the agent's knowledge is brought forward to subsequent challenge levels ensuring that the agent's skills and challenge complexity are maintained in balance (in a flow-zone) throughout the process.

The attributes of the RL model are as presented in Table 1. Greedy learning was used for all compared models for a fair comparison. The values for the decay constants ε, λ_1, and λ_2 were decided after a sensitivity analysis with the CCRL model using accumulated rewards for ten values each in the range of [0,1]. All 3 reward schemes were tested with CCRL, and accumulated reward scheme was used for the compared models, since this showed the best performance with CCRL as discussed in Sect. 4.2.

4.2   Results

The experiments with all models were repeated for 20 different seeds each and the aggregated result was used for the comparisons. Two-sample t-test was used for comparisons of two groups, whereas one-way ANOVA was used for comparisons across more than two groups. The statistical significance level is set at $p = 0.05$.

Table 1. Attributes for the setup of CCRL model

Attribute	Value
Novelty Threshold (φ)	-1
Decay Constant (ε) Initial: for epsilon-greedy q-learning	0.4
Decay Constant (ε) Update	$\varepsilon * 0.9$
Decay Constant (λ_1)	0.9
Decay Constant (λ_2)	0.5

Figure 1 illustrates the results for the analysis across 50 challenge levels for all three reward schemes for CCRL: accumulated, immediate, delayed; Flow-based RL; incremental RL, and traditional RL models. Total rewards collected by each approach across the challenge levels is given in Fig. 1a, and total time taken (determined by the total number of state, action pairs visited across all episodes) for learning across the challenge levels is given in Fig. 1b.

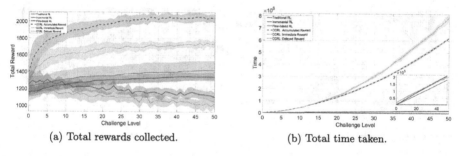

(a) Total rewards collected.   (b) Total time taken.

Fig. 1. Analysis of the reward collection or each challenge level. The experimental results are averaged across 50 runs each and the shaded areas depict the standard deviation.

Results show that the CCRL model with the accumulated reward scheme achieves the best performance with a significant improvement over the other models ($p < 0.05$). The analysis considers the reward collected by the agent that won the competition in all three CCRL models. Starting from an average of 1400 total reward points in the initial challenge levels, CCRL-A (accumulated rewards) collects over 2000 points by the 50th level. Therefore, as the complexity of the task increases, the CCRL model is still capable of identifying the best route while minimising the associated switching costs. CCRL-I (immediate rewards) follows next with the ability to collect more than 1700 total reward points by the 50th level, whereas Flow-based and incremental learning models are in the range of 1200–1400 points on average. Even with the advantage of previously learnt knowledge being transferred to the next complexity levels during training, the Flow and incremental models are not capable of surpassing the two CCRL

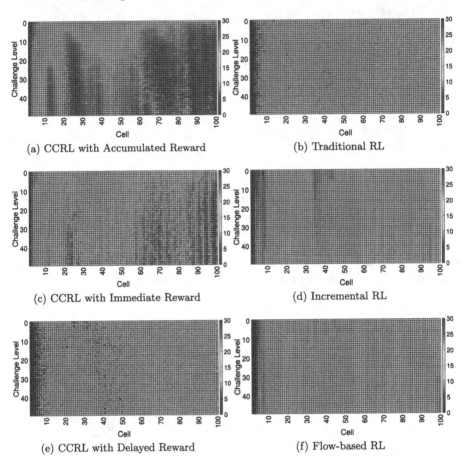

Fig. 2. Individual rewards reaped at each cell across 100 steps by the 3 CCRL models, for 50 challenge levels. The results are averaged across 20 runs each.

models. Both the traditional RL model and the CCRL-D (delayed rewards) perform poorly compared to the others and show a decrease in the rewards they capture across the challenge levels. This shows that they are not capable of learning to avoid the channel switching costs which become more significant as the complexity of the task increases. The results suggest that rewarding the agent only at the end of the task is not sufficient even with the introduction of competitive collaboration characteristics into the system.

As CCRL use two competing agents, the time values illustrated in Fig. 1b for all three CCRL models show the combined time taken by both agents for a fair evaluation. The cumulative time to learn all challenge levels up-to each specific challenge is considered for both incremental and Flow models as they carry forward their previous knowledge. The traditional model considers the time taken to learn each specific challenge independently. The comparison results

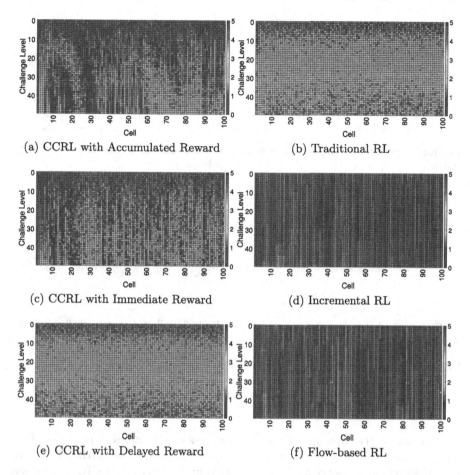

Fig. 3. Individual costs incurred at each cell across 100 steps by the three CCRL models, for all 50 challenge levels. The results are averaged across 20 runs each.

show that the significant performance improvements of CCRL with accumulated and immediate rewards come at a cost. CCRL-I take the longest time to train followed closely by CCRL-A. This can be attributed to the time taken to compete and learn to avoid channel switching costs while collecting rewards. Considering CCRL-A yielded significantly better results and takes less time than CCRL-I, it stands out as the best performing model. CCRL-D uses a similar computational time cost as the incremental learner, although the performance results are worse. Traditional model takes the least time to train, however, is not intelligent enough to learn the balance between cost avoidance and reward collection as discussed.

Figures 2 and 3 illustrate the distribution of rewards collected and costs incurred at each of the 100 cell steps across the 50 challenge levels, respectively. CCRL-A collects very high rewards compared to the other models throughout the 100 steps. It is more evident as the complexity of the task increases. Figure 3a

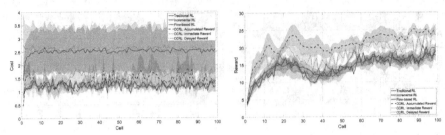

(a) Cost incurred at each cell across all challenge levels.

(b) Rewards collected at each cell across all challenge levels.

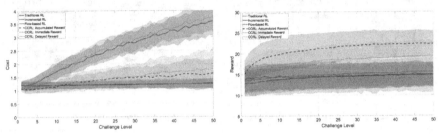

(c) Cost incurred at each challenge level across all cells.

(d) Rewards collected at each challenge level across all cells.

Fig. 4. Total cost and rewards at each cell and during each challenge level for the three CCRL models, and traditional, incremental, and Flow-based RL models. Results are averaged across 20 runs each and the shaded areas depict the standard deviation.

illustrates that the costs also relatively increase as the complexity increases. This suggests that CCRL-A model is capable of developing a policy that learns to take risks when necessary for higher rewards in the long run. CCRL-I follows a similar approach, however, it is not capable of collecting as higher rewards as CCRL-A. Both Flow and incremental learning models follow a safer approach to collect rewards. As seen from Figs. 3d and 3f, the costs are quite low for the two models which suggest that they limit their movement to only the nearby channels at a time without risking drastic channel changes which incur higher costs. The knowledge gathered from previous challenge levels may encourage this behaviour. However, this also limits their capacity to achieve higher rewards that could have resulted from switches to channels further apart. Conversely, both traditional and CCRL-D are not capable of developing an intelligent policy to balance the rewards and costs resulting in drastic channel switches with very high costs as depicted in Figs. 3e and 3b without any long or short term benefits of higher reward points.

These observations are confirmed by the cost and reward analysis presented in Fig. 4. According to Figs. 4a and 4c, both traditional and CCRL-D incur a higher cost (2.5 on average) across all cells and it increases with challenge levels, proving that they cannot balance rewards versus costs. All other models incur a cost in the range of 1–1.5 on average which are not significantly different ($p > 0.05$). However, the average cost for both CCRL-A and CCRL-I slightly increase over the challenge levels. These two models can adjust their policies to maintain the same cost range across cells, but distribute this weight across the challenge levels such that higher costs are incurred in higher complexity levels which also lead to better rewards in the long run. The competitive collaboration characteristics introduced are further capable of improving the potential of the agents to collect higher rewards as evident from Figs. 4b and 4d.

5   Conclusion and Future Work

This paper introduced a novel competitive collaboration-based RL model for learning complex goal oriented tasks. Most existing approaches that use competition and collaboration for performance optimisation are focused on multiple distributable objectives and/or heterogeneous interactions [17]. The experimental evaluations presented here explore competitive collaboration for tasks that have homogeneous agent interactions. In comparison to a state-of-the-art Flow-based model, an incremental learning model, and a traditional RL model, the proposed CCRL model performs significantly better.

Both Flow-based and incremental models displayed the adoption of a sub-optimal strategy which is to navigate the environment by moving on to successive channels reducing the associated costs when collecting rewards. This strategy, while better than the traditional RL model, still hinders the full potential of the agent for collection of more rewards. In contrast, CCRL-A and CCRL-I were more open to risk taking and managed to find policies that strategically make drastic channel switches targeting higher rewards in the long run. The competition between the two agents drives the behaviours towards exploring riskier strategies to beat the opponent, while the collaboratively updated common knowledge pool further facilitates identification of high-rewarding strategies. However, CCRL-D performs even poorly than the traditional model which lends evidence to support the significance of the delayed reward problem [13] that is persistent in the RL domain. This signifies that despite the learning strategy adopted, the reward mechanism in a RL training environment is pertinent for its performance.

Given the potential of CCRL-A and CCRL-I in significantly surpassing state-of-the-art RL models, it opens new research avenues to investigate its use in more dynamic problems. Evaluations on the impact of increasing the number of competitors on performance and computational resource requirement would be interesting to determine the model's capacity. Further, the CCRL approach evaluated here does not carry forward the knowledge from previous levels to learn more complex challenges. It would be interesting to combine CCRL with a model such as Flow or incremental learning to understand how the agents can better utilise this knowledge for future task learning. Finally, the proposed CCRL model was applied on a Q-learning strategy which can be limiting when the size of the learning environment increases, requiring more computational time [8]. It can be incorporated into more advanced learning models which may yield interesting insights into performance in complex environments.

References

1. Cruz, D.L., Yu, W.: Path planning of multi-agent systems in unknown environment with neural kernel smoothing and reinforcement learning. Neurocomputing **233**, 34–42 (2017)
2. Ding, L., Lin, Z., Shi, X., Yan, G.: Target-value-competition-based multi-agent deep reinforcement learning algorithm for distributed nonconvex economic dispatch. IEEE Trans. Power Syst. 1 (2022). https://doi.org/10.1109/TPWRS.2022.3159825
3. Esteban, P.G., Insua, D.R.: Supporting an autonomous social agent within a competitive environment. Cybern. Syst. **45**(3), 241–253 (2014)
4. Goldberg, D., Matarić, M.J.: Maximizing reward in a non-stationary mobile robot environment. Auton. Agent. Multi-Agent Syst. **6**, 287–316 (2003)
5. Jiang, Y., Ji, L., Liu, Q., Yang, S., Liao, X.: Couple-group consensus for discrete-time heterogeneous multiagent systems with cooperative-competitive interactions and time delays. Neurocomputing **319**, 92–101 (2018). https://doi.org/10.1016/j.neucom.2018.08.048
6. Jin, D., Kannengießer, N., Sturm, B., Sunyaev, A.: Tackling challenges of robustness measures for autonomous agent collaboration in open multi-agent systems. In: HICSS, pp. 1–10 (2022)
7. Kouka, N., BenSaid, F., Fdhila, R., Fourati, R., Hussain, A., Alimi, A.M.: A novel approach of many-objective particle swarm optimization with cooperative agents based on an inverted generational distance indicator. Inf. Sci. **623**, 220–241 (2023)
8. Low, E.S., Ong, P., Cheah, K.C.: Solving the optimal path planning of a mobile robot using improved Q-learning. Robot. Auton. Syst. **115**, 143–161 (2019)
9. Lu, C.X., Sun, Z.Y., Shi, Z.Z., Cao, B.X.: Using emotions as intrinsic motivation to accelerate classic reinforcement learning. In: 2016 International Conference on Information System and Artificial Intelligence (ISAI), pp. 332–337 (2016). https://doi.org/10.1109/ISAI.2016.0077
10. Mukherjee, D., Gupta, K., Chang, L.H., Najjaran, H.: A survey of robot learning strategies for human-robot collaboration in industrial settings. Robot. Comput.-Integr. Manuf. **73**, 102231 (2022). https://doi.org/10.1016/j.rcim.2021.102231
11. Samarasinghe, D., Barlow, M., Lakshika, E.: Flow-based reinforcement learning. IEEE Access **10**, 102247–102265 (2022). https://doi.org/10.1109/ACCESS.2022.3209260
12. Samarasinghe, D., Barlow, M., Lakshika, E., Kasmarik, K.: Grammar-based cooperative learning for evolving collective behaviours in multi-agent systems. Swarm Evol. Comput. **69**, 101017 (2022). https://doi.org/10.1016/j.swevo.2021.101017
13. Shen, S., Chi, M.: Reinforcement learning: the sooner the better, or the later the better? In: Proceedings of the 2016 Conference on User Modeling Adaptation and Personalization, pp. 37–44. UMAP '16, Association for Computing Machinery, New York, NY, USA (2016)
14. Shen, Z., Miao, C., Tao, X., Gay, R.: Goal oriented modeling for intelligent software agents. In: Proceedings. IEEE/WIC/ACM International Conference on Intelligent Agent Technology, 2004. (IAT 2004), pp. 540–543 (2004). https://doi.org/10.1109/IAT.2004.1343014

15. Sutton, R.S., Barto, A.G.: Reinforcement Learning: An Introduction. MIT Press, Cambridge (2018)
16. Tadewos, T.G., Shamgah, L., Karimoddini, A.: Automatic decentralized behavior tree synthesis and execution for coordination of intelligent vehicles. Knowl.-Based Syst. **260**, 110181 (2023). https://doi.org/10.1016/j.knosys.2022.110181
17. Wang, J., et al.: Cooperative and competitive multi-agent systems: from optimization to games. IEEE/CAA J. Autom. Sin. **9**(5), 763–783 (2022). https://doi.org/10.1109/JAS.2022.105506
18. Watkins, C.J., Dayan, P.: Q-learning. Mach. Learn. **8**, 279–292 (1992)
19. Zeng, S., Chen, T., Garcia, A., Hong, M.: Learning to coordinate in multi-agent systems: a coordinated actor-critic algorithm and finite-time guarantees. In: Firoozi, R., et al. (eds.) Proceedings of The 4th Annual Learning for Dynamics and Control Conference. Proceedings of Machine Learning Research, vol. 168, pp. 278–290. PMLR, 23–24 June 2022
20. Zhang, Y., Zhang, C., Liu, X.: Dynamic scholarly collaborator recommendation via competitive multi-agent reinforcement learning. In: Proceedings of the Eleventh ACM Conference on Recommender Systems, pp. 331–335. RecSys '17, Association for Computing Machinery, New York, NY, USA (2017). https://doi.org/10.1145/3109859.3109914
21. Zhou, Z., Xu, H.: Mean field game and decentralized intelligent adaptive pursuit evasion strategy for massive multi-agent system under uncertain environment. In: 2020 American Control Conference (ACC), pp. 5382–5387 (2020)

Limiting Inequalities in
Repeated House and Task Allocation

Martin Aleksandrov[(✉)][iD]

Freie Universität Berlin, Berlin, Germany
martin.aleksandrov@fu-berlin.de

Abstract. We consider house and task allocation markets over multiple rounds, where agents have endowments and valuations at each given round. The endowments encode allocations of agents at the previous rounds. The valuations encode the preferences of agents at the current round. For these problems, we define novel axiomatic norms, denoted as JFXRC, JFXRR, JF1RC, JF1RR, and JF1B, that limit inequalities in allocations gradually. When the endowments are equal, we prove that computing JFXRC and JFXRR allocations may take exponential time whereas computing JF1RC and JF1RR allocations takes polynomial time. However, when the endowments are unequal, JF1RC or JF1RR allocations may not exist whereas computing JF1B allocations takes polynomial time. Finally, our work offers a number of polynomial-time algorithms for limiting inequalities in repeated house and task allocation markets.

Keywords: Social Good · Individual Well-being · Algorithmic Decisions

1 Introduction

Limiting inequalities and ensuring that no one is left behind are integral to achieving the United Nations' Sustainable Development Goals. Despite some positive signs toward limiting inequality in some dimensions, such as limiting relative income inequality in some countries and preferential trade status benefiting lower-income countries, inequality still persists. Inequalities are also deepening for vulnerable populations in countries with weaker health systems and those facing existing humanitarian crises. Refugees and migrants, as well as indigenous peoples, older persons, and people with disabilities, are particularly at risk of facing increasing inequalities worldwide. Many of these people live in poor houses but have an increasing number of tasks in their daily routines. Hence, limiting inequalities in the associated house allocation [1] and task allocation [13] markets may promote social integration and life stability for those people.

Indeed, we expect that those with bigger houses would also have to do more tasks for maintaining those houses from one day to another. We study thus limiting inequalities in repeated settings, where agents have general valuations for houses and tasks. Unlike static settings [4], *repeated settings* enable us to balance inequalities not only within a fixed day but also over multiple days. For example, we might want to give

The work was supported by the DFG Individual Research Grant on "Fairness and Efficiency in Emerging Vehicle Routing Problems" (497791398).

greater priority tomorrow to those who could not get houses today. Furthermore, unlike additive valuations where agents enjoy summing up their valuations for items in bundles [2], *general valuations* do not rely on assumptions about the function agents use to aggregate bundle valuations. For example, people might have negative time valuations for individual tasks, but they might prefer avoiding doing multiple tasks because this takes additional time as they get tired, in which case their valuations for bundles of tasks might be strictly lower than the sum of their valuations for the individual bundle tasks.

Social goods (e.g. items liked by everyone such as houses) and *social bads* (e.g. items disliked by everyone such as tasks) received some research attention [3,5]. If there were money as a divisible resource, we could use it to reduce potential inequalities over houses and tasks by paying more to agents who receive poorer houses but have more tasks. For this reason, we consider the more challenging case when there is no money and all items are *indivisible*. As a result, *jealousy*, caused by interpersonal comparisons of well-being among agents [10], is unavoidable. For example, if our neighbors have bigger houses than us, we might be jealous of them, but not if we know that they work much harder than us. Perceiving lower inequalities relates, therefore, to how well we balance our lives and how comfortable we feel living in a given area. As a consequence, limiting inequalities could improve our well-being!

2   Overview and Contributions

In Sect. 3, we present related works. In Sect. 4, we give the formal preliminaries. We allocate social goods and bads over multiple rounds. Pick a given allocation round. If we allocate tasks and their associated wages, we might be jealous of employees who get fewer tasks and higher wages than us but we get more tasks and lower wages than them. However, we might be able to eliminate this jealousy if they were getting lower wages or if we gave them some of our tasks. If we allocate houses and their associated taxes, we might be jealous of neighbors who get bigger houses than us because we also like bigger houses but get smaller houses. However, we might be able to eliminate this jealousy if they were getting smaller houses or if we did not have to pay taxes. In Sect. 5, we first define two novel axiomatic norms that reflect these observations:

1 *Jealousy-Freeness Limited To Every Non-zero Removed Good and Copied Bad (JFXRC)* requires that an agent's endowment plus valuation is not lower than any other agent's endowment plus valuation after any non-zero individual bad is *copied* from the former agent's bundle into the latter agent's bundle, and any non-zero individual good is *removed* from the latter agent's bundle. With zero endowments (i.e. agents are not allocated anything prior to the current round), we show that JFXRC allocations exist in problems with goods and bads (Theorem 1 in Sub-sect. 5.1).

2 *Jealousy-Freeness Limited To Every Non-zero Removed Good and Removed Bad (JFXRR)* requires that an agent's endowment plus valuation is at least as much as any other agent's endowment plus valuation after any given non-zero individual bad is *removed* from the former agent's bundle, and any given non-zero individual good is *removed* from the latter agent's bundle. When the endowments are zeros, we prove that JFXRR allocations exist in problems with either goods or bads (Theorem 2 in Sub-sect. 5.1).

Furthermore, we observe that computing JFXRC and JFXRR allocations might exhibit high computational complexity. For this reason, we weaken these norms and define two other novel axiomatic norms that also limit any agent-pairwise jealousy in allocations, but can be satisfied in polynomial time:

3 *Jealousy-Freeness Limited To One Removed Good or Copied Bad (JF1RC)* requires that the removal and copying operations in the formulation of JFXRC concern not every non-zero valued good and bad, but at least one, possibly zero-valued, item. With zero endowments, we prove that JF1RC allocations can be computed in polynomial time (Theorem 3 in Sub-sect. 5.2).
4 *Jealousy-Freeness Limited To One Removed Good or Removed Bad (JF1RR)* requires that the two removal operations in the formulation of JFXRR concern not every non-zero valued good and bad, but at least one, possibly zero-valued, item. With zero endowments, we also prove that JF1RR allocations can be computed in polynomial time (Theorem 4 in Sub-sect. 5.2).

For the case when the endowments are not necessarily zeros, we make the following two contributions in a given allocation round.

5 When the endowments are equal (i.e. agents are allocated items of equal, possibly non-zero, utility prior to the current round), we prove that JFXRC, JFXRR, JF1RC, and JF1RR allocations exist (Theorem 5 in Sect. 6).
6 When the endowments are unequal, we show that JF1RC or JF1RR allocations might not always exist (Theorem 6 in Sect. 6). As a response, we relax these properties and define an axiomatic norm for bounding agent pairwise jealousy in up to one item fashion by the corresponding absolute endowment difference (JF1B). We prove that JF1B allocations can be computed in polynomial time (Theorem 7 in Sect. 6).

Finally, we conclude and highlight several future directions in Sect. 7.

3   Related Works

Gourvès et al. [10] defined a notion called *near jealousy-freeness* in static instances without endowments and additive valuations for goods. Also, Freeman et al. [8,9] defined a notion called *equitability up to one item* in static instances without endowments and additive valuations for either goods or chores. Hence, neither near jealousy-freeness nor equitability up to one item is well-defined for repeated house and task allocations with endowments and general valuations. In fact, we are not aware of any prior work that studies limiting inequalities over multiple rounds. We are nevertheless aware of a very recent work [12] that studies limiting *envy*, another fairness metric of intrapersonal comparison of different consumption bundles [7], over multiple rounds. However, Herreiner and Puppe [11] found out through an experiment that people preferred allocations of fewer inequalities to allocations, say those limiting envy, that normally induce more inequalities. Our normative design reflects this experiment because JFXRC and JFXRR limit inequalities more than JF1RC and JF1RR, as well as JF1RC and JF1RR limit inequalities more than JF1B.

4   Formal Preliminaries

For given $p \in \mathbb{N}_{\geq 1}$, we let $[p] = \{1, \ldots, p\}$. We consider *repeated fair division instances* over rounds 1 to $t \in \mathbb{N}_{\geq 1}$. At each $\tau \in [t]$, we consider items from M_τ that must be allocated to agents from N_τ. We let $n_\tau = |N_\tau|$ and $m_\tau = |M_\tau|$ hold. We note that the sets of agents and items across different rounds might be different.

Pick τ. For each $a \in N_\tau$ and each bundle $M \subseteq M_\tau$, we let $v_a^\tau(M) \in \mathbb{R}$ denote their *general valuation* for M. We let $v_a^\tau(\emptyset) = 0$ hold. We write $v_a^\tau(o)$ for $v_a^\tau(\{o\})$. We say that $v_a^\tau(M)$ is *additive* if $v_a^\tau(M) = \sum_{o \in M} v_a^\tau(o)$ holds. Valuations can be elicited from agents in an experiment (see e.g. [11]), through a website (see e.g. [6]), or via an oracle (see e.g. [14]).

Pick $o \in M_\tau$. o is *good* for $a \in N_\tau$ with respect to (WRT) $M \subseteq M_\tau$ if $v_a^\tau(M \cup \{o\}) \geq v_a^\tau(M)$ holds, and *pure good* if $v_a^\tau(M \cup \{o\}) > v_a^\tau(M)$ holds. Also, o is *bad* for a WRT M if $v_a^\tau(M \cup \{o\}) \leq v_a^\tau(M)$ holds, and *pure bad* if $v_a^\tau(M \cup \{o\}) < v_a^\tau(M)$ holds. As we mentioned, we consider (social) goods and (social) bads. Hence, we can partition M_τ into $G_\tau = \{o \in M_\tau | \forall a \in N_\tau \forall M \subseteq M_\tau : v_a^\tau(M \cup \{o\}) \geq v_a^\tau(M)\}$ and $B_\tau = \{o \in M_\tau | \forall a \in N_\tau \forall M \subseteq M_\tau : v_a^\tau(M \cup \{o\}) \leq v_a^\tau(M)\}$.

A complete *allocation* at τ is $A^\tau = (A_1^\tau, \ldots, A_{n_\tau}^\tau)$, where (1) A_a^τ is the bundle of agent $a \in N_\tau$, (2) $\bigcup_{a \in N_\tau} A_a^\tau = M_\tau$ holds, and (3) $A_a^\tau \cap A_b^\tau = \emptyset$ holds for each $a, b \in N_\tau$ with $a \neq b$. We note that some agents might be allocated items across multiple rounds.

We trace therefore the valuations of agents over rounds. For this reason, we make use of *endowment* $e_a^\tau \in \mathbb{R}$ for each $a \in N_\tau$ that depends on the items allocated to a prior to τ. Thus, for $\tau = 1$, we let $e_a^1 = 0$ hold. But, for $\tau > 1$, the value of e_a^τ could be equal to the additive item number up to τ, i.e. $e_a^\tau = \sum_{\kappa \in 1:(\tau-1)} |A_a^\kappa|$, or the additive valuation sum up to τ, i.e. $e_a^\tau = \sum_{\kappa \in 1:(\tau-1)} v_a^\kappa(A_a^\kappa)$.

Thus, at τ, we let a receive *general utility* $u_a^\tau(A_a^\tau) = (e_a^\tau + v_a^\tau(A_a^\tau))$. With zero endowments at τ, we note that $u_a^\tau(A_a^\tau) = v_a^\tau(A_a^\tau)$ holds. We say that $a \in N_\tau$ is *jealousy-free* of $b \in N_\tau$ in A^τ if $u_a^\tau(A_a^\tau) \geq u_a^\tau(A_b^\tau)$ holds. Thus, A^τ is *jealousy-free* if, for each $a, b \in N_\tau$, a is jealousy-free of b in A^τ. With one round and zero endowments, giving one pure good to an agent makes any other agent feel jealous. Hence, it might be impossible to guarantee jealousy-freeness. For this reason, we propose to relax it.

5   One-Round Instances

We begin with *one-round fair division instances* which are repeated fair division instances that repeat one round. That is, in such instances, we have that $t = 1$ holds. In this section, for simplicity, we **omit** the subscripts $_\tau$ and superscripts $^\tau$ from the model notation because $\tau = 1$ holds. Also, as there are no repetitions of the instance prior to round one, no agent $a \in N$ has been allocated anything prior to this round and, for this reason, we have that $e_a = 0$ hold. For example, when allocating tasks to employees today, we might want to make use of all employees equally regardless of whether they serviced unequal numbers of tasks in the previous days.

5.1 Jealousy Freeness Up to Every Non-zero Item

As jealousy-freeness might be too demanding, we next relax it. We say that *agent a is JFXRC of agent b* whenever a's utility is at least as much as b's utility, after hypothetically *coping* any given non-zero individual bad from a's bundle into b's bundle, and hypothetically *removing* any given non-zero individual good from b's bundle. Also, we say that *agent a is JFXRR of agent b* whenever a's utility is at least as much as b's utility, after hypothetically *removing* any given non-zero individual bad from a's bundle, and hypothetically *removing* any given non-zero individual good from b's bundle.

Definition 1 (JFXRC). *Allocation A is JFXRC if, $\forall a, b \in N$ such that a is not jealousy-free of b, (1) $\forall o \in A_a$ such that $u_a(A_a) < u_a(A_a \setminus \{o\})$: $u_a(A_a) \geq u_b(A_b \cup \{o\})$ and (2) $\forall o \in A_b$ such that $u_b(A_b) > u_b(A_b \setminus \{o\})$: $u_a(A_a) \geq u_b(A_b \setminus \{o\})$.*

Definition 2 (JFXRR). *Allocation A is JFXRR if, $\forall a, b \in N$ such that a is not jealousy-free of b, (1) $\forall o \in A_a$ such that $u_a(A_a) < u_a(A_a \setminus \{o\})$: $u_a(A_a \setminus \{o\}) \geq u_b(A_b)$ and (2) $\forall o \in A_b$ such that $u_b(A_b) > u_b(A_b \setminus \{o\})$: $u_a(A_a) \geq u_b(A_b \setminus \{o\})$.*

By definition, jealousy-freeness is *stronger* than JFXRC and JFXRR as such allocations satisfy JFXRC and JFXRR. However, unlike jealousy-free allocations that may not exist in some instances with pure goods, we prove that JFXRC and JFXRR allocations exist in every such instance. However, we first observe similarities between them.

When allocating n fixed-price houses among n agents, any allocation that gives one house to each agent is JFXRC and JFXRR. Furthermore, when allocating n fixed-time tasks among n agents, any allocation that gives one task to each agent is JFXRC and JFXRR. At the same time, there are fundamental differences between these two norms.

For example, achieving JFXRC might require giving unequal numbers of tasks to the agents, but thus optimize various welfares such as the Nash welfare and utilitarian welfare [6], whereas achieving JFXRR might require giving equal numbers of tasks to agents, but still may not optimize such welfares. We demonstrate this in Example 1.

Example 1. *Let us consider agents 1, 2, and 3 with zero endowments and valuations $(-\epsilon, -\epsilon, -\epsilon)$, $(-\epsilon, -\epsilon, -\epsilon)$, and $(-1, -1, -1)$, respectively, for three tasks. We let $\epsilon \in (0, 1/3)$ hold. Each JFXRR allocation gives one item to each agent and: minimizes the product of agents' valuations (i.e. Nash welfare) to $-\epsilon^2$; achieves a sum of agents' valuations (i.e. the utilitarian welfare) of $(-1 - 2\epsilon)$; returns a difference between the maximum and minimum agent's valuations (i.e. the inequality variance) of $(1 - \epsilon)$. By comparison, each JFXRC allocation shares all items only among agents 1 and 2 and: maximizes the Nash welfare to $0 > -\epsilon^2$; maximizes the utilitarian welfare to $-3\epsilon > (-1 - 2\epsilon)$; minimizes the inequality variance to $2\epsilon < (1 - \epsilon)$. Hence, we might prefer achieving JFXRC to achieving JFXRR, especially if we care about maximizing welfare and minimizing variance as secondary objectives.*

By Example 1, it follows that we might prefer achieving JFXRC to achieving JFXRR. For this reason, we start with JFXRC. We prove that JFXRC allocations are guaranteed to exist in every instance with goods and bads. Such allocations are returned by the *leximin++* solution [16].

To define this solution, we let $\overrightarrow{u}(A) \in \mathbb{R}^n$ denote the vector of agents' utilities in A, which are (re-)arranged in some non-decreasing order. We next write $A \succ_{++} B$ if there exists an index $i \leq n$ such that $\overrightarrow{u}(A)_j = \overrightarrow{u}(B)_j$ and $|A_j| = |B_j|$ for each $1 \leq j < i$, and either $\overrightarrow{u}(A)_i > \overrightarrow{u}(B)_i$ or, $\overrightarrow{u}(A)_i = \overrightarrow{u}(B)_i$ and $|A_i| > |B_i|$, hold. Thus, the leximin++ solution is defined as a maximal element under $\succ_{++}$. Informally, the leximin++ solution maximizes the least agent's utility, then maximizes the bundle's size of an agent with the least utility, before it maximizes the second least agent's utility and the bundle's size of an agent with the second least utility, and so on.

Theorem 1. *Pick a one-round instance. With zero endowments and general valuations for goods and bads, the leximin $++$ solution returns JFXRC allocations.*

We proceed with JFXRR. By definition, in instances with only goods, an allocation is JFXRR if and only if it is JFXRC. Hence, by Theorem 1, each leximin++ allocation satisfies JFXRR in instances with only goods. However, in instances with only bads, there is another solution for returning JFXRR allocations. This one is our Algorithm 1.

Algorithm 1 allocates the items one by one in rounds. At each round, Algorithm 1 picks the *maximum* utility *agent* given the current partial allocation. After allocating the current item to them, if the new partial allocation remains JFXRR then Algorithm 1 moves to the next round and, otherwise, Algorithm 1 picks an inclusion-wise *minimal* strict *subset* of the bundle of the selected agent, assigns this subset to them, and returns the remaining items from their bundle to the pool of unassigned items. Algorithm 1 runs in pseudo-polynomial time. This run-time complexity could be useful in practice whenever the utilities are specified in unary. This has been observed in settings where people divide service fares and house rent [6].

Algorithm 1. JFXRR in round instances with zero endowments and general valuations for bads.

1: **procedure** MAXAGENTMINSUBSET($N, M, (u_a)_n$)
2:     $O \leftarrow M, \forall a \in N : A_a \leftarrow \emptyset$
3:     **while** $O \neq \emptyset$ **do**
4:         $o \leftarrow$ an unallocated bad from O, $b \leftarrow \arg\max_{a \in N} u_a(A_a)$
5:         **if** $(A_1, \ldots, A_b \cup \{o\}, \ldots, A_n)$ is JFXRR **then**
6:             $O \leftarrow O \setminus \{o\}, A_b \leftarrow A_b \cup \{o\}$
7:         **else**
8:             $X \leftarrow$ inclusion-wise minimal strict subset of $A_b \cup \{o\}$ s.t. $u_b(X) < u_a(A_a)$ for some $a \in N, O \leftarrow O \cup [A_b \cup \{o\} \setminus X], A_b \leftarrow X$
9:     **return** A

Theorem 2. *Pick a one-round instance. (1) With zero endowments and general valuations for goods, the leximin++ solution returns JFXRR allocations. (2) With zero endowments and general valuations for bads, Algorithm 1 returns JFXRR allocations.*

Proof. For instances with goods, the result for the leximin $++$ follows by the proof of Theorem 1. For instances with bads, the argument for Algorithm 1 is inductive. Let A denote the partial allocation. Suppose that A is JFXRR. We argue that the allocation after the current round remains JFXRR. If $(A_1, \ldots, A_b \cup \{o\}, \ldots, A_n)$ is JFXRR, then the proof is done. Otherwise, we claim that $(A_1, \ldots, X, \ldots, A_n)$ is JFXRR.

For this claim, we need to show that agent b and agent $c \in N \setminus \{b\}$ are JFXRR of each other. Every two agents where agent b is not involved are clearly JFXRR of each other because their allocations remain unchanged at the end of the current iteration.

Indeed, as $(A_1, \ldots, A_b \cup \{o\}, \ldots, A_n)$ violates JFXRR, $u_b(A_b \cup \{o\} \setminus \{g^-\}) < u_a(A_a)$ holds for some $g^- \in A_b, a \in N$. Thus, we can find an inclusion-wise minimal subset $X \subset A_b \cup \{o\}$ such that $u_b(X) < u_a(A_a)$ holds, e.g. $X = A_b \cup \{o\} \setminus \{g^-\}$. Even more, for a fixed $Y \subset X$, we have $u_b(Y) \geq u_c(A_c)$ for each $c \in N$.

Since X is a minimal subset of $A_b \cup \{o\}$ such that $u_b(X) < u_a(A_a)$ holds for some $a \in N$, we have $u_b(X \setminus \{g^-\}) \geq u_c(A_c)$ for each $g^- \in X$ and each $c \in N$. Therefore, agent b remains indeed JFXRR of agent $c \in N \setminus \{b\}$. We next also show that agent $c \in N \setminus \{b\}$ remains JFXRR of agent b.

As $u_b(X) < u_a(A_a)$ holds for some $a \in N$ and $u_c(A_c) \leq u_b(A_b)$ holds for each $c \in N$ by the choice of b, we conclude that $u_b(X) < u_b(A_b)$ holds. Thus, we derive $u_c(A_c \setminus \{g^-\}) \geq u_b(A_b) > u_b(X)$ for all $g^- \in A_c$, where the inequality $u_c(A_c \setminus \{g^-\}) \geq u_b(A_b)$ follows from A being JFXRR. The argument concludes.

At each iteration of the algorithm, either (1) one bad is allocated or (2) some bads are returned to the pool of currently unallocated bads but the overall sum of agents' utilities strictly decreases. As the number of bads is bounded from above by m and the minimum overall sum of agents' utilities is also bounded from below, the algorithm is guaranteed to terminate in a finite number of rounds.

Indeed, there can be at most m rounds where (1) holds. After that, either all bads are allocated or there is a round where (2) holds. We let U denote the maximum value of $|\sum_{a \in [n]} u_a(A_a)|$ and u denote the minimum value of $|u_a(X) - u_b(Y)| > 0$ for each $a, b \in N$ and each $X \subseteq M, Y \subseteq M$. After $O(m\frac{U}{u})$ rounds, either the sum of agents' utilities reaches $-U$ or all bads are allocated.   □

By Theorem 2, JFXRR allocations exist in instances with general utilities for either goods *or* bads. We hoped to prove that JFXRR allocations exist in instances with general utilities for goods *and* bads. Despite our efforts, we could not come up with a solution that satisfies JFXRR in this case. For this reason, we leave it as an *open* problem.

As a result, we might prefer JFXRC to JFXRR because JFXRC allocations exist in instances with general utilities for goods and bads, whereas JFXRR allocations exist in instances with general utilities for either goods or bads, and it remains unclear whether JFXRR allocations exist in instances with general utilities for goods and bads.

5.2 Jealousy Freeness Up to One Item

Each leximin++ allocation satisfies JFXRC. But, it can be computed in $O(n^m)$ time [16]. This run-time complexity might be fine for a constant value of m. However, m can be much larger than n in practice. For this reason, we propose to relax JFXRC.

Likewise, as we mentioned previously, we do not know whether JFXRR allocations exist in instances with general utilities for goods and bads. Nevertheless, we expect that a possible solution will inherit the complexity of computing leximin++ allocations. For this reason, we also propose to relax JFXRR.

Respectively, we say that *agent a is JF1RC or JF1RR of agent b* whenever a is not JFXRC or JFXRR, but the JFXRC or JFXRR conditions hold for at least one individual bad in a's bundle or at least one individual good in b's bundle.

Definition 3 (JF1RC). *A is* JF1RC *if,* $\forall a, b \in N$ *such that a is not JFXRC of b, (1)* $\exists o \in A_a$ *such that* $u_a(A_a) \geq u_b(A_b \cup \{o\})$ *or (2)* $\exists o \in A_b$ *such that* $u_a(A_a) \geq u_b(A_b \setminus \{o\})$.

Definition 4 (JF1RR). *A is* JF1RR *if,* $\forall a, b \in N$ *such that a is not JFXRR of b, (1)* $\exists o \in A_a$ *such that* $u_a(A_a \setminus \{o\}) \geq u_b(A_b)$ *or (2)* $\exists o \in A_b$ *such that* $u_a(A_a) \geq u_b(A_b \setminus \{o\})$.

By definition, JFXRC is stronger than JF1RC and JFXRR is stronger than JF1RR. Like JFXRC and JFXRR, JF1RC and JF1RR could exhibit substantial differences. To see this, let us recall Example 1. It is easy to observe that an allocation is JFXRC if and only if JF1RC; JFXRR if and only if JF1RR. Hence, in Example 1, each JF1RC allocation maximizes the Nash welfare and utilitarian welfare, whereas each JF1RR allocation does not do that.

However, unlike JFXRC and JFXRR whose satisfiability exhibits respectively exponential and pseudo-polynomial solutions, computing JF1RC and JF1RR allocations can be done in polynomial time by using our novel Algorithms 2 and 3, respectively.

Algorithm 2 allocates the items one by one. If the current item is good, then the algorithm gives it to the *min*imum utility *agent*. Otherwise, the algorithm runs a *pseudo* experiment, where a copy of the bad is allocated to every agent, and actually gives the bad to the *max*imum utility *agent* in this experiment.

By comparison, Algorithm 3 allocates the items one by one. If the current item is good, then the algorithm gives it to the *min*imum utility *agent*. Otherwise, the algorithm gives it to the *max*imum utility *agent*.

When allocating houses, both Algorithms 2 and 3 give priority to the more impacted people of the current unfair allocations. When allocating tasks, they give priority to the least impacted people of the current unfair allocations. These decision criteria could be used to explain to people how the algorithms limit inequalities.

Algorithm 2. JF1RC in round instances with zero endowments and general valuations.

1: **procedure** MINAGENTPSEUDOMAXAGENT($N, M, (u_a)_n$)
2:     $O \leftarrow M, \forall a \in N : A_a \leftarrow \emptyset$
3:     **while** $O \neq \emptyset$ **do**
4:         $o \leftarrow$ an item from O
5:         **if** o is social good **then**
6:             $a \leftarrow \arg\min_{b \in N} u_b(A_b)$
7:         **else**   ▷ i.e. o is social bad
8:             $a \leftarrow \arg\max_{b \in N} u_b(A_b \cup \{o\})$
9:         $A_a \leftarrow A_a \cup \{o\}, O \leftarrow O \setminus \{o\}$
10:    **return** A

Algorithm 3. JF1RR in round instances with zero endowments and general valuations.

1: **procedure** MINAGENTMAXAGENT($N, M, (u_a)_n$)
2:     Copy lines 2-7 from Algorithm 2.
3:         $a \leftarrow \arg\max_{b \in N} u_b(A_b)$
4:     Copy lines 9-10 from Algorithm 2.

Theorem 3. *Pick a one-round instance. With zero endowments and general valuations for goods and bads, Algorithm 2 returns JF1RC allocations.*

Theorem 4. *Pick a one-round instance. With zero endowments and general valuations for goods and bads, Algorithm 3 returns JF1RR allocations.*

Proof. The proof is inductive. In the base case, no items are allocated. This allocation is JF1RR. In the hypothesis, we let A denote the partial allocation and assume that A is JF1RR. In the step case, we show that allocating o to agent a preserves JF1RR. By the hypothesis, agents $b \neq a$ and $c \neq a$ remain JF1RR after o is allocated to a because their allocations remain intact. We thus consider two cases.

Case 1: Let o be a social good. In this case, the proof is identical to the proof of Case 1 from Theorem 3. However, the conclusions are that: agent c remains JF1RR of agent a and agent a remains JF1RR of agent c.

Case 2: Let o be a social bad. In this case, $u_b(A_b \cup \{o\}) \leq u_b(A_b)$ and $u_b(A_b) \leq u_a(A_a)$ for each $b \in N$. Let us consider some agent $c \neq a$. If $u_a(A_a \cup \{o\}) < u_a(A_a)$, it follows that $u_c(A_c) \leq u_a(A_a \cup \{o\} \setminus \{o\})$ holds. If $u_a(A_a \cup \{o\}) = u_a(A_a)$, it follows that $u_c(A_c) \leq u_a(A_a \cup \{o\})$ holds. Agent a remains JF1RR of agent c.

In the opposite direction, as A is JF1RR, we conclude $u_c(A_c) \geq u_a(A_a \setminus \{g^+\})$ for some non-zero marginal good $g^+ \in A_a$ or $u_c(A_c \setminus \{g^-\}) \geq u_a(A_a)$ for some non-zero marginal bad $g^- \in A_c$. Thus, as item o is social bad, we derive $u_a(A_a \cup \{o\}) \leq u_a(A_a)$ and $u_a(A_a \cup \{o\} \setminus \{g^+\}) \leq u_a(A_a \setminus \{g^+\})$. Agent c remains JF1RR of agent a.   □

The input of both Algorithms 2 and 3 is bounded by $O(mn)$. Their running times are dominated by computing a sorting of the n agents' bundle utilities for each of the m items, i.e. $O(m \cdot n)$.

To sum up, as both JF1RC and JF1RR allocations can be computed in polynomial time, we may be indifferent between such allocations or insist on a secondary objective such as envy-freeness [7] or Pareto optimality [15].

6   Repeated Instances

We end with repeated fair division instances that repeat at least one round. That is, in such instances, we have that $t > 1$ holds. Pick round $\tau \in [t]$. If $\tau = 1$, then $e_a^1 = 0$ holds for each $a \in N_1$. However, if $\tau > 1$, some agents might have been allocated items in previous rounds and, therefore, $e_a^\tau \neq 0$ might hold for some $a \in N_\tau$ at round τ. A special case is when the agent endowments are *equal* at round τ, i.e. $e_a^\tau = e^\tau$ holds for each $a \in N_\tau$. In this case, we can use the previously proposed solutions to return almost jealousy-free allocations.

Theorem 5. *Pick a repeated instance and round τ in it. (A) With equal endowments and general valuations for goods and bads at round τ, the leximin++ solution returns JFXRC allocations at τ. (B) With equal endowments and general valuations for goods at τ, the leximin++ solution returns JFXRR allocations at τ. With equal endowments and general valuations for bads at τ, Algorithm 1 returns JFXRR allocations at τ. (C) With equal endowments and general valuations for goods and bads at τ, Algorithm 2 returns JF1RC allocations at τ. (D) With equal endowments and general valuations for goods and bads at τ, Algorithm 3 returns JF1RR allocations at τ.*

Proof. The key argument is that, in a repeated instance and round τ in it, we can effectively reduce the round instance and allocations in it at τ where all endowments are the same to a one-round instance and allocations in it where all endowments are zeros. Then, by the definitions of the axiomatic norms, statement (A) follows by Theorem 1, statement (B) follows by Theorem 2, statement (C) follows by Theorem 3, and statement (D) follows by Theorem 4.   □

As soon as two endowments are not equal at round $\tau > 1$ (i.e. $e_a^\tau \neq e_b^\tau$ holds for some $a, b \in N_\tau$ with $a \neq b$), we may no longer be able to achieve JF1RC or JF1RR.

Theorem 6. *Pick a repeated instance and round $\tau > 1$ in it. With two agents, unequal endowments, and additive valuations at round τ, it might be the case that* no *allocation at round τ satisfies JF1RC or JF1RR.*

Proof. Let us consider two goods g_1 and g_2, each giving a unit valuation to every agent. Further, let us define the endowments of agents a and b as $e_a^\tau = 4$ and $e_b^\tau = 1$, respectively. The result follows because there are no bads in the instance at round τ and, in each allocation $A^\tau = (A_a^\tau, A_b^\tau)$, we cannot eliminate the jealousy of b due to the fact that the following inequalities hold: $u_b^\tau(A_b^\tau) \leq u_b^\tau(\{g_1, g_2\}) = e_b^\tau + v_b^\tau(g_1) + v_b^\tau(g_2) = 3 < 4 = e_a^\tau + v_a^\tau(\emptyset) = e_a^\tau = u_a^\tau(\emptyset) \leq u_a^\tau(A_a^\tau \setminus \{g_1\}) = u_a^\tau(A_a^\tau \setminus \{g_2\})$.   □

In response, we propose to place a bound on the agent-pairwise utility difference in a given allocation. Thus, for every pair of agents $a, b \in N_\tau$ where a is neither JF1RC nor JF1RR of b, we require that a *is JF1B of* b, i.e. the utility of b minus the utility of a is bounded from above by their absolute endowment difference plus the maximum marginal valuation for any meaningful item move, that either increases the utility of a or decreases the utility of b, within the bundles of a and b. We next formally extend this property to allocations.

Definition 5 (JF1B). *Allocation A^τ is* bounded jealousy-free *if, $\forall a, b \in N$ such that a is not JF1RC or JF1RR of b, $u_b^\tau(A_b^\tau) - u_a^\tau(A_a^\tau) \leq |e_b^\tau - e_a^\tau| + \max\{J_a^-, J_a^+, JF_b^-, JF_b^+\}$, where we have (a) $J_a^- = \max_{\forall o \in A_a^\tau : [v_a^\tau(A_a^\tau \setminus \{o\}) - v_a^\tau(A_a^\tau)] > 0} [v_a^\tau(A_a^\tau \setminus \{o\}) - v_a^\tau(A_a^\tau)]$ and (b) $J_a^+ = \max_{\forall o \in A_a^\tau : [v_a^\tau(A_a^\tau \cup \{o\}) - v_a^\tau(A_a^\tau)] > 0} [v_a^\tau(A_a^\tau \cup \{o\}) - v_a^\tau(A_a^\tau)]$, and (c) $JF_b^- = \max_{\forall o \in A_b^\tau : [v_b^\tau(A_b^\tau) - v_b^\tau(A_b^\tau \cup \{o\})] > 0} [v_b^\tau(A_b^\tau) - v_b^\tau(A_b^\tau \cup \{o\})]$ and (d) $JF_b^+ = \max_{\forall o \in A_b^\tau : [v_b^\tau(A_b^\tau) - v_b^\tau(A_b^\tau \setminus \{o\})] > 0} [v_b^\tau(A_b^\tau) - v_b^\tau(A_b^\tau \setminus \{o\})]$.*

By definition, if an allocation at round τ is JF1RC or JF1RR then it is also bounded jealousy-free. However, the reversed implication may not always be true. For example, in the allocation from Theorem 6, where the jealousy-free agent a gets no goods and the jealous agent b gets all goods, agent b is not JF1RC or JF1RR of agent a, but agent b is JF1B of agent a because $1 = u_a^\tau(\emptyset) - u_b^\tau(\{g_1, g_2\}) \leq |e_a^\tau - e_b^\tau| + \max\{0, 0, 0, 0\} = 3$ hold. Interestingly, we can compute JF1B allocations at round τ in polynomial time. For this purpose, we can use Algorithms 2 and 3.

Theorem 7. *Pick a repeated instance and round τ in it. With general endowments and general valuations for goods and bads, Algorithms 2 and 3 return JF1B allocations.*

Proof. The proof is inductive. In the base case, we let no items be allocated. This partial allocation is JF1B. In the hypothesis, we let some but not all items be allocated. We assume that the current allocation is JF1B. In the step case, we let the current item o be allocated to agent a. Thus, as only the bundle of agent a is changed, we only need to show that agents a and $b \neq a$ remain JF1B of each other because every other two agents remain JF1B of each other by the hypothesis.

Case 1 for Algorithms 2 and 3: Let o be a social good. Immediately after o is allocated to a, the utility of a does not decrease, and, hence, a remains JF1B of every b by the hypothesis. Also, immediately before o is allocated to a, every b is JF1RC and JF1RR of a because a is the minimum utility agent. Hence, immediately after o is allocated to a, every b that becomes neither JF1RC nor JF1RR of a is such that the utility of a minus the utility of b is at most $|e_a^\tau - e_b^\tau|$ plus the valuation of a with o minus the valuation of a without o, which is at most $|e_a^\tau - e_b^\tau|$ plus JF_a^+.

Case 2 for Algorithm 2: Let o be a social bad. As soon as o is allocated to a, the utility of a does not increase, and, hence, any other b remains JF1B of a by the hypothesis. Also, immediately before o is allocated to a, the utility of b with o is at most the utility of a with o because a is a pseudo maximum utility agent. Hence, immediately after o is allocated to a, every b for which a becomes neither JF1RC nor JF1RR of b is such that the utility of b minus the utility of a with o is at most $|e_b^\tau - e_a^\tau|$ plus the valuation of b without o minus the valuation of b with o, which is at most $|e_b^\tau - e_a^\tau|$ plus JF_b^-.

Case 2 for Algorithm 3: Let o be a social bad. As soon as o is allocated to a, the utility of a does not increase, and, hence, every b remains JF1B of a by the hypothesis. Also, immediately before o is allocated to a, a is JF1RC and JF1RR of every b because a is the maximum utility agent. Hence, immediately after o is allocated to a, every b for which a becomes neither JF1RC nor JF1RR of b is such that the utility of b minus the utility of a is at most $|e_b^\tau - e_a^\tau|$ plus the valuation of a without o minus the valuation of a with o, which is at most $|e_b^\tau - e_a^\tau|$ plus J_a^-.   □

7   Conclusions and Future Directions

We considered limiting inequalities in repeated house and task allocation problems. We looked at these problems because every individual in our society has the right to have better living and working conditions. For such problems, we proposed five axiomatic norms, denoted as JFXRC, JFXRR, JF1RC, JF1RR, and JF1B, that limit inequalities in allocations gradually, namely people prefer JFXRC and JFXRR to JF1RC and JF1RR, and JF1RC and JF1RR to JF1B. Our conclusions are two-fold: (1) computing JFXRC and JFXRR allocations might take time longer than the time people can wait to be allocated houses and tasks; (2) computing JF1RC, JF1RR, and JF1B allocations is faster but JF1RC and JF1RR allocations may not always exist, in which case people may accept JF1B allocations that limit inequalities over multiple rounds. In the future, we will run experiments and estimate the performance of the proposed solutions in practice in terms of their scalability. Finally, we will surely also extend our work to include items that could be good for some individuals and bad for other individuals.

References

1. Abdulkadiroğlu, A., Sönmez, T.: House allocation with existing tenants. J. Econ. Theory **88**(2), 233–260 (1999). https://doi.org/10.1006/jeth.1999.2553, https://www.sciencedirect.com/science/article/pii/S002205319992553X

2. Aziz, H., Moulin, H., Sandomirskiy, F.: A polynomial-time algorithm for computing a Pareto optimal and almost proportional allocation. Oper. Res. Lett. **48**(5), 573–578 (2020). https://doi.org/10.1016/j.orl.2020.07.005

3. Aziz, H., Rey, S.: Almost group envy-free allocation of indivisible goods and chores. In: Proceedings of the Twenty-Ninth International Joint Conference on Artificial Intelligence, pp. 39–45. IJCAI'20, IJCAI Press (2021). https://dl.acm.org/doi/abs/10.5555/3491440.3491446

4. Brams, S.J., Taylor, A.D.: Fair Division - From Cake-Cutting to Dispute Resolution. Cambridge University Press, Cambridge (1996). https://doi.org/10.1017/CBO9780511598975

5. Caragiannis, I., Kaklamanis, C., Kanellopoulos, P., Kyropoulou, M.: The efficiency of fair division. Theory Comput. Syst. **50**(4), 589–610 (2012). https://doi.org/10.1007/s00224-011-9359-y

6. Caragiannis, I., Kurokawa, D., Moulin, H., Procaccia, A.D., Shah, N., Wang, J.: The unreasonable fairness of maximum Nash welfare. ACM Trans. Econ. Comput. **7**(3) (2019). https://doi.org/10.1145/3355902

7. Foley, D.K.: Resource allocation and the public sector. Yale Econ. Essays **7**(1), 45–98 (1967). https://www.proquest.com/docview/302230213?pq-origsite=gscholar&fromopenview=true

8. Freeman, R., Sikdar, S., Vaish, R., Xia, L.: Equitable allocations of indivisible goods. In: Proceedings of the 28th International Joint Conference on Artificial Intelligence, pp. 280–286. IJCAI'19, AAAI Press (2019). https://dl.acm.org/doi/10.5555/3367032.3367073

9. Freeman, R., Sikdar, S., Vaish, R., Xia, L.: Equitable allocations of indivisible chores. In: Proceedings of the 19th International Conference on Autonomous Agents and MultiAgent Systems, pp. 384–392. AAMAS '20, IFAAMAS, Richland, SC (2020). https://dl.acm.org/doi/10.5555/3398761.3398810

10. Gourvès, L., Monnot, J., Tlilane, L.: Near fairness in matroids. In: Proceedings of the Twenty-First European Conference on Artificial Intelligence, pp. 393–398. ECAI'14, IOS Press, NLD (2014). https://dl.acm.org/doi/10.5555/3006652.3006719

11. Herreiner, D., Puppe, C.: Envy freeness in experimental fair division problems. Theory Decis. **67** (2005). https://doi.org/10.1007/s11238-007-9069-8

12. Igarashi, A., Lackner, M., Nardi, O., Novaro, A.: Repeated fair allocation of indivisible items (2023). https://arxiv.org/abs/2304.01644

13. Krynke, M., Mielczarek, K., Vaško, A.: Analysis of the problem of staff allocation to work stations. In: Conference Quality Production Improvement - CQPI, vol. 1, no. 1, pp. 545–550 (2019). https://doi.org/10.2478/cqpi-2019-0073

14. Lipton, R.J., Markakis, E., Mossel, E., Saberi, A.: On approximately fair allocations of indivisible goods. In: Proceedings of the 5th ACM Conference on Electronic Commerce, New York, USA, 17–20 May 2004, pp. 125–131 (2004). https://dl.acm.org/doi/10.1145/988772.988792

15. Pareto, V.: Cours d'Économie politique. Professeur à l'Université de Lausanne, Vol. I, pp. 430. 1896. Vol. II, pp. 426. 1897, F. Rouge, Lausanne (1897). https://www.cairn.info/cours-d-economie-politique-tomes-1-et-2-9782600040143.htm

16. Plaut, B., Roughgarden, T.: Almost envy-freeness with general valuations. SIAM J. Discret. Math. **34**(2), 1039–1068 (2020). https://doi.org/10.1137/19M124397X

Non-stationarity Detection in Model-Free Reinforcement Learning via Value Function Monitoring

Maryem Hussein[1] , Marwa Keshk[2] , and Aya Hussein[3]([✉])

[1] Faculty of Engineering, Cairo University, Giza, Egypt
[2] School of Professional Studies, University of New South Wales, Canberra, Australia
[3] School of Systems and Computing, University of New South Wales, Canberra, Australia
a.hussein@adfa.edu.au

Abstract. The remarkable success achieved by Reinforcement learning (RL) in recent years is mostly confined to stationary environments. In realistic settings, RL agents can encounter non-stationarity when the environmental dynamics change over time. Detecting when this change occurs is crucial for activating adaptation mechanisms at the right time. Existing research on change detection mostly relies on model-based techniques which are challenging for tasks with large state and action spaces. In this paper, we propose a model-free, low-cost approach based on value functions (V or Q) for detecting non-stationarity. The proposed approach calculates the change in the value function (ΔV or ΔQ) and monitors the distribution of this change over time. Statistical hypothesis testing is used to detect if the distribution of ΔV or ΔQ changes significantly over time, reflecting non-stationarity. We evaluate the proposed approach in three benchmark RL environments and show that it can successfully detect non-stationarity when changes in the environmental dynamics are introduced at different magnitudes and speeds. Our experiments also show that changes in ΔV or ΔQ can be used for context identification leading to a classification accuracy of up to 88%.

Keywords: Deep Reinforcement Learning · Context Detection · Non-stationarity

1 Introduction

Reinforcement learning (RL) has greatly advanced in the last decade showing remarkable success across a wide variety of domains from video games [12] to serious defence applications [16]. The aim of an RL agent is to learn an optimal policy for a given task through trial and error. By repeatedly interacting with the environment (via observing its state, executing actions, and receiving rewards), the agent learns a policy that specifies how actions should be selected in different states to maximise the accumulated reward. State-of-the-art RL algorithms

T. Liu et al. (Eds.): AI 2023, LNAI 14472, pp. 350–362, 2024.
https://doi.org/10.1007/978-981-99-8391-9_28

have demonstrated their ability to address increasingly complex tasks and have outperformed experienced human beings across multiple domains [4,16].

However, much of the progress achieved in RL is confined to tasks with stationary environments in which the dynamics are mostly fixed throughout the learning process. In many real-world settings, the assumption of stationarity does not hold. In contrast, the task at hand can exhibit significant changes in its dynamics rendering previously learnt policy insufficient to deal with the new dynamics. An example of non-stationary environments can be found in the control of traffic lights in a road network where the dynamics can change based on the time of the day or the presence of big events. Non-stationarity can also arise from the agent itself. For instance, in a robotic-embodied RL agent, the accuracy of the robot's sensors and actuators can change due to aging effects leading to a change in the dynamics [17].

Non-stationarity is a key challenge faced by RL algorithms [5]. Past studies have shown that RL agents suffer from significant performance drops in response to changes in the environment dynamics [10]. Several research directions, for instance transfer learning [20] and meta learning [2], have focused on developing adaptation techniques which can be activated when the agent moves between different tasks. However, these techniques rely on an external signal for defining task boundaries. Equipping RL agents with the ability to autonomously detect changes in the dynamics is thus desirable for increasing agent autonomy.

Most existing methods for detecting non-stationarity learn a model of the environment such that if the current dynamics diverge from the learnt model, a change is detected. However, most state-of-the-art RL algorithms are model free. Leaning a model for these algorithms would incur a significant computational cost, particularly in tasks with high-dimensional state and/or action spaces. This paper aims to address this gap by proposing a low-cost approach for change detection in model-free RL algorithms that use value functions. The proposed approach is based on the following rationale: given a fixed policy, the rate at which state utility improves in a given environment is determined by the environment dynamics. Monitoring this rate can simply be done by monitoring the change in value function. When the distribution of this rate changes, this reflects a change in the dynamics. The next section gives a background on the concepts used in this paper.

2 Preliminaries

2.1 RL Formulation

RL is typically formulated as a Markov Decision Process (MDP) described in terms of the tuple $(S, A, R, T, \rho, \gamma)$ such that S is the set of possible states, A is the set of allowable actions, $T : S \times A \rightarrow p(s'|s, a)$ is the transition function determining the probability of transitioning to state s' from state s when performing action a, $R : S \rightarrow \mathbb{R}$ is the reward function that gives an immediate reward for transitioning to state s, ρ is the distribution of initial states, and $\gamma \in [0, 1]$ is the discounting factor that describes the importance of delayed reward relative to immediate ones. The accumulated reward (or return) for an agent is calculated as: $G = \Sigma_{t=0}^{H-1} \gamma^t \cdot R(s_t)$ such that H is the task horizon.

The action strategy of an RL agent is given by its policy $\pi : S \rightarrow p(a|s)$ which defines the mapping from states to action probabilities. RL algorithms aim to learn an optimal policy π^* that maximises the expected long-term reward, i.e. $\pi^* = \underset{\pi}{\mathrm{argmax}}\ \mathbb{E}[G|\pi]$. A wide variety of algorithms have been proposed to search for π^*. The focus of the current paper is on non-stationarity detection in model-free RL algorithms that use value functions. As such, the rest of this section is devoted to explaining the concept of model-free algorithms and defining the role of value functions in RL.

2.2   Model-Based Versus Model-Free RL Algorithms

RL algorithms can be model free or model based [15]. *Model-based* algorithms are those in which the RL agent has access to or explicitly learns a model of the environment and its dynamics (i.e. the transition T and reward R functions). Such a model allows the agent to plan ahead by estimating the consequences of its actions before deciding on which action to execute. This results in significant speedup in the learning. However, a model of the environment is usually not available and the model learning process comes with significant challenges that make it difficult for the model-based algorithms to pay off. On the other hand, *model-free* algorithms learn directly without relying on a model of the environment which makes these algorithms easier to use [1]. Advances in deep learning have furthered the impact of model-free RL by enhancing its ability to deal with large state spaces [3]. As such, model-free algorithms have received an extensive deal of attention and have been successfully applied to various domains. The rest of this section presents the key approaches to model-free RL.

2.3   Value-Based, Policy-Based, and Actor-Critic Algorithms

There are three approaches to model-free RL based on whether the algorithm uses value functions, policy search, or both for finding an effective policy. The first approach, referred to as *value-based* algorithms, uses a learnable value function to estimate how good it is to be in a given state. Two types of value functions have been used: state value function $V(s)$ and state-action value function (or quality function) $Q(s, a)$. The state value function $V(s) : S \rightarrow \mathbb{R}$ approximates the expected return when starting at state s and then following π afterwards.

$$V(s) = R(s) + \gamma \cdot \sum_{s'} p(s'|s, \pi(s)) \cdot V(s') \tag{1}$$

V(s) contains useful information about the utility of a state, however this information is not sufficient for determining the actions that lead to the desirable states. On the other hand, the state-action value function $Q(s, a) : S \times A \rightarrow \mathbb{R}$

approximates the expected return for performing action a at state s and then following π afterwards.

$$Q(s,a) = R(s) + \gamma \cdot \sum_{s'} p(s'|s,a) \cdot V(s') \tag{2}$$

The Q function holds sufficient information for differentiating between the available actions, and is thus used in value-based algorithms to infer the policy $\pi(s) = \arg\max_a Q(s,a)$. The main weakness of value-based algorithms is that they suffer from convergence issues in many settings [13].

The second approach to model-free RL is used in *policy-based* algorithms which aim to directly learn a near-optimal policy π by updating the policy parameters to maximise the expected return. Policy-based algorithms are effective for high-dimensional or continuous action spaces. However, they suffer from high variance in their gradient estimator and they are prone to local minima, particularly for large number of parameters [13]. The third approach, *actor-critic*, aims to combine the advantages of the first two approaches by learning both an explicit policy (or actor) and a value function (or critic). The actor selects actions, whereas the critic estimates the value function to evaluate the performance of the policy.

3 Related Work

Non-stationarity is one of the key challenges faced by RL. Several studies have proposed solutions to deal with non-stationarity in RL domains as discussed in this section.

3.1 Context Detection

Among the early studies on modeling non-stationary environments is the work by Choi et al. [6] who extended MDP to hidden-mode MDP (HM-MDP). HM-MDP comprises a set of MDPs, each represents a mode of the environment. The MDPs share the same state and action spaces, but have different transition and/or reward functions. Choi et al. proposed a variant of the Baum-Welch algorithm to learn the HM-MDP model of a given environment. Once learnt, model-based RL can be used for finding an optimal policy. A key limitation of this work is that it assumes the number of modes is small and known a priori and that the probability of transition between modes is fixed.

Da Silva et al. [7] proposed an algorithm for dealing with unknown number of modes. The algorithm deploys multiple partial models of the environment, each represents a single mode. Mode change detection is performed by monitoring the prediction error of the active model. When the error crosses a threshold λ, a change is detected and the model with the lowest error is activated. If the error of all the existing models exceeds λ, a new model is created and learnt. Recently, Hadoux et al. [11] adapted the algorithm in [7] by replacing the change detection component by a component based on statistical tests for sequential change point detection. Their algorithm is theoretically based and it requires less parameter tuning than [7].

All the above–mentioned solutions need to learn partial models of the environments. However, model learning is a challenging task, as discussed in Sect. 2.2. This is exacerbated by non-stationarity due to the need to learn several models in addition to deciding which model should be updated at each time frame. Another limitation of these techniques is the inherent assumption that mode transitions are abrupt which is not the case in many real-life environments where changes occur gradually.

3.2   Model-Free Non-stationarity Detection

Model-free RL techniques are widely used across various domains due to their great success and ease of use. Most existing techniques for non-stationarity detection need to learn a model for each environment mode, which represents a significant computation overhead for model-free algorithms. Canonaco et al. [5] proposed a method for non-stationarity detection for policy-based RL algorithms. Their method aims to detect changes between any two training iterations i and $i + k$ through the use of a two-level module. The lower level applies importance sampling on trajectories collected at the two iterations i, $i + k$. This allows for statistically testing whether the expected value of the reward at these iterations are similar given a common policy π_μ where π_μ is selected via an optimisation step. The outcome of this statistical test is then fed into the higher-level module which performs sequential analysis by means of a CUSUM approach to detect non-stationarity. This algorithm has been shown to effectively detect changes with different magnitudes and speeds. However, it is only limited to policy-gradient RL algorithms. Our proposed technique is similar to this approach as they both are model-free. However, our technique is targeted to value-based and actor-critic RL algorithms.

3.3   Multi-task Learning and Meta-learning

Multi-task learning and meta-learning are active areas in RL that deal with learning under non-stationarity. In multi-task learning [21], an agent is required to optimise its performance on a set of related tasks that might have different transition or reward functions. The underlying assumption in multi-task learning is that tasks have a common structure such that sharing experience between tasks leads to more efficient learning than learning each task in isolation. On the other hand, in meta-learning [8] an agent is exposed to a set of tasks sampled from a fixed distribution. The goal of meta-learning is to use these training tasks to learn how to quickly adapt to previously unseen tasks sampled from the same distribution. Both multi-task learning and meta-learning assume that task boundaries are defined and hence do not require specific techniques for non-stationarity detection. Instead, they focus on the adaptation techniques themselves.

4 Non-stationarity Detection via Value Function Monitoring

As discussed in Sect. 2.3, the primary usage of value functions in RL is either to determine action selection or to provide a critique for the policy to improve its performance. In both cases, value functions perform their role by giving crucial information about the expected return of a state or of performing an action at a given state. We hypothesise that after convergence this information could also be useful for detecting non-stationarity. This is because when the policy is fixed, the probabilities of state trajectories in a given environment are determined by the environment dynamics. Consequently, given a fixed policy, the environment dynamics also determine the distributions of the rate at which state utility improves within an episode. Monitoring this rate can simply be done by monitoring the change in value function. When the distribution of this rate changes, this reflects a change in the dynamics. We use V in the following formulation, however very similar formulation can be replicated for Q. Assume that the state-value function V of an RL agent has been trained till convergence under a given stationary context. Let $\mathbb{E}[\Delta V^w|_{\rho,T}]$ denote the expected change in V after time window w from the start of an episode, given the initial state distribution ρ and the transition function T. After one time step ($w = 1$), the expected change in V is given by the equation:

$$\mathbb{E}[\Delta V^1_{\rho,T}] = \sum_{s_j} p(s_j) \cdot \left(\sum_{s_i} p(s_i|s_j, \pi(s_j)) \cdot V(s_i) - V(s_j) \right) \qquad (3)$$

Such that $p(s_j)$ is the probability that s_j is the initial state and $p(s_i|s_j, \pi(s_j))$ is the probability of transitioning from s_j to s_i when following the policy π. As the values of $V(s_i), V(s_j), \pi(s_j)$ in Eq. 3 are constant after convergence, $\mathbb{E}[\Delta V^1]$ is determined by the distribution of initial states and the transition function.

4.1 Change in Initial State Distribution

First, consider a change in the initial state distribution from ρ to ρ' such that the new initial state probabilities can be expressed as:

$$p'(s) = p(s) + \delta_s \quad s.t. \sum_{s \in S} \delta_s = 0, \ \exists \delta_s \neq 0 \qquad (4)$$

In this case, the distribution of ΔV for $w = 1$ changes and its expectation can be obtained by replacing $p(s_j)$ by $p(s_j) + \delta_{sj}$ in Eq. 3, as follows:

$$\mathbb{E}[\Delta V^1_{\rho',T}] = \sum_{s_j} (p(s_j) + \delta_{sj}) \cdot \left(\sum_{s_i} p(s_i|s_j, \pi(s_j)) \cdot V(s_i) - V(s_j) \right) \qquad (5)$$

By re-organising the terms of Eq. 5, $\mathbb{E}[\Delta V^1_{\rho',T}]$ can be expressed as:

$$\mathbb{E}[\Delta V^1_{\rho',T}] = \mathbb{E}[\Delta V^1_{\rho,T}] + \sum_{s_j} \delta_{sj} \cdot \left(\sum_{s_i} p(s_i|s_j, \pi(s_j)) \cdot V(s_i) - V(s_j) \right) \qquad (6)$$

The difference between $\mathbb{E}[\Delta V^1_{\rho,T}]$ and $\mathbb{E}[\Delta V^1_{\rho',T}]$ is generally non-zero as δ_{sj} is non-zero for at least a subset of states. For the generic case of $w = \omega$, the expected change in value function is calculated as:

$$\mathbb{E}[\Delta V^{\omega}_{\rho',T}] = \sum_{s_j} \left(p(s_j) + \delta_{sj}\right) \cdot \left(\sum_{s_i} p(s(\omega) = s_i | s(0) = s_j, \pi) \cdot V(s_i) - V(s_j)\right)$$

$$(7)$$

such that $p(s(\omega) = s_i | s(0) = s_j, \pi)$ is the probability of transitioning to state s_i at time ω when starting at state s_j and following π. This probability is independent of the initial state distribution. Hence, Eq. 7 can be rewritten as:

$$\mathbb{E}[\Delta V^{\omega}_{\rho',T}] = \mathbb{E}[\Delta V^{\omega}_{\rho,T}] + \sum_{s_j} \delta(s_j) \cdot \left(\sum_{s_i} p(s(\omega) = s_i | s(0) = s_j, \pi) \cdot V(s_i)\right.$$

$$\left. - V(s_j)\right) \qquad (8)$$

The second term in Eq. 8 is zero only if the summed terms cancel each other, that is only if the change from ρ to ρ' causes no change to the average performance.

4.2   Change in the Transition Function

Now, consider the case when the transition function changes into T'. This can be expressed as follows:

$$p'(s_i|s_j, a) = p(s_i|s_j, a) + \delta^1_{si,sj,\pi}$$

$$\text{s.t. } \exists s_i, s_j \in S \,, \exists a \in A \; \delta^1_{si,sj,a} \neq 0 \qquad (9)$$

In this case, $\mathbb{E}[\Delta V^1_{\rho,T'}]$ can be calculated by replacing $p(s_i|s_j, \pi(s_j))$ in Eq. 3 by $p(s_i|s_j, \pi(s_j)) + \delta^1_{si,sj,\pi}$ as follows:

$$\mathbb{E}[\Delta V^1_{\rho,T'}] = \sum_{s_j} p(s_j) \cdot \left(\sum_{s_i} \left(p(s_i|s_j, \pi(s_j)) + \delta^1_{si,sj,\pi}\right) \cdot V(s_i) - V(s_j)\right) \quad (10)$$

Re-organising the terms yields:

$$\mathbb{E}[\Delta V^1_{\rho,T'}] = \sum_{s_j} p(s_j) \cdot \left(\sum_{s_i} p(s_i|s_j, \pi(s_j)) \cdot V(s_i) - V(s_j)\right) +$$

$$\sum_{s_j} p(s_j) \cdot \sum_{s_i} \delta^1_{si,sj,\pi} \cdot V(s_i) \qquad (11)$$

$$\mathbb{E}[\Delta V^1_{\rho,T'}] = \mathbb{E}[\Delta V^1_{\rho,T}] + \sum_{s_j} p(s_j) \cdot \sum_{s_i} \delta^1_{si,sj,\pi} \cdot V(s_i) \qquad (12)$$

For $w = \omega$, the expectation of ΔV is calculated as:

$$\mathbb{E}[\Delta V^{\omega}_{\rho,T'}] = \sum_{s_j} p(s_j) \cdot \left(\sum_{s_i} p'(s(\omega) = s_i | s(0) = s_j, \pi) \cdot V(s_i) - V(s_j) \right) \quad (13)$$

This can be written in terms of $\mathbb{E}[\Delta V^{\omega}_{\rho,T}]$ as follows:

$$\mathbb{E}[\Delta V^{\omega}_{\rho,T'}] = \mathbb{E}[\Delta V^{\omega}_{\rho,T}] + \sum_{s_j} p(s_j) \cdot \left(\sum_{s_i} \delta^{\omega}_{si,sj,\pi} \cdot V(s_i) \right) \quad (14)$$

such that $\delta^{\omega}_{si,sj,\pi} = p'(s(\omega) = s_i | s(0) = s_j, \pi) - p(s(\omega) = s_i | s(0) = s_j, \pi)$. The second term of Eq. 14 is non-zero except when the summed terms cancel each other which is the case when the change in transition function leads to no change to the average performance. As such, monitoring the distribution of ΔV^{ω} can be used to detect changes in the dynamics of the environment.

4.3   Non-stationarity Detection Algorithm

Given that the distribution of ΔV^{ω} is sensitive to changes in both ρ and T, non-stationarity can be revealed by monitoring the distribution of ΔV^{ω} across different time frames. To do this, we use a fixed-length buffer B to store the values of $\Delta V / episode\,length$ from the most recent episodes. We store $\Delta V / episode\,length$ rather than ΔV to allow for environments with variable-length episodes. The last n elements in the buffer are compared to the rest of the elements by statistically testing the hypothesis that both sets of elements have similar distributions. T-test for independent samples is used for hypothesis testing such that if the $p - value < \alpha$, the hypothesis is rejected and the change detection alarm is activated. Algorithm 1 provides the steps used for detecting non-stationarity. The parameters used are $\alpha = .05$, $length = 100$, and $n = 10$.

Algorithm 1. Non-stationarity detection algorithm.

1: **INPUT** B, V_{start}, V_{end}, $episode\_length$
2: **Constant** $length$, n, α
3: $alarm = false$
4: $delta = (V_{end} - V_{start})/episode\_length$
5: $B = append(B, delta)$
6: **if** length(B)$< length$ **then**
7:     **return** $alarm$, B
8: **end if**
9: $p\_value = t - test(B[1 : length - n], B[length - n : n])$
10: **if** $p\_value < \alpha$ **then**
11:     $alarm = true$
12: **end if**
13: $B = remove\_at(B, 1)$
14: **return** $alarm$, B

---

5   Test Environments

This section presents the three benchmark environments used for evaluation and describes how non-stationarity is introduced into them. These environments have been selected from the *OpenAI Gym* kit. They are *Pendulum-v0*, *MountainCar-v0*, and *LunarLanderContinuous-v2*.

Pendulum-v0 is a classic inverted pendulum problem where a frictionless pendulum starts at a random position. The goal is to swing the pendulum up so that it stays upright. The environment has three observations $[sin(\theta), cos(\theta), \theta\cdot]$ representing the angle and angular velocity of the pendulum. A one-dimensional action $a \in [-2, 2]$ is used representing the force applied on the pendulum. The reward function R(s) is $-(\theta^2 + 0.1 \cdot (\theta\cdot)^2 + 0.001 \cdot a^2)$. The environment has fixed-length episodes with its horizon set to $H = 200$ time steps.

MountainCar-v0 is a standard RL problem where an under-powered car is positioned in a valley between two mountains. The goal is to drive up the mountain on the right. As the car engine is not powerful enough to make it accelerate up the steep slope, the car needs to learn to drive back and forth to build up momentum. The state vector consists of the car position and velocity $[x, x\cdot]$. The car has three discrete actions: accelerate to the left, do not accelerate, and accelerate to the right. A reward of -1 is given every time step spent in the task before reaching the goal. When the goal is reached, a reward of 0 is given and the episode terminates. This environment has variable-length episodes with its maximum horizon set to $H = 1000$ time steps per episode.

LunarLanderContinuous-v2 consists of a lander and a landing pad situated between two flags at specific locations in an environment characterised by low gravity. The goal is to land at the landing pad as softly and fuel-efficiently as possible using the three engines available on the lander. The episode ends when the lander comes to rest at the landing pad or when it crashes. The input state is an 8-dimensional vector representing the position x, y, velocity $x\cdot, y\cdot$, angle θ, angular velocity $\theta\cdot$, and two binary variables L, R representing whether the left and right leg have contact with the ground, respectively. The action is a 2-dimensional vector of float numbers $[a_{main}, a_{side}]$ such that a_{main} is the power applied to the main engine, and a_{side} is the power applied to one of the side engines (if $a_{side} < 0$ then left engine, otherwise right engine). The reward function is a linear combination of the following terms: $-0.3 \cdot a_{main}$, $-0.03 \cdot a_{side}$, 100 for resting at the landing pad, -100 for crashing, and a shaping component. This environment has variable-length episodes with its maximum horizon set to $H = 1000$ time steps per episode.

These three environments are originally stationary. To introduce non-stationarity, we changes task dynamics in a way similar to [5]. Let i^* denote the episode at which non-stationary starts. The magnitude of the change and the duration of its transient component are controlled by two variables e and c such that the change increases linearly from 0 at episode i^* to its maximum value of $e \cdot c$ at episode $i^* + c$. The pendulum environment is modified to include non-stationarity by changing its state vector to $[sin(\theta), cos(\theta), \theta + v]$, where v is calculated as:

$$v = \begin{cases} 0 & \text{if } i < i^* \\ e \cdot (i - i^*) & \text{if } i^* \leq i < c \\ e \cdot c & \text{otherwise} \end{cases}$$

The mountain car environment is modified by changing its state vector to $[x, x - v]$. The lunar lander environment is modified by changing the action applied to the main engine to $a_{main} + v$.

6   Experimental Evaluation

Three state-of-the-art RL agents have been used for experimentation: Deep Q-Network (DQN) [14], Proximal Policy Optimisation (PPO) [19], and Twin Delayed Deep Deterministic Policy Gradients (TD3) [9]. DQN is an example of value-based algorithms, PPO is an example of on-policy actor-critic algorithms, and TD3 is an example of off-policy actor-critic algorithms. The implementation of these RL agents is based on *Stable-Baselines3* [18] with some modifications to allow for non-stationarity detection. The parameters for these RL algorithms are set to their default values in *Stable-Baselines 3*. DQN has been used for the mountain car environment, TD3 has been used for pendulum, and PPO has been used for lunar lander.

Table 1. Results of the proposed algorithms in the three evaluation environments. For DD, the mean and standard deviation are reported.

Environment	Change parameters	FPR	FNR	DD
Pendulum	$e = 0.025, c = 20$	.016	0	8.3 (4.24)
	$e = 0.025, c = 30$	.024	0	10.9 (4.57)
	$e = 0.05, c = 20$	.008	0	7.8 (3.34)
	$e = 0.05, c = 30$	.12	0	5.78 (4.61)
Lunar lander	$e = 0.04, c = 20$	.012	0	18 (7.89)
	$e = 0.04, c = 30$	.014	0	18.3 (5.92)
	$e = 0.06, c = 20$	.046	0	15.8 (7.24)
	$e = 0.06, c = 30$	.044	0	15.9 (6.35)
Mountain car	$e = 2.5 \cdot 10^{-5}, c = 20$	.018	0	8.4 (2.33)
	$e = 2.5 \cdot 10^{-5}, c = 30$	.016	0	7.5 (2.87)
	$e = 5 \cdot 10^{-5}, c = 20$	.008	0	5.6 (1.69)
	$e = 5 \cdot 10^{-5}, c = 30$	.014	0	5.7 (1.62)

The experiments comprise two stages: initial training and continual training. During the initial training, each RL agent has been trained till convergence in its environment under stationarity by setting $e = 0$ and $c = 0$ in Eq. 5. The resulting RL models have been saved. Then, in the continual training stage, the pre-trained models have been loaded and training is continued with the episode counter reset to $i = 0$. The algorithm for non-stationarity detection has been activated at episode $\tilde{i} = 250$. A change in the environment dynamics is introduced at episode $i^* = 300$. Different combinations of e and c have been used for experimentation with ten random seeds for each experiment.

False positive rate (FPR), false negative rate (FNR), and detection delay (DD) are used for evaluation. A false positive event happens whenever an alarm is raised at $i < i^*$, whereas a false negative event happens if no alarm is raised at any episode i where $i^* < i < i^* + 50$. When correct detection first happens at episode $i > i^*$, detection delay is calculated as $i - i^*$. Table 1 lists the results of the change detection algorithm in the three test environments. The results show that false positives happened at very low rates whereas false negatives never happened in any of the experiments. Across all test environments, the length of the transition phase c did not have a statistically significant effect on the detection delay. The increase in the initial change magnitude e resulted in a decrease in the detection delay in all the environments such that the decrease was significant ($p < .05$) in the mountain car environment.

We also investigated the potential of using information from the value functions to differentiate between different contexts. Context identification enables the agent to determine if the current context has been previously encountered and if so, to activate the corresponding learnt policy. To enable this, we used supervised leaning with a feed forward neural network consisting of a single hidden layer consisting of 60 neurons. The network input is a 2D vector containing the initial value function (V_{start} or Q_{start}) and the rate of change in value function (ΔV or ΔQ). The output layer consists of four neurons to classify the four different contexts for each of the three environments, as shown in Table 1.

The training results are provided in Fig. 1. The classification accuracy varied between the three environments. The highest accuracy of 88.2% was recorded for the pendulum environment, followed by an accuracy of 83.3% for the mountain car environment, and lastly the accuracy for the lunar lander environment was only 54.9%. The results are consistent with the context detection results shown in Table 1 where the lunar lander environment recorded the highest DD and FPR reflecting the relative difficulty of detecting changes in this environment. It is worth mentioning that event the lowest identification accuracy of 54.9% represents a considerable improvement over mere context detection where a random guess would lead to a 25% accuracy.

7   Summary and Future Work

Non-stationarity is a key challenge that RL has to overcome before being used in many real environments. The ability to correctly detect when the environmental dynamics change is crucial for activating adaptation mechanisms at the right time. Most previous work in change detection used model-based techniques which incur significant computational costs for model-free RL. This paper proposes a

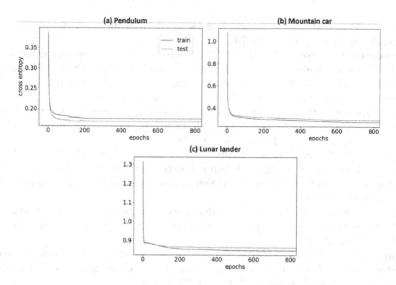

Fig. 1. Context identification training results.

low-cost approach for change detection which can be used in RL models that use value functions. The proposed approach monitors the distribution of ΔV (or ΔQ) over time and applies statistical test to detect when the distribution changes. The proposed approach has been applied to three benchmark RL environments and has demonstrated its ability to detect different levels of changes that are being introduced at different speeds.

The experimental results also showed that different changes in the dynamics can have distinct effects on the distribution of ΔV (or ΔQ). This suggests that the proposed algorithm can be extended to perform context identification. This is useful in scenarios where different policies can be learnt and saved for different contexts. Context identification in this case will be used to select which policy should be activated at a given time. Our future work will investigate the use of the proposed approach for context identification.

Acknowledgement. This work was funded by the Department of Defence and the Office of National Intelligence under the AI for Decision Making Program, delivered in partnership with the NSW Defence Innovation Network Grant Number RG213520.

References

1. Achiam, J.: Spinning up in deep reinforcement learning. GitHub repository (2018)
2. Al-Shedivat, M., Bansal, T., Burda, Y., Sutskever, I., Mordatch, I., Abbeel, P.: Continuous adaptation via meta-learning in nonstationary and competitive environments. In: International Conference on Learning Representations (2018)
3. Arulkumaran, K., Deisenroth, M.P., Brundage, M., Bharath, A.A.: Deep reinforcement learning: a brief survey. IEEE Sig. Process. Mag. **34**(6), 26–38 (2017)

4. Berner, C., et al.: Dota 2 with large scale deep reinforcement learning. arXiv preprint arXiv:1912.06680 (2019)
5. Canonaco, G., Restelli, M., Roveri, M.: Model-free non-stationarity detection and adaptation in reinforcement learning. In: Proceedings of the Twenty-Fourth European Conference on Artificial Intelligence, ECAI, pp. 1047–1054. IOS Press (2020)
6. Choi, S.P.M., Yeung, D.-Y., Zhang, N.L.: Hidden-mode Markov decision processes for nonstationary sequential decision making. In: Sun, R., Giles, C.L. (eds.) Sequence Learning. LNCS (LNAI), vol. 1828, pp. 264–287. Springer, Heidelberg (2000). https://doi.org/10.1007/3-540-44565-X_12
7. Da Silva, B.C., Basso, E.W., Perotto, F.S., Bazzan, A.L.C., Engel, P.M.: Improving reinforcement learning with context detection. In: Proceedings of the Fifth International Joint Conference on Autonomous Agents and Multiagent systems, pp. 810–812 (2006)
8. Finn, C., Abbeel, P., Levine, S.: Model-agnostic meta-learning for fast adaptation of deep networks. In: International Conference on Machine Learning, pp. 1126–1135. PMLR (2017)
9. Fujimoto, S., Hoof, H., Meger, D.: Addressing function approximation error in actor-critic methods. In: Proceedings of the Thirty-Fifth International Conference on Machine Learning, pp. 1587–1596. PMLR (2018)
10. Gamrian, S., Goldberg, Y.: Transfer learning for related reinforcement learning tasks via image-to-image translation. In: International Conference on Machine Learning, pp. 2063–2072. PMLR (2019)
11. Hadoux, E., Beynier, A., Weng, P.: Sequential decision-making under non-stationary environments via sequential change-point detection. In: Learning Over Multiple Contexts (LMCE) (2014)
12. Jaderberg, M., et al.: Human-level performance in 3D multiplayer games with population-based reinforcement learning. Science **364**(6443), 859–865 (2019)
13. Konda, V., Tsitsiklis, J.: Actor-critic algorithms. In: Advances in Neural Information Processing Systems, vol. 12 (1999)
14. Mnih, V., et al.: Playing atari with deep reinforcement learning. In: NIPS Deep Learning Workshop (2013)
15. van Otterlo, M., Wiering, M.: Reinforcement learning and Markov decision processes. In: Wiering, M., van Otterlo, M. (eds.) Reinforcement Learning. ALO, vol. 12, pp. 3–42. Springer, Heidelberg (2012). https://doi.org/10.1007/978-3-642-27645-3_1
16. Pope, A.P., et al.: Hierarchical reinforcement learning for air-to-air combat. arXiv preprint arXiv:2105.00990 (2021)
17. Qiao, G., Weiss, B.A.: Quick health assessment for industrial robot health degradation and the supporting advanced sensing development. J. Manuf. Syst. **48**, 51–59 (2018). Special Issue on Smart Manufacturing
18. Raffin, A., Hill, A., Gleave, A., Kanervisto, A., Ernestus, M., Dormann, N.: Stable-baselines3: reliable reinforcement learning implementations. J. Mach. Learn. Res. **22**(268), 1–8 (2021)
19. Schulman, J., Wolski, F., Dhariwal, P., Radford, A., Klimov, O.: Proximal policy optimization algorithms. arXiv preprint arXiv:1707.06347 (2017)
20. Taylor, M.E., Stone, P.: Transfer learning for reinforcement learning domains: a survey. J. Mach. Learn. Res. **10**(7), 1633–1685 (2009)
21. Yu, T., Kumar, S., Gupta, A., Levine, S., Hausman, K., Finn, C.: Gradient surgery for multi-task learning. In: Larochelle, H., Ranzato, M., Hadsell, R., Balcan, M.F., Lin, H. (eds.) Advances in Neural Information Processing Systems, vol. 33, pp. 5824–5836. Curran Associates, Inc. (2020)

Toward a Unified Framework for RGB and RGB-D Visual Navigation

Heming Du, Zi Huang, Scott Chapman, and Xin Yu[✉]

The University of Queensland, Brisbane, Australia
xin.yu@uq.edu.au

Abstract. In object-goal navigation, an agent is steered toward a target object based on its observations. The solution pipeline is usually composed of scene representation learning and navigation policy learning: the former reflects the agent observation, and the latter determines the navigation action. To this end, this article proposes a unified visual navigation framework, dubbed VTP, which can employ either RGB or RGB-D observations for object-goal visual navigation. Using a unified Visual Transformer Navigation network (VTN), the agent analyzes image areas in relation to specific objects, producing visual representations that capture both instance-to-instance relationships and instance-to-region relationships. Meanwhile, we utilize depth maps to explore the spatial relationship between instances and the agent. Additionally, we develop a pre-training scheme to associate visual representations with navigation signals. Furthermore, we adopt Tentative Policy Learning (TPL) to guide an agent to escape from deadlocks. When an agent is detected as being in deadlock states in training, we utilize Tentative Imitation learning (TIL) to provide the agent expert demonstrations for deadlock escape and such demonstrations are learned in a separate Tentative Policy Network (TPN). In testing, under deadlocks, estimated expert demonstrations are given to the policy network to find an escape action. Our system outperforms other methods in both iTHOR and RoboTHOR.

Keywords: Visual Navigation · Reinforcement Learning · Vision Transformer

1 Introduction

The goal of target-driven visual navigation, or object-goal navigation, is to guide an agent to reach instances of a given target category based on visual observations. A typical object-goal navigation system has two major components, *i.e.*, visual representation learning and navigation policy learning. An agent first extracts a visual representation from the observation and then passes the visual representation and experienced states through a navigation policy for an action prediction, *e.g.*, `MoveAhead` and `RotateLeft`.

In order to accomplish this task with effectiveness and efficiency, it is highly desirable to extract informative visual representations correlated to directional

© The Author(s), under exclusive license to Springer Nature Singapore Pte Ltd. 2024
T. Liu et al. (Eds.): AI 2023, LNAI 14472, pp. 363–375, 2024.
https://doi.org/10.1007/978-981-99-8391-9_29

navigation signals from observations. We observe that the agent may fail to move toward the target due to being stuck in deadlocks, *e.g.*, continuously repeating the same action, in complex environments, and thus a deadlock-free navigation policy is critical for the agent. Meanwhile, given that a successful episode requires the agent to be within 1.5 m of the target, some agents do not achieve this requirement because of imprecise estimations of the distance between them and the target. Therefore, it is vital to explore the spatial relationships between the agent and the target instances.

In this paper, we propose a unified visual navigation system named VTP that can steer agents based on both RGB and RGB-D observations to target objects effectively and efficiently. In VTP, we first introduce a unified Visual Transformer Navigation network (VTN) to generate expressive visual representations that embody two key properties: (i) relationships among all instances in a scene and (ii) the spatial locations of objects and image regions. Furthermore, we aim to encode the relationships between the agent and all observed instances by leveraging the depth input. Then, we present a pre-training scheme to facilitate navigation policy learning by associating the visual representation with directional navigation signals. To be specific, VTN exploits two spatial-aware descriptors as the key and query (*i.e.*, a spatial-enhanced local descriptor and a positional global descriptor) and then encode the descriptors to obtain a comprehensive visual representation.

In order to capitalize on all detected instances, we encode instances by spatial-enhanced local descriptors. The spatial-enhanced local descriptors are designed to explore spatial and category relationships among instances. Here, we employ an object detector DETR [1] to detect instances from the given RGB observations. DETR features not only encode object-specific information, such as class labels and bounding boxes, but also contain the relations between instances and global contexts. Considering object positions cannot be explicitly decoded without the feed-forward layer of DETR, we enhance all the detected instance features with their locations to obtain spatial-enhanced local descriptors. Additionally, when the agent receives a depth observation, we encode the distance between instances and the agent into the spatial-enhanced local descriptors. Therefore, the agent can explore the relationships between instances and agents for effective navigation. Then, we take all the spatial-enhanced local descriptors as the key of the VTN encoder to model the relationships among detected instances, such as category concurrence and spatial correlations.

Furthermore, we apply a positional global descriptor as the query of the VTN decoder to encode the spatial locations of objects and image regions. Specifically, we divide a global observation into multiple regions based on spatial layouts and assign a positional embedding to each regional feature for a positional global descriptor. It facilitates the exploration of the correspondences between navigation actions and image regions. Then, we attend the spatial-enhanced local descriptor to the positional global descriptor query to learn the spatial relationships between instances and observation regions via the VTN decoder. Subsequently, the output of the VTN decoder is flattened as a visual representation.

After obtaining an expressive visual representation, we employ the Asynchronous Advantage Actor Critic (A3C) algorithm [2] with a standard Long Short Term Memory (LSTM) network [3] to learn a navigation policy. However, we found that directly training our VTN with a navigation policy network fails to converge due to the training difficulty of Transformers [4]. Therefore, we present a pre-training scheme to endow our VTN with the capability of encoding directional navigation signals. Specifically, the pre-training scheme pushes the system to make moves even if it might take longer, ensuring better outcomes. After warming-up through expert experience, VTN provides instructional representations for navigation.

Given a visual representation encoded with directional navigation signals, an agent aims at learning a navigation policy network to map the current visual representation and previous states to action toward target instances. To achieve this goal, the agent will be punished as the number of actions increases and rewarded when it reaches a target, thereby optimizing the navigation policy network for a high reward in each episode. However, we notice that an agent might fail to reach a target and is stuck in a deadlock state, e.g., endlessly repeating the same actions. Only using reinforcement learning in training cannot solve this problem since the reward does not provide explicit guidance on leaving deadlock states to an agent. Thus, explicit instruction is required to provide when an agent is trapped in a deadlock.

Therefore, we introduce Tentative Policy Learning (TPL) to provide explicit deadlock escape guidance to agents duly. TPL consists of two modules, i.e., Trial-driven Imitation Learning (TIL) and a memory-augmented Tentative Policy Network (TPN). TIL is inspired by human trial-and-practice behaviors and designed to guide an agent to escape deadlock in training. Specifically, if an agent gets stuck, like going in circles, we give it clear action instructions (i.e., expert experience) to follow.

However, in unseen environments, TIL is not available to an agent, and the agent may still fall into a deadlock due to the complexity of the scenes. In order to enable an agent to avoid deadlock states in testing, we develop a memory-augmented TPN. TPN employs an external memory to spot when agents might be stuck by looking at past visuals. Then, TPN utilizes an internal memory that stores the past state and action pairs to generate explicit instructions for the agent, allowing it to leave a deadlock in testing. Different from the previous work [5] providing a scalar reward in testing, TPN updates the navigation policy network based on explicit action instructions in testing. Thus, our method obtains an efficient and effective navigation policy.

In the popular widely-used navigation environments iTHOR and RoboTHOR [6], our method significantly outperforms the state-of-the-art. Experimental results demonstrate that our visual navigation system outperforms comparable methods in the Success Rate under RGB and RGB-D inputs, respectively. Our contributions are summarized as follows:

- We design a unified visual navigation system, called VTP, to approach target instances with either RGB or RGB-D images effectively and efficiently.

– Experimental results demonstrate that our method significantly improves the baseline visual navigation systems in unseen environments with both RGB and RGB-D observations.

2   Related Works

Visual navigation has grown a lot lately as an important task in robotics. Previous methods [7,8] use maps to navigate with three main steps: making maps, finding places, and making routes. Some methods [9,10] use maps to obviate obstructions. Dissanayake et al. [11] use Simultaneous Localization and Mapping (SLAM) to figure out robot positions.

With the progress of Deep Neural Networks [4,12], Gupta et al. [13] build maps and make routes via Cognitive Mapping and Planning (CMP). Meanwhile, Reinforcement Learning [14] has been used in navigation. Mirowski et al. [15] design a reinforcement learning based method to navigate 3D mazes. Meanwhile, [16,17] focus on avoiding collision in mazes. There are also methods [18,19] utilizing both visual features and the topological guidance of scenes to guide navigation.

Some works et al. [17,20] propose a Transformer-based method to handle navigation. Kahn et al. [21] design a self-driven method to understand surroundings. Wu et al. [22] focus on room layout exploration instead of navigation. Shen et al. [23] adopt multiple visual representations to decide on actions. Furthermore, Zhu et al. [24] aim to navigate toward a given image. Yang et al. [25] try to associate different object categories based on external knowledge. Mousavian et al. [26] use visual clues via semantic segmentation and object detection for navigation guidance. Wortsman et al. [5] use word hints to show target types and create a guiding system. Moreover, Du et al. [27] introduce a graph-based network called ORG to explore relationships among instances. Zhang's team [28] suggests a system, HOZ, to give agents small goals as they move toward the main target. Lastly, Khandelwal [29] adopt CLIP [30] to extract visual representations for navigation policy.

3   VTP: A Unified Visual Navigation System

The goal is to introduce an informative visual representation and a robust navigation policy for a unified visual navigation system. As illustrated in Fig. 1, VTP includes three parts: (i) a visual representation extracted from observations; (ii) a navigation policy based on the visual representations and previous states; (iii) a tentative policy utilizing historical states and actions.

3.1   Task Definition and Setup

In the object goal visual navigation task, an agent must navigate to a target object in an unknown environment. The agent has no prior knowledge

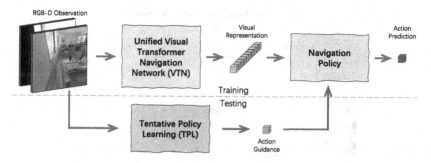

Fig. 1. Overview of VTP. Our VTP involves a unified Visual Transformer Navigation network (VTN) with a pre-training scheme, Tentative Policy Learning (TPL) and a navigation policy network.

about the environment, and the only information it has is the current view and the previous state. The agent can move between grid points using six different actions: MoveAhead, RotateLeft, RotateRight, LookUp, LookDown, Done. Following the works [5,27], an episode is considered successful if the agent reaches the target object within the allowed number of steps. Meanwhile, the target should be in the observation of the agent and the distance between the agent and the target is less than 1.5 m.

At the beginning of each episode, a target object and a start state are randomly generated. At each timestamp, the agent records an image from its vision sensor. The agent then uses this image and the previous state to generate a policy that predicts the distribution of actions at the current time step. The agent selects the action with the highest probability for navigation.

3.2 Visual Navigation Pipeline

Visual navigation frameworks typically use two components to navigate in complex environments: visual representation learning and navigation policy prediction. Visual representation learning extracts features from images that are relevant to navigation. Meanwhile, navigation policy adopts extracted visual representations to predict the best action to take at a given time. This can be done using reinforcement learning algorithms, such as Asynchronous Advantage Actor-Critic (A3C) architecture [2]. The navigation policy network takes the combination of the current visual representation, the previous action, and the state embedding as input. Then the policy outputs the action distribution and value. The agent selects the most probable actions from the predicted policy and uses the predicted value to train the navigation policy network.

3.3 Unified Visual Transformer Navigation Network

In our VTP, we adopt a unified Visual Transformer Navigation Network (VTN) to explore the relationship among all objects and their spatial correlations.

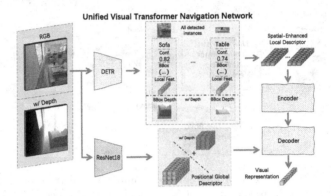

Fig. 2. Illustration of VTN. We combine the detection results and instance features into a spatial-enhanced local descriptor. Meanwhile, we extract a positional global descriptor from the observation. In both branches, we can easily encode spatial relationships by exploiting depth maps. The spatial-enhanced local and positional global descriptors are fused by the VTN encoder and then the visual representation is output by our VTN decoder.

As demonstrated in Fig. 2, we extract two spatial-aware descriptors: a spatial-enhanced local descriptor and a positional global descriptor. We fuse these two types of descriptors with a multi-head attention operation to produce visual representations. Furthermore, we enforce visual representations to be highly correlated to navigation signals via a pre-training scheme for navigation effectiveness.

Spatial-Enhanced Local Descriptor. We first employ the DETR [1] for object instance identification and localization. Meanwhile, we extract features from the last layer of DETR. These features are originally used to predict detection results and are rich in information about the objects in the image. We use these features as the local spatial features in our navigation system. This allows us to take into account the spatial relationships between objects, which is important for navigation. Furthermore, local spatial features are derived by integrating normalized bounding box parameters, associated confidence levels, and predominant semantic labels for each object. A one-hot encoded target vector is concatenated with the spatial feature to specify a target class. Moreover, we project bounding boxes onto the corresponding depth map to encode the distance between the agent and each detected object. Then we calculate the average depth value for each detected object and attach the depth value to the local spatial feature. Finally, we fuse local spatial features with an MLP to obtain a spatial-enhanced local descriptor, which is used as the key of our VTN encoder.

Positional Global Descriptor. Agents integrate a spatial-enhanced local descriptor with a global feature to represent the surrounding environment. We use an ImageNet [31] pretrained ResNet18 [12] to extract global features. Meanwhile, we apply non-overlapping sliding windows on the depth map to yield global depth representations and then concatenate the depth representation with the global feature. To align with the spatial-enhanced local descriptor, we then project the global feature into d dimension via a 1×1 convolution.

Furthermore, we adopt a more informative representation of the global scene, called a positional global descriptor. Specifically, we add a positional embedding on each region feature to emphasize the location of each region in the image. Let u and v represent the row and column indexes of an image region, respectively. Our positional embedding is expressed as:

$$
\begin{aligned}
PE_{2i}(u,v) &= \begin{cases} \sin(\frac{u}{1e4^{2i/d}}), & 0 < i \le \frac{d}{2} \\ \sin(\frac{v}{1e4^{2i/d}}), & \frac{d}{2} < i \le d \end{cases} \\
PE_{2i+1}(u,v) &= \begin{cases} \cos(\frac{u}{1e4^{2i/d}}), & 0 < i \le \frac{d}{2} \\ \cos(\frac{v}{1e4^{2i/d}}). & \frac{d}{2} < i \le d \end{cases}
\end{aligned}
\tag{1}
$$

Therefore, each global feature represents one particular region. We then reshape the features into the positional global descriptor.

Visual Transformer. After obtaining spatial-enhanced and positional global descriptors, we use a visual transformer. To leverage the spatial relationships among detected instances and observed regions, we apply attention between spatial-enhanced local and positional global descriptors via a transformer mechanism. Following the transformer architecture [4,32,33], spatial-enhanced local descriptors serve as both keys and values, undergoing multi-head self-attention. Furthermore, guided by intuitive navigation behavior patterns, the decoder is designed to explore correlations between observation regions and navigation actions. Then, the positional global descriptors are used as the location queries and fed into the decoder.

Pre-training Visual Transformer. We found that the agent cannot learn to navigate properly when we adopt VTN directly in training. This is mainly because training a transformer based on guiding signals from reinforcement learning is hard. The information it produced might be unclear and mislead the system. As a result, the agent often chooses to stop quickly instead of moving.

To solve this problem, we apply a pre-training scheme for VTN. We enforce information it produced was clear and useful by teaching it through a process known as imitation learning, as shown in Fig. 1. We used a well-known method called Dijkstra's Shortest Path First algorithm to create navigation instructions. Then VTN learned to follow these instructions.

During this special pre-training scheme, we do not use our usual navigation system or consider previous actions or states. Instead, we used an MLP to predict actions based only on the current visual information. We adopt cross-entropy loss to guide the pre-training of our VTN. After pre-training, VTN is better at understanding directions. Therefore, the pre-training scheme facilitates navigation network training.

Table 1. Comparison with other RGB methods on iTHOR and RoboTHOR. We report the average success rate (%) and SPL, by repeating experiments five times. $L > 5$ represents the episodes that require at least 5 steps.

Method	iTHOR				RoboTHOR			
	ALL		$L \geq 5$		ALL		$L \geq 5$	
	Success	SPL	Success	SPL	Success	SPL	Success	SPL
Random	8.0	0.036	0.3	0.001	4.0	0.016	0.2	0.1
WE	33.0	0.147	21.4	0.117	7.0	0.022	4.7	0.017
SP [25]	35.1	0.155	22.2	0.114	10.9	0.042	7.2	0.032
SAVN [5]	40.8	0.161	28.7	0.139	10.2	0.039	6.9	0.028
Baseline	62.6	0.364	51.5	0.345	46.5	0.233	41.4	0.207
ORG	65.3	0.375	54.8	0.361	45.4	0.212	39.6	0.185
VTN	72.2	0.449	63.4	0.440	53.2	0.275	47.0	0.233
VTN + TIL	75.9	0.455	69.9	**0.445**	58.8	0.305	54.2	0.263
VTP [27]	**77.4**	**0.460**	**70.6**	**0.445**	**63.6**	**0.341**	**59.5**	**0.307**

3.4 Tentative Policy Learning

In complex simulations, agents can sometimes get stuck in challenging situations. However, reinforcement learning rewards cannot give the agents clear instructions to get out of these spots. Therefore, we adopt a Tentative Policy Network (TPN) to guide the agent using past experiences. It is designed to explore stuck situations and uses previous successful experiences to suggest actions.

Trial-Driven Imitation Learning. Here, we first teach the agent using expert strategies, specifically using Dijkstra's algorithm, via Trial-driven Imitation Learning (TIL). This helps the agent learn the best moves to avoid being stuck. We supervise action predictions via cross-entropy loss. Furthermore, with limited training data, there is a risk of the agent learning only the specific examples and not generalizing unseen environments well. We aim to improve this by guiding the agent only in tricky situations rather than all the time. This method lets the agent continue learning without getting stuck often.

Memory-Augmented Tentative Policy Network. In the testing phases, agents do not have access to expert advice. Thus, we provide the agent with an external memory tool to identify and escape from challenging situations. The memory keeps track of previous observations. We assume the agent is stuck if the agent notices something similar to the memory. Furthermore, we use another internal memory to store the historical states and actions for instruction generation. Each memory slot includes a historical state representation and its corresponding action distribution.

The agent utilizes imitation learning to guide TPN when the agent is stuck. Then, TPN helps the agent decide on the escaping action in testing. Thus, TPN benefits the agent in both navigation efficiency and effectiveness.

4   Experiments

4.1   Protocols and Experimental Details

Dataset. We test our methods on two environments: iTHOR and RoboTHOR. **iTHOR** has four types of rooms: kitchens, living rooms, bedrooms, and bathrooms. Each type contains 30 different room layouts. We select targets from 22 types of objects, such as Book. We use 80 rooms for training and split the remaining 40 rooms for validation and testing. **RoboTHOR** contains 75 rooms, each with multiple rooms. We choose our targets from 13 types of objects, such as AlarmClock. We use 55 scenes for training, 5 scenes for validation, and the reset 15 scenes for testing. For our results, we report the model that achieves the best performance on the validation set and test it.

Table 2. Comparison with other RGB-D methods on iTHOR and RoboTHOR.

Method	iTHOR				RoboTHOR			
	ALL		$L \geq 5$		ALL		$L \geq 5$	
	Success	SPL	Success	SPL	Success	SPL	Success	SPL
Baseline	73.2	0.410	64.2	0.398	51.5	0.233	46.6	0.200
ORG	69.0	0.322	61.9	0.317	49.2	0.225	44.5	0.206
VTN	76.7	0.478	69.8	0.461	58.5	0.331	54.1	0.296
VTN + TIL	78.1	0.496	71.1	0.491	63.1	0.355	59.9	0.325
VTP [27]	**77.5**	**0.474**	**71.1**	**0.462**	**72.2**	**0.383**	**70.0**	**0.359**

Evaluation Metrics. We assess our model using two metrics: Success Rate and Success Weighted by Path Length (SPL). The success rate gauges how effectively our model navigates based on the number of successful episodes. Meanwhile, SPL evaluates how efficiently the model navigates by comparing the length of the episode to the optimal path length.

Training Details. We use a three-step training approach. First, we train the visual transformer for 20 epochs using ideal action guides to associate observations with actions. Next, we train our navigation method for 6M episodes with 16 asynchronous agents. Finally, we train TPN for 2M episodes with 12 agents, using the best-performing navigation model as a base. We penalize each action slightly and reward successful tasks. We use the Adam optimizer [34] with a learning rate 10^{-4} to update our models.

4.2   Competing Methods

We compare our framework against several others: **Random policy** just picks actions by chance. **Scene Prior (SP)** [25] uses a special network to understand scenes and how things relate. **Word Embedding (WE)** uses GloVe embedding [35] to recognize objects not by seeing but by their names. **Self-adaptive Visual Navigation (SAVN)** [5] helps an agent learn to work in new places via a met reinforcement learning method. **Object Relationship Graph (ORG)** [27] aims to understand how categories relationships. **Baseline** is a basic version of VTN. We feed the concatenation of the local instance features from DETR and the global feature to the policy network for navigation.

4.3   Evaluation Results

Table 1 demonstrates that VTP is much better than the baseline in terms of success rate and SPL score. The baseline only uses basic features to predict navigation action. VTP, on the other hand, uses informative visual representations for navigation to make better decisions. This means VTP makes our system work more effectively and quickly.

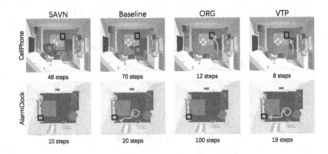

Fig. 3. Visual results of four different models in testing environments. The target objects (*i.e.*, *CellPhone*) are highlighted by the blue boxes. Green and red curves represent success and failure cases, respectively. The episode produced by VTP is successful in reaching the target with the shortest steps. (Color figure online)

When comparing VTP to other methods like SP and SAVN, VTP also stands out. SP and SAVN use word cues to guide them. On the contrary, VTN uses a VT to exploit all the detected instances to deduce the relationships among objects. Furthermore, VTP matches well with navigation actions, making the system learn faster. ORG focuses on one thing at a time, but VTP sees the whole picture. VTP can understand how different objects in the image relate to each other, especially with the help of DETR.

Table 2 further shows the dominance if VTP over other frameworks in RGB-D scenarios. This suggests the adeptness of VTP at integrating depth map data. Moreover, VTP with RGB-D observations surpasses its performance with just

RGB, highlighting the importance of incorporating depth map data into visual interpretations.

Case Study. Figure 3 shows that both SAVN and Baseline try to reach the target. However, they stop early and do not arrive at the target. This might mean that SAVN and Baseline are not good at understanding how things relate to each other. On the other hand, both ORG and VTN do reach the target. Meanwhile, with clear direction signals encoded in our visual representations, VTP gets to the object faster than the others with the fewest steps.

5 Conclusion

In this paper, we proposed a unified visual navigation framework named VTP for both RGB and RGB-D observations. In VTP, we adopt a VTN to encode visual observations. VTN leverages depth maps and two spatial-aware descriptors, *i.e.*, a spatial-enhanced local object descriptor and a positional global descriptor, to explore the instance-to-instance, instance-to-region and instance-to-agent relationships. Furthermore, we also adopt TPN to avoid getting stuck. Extensive results demonstrate that VTP outperforms other methods on both iTHOR and RoboTHOR in terms of effectiveness and efficiency.

Acknowledgements. This research is funded in part by ARC-Discovery grant (DP220100800 to XY) and ARC-DECRA grant (DE230100477 to XY). We thank all anonymous reviewers and ACs for their constructive suggestions.

References

1. Carion, N., Massa, F., Synnaeve, G., Usunier, N., Kirillov, A., Zagoruyko, S.: End-to-end object detection with transformers. arXiv preprint arXiv:2005.12872 (2020)
2. Mnih, V., et al.: Asynchronous methods for deep reinforcement learning. In: International Conference on Machine Learning, pp. 1928–1937 (2016)
3. Hochreiter, S., Schmidhuber, J.: Long short-term memory. Neural Comput. 9(8), 1735–1780 (1997)
4. Vaswani, A., et al.: Attention is all you need. In: Advances in Neural Information Processing Systems, pp. 5998–6008 (2017)
5. Wortsman, M., Ehsani, K., Rastegari, M., Farhadi, A., Mottaghi, R.: Learning to learn how to learn: self-adaptive visual navigation using meta-learning. In: Proceedings of the IEEE Conference on Computer Vision and Pattern Recognition, pp. 6750–6759 (2019)
6. Kolve, E., et al.: AI2-THOR: an interactive 3D environment for visual AI. arXiv (2017)
7. Oriolo, G., Vendittelli, M., Ulivi, G.: On-line map building and navigation for autonomous mobile robots. In: Proceedings of 1995 IEEE International Conference on Robotics and Automation, vol. 3, pp. 2900–2906. IEEE (1995)
8. Kortenkamp, D., Weymouth, T.: Topological mapping for mobile robots using a combination of sonar and vision sensing. In: AAAI, vol. 94, pp. 979–984 (1994)

9. Borenstein, J., Koren, Y.: Real-time obstacle avoidance for fast mobile robots. IEEE Trans. Syst. Man Cybern. **19**(5), 1179–1187 (1989)

10. Borenstein, J., Koren, Y.: The vector field histogram-fast obstacle avoidance for mobile robots. IEEE Trans. Robot. Autom. **7**(3), 278–288 (1991)

11. Dissanayake, M.G., Newman, P., Clark, S., Durrant-Whyte, H.F., Csorba, M.: A solution to the simultaneous localization and map building (SLAM) problem. IEEE Trans. Robot. Autom. **17**(3), 229–241 (2001)

12. He, K., Zhang, X., Ren, S., Sun, J.: Deep residual learning for image recognition. In: Proceedings of the IEEE Conference on Computer Vision and Pattern Recognition, pp. 770–778 (2016)

13. Gupta, S., Davidson, J., Levine, S., Sukthankar, R., Malik, J.: Cognitive mapping and planning for visual navigation. In: Proceedings of the IEEE Conference on Computer Vision and Pattern Recognition, pp. 2616–2625 (2017)

14. Sutton, R.S., Barto, A.G.: Reinforcement Learning: An Introduction. MIT Press, Cambridge (2018)

15. Mirowski, P., et al.: Learning to navigate in complex environments. arXiv preprint arXiv:1611.03673 (2016)

16. Chen, T., Gupta, S., Gupta, A.: Learning exploration policies for navigation. arXiv preprint arXiv:1903.01959 (2019)

17. Fang, K., Toshev, A., Fei-Fei, L., Savarese, S.: Scene memory transformer for embodied agents in long-horizon tasks. In: Proceedings of the IEEE Conference on Computer Vision and Pattern Recognition, pp. 538–547 (2019)

18. Sepulveda, G., Niebles, J.C., Soto, A.: A deep learning based behavioral approach to indoor autonomous navigation. In: 2018 IEEE International Conference on Robotics and Automation (ICRA), pp. 4646–4653. IEEE (2018)

19. Chen, K., et al.: A behavioral approach to visual navigation with graph localization networks. arXiv preprint arXiv:1903.00445 (2019)

20. Hao, W., Li, C., Li, X., Carin, L., Gao, J.: Towards learning a generic agent for vision-and-language navigation via pre-training. In: Proceedings of the IEEE/CVF Conference on Computer Vision and Pattern Recognition, pp. 13137–13146 (2020)

21. Kahn, G., Villaflor, A., Ding, B., Abbeel, P., Levine, S.: Self-supervised deep reinforcement learning with generalized computation graphs for robot navigation. In: 2018 IEEE International Conference on Robotics and Automation (ICRA), pp. 1–8. IEEE (2018)

22. Wu, Y., Wu, Y., Tamar, A., Russell, S., Gkioxari, G., Tian, Y.: Bayesian relational memory for semantic visual navigation. In: Proceedings of the IEEE International Conference on Computer Vision, pp. 2769–2779 (2019)

23. Shen, W.B., Xu, D., Zhu, Y., Guibas, L.J., Fei-Fei, L., Savarese, S.: Situational fusion of visual representation for visual navigation. arXiv preprint arXiv:1908.09073 (2019)

24. Zhu, Y., et al.: Target-driven visual navigation in indoor scenes using deep reinforcement learning. In: IEEE International Conference on Robotics and Automation (ICRA), pp. 3357–3364. IEEE (2017)

25. Yang, W., Wang, X., Farhadi, A., Gupta, A., Mottaghi, R.: Visual semantic navigation using scene priors. arXiv preprint arXiv:1810.06543 (2018)

26. Mousavian, A., Toshev, A., Fišer, M., Košecká, J., Wahid, A., Davidson, J.: Visual representations for semantic target driven navigation. In: 2019 International Conference on Robotics and Automation (ICRA), pp. 8846–8852. IEEE (2019)

27. Du, H., Yu, X., Zheng, L.: Learning object relation graph and tentative policy for visual navigation. arXiv preprint arXiv:2007.11018 (2020)

28. Zhang, S., Song, X., Bai, Y., Li, W., Chu, Y., Jiang, S.: Hierarchical object-to-zone graph for object navigation. In: Proceedings of the IEEE/CVF International Conference on Computer Vision, pp. 15130–15140 (2021)
29. Khandelwal, A., Weihs, L., Mottaghi, R., Kembhavi, A.: Simple but effective: clip embeddings for embodied AI. In: Proceedings of the IEEE/CVF Conference on Computer Vision and Pattern Recognition, pp. 14829–14838 (2022)
30. Radford, A., et al.: Learning transferable visual models from natural language supervision (2021)
31. Deng, J., Dong, W., Socher, R., Li, L.-J., Li, K., Fei-Fei, L.: ImageNet: a large-scale hierarchical image database. In: IEEE Conference on Computer Vision and Pattern Recognition, pp. 248–255. IEEE (2009)
32. Khan, S., Naseer, M., Hayat, M., Zamir, S.W., Khan, F.S., Shah, M.: Transformers in vision: a survey. arXiv preprint arXiv:2101.01169 (2021)
33. Song, H., et al.: ViDT: an efficient and effective fully transformer-based object detector. arXiv preprint arXiv:2110.03921 (2021)
34. Kingma, D.P., Ba, J.: Adam: a method for stochastic optimization. arXiv preprint arXiv:1412.6980 (2014)
35. Pennington, J., Socher, R., Manning, C.: GloVe: global vectors for word representation. In: Proceedings of the 2014 Conference on Empirical Methods in Natural Language Processing (EMNLP), pp. 1532–1543 (2014)

Improving CCA Algorithms on SSVEP Classification with Reinforcement Learning Based Temporal Filtering

Liang Ou[✉], Thomas Do, Xuan-The Tran, Daniel Leong, Yu-Cheng Chang, Yu-Kai Wang, and Chin-Teng Lin

GrapheneX-UTS Human-Centric Artificial Intelligence Centre, University of Technology Sydney, Ultimo, NSW 2000, Australia
liang.ou-1@student.uts.edu.au

Abstract. Canonical Correlation Analysis (CCA) has been widely used in Steady-State Visually Evoked Potential (SSVEP) analysis, but there are still challenges in this research area, specifically regarding data quality and insufficiency. In contrast to most previous studies that primarily concentrate on the development of spatial or spectral templates for SSVEP data, this paper proposes a novel temporal filtering method based on a reinforcement learning (RL) algorithm for CCA on SSVEP data. The proposed method leverages RL to automatically and precisely detect and filter low-quality segments in the SSVEP data, thereby improving the accuracy of CCA. Additionally, the proposed RL-based Temporal Filtering is algorithm-independent and compatible with various CCA algorithms. The RL-based Temporal Filtering is evaluated using a wearable dataset consisting of 102 subjects. The experimental results demonstrate significant advancements in CCA accuracy, particularly when combined with the extended CCA (ECCA) algorithm. In addition to performance enhancement, the RL-based Temporal Filtering method provides visualizable filters, which can ensure the transparency of the filtering process and the reliability of the obtained results. By addressing data quality and insufficiency concerns, this novel RL-based Temporal Filtering approach demonstrates promise in advancing SSVEP analysis for various applications.

Keywords: Canonical Correlation Analysis (CCA) · Steady-State Visually Evoked Potential (SSVEP) · Reinforcement Learning (RL) · Temporal Filter

1 Introduction

Steady-State Visually Evoked Potential (SSVEP) is an electrophysiological brain response to visual stimuli presented at a constant frequency, often involving flickering or flashing stimuli. The main objective of SSVEP studies is to identify targets by representing them as various flickering stimuli. Using Canonical

T. Liu et al. (Eds.): AI 2023, LNAI 14472, pp. 376–386, 2024.
https://doi.org/10.1007/978-981-99-8391-9_30

Correlation Analysis (CCA) [1] for SSVEP signals analysis has enjoyed major attention from researchers working on SSVEP data processing. CCA considers the frequency and phase of the flickering stimuli as reference signals and computes the correlation between the subject's EEG signal when they focus on a particular flickering stimulus and the reference signal associated with that stimulus. Although traditional CCA takes advantage of EEG's high-temporal resolution, leading to the high classification accuracy of above 90% [1], it is still limited by its sensitivity to signal noise ratio (SNR) and biases associated with subjects and sensor spatial positions. In attempts to improve CCA classification accuracy and minimize the duration of flickering stimuli, several studies have explored solutions focusing on (1) enhancing the SNR of EEG signals through various signal processing and filtering techniques, (2) improving EEG signal spatial accuracy via spatial filtering methods, and (3) employing individual templates as calibration data to mitigate subject-related biases in the EEG signals.

In [2], Poryzala utilizes the cluster analysis of CCA coefficients (CACC) method, which enables asynchronous SSVEP-based target identification through k-means cluster analysis to distinguish detection and idle states. On the other hand, the multi-way CCA approach [3] enhances target identification accuracy by utilizing optimization mathematics to correlate stimulus reference signals and the subject's EEG data. Similarly, in [4], the authors further optimize the correlation between the reference and EEG signals by applying L1-regularization to penalize the correlation in incorrect trials. To address the limitation concerning EEG spatial accuracy, Zhang [5] proposed multi-set CCA methods, which utilize joint spatial filtering of multiple subject EEG training sets to derive an optimization function that maximizes the correlation among these sets. Additionally, to mitigate personally biased EEG signals, the Individual Template Based CCA (IT-CCA) [6] approach replaces the reference signals used in traditional CCA with individual templates. These templates are obtained by averaging subject EEG signals during multiple training trials with various flickering stimuli. Nakanishi [7] conducted a comprehensive comparison of various CCA methods, assessing target classification accuracy using different evaluation metrics, including stimuli duration, number of EEG channels, and number of trials. The study found that CCA performance stabilizes after 3 s of stimuli duration. Moreover, increasing the number of EEG channels and trials positively impacts CCA classification performance, and the individual template approach demonstrated notable advantages in enhancing CCA performance.

Researchers have made significant efforts to improve the performance of CCA by incorporating spatial filters. One typical example is the extended CCA (ECCA) [8], which defines three types of spatial filters for each trial. Additionally, the Sum of Squared Correlations (SSCOR) and ensemble sum of Squared Correlations (eSSCOR) methods [9] adopt spatial filters but break them down into stimulus levels. Recent research [10–14] continues to focus on defining better spatial filters to enhance CCA accuracy, with some novel attempts emerging. For instance, the Spatio-Spectral CCA (SS-CCA) [15] includes both spatial and spectral filters, broadening the scope of improvements. Similarly, the

Time-Weighting Canonical Correlation Analysis (TWCCA) [16] introduces a time-dimension weight to differentiate time periods during the analysis. In our research, we address data quality enhancement through the inclusion of a temporal filter. This idea arises from the observation that CCA tends to favour temporal aspects over spectral ones [7]. To implement this temporal filtering, we employ a Reinforcement Learning (RL) agent, allowing the filter to learn and adapt to the unique characteristics of the data. The advantage of using RL is that it enables collaboration with any CCA algorithm, making our proposed method applicable and compatible with various CCA algorithms.

Another recent trend in SSVEP classification involves the adoption of machine learning methods for direct flicker class prediction. There are three major types of machine learning models commonly used for SSVEP classification. The first type is LSTM/RNN [17,18], which effectively encodes temporal correlations of SSVEP signals. The second type, also the most popular one, is CNN [19–21], particularly EEGNet-based models. This method uses convolutional kernels to extract both spatial and temporal information. The last type, and the newest one, is the transformer-based method [22,23]. This approach leverages the self-attention mechanism to extract correlations from both the temporal and spatial domains of SSVEP data. In this paper, our RL agent is built upon an EEGNet-based deep model, following the results of our experiments. In contrast to the aforementioned deep models, our proposed method combines deep learning with CCA algorithms to leverage information from the reference signal of CCAs. This design is also found in CCA-CWT-SVM [24] and [25]. While CCA-CWT-SVM combines a support vector machine (SVM) with CCA, [25] models both the reference signal and SSVEP data with CNN models. However, our proposed method is independent of CCA, allowing the RL-based Temporal filter to collaborate with any CCA algorithm, potentially enhancing their classification accuracy.

In summary, our contributions are as follows:

1. We introduce a unique filtering method that treats EEG data segments as an RL environment. By allowing an RL agent to explore this environment, we can intelligently determine the quality of each segment, making informed decisions about retaining or discarding specific parts of the data.

2. Compared to other deep CCA methods, our proposed approach is more explainable, providing a meaningful filter that can be reviewed, understood, and validated by domain experts, thus promoting reliability in the results.

3. To address the subject difference problem, we propose an EEGNet-based quality classifier. This classifier can accurately identify whether a subject's data quality is potentially risky or not, allowing us to decide whether data filtering is necessary for each individual subject.

4. Through our proposed method, we significantly enhance the tolerance of SSVEP data quality. Even with low-quality data, our advanced data filtering process ensures that the remaining information is sufficient for generating reliable outcomes, thus reducing data waste and increasing the efficiency of data analysis.

2 Method

Our proposed method consists of two major components: the EEGNet [26]-based quality classifier and the RL-based Temporal filtering (see Fig. 1). These components work in synergy to ensure that only high-quality data is fed into the subsequent CCA matching process.

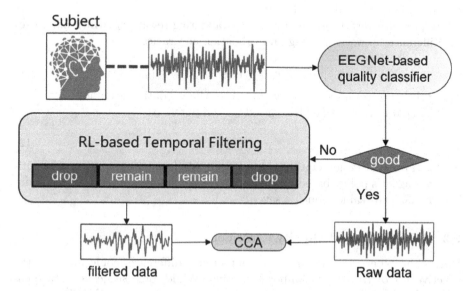

Fig. 1. The workflow of the proposed method. The EEGNet-based quality classifier and the RL-based Temporal Filtering components work together to enhance the classification accuracy of SSVEP signals.

In the initial step, the subject's SSVEP data is passed through the EEGNet-based quality classifier. This classifier evaluates the data quality and determines if it meets the required standards. If the data is deemed of good quality, it is directly used for CCA matching. However, if the quality is suboptimal, the RL-based Temporal filtering process is activated to identify and exclude problematic segments in the time domain. The filtered data is then used for CCA matching, ensuring that only relevant and high-quality information is considered.

2.1 RL-Based Temporal Filtering

The RL-based Temporal Filtering component is at the core of our proposed method, aiming to enhance SSVEP data quality. We assume that during the experiment, subjects may get distracted or focus on the wrong flicker, leading to the inclusion of irrelevant segments in the recorded data. The RL agent is designed to explore the SSVEP data as a RL environment, deciding which segments to retain and which to discard, based on their contribution to CCA accuracy.

For each SSVEP data trial, represented as a 2D matrix with channel and time sample dimensions, the RL agent begins exploration at the first time sample and examines a window of time samples (state s). The agent then takes actions a (filtering options) at each step, determining whether to retain or drop segments. After fully exploring the training environment, the remaining parts of the data trail, combined with a reference signal under the same filter, are fed to the CCA algorithm. The agent's performance is evaluated based on CCA accuracy, and a reward r_a is calculated accordingly.

To prevent overfitting, we introduce a validating reward r_v, which minimizes the difference between training and validation accuracies:

$$r_v = |A_t - A_v|, \tag{1}$$

where A_t is the training accuracy, and A_v is the validation accuracy.

The final reward for the RL episode is computed as:

$$r = r_a + r_v. \tag{2}$$

The Proximal Policy Optimization (PPO) [27] framework is utilized to optimize the agent's policy based on the reward r. After several rounds of RL training, the final workable model is saved.

2.2   EEGNet-Based Quality Classifier

During the experiment, we observed that subject differences could influence the effectiveness of RL-based Temporal filtering. While some subjects' data quality can be substantially improved through RL-based Temporal Filtering, others already exhibit good enough quality without requiring such filtering. To address this variability and ensure CCA accuracy across all subjects, we introduce an EEGNet-based quality classifier.

The EEGNet-based quality classifier is a binary classifier, fed with one SSVEP data trial at a time, producing a binary decision of good or bad quality. We employ the EEGNet architecture for this classifier and train it using the results from the RL agent's performance. Subjects whose data can be effectively improved by the RL method are labelled as "bad", while others are labelled as "good". After several epochs of training, the classifier achieves an accuracy of over 99%, allowing us to make informed decisions about whether data filtering is necessary for each subject. This approach ensures that the method generalizes well across diverse datasets and guarantees the quality of data delivered to the subsequent CCA analysis.

2.3   Learning Scheme

In our proposed method, both major components rely on deep models that require training, and the classification head depends on the labelling results from RL-based Temporal Filtering. To facilitate this process, we have defined a 5-step learning scheme outlined in Algorithm 1.

Algorithm 1. (Learning Scheme)

Input:	T (all subjects trails data), N (number of subjects).
	for $i \leftarrow 1$ to N **do**
Step 1:	select t_i from $T_i \in T$, train a RL agent on t_i
Step 2:	$l_i \leftarrow q$ (q is the subject quality found from RL agent)
	$L \leftarrow l_i$, store subject label in L
	end for
Step 3:	Train the EEGNet-based classifier H with T and L
Step 4:	**for** $i \leftarrow 1$ to N **do**
	if l_i="Bad" **then**
	Train a RL agent A_i on T_i
	$\mathcal{A} \leftarrow A_i$, save agent
	end if
	end for
Step 5:	Combine saved models: $\{H, \mathcal{A}\}$

The learning process begins with labelling the data quality of subjects. We achieve this by running agents in RL environments with a small amount (10%) of trial data, which provides an overview of subjects' data quality. This overview is then used to label the subjects in the second step. In the third step, we train a reliable distinguisher, an EEGNet-based quality classifier, for subjects' data quality. Our experiments show that this classifier can easily achieve 99% accuracy with just a few epochs of training. With the help of the pre-trained quality classifier, we can now use all data trials for those labelled as "Bad" quality to train dedicated RL agents in step four. This approach avoids unnecessary training for subjects who are already determined to have good data quality, which provides a solution for the subject difference issue in SSVEP experiments. Once all RL agents are optimized, we integrate the entire architecture of the method, comprising one shareable classifier and multiple RL agents dedicated to different subjects. This integrated model can then be used for unseen data.

3   Experiment Results

To evaluate and demonstrate the advantages of our proposed method, we conducted experiments using a wet wearable dataset kindly provided by 2020 International brain–computer interface competition [28]. This dataset comprises SSVEP data from 102 subjects, each recorded with 8 EEG channels: POz, PO3, PO4, PO5, PO6, Oz, O1, and O2. The data was recorded with a sampling rate of 1000 Hz and then down-sampled to 250 Hz without any other processing. Each subject's data consists of 120 trials, with 12 targets/classes and 10 blocks. For each trial, we collected 500 time samples, which were treated as individual runs of the RL environment. To process the data and facilitate RL exploration, we set the RL agent with a window size of 50 time samples. In other words, the input provided to the RL agent's EEGNet-based deep model is an 8×50 matrix, representing the 8 EEG channels over a segment of 50 time samples."

Table 1. Network Architecture Summary and Learning Settings of PPO.

	Architecture/Settings
EEGNet	CNN kernel size 1×8, 2×1, 1×15
	Activation: Exponential Linear Unit (ELU)
PPO settings	Total learning step: 10000
	Steps of update: 128
	Feature extractor: EEGNet (with 128 latent dimensions)
	Policy network: 2 layers of FFN
	Critic network: 1 layer of FFN

To clarify our experimental settings and the design of EEGNet, which was utilized both as the Quality Classifier and the PPO feature extractor, we have meticulously outlined all relevant configurations in Table 1. To tailor the EEGNet architecture to our wearable dataset, we made adjustments to the kernel size and activation functions. Furthermore, we imposed specific constraints on the PPO training process: the total training steps were capped at 10,000, and updates were performed every 128 steps. This deliberate approach was chosen to ensure a swift and efficient learning trajectory for the PPO agent. In the subsequent subsection, we include a discussion of the outcomes derived from our experiment.

3.1   Performance Comparisons

In order to demonstrate the advantages of our method, we conducted a comprehensive comparison with several existing CCA algorithms, namely ECCA, SSCOR, and ESSCOR. The results are summarized in the following table, which presents the improvements achieved by our RL-based Temporal Filter:

Table 2. Accuracy improvement summarizing of adopting the proposed method. This experiment includes three CCA algorithms, which are ECCA, SSCOR, and ESSCOR.

CCA Algorithms	#Improved	Acc without filter	Acc with filter	Acc lift
ECCA	44	53.1%	76.7%	23.6%
SSCOR	5	0.0%	8.8%	8.8%
ESSCOR	2	0.0%	17.5%	17.5%

The first column of the table indicates the number of subjects (out of 102) whose performance improved using our approach. The second and third columns compare the average accuracy of these improved subjects before and after applying our method, respectively. The last column summarizes the improvement percentages achieved with our RL-based Temporal Filter. One key advantage of our method is its ability to selectively target problematic subjects. The classification head accurately identifies subjects that stand to benefit from the RL filter's intervention, allowing us to consistently enhance CCA accuracy. This adaptability ensures that our method can improve CCA accuracy across varying scenarios and datasets. In particular, we observed a significant improvement when combining our method with ECCA. Our filter yields enhancements in nearly half of the subjects' data, with an impressive average improvement of 23.6%. Conversely, SSCOR and ESSCOR showed limited improvements with our approach. The 0.0% accuracy for some subjects indicates that SSCOR and ESSCOR struggle to make correct predictions for all trial data. However, our method still demonstrates we can utilize 8.8% and 17.5% of the subjects' trials to yield correct predictions (Table 2).

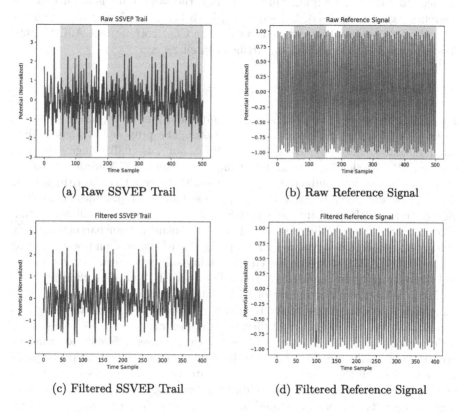

(a) Raw SSVEP Trail   (b) Raw Reference Signal

(c) Filtered SSVEP Trail   (d) Filtered Reference Signal

Fig. 2. Example of one trial data applied with the RL-based Temporal Filter, visualizing the Oz channel of the SSVEP trial along with the ground truth of the reference signal.

3.2   Filter Explanation and Visualizations

We present a data filtering technique to elucidate the mechanics of our proposed approach. In Fig. 2, we have chosen a single trial of SSVEP data from a subject to analyze its CCA classification results. In Fig. 2(a), the displayed SSVEP trial data exhibits low quality, and when used with its corresponding reference signal (Fig. 2(b)), the CCA classification yields an incorrect result. In this instance, CCA predicts a label of 8, while the true label indicated by the reference signal is 9.

To address this issue, we introduce our RL-based filter, as detailed in Sect. 2.1. The RL filter systematically scans the problematic trial and identifies segments that require retraining (depicted in grey). Upon applying the RL filter, the trial data transforms filtered trial data, as illustrated in Fig. 2(c), from 500 time samples to 400 time samples. Simultaneously, the same filter is applied to the reference signal (Fig. 2(d)). Subsequently, both the filtered trial data and the filtered reference signal are utilized to calculate CCA correlations. AJCAI 2023: Australasian Joint Conference on Artificial Intelligence.

4   Conclusion

This paper presents a novel RL-based Temporal Filtering approach that can be used in CCA for SSVEP classification tasks. The proposed RL-based Temporal Filtering effectively identifies and filters out low-quality segments from raw SSVEP data, leading to significant improvements in CCA accuracy. To validate its efficacy, we conducted experiments with three CCA algorithms using the wearable datasets. Additionally, the proposed approach provides an intuitive view of how the RL agent interacts with SSVEP data through visualizations that allow users to interpret the results of filters, thus ensuring transparency and reliability. An important advantage of adopting the RL-based Temporal Filtering method is its ability to leverage even low-quality subjects' data. This holds critical importance when considering the challenge of EEG data insufficiency. This challenge arises from the nature of EEG data collection, in which experimental settings vary from one study to another. Variations include the length of SSVEP flickers, the frequency of flickers, sampling rate, and noise levels, all of which render EEG data incompatible for sharing across different experiments, even if they involve similar tasks. Additionally, the variability in participants necessitates machine learning research to create dedicated models for each individual, further limiting the available data for training. By harnessing more data from low-quality subjects, our method becomes more robust and practical for real-world applications. In addition, the novel combination of RL and CCA provides additional explainability and a solution for the subject difference issue. Our future research objectives involve conducting in-depth analyses of the behaviour of these filters. It aims to uncover the underlying reasons behind incorrect CCA classifications on SSVEP data, leading to further advancements in our understanding of SSVEP signals and refining our proposed approach. We would also like to extend our experiment with more SSVEP data without quality selection.

When conducting online SSVEP classification, the tolerance of low-quality data plays a more important role than offline analysis.

References

1. Lin, Z., Zhang, C., Wu, W., Gao, X.: Frequency recognition based on canonical correlation analysis for SSVEP-based BCIs. IEEE Trans. Biomed. Eng. **53**(12), 2610–2614 (2006)
2. Poryzala, P., Materka, A.: Cluster analysis of CCA coefficients for robust detection of the asynchronous SSVEPs in brain-computer interfaces. Biomed. Sig. Process. Control **10**, 201–208 (2014)
3. Zhang, Yu., et al.: Multiway canonical correlation analysis for frequency components recognition in SSVEP-based BCIs. In: Lu, B.-L., Zhang, L., Kwok, J. (eds.) ICONIP 2011. LNCS, vol. 7062, pp. 287–295. Springer, Heidelberg (2011). https://doi.org/10.1007/978-3-642-24955-6_35
4. Zhang, Y., Zhou, G., Jin, J., Wang, M., Wang, X., Cichocki, A.: L1-regularized multiway canonical correlation analysis for SSVEP-based BCI. IEEE Trans. Neural Syst. Rehabil. Eng. **21**(6), 887–896 (2013)
5. Zhang, Y., Zhou, G., Jin, J., Wang, X., Cichocki, A.: Frequency recognition in SSVEP-based BCI using multiset canonical correlation analysis. Int. J. Neural Syst. **24**(04), 1450013 (2014)
6. Bin, G., Gao, X., Wang, Y., Li, Y., Hong, B., Gao, S.: A high-speed BCI based on code modulation VEP. J. Neural Eng. **8**(2), 025015 (2011)
7. Nakanishi, M., Wang, Y., Wang, Y.T., Jung, T.P.: A comparison study of canonical correlation analysis based methods for detecting steady-state visual evoked potentials. PLoS ONE **10**(10), e0140703 (2015)
8. Chen, X., Wang, Y., Nakanishi, M., Gao, X., Jung, T.P., Gao, S.: High-speed spelling with a noninvasive brain-computer interface. Proc. Natl. Acad. Sci. **112**(44), E6058–E6067 (2015)
9. Kumar, G.K., Reddy, M.R.: Designing a sum of squared correlations framework for enhancing SSVEP-based BCIs. IEEE Trans. Neural Syst. Rehabil. Eng. **27**(10), 2044–2050 (2019)
10. Yuan, X., Sun, Q., Zhang, L., Wang, H.: Enhancing detection of SSVEP-based BCIs via a novel CCA-based method. Biomed. Sig. Process. Control **74**, 103482 (2022)
11. Peng, F., Li, M., Zhao, S.N., Xu, Q., Xu, J., Wu, H.: Control of a robotic arm with an optimized common template-based CCA method for SSVEP-based BCI. Front. Neurorobot. **16**, 855825 (2022)
12. Kordmahale, S.N.A., Kilani, S., Ghassemlooy, Z., Wu, Q., Maleki, A.: A novel artifact removal method for the SSVEP signal using hybrid CCA-DWT and comparative analysis for feature selection and classification in the P300 signal. In: 2022 13th International Symposium on Communication Systems, Networks and Digital Signal Processing (CSNDSP), pp. 390–394. IEEE (2022)
13. Wang, S., Ji, B., Shao, D., Chen, W., Gao, K.: A methodology for enhancing SSVEP features using adaptive filtering based on the spatial distribution of EEG signals. Micromachines **14**(5), 976 (2023)
14. Yan, W., Wu, Y., Du, C., Xu, G.: Cross-subject spatial filter transfer method for SSVEP-EEG feature recognition. J. Neural Eng. **19**(3), 036008 (2022)

15. Cherloo, M.N., Amiri, H.K., Daliri, M.R.: Spatio-spectral CCA (SS-CCA): a novel approach for frequency recognition in SSVEP-based BCI. J. Neurosci. Methods **371**, 109499 (2022)

16. Sun, Y., et al.: Cross-subject fusion based on time-weighting canonical correlation analysis in SSVEP-BCIs. Measurement **199**, 111524 (2022)

17. Thomas, J., Maszczyk, T., Sinha, N., Kluge, T., Dauwels, J.: Deep learning-based classification for brain-computer interfaces. In: 2017 IEEE International Conference on Systems, Man, and Cybernetics (SMC), pp. 234–239. IEEE (2017)

18. Kobayashi, N., Ishizuka, K.: LSTM-based classification of multiflicker-SSVEP in single channel dry-EEG for low-power/high-accuracy quadcopter-BMI system. In: 2019 IEEE International Conference on Systems, Man and Cybernetics (SMC), pp. 2160–2165. IEEE (2019)

19. Kwak, N.S., Müller, K.R., Lee, S.W.: A convolutional neural network for steady state visual evoked potential classification under ambulatory environment. PLoS ONE **12**(2), e0172578 (2017)

20. Waytowich, N., et al.: Compact convolutional neural networks for classification of asynchronous steady-state visual evoked potentials. J. Neural Eng. **15**(6), 066031 (2018)

21. Li, M., Ma, C., Dang, W., Wang, R., Liu, Y., Gao, Z.: DSCNN: dilated shuffle CNN model for SSVEP signal classification. IEEE Sens. J. **22**(12), 12036–12043 (2022)

22. Wan, Z., Li, M., Liu, S., Huang, J., Tan, H., Duan, W.: EEGformer: a transformer-based brain activity classification method using EEG signal. Front. Neurosci. **17**, 1148855 (2023)

23. Zhang, C., Han, S., Zhang, M.: Single-channel EEG completion using Cascade Transformer. In: 2022 IEEE Biomedical Circuits and Systems Conference (BioCAS), pp. 605–609. IEEE (2022)

24. Ma, P., et al.: A classification algorithm of an SSVEP brain-computer interface based on CCA fusion wavelet coefficients. J. Neurosci. Methods **371**, 109502 (2022)

25. Li, Y., Xiang, J., Kesavadas, T.: Convolutional correlation analysis for enhancing the performance of SSVEP-based brain-computer interface. IEEE Trans. Neural Syst. Rehabil. Eng. **28**(12), 2681–2690 (2020)

26. Lawhern, V.J., Solon, A.J., Waytowich, N.R., Gordon, S.M., Hung, C.P., Lance, B.J.: EEGNet: a compact convolutional neural network for EEG-based brain-computer interfaces. J. Neural Eng. **15**(5), 056013 (2018)

27. Schulman, J., Wolski, F., Dhariwal, P., Radford, A., Klimov, O.: Proximal policy optimization algorithms. arXiv preprint arXiv:1707.06347 (2017)

28. Jeong, J.H., et al.: 2020 international brain-computer interface competition: a review. Front. Hum. Neurosci. **16**, 898300 (2022)

Evolving Epidemic Management Rules Using Deep Neuroevolution: A Novel Approach to Inspection Scheduling and Outbreak Minimization

Victoria Huang[1]([✉]) [iD], Chen Wang[1] [iD], Samik Datta[1] [iD], Bryce Chen[1],
Gang Chen[2] [iD], and Hui Ma[2] [iD]

[1] The National Institute of Water and Atmospheric Research, Wellington 6021,
New Zealand
{victoria.huang,chen.wang,samik.datta,bryce.chen}@niwa.co.nz
[2] Centre for Data Science and Artificial Intelligence & School of Engineering and
Computer Science, Victoria University of Wellington, Wellington 6041, New Zealand
{aaron.chen,hui.ma}@ecs.vuw.ac.nz

Abstract. In epidemic management, the unpredictable dynamics of outbreaks, the constraints of available resources, and the complexity of variable interactions pose a significant challenge in designing effective management rules. While traditional mathematical models offer insights into the spread of the outbreaks, they are not effective due to the lack of available information on the outbreaks. Moreover, they are valuable for the evaluation of different management rules but do not provide management rules. In this paper, we propose a novel approach that leverages Reinforcement Learning (RL) to automatically design epidemic management rules. By using RL, we formulate the problem as determining the optimal daily inspection frequency during an outbreak with the goal of minimizing both the inspection cost and the epidemic size. A management rule is trained using a deep neuroevolution algorithm called Evolutionary Strategy (ES). An epidemic simulator is developed to provide a realistic and dynamic environment for evaluating our proposed approach's effectiveness. Extensive experiments have shown that our RL-based approach outperformed conventional methods and can better balance the trade-off between the inspection cost and epidemic size. The findings from this work provide not only a framework for autonomous learning of epidemic management rules, but also potential implications for real-world epidemic control and policy-making.

Keywords: Epidemic Management · Reinforcement Learning

1 Introduction

In an epidemic, policymakers need to conduct effective epidemic management to rapidly control the spread of the epidemic with limited available management resources and minimize its impact. Mathematical models and simulations,

T. Liu et al. (Eds.): AI 2023, LNAI 14472, pp. 387–399, 2024.
https://doi.org/10.1007/978-981-99-8391-9_31

such as the spatial SIR (Susceptibl-Infected-Removed) model [17], have been widely used to evaluate different management rules and forecast outbreak severity [5,7,11]. However, the accuracy of these methods relies on the estimation of model parameters. During the early stage of an epidemic, only limited information/data is available, leading to increased uncertainty in model parameter estimation. Furthermore, mathematical models and simulations are valuable for the evaluation on different management rules, but they fail to address the challenges of designing an effective management rule.

It is challenging to design epidemic management rules [6]. The complexity comes from several factors that impact the size of the epidemic, such as the nature of the disease, its transmission rate, and the population's susceptibility. Moreover, the necessity of balancing practical limitations, such as staff availability and budget, against the optimum number of inspections needed for effective outbreak control poses a considerable challenge. Additionally, the rapidly evolving dynamics of an epidemic demand continuous reassessment and modification of this rule. These rules also need to be flexible, adapting to diverse contexts and outbreak stages. Therefore, crafting such a management rule requires a scientifically rigorous approach that meticulously balances the size of the epidemic and the number of inspections while maintaining adaptability.

Reinforcement Learning (RL) is a promising approach to address the challenges in devising a robust, adaptable, and efficient management rule [3]. RL, characterized by its ability to learn optimal policies through interactions with a complex and dynamic environment, is uniquely suited to manage epidemics. It holds the potential to dynamically optimize daily inspection frequency and the size of the epidemic, adapting to changing epidemic conditions. To tackle the task of designing an epidemic management rule, we propose an RL-based approach in this paper to determine the daily inspections during epidemic outbreaks. This approach learns the consequences of each decision, thereby progressively enhancing its policy over time. Moreover, our RL approach can simultaneously handle multiple objectives, such as minimizing inspection numbers while curtailing the epidemic size – an essential capability for efficiently managing limited resources yet effectively controlling epidemic spread.

The aim of this paper is to propose an RL-based approach to autonomously and effectively learn a management rule that provides the frequency of daily inspections amidst an epidemic outbreak. In this paper, we conceptualize the creation of a management rule to minimize the size of the epidemic and the number of inspections. Specifically, we first introduce a new problem model that formulates the complexity and dynamics of epidemic management. We then model a neural network-based management rule with designed epidemic status and inspection queue inputs. The rule is effectively trained via a deep neuroevolution RL algorithm, i.e., OpenAI Evolutionary Strategy (ES) [18]. To facilitate experimentation, an epidemic simulator is developed to accurately simulate the dynamic process of epidemic outbreaks and keep track of infection status and rewards. Extensive experiments have been conducted to demonstrate that the RL-learned management rule can significantly outperform existing approaches.

2   Related Work

Existing works on epidemic management can be classified into two categories: (1) mathematical models to simulate the dynamics of epidemics, and (2) optimization methods to optimize the design of management rules.

Mathematical models have been widely used to simulate the spread of epidemics [5,7,11,12]. For example, Datta et al. [5] adopted a statistical Markov chain Monte Carlo (MCMC) model to simulate the spatial spread of disease in American honeybees and evaluate the effectiveness of control strategies using a stochastic Bayesian method. The disease control strategies adopted are standard practices specific to honeybees. Similarly, Leung et al. [12] modelled the non-indigenous species invasion using statistical and distance-based kernel, and simulated control algorithms for detecting and delimiting the invasion. But the performance of the proposed method highly relies on the availability of data. Although these proposed mathematical models can be used to compare different management strategies, the design itself remains an open question. Meanwhile, at the brink of an epidemic outbreak, limited data is available to estimate the model parameters, which limits the predictive performance of the model.

In addition to the modelling efforts reviewed above, optimization methods have also been developed successfully for effective epidemic management. These methods can be divided into exact methods [1,16,19], heuristic methods [8,14], and machine learning methods [2,15]. For exact methods, Sepulveda et al. [19] formulated an optimal control mathematical model to manage the dengue outbreak under a limited budget of insecticides. For heuristic approaches, Hazard et al. [8] proposed a tabu heuristic search strategy to determine the optimal vaccine plan that minimizes the number of affected people for COVID-19. We have also seen applications of RL on disease control such as [15] that proposed an optimal control policy for the allocation of test kits and vaccine doses on COVID-19 cases. However, this research does not tackle the disease control problem directly. Bushaj et al. [3] also applied RL to simulate epidemic control using COVID-19 cases to serve different epidemic control objectives, without considering the importance of budget constraints and practical limitations. The aforementioned optimization methods cannot be applied to solve our problem directly because they mostly rely on their problems and require rigorous mathematical formulation to fit into different policy search problems. For example, transferring the disease control model from mosquito to another context such as invasive species control requires a reformulation of problems, including objectives, variables, constraints and parameters related to dynamics.

This paper focuses on optimization methods for epidemic control and management using RL as it has achieved promising results compared with traditional methods in terms of scalability and effectiveness. Based on this, we will develop an RL-based approach to train our designed management policy with the aim of minimizing the number of inspections and the size of an epidemic.

3   Problem Formulation

We follow the SIR (Susceptible-Infected-Recovered) model and assume that an epidemic takes place within a $N \times N$ grid where a population of I individuals randomly reside. The epidemic starts with the introduction of an initial infection of a randomly selected individual i. The epidemic then progresses at discrete timesteps. The whole epidemic lasts for T timesteps, where every timestep t is defined as a day in this paper. A matrix $\mathcal{X} \in \mathbb{R}^{I \times T}$ is used to record the population infection status where each element $x_{i,t}$ presents the infection status of individual i at timestep t. Specifically,

$$x_{i,t} = \begin{cases} 1, & \text{if individual } i \text{ is infected at timestep } t \\ 0, & \text{otherwise} \end{cases}$$

At each timestep t, new infections can happen. In particular, nearby individuals may be infected by their infected neighbours. The spread to neighbouring individuals is a stochastic process, depending on the transmission rate and the distance exponent. Specifically, the disease transmission rate between an infected individual i to a susceptible individual j can be formulated as [5]

$$r_{i,j} = \frac{\beta}{1 + \left(\frac{d_{i,j}}{\alpha}\right)^2} \tag{1}$$

where β is the transmission rate, which signifies how frequently the disease passes from an infected individual to a susceptible one. $d_{i,j}$ is the distance between individuals i and j. α is the distance exponent which models the rate at which the transmission of the disease decreases with distance.

We assume that each infected individual can be detected (e.g., infected livestock detected by farmers), which follows a probability

$$1 - e^{-\gamma dt} \tag{2}$$

where γ is a detection parameter, which indicates the probability at which infected individuals can be detected. The detected individuals will be added into an inspection queue with assigned priorities (see Sect. 4.2 for inspection queue details). Meanwhile, whenever an infected individual is detected, a radial check will be applied which adds neighbouring individuals within a certain radius of the discovered infected individual to the inspection queue.

Given the latest epidemic observation, policymakers need to decide the number of inspections a_t at timestep t given the daily inspection limit L. During the inspections, a_t individuals with the highest priorities will be selected from the inspection queue. The detected individuals will be inspected first. If there are still remaining inspection quotas, the earliest neighbouring individuals added to the queue will be inspected. During the inspection, any infected individual i will be recovered and $x_{i,t} = 0$. Its neighbouring individuals will be added to the inspection queue. After that, the epidemic proceeds to the next timestep $t + 1$.

Our ultimate aim is to produce an optimized management rule to jointly optimize the following: (1) to reduce the number of inspections to the minimum required for effective control, and (2) to limit the overall size of the epidemic to minimize the disease impact. We have modelled the objective as minimizing the cumulative total of two factors: the total number of inspections conducted, and the total number of infected days across the entire population, measured as the sum of timesteps each individual has been infected.

$$\min_{\{a_t\}} \sum_t^T a_t + \sum_i^I \sum_t^T x_{i,t} \quad \text{s.t. } a_t \leq L \text{ and } t = 1, ..., T \quad (3)$$

Specifically, we assume that the disease characteristics (e.g., the dispersal distance and transmission rate) are unknown.

4   Design of Epidemic Management Rule

4.1   Rule Representation

In this paper, we model the epidemic management rule π_θ in the form of a Neural Network (NN) parameterized by θ, as shown in Fig. 1. The NN, trained through evolutionary strategy, is designed to balance the two objectives defined in Eq. (3) based on the current state of the epidemic and the inspection resources available. By considering the number and status of detected and recovered individuals, as well as those in the inspection queue, the NN can dynamically adjust the inspection frequency to efficiently manage the epidemic. This approach offers a promising pathway for the development of flexible and effective management policies for epidemic outbreaks.

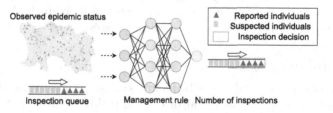

Fig. 1. Design of the epidemic management rule.

The decision-making process of this management rule requires two key pieces of information. The first one is individual status: detected individuals who have been identified but not yet recovered and recovered individuals who have successfully overcome the infection through the inspection process. The individual status can be represented by a list $\eta \in \mathbb{R}^I$. The second one is the inspection queue $\mu \in \mathbb{R}^I$: individuals on the inspection queue waiting to be inspected, which includes those within the infection radius as well as the reported infected individuals who have not yet recovered. Therefore, at every timestep t, the above key information s_t can be expressed as:

$$s_t = [\eta, \mu] \tag{4}$$

The management rule π_θ outputs the number of daily inspections a_t:

$$a_t = \pi_\theta(s_t) \tag{5}$$

The individuals chosen for inspection are determined by the inspection queue, with the rule being executed once per timestep.

4.2   Inspection Queue

To effectively manage and respond to an epidemic, we propose and manage an inspection queue that keeps track of the detected individuals and neighbouring exposed individuals. Within the queue, individuals are prioritized in a descending order based on a priority score $p_{i,t}$ defined in Eq. (6). $p_{i,t}$ is calculated based on when individual i was added to the queue and why i was added to the queue (e.g., i is a detected infection or it is suspected to be infected due to its exposure to an infection).

$$p_{i,t} = \begin{cases} C - t, & \text{if individual } i \text{ is infected at timestep } t \\ -t, & \text{if individual } i \text{ is within the radius at timestep } t \end{cases} \tag{6}$$

where C is a positive constant and $C > T_{\max}$. This priority score guarantees that immediate actions are directed toward confirmed cases, especially the ones entering the queue earlier. Additionally, it recognizes the potential threat from exposed individuals and ranks them based on their time of exposure (i.e., the longer the exposure is, the higher the rank is), ensuring a timely response. Note that scores for infected individuals should always be positive, positioning them ahead in the queue. Conversely, scores from exposed individuals should always be negative, placing them behind the confirmed cases.

4.3   Algorithm Design

Learning a management rule described above can be achieved through RL. Yet, the performance of numerous RL algorithms is highly sensitive to their hyper-parameter settings [4,9]. Adjusting these hyper-parameters, however, poses its own set of challenges [10,13]. In this work, we adopt a robust Evolutionary Strategy (ES) [18] algorithm by OpenAI to train our management rule.

The overall training process is summarized in Fig. 2. It starts with a management rule π_θ with randomly initialized NN parameters θ. Since ES is a population-based policy search method, at each iteration, the algorithm generates a set of parameters $[\hat{\theta}_j]_{j=1,...,M}$ for M NNs through introducing variations of the current parameters θ. Each set of NN parameters $\hat{\theta}_j$ follows:

$$\hat{\theta}_j = \theta + \sigma\epsilon_j, \epsilon_j \sim \mathcal{N}(0, I) \tag{7}$$

where σ is a hyper-parameter that controls the scale of noise that gives rise to the diversity of the generated NNs.

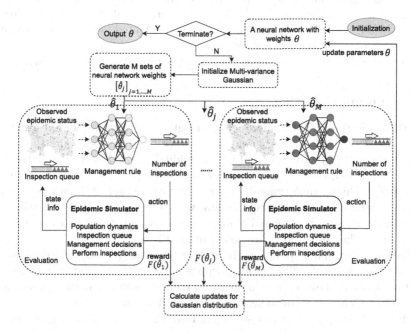

Fig. 2. Training the epidemic management rule.

Each NN is independently executed within an environment (i.e., the simulator in Sect. 5.1) as described in Sect. 4. The performance of each NN parameterized by $\hat{\theta}_j$ (i.e., "reward") F is measured by the objective function in Eq. (3):

$$F(\hat{\theta}_j) = F(\theta + \sigma\epsilon_j) = \sum_t^T a_t + \sum_i^I \sum_t^T x_{i,t}$$

To minimize Eq. (3), ES updates θ following:

$$\theta \leftarrow \theta + \delta\nabla_\theta \mathbb{E}_{\theta \sim p_\psi} F(\theta) = \theta + \delta\nabla_\theta \mathbb{E}_{\epsilon \sim \mathcal{N}(0,I)} F(\theta + \sigma\epsilon)$$

$$= \theta + \frac{\delta}{\sigma}\mathbb{E}_{\epsilon \sim \mathcal{N}(0,I)}\left[F(\theta + \sigma\epsilon)\epsilon\right] \approx \theta + \frac{\delta}{M\sigma}\sum_{j=1}^M \left[F(\theta + \sigma\epsilon_j)\epsilon_j\right] \qquad (8)$$

5   Experiment

5.1   Epidemic Simulation

As part of our contributions, we have developed a simulator to model the evolving dynamics of an epidemic among individuals within a given space. The simulation process begins by randomly positioning individuals within this space and introducing an infection to one of them. The simulation then proceeds forward at discrete timesteps to capture the dynamics of the disease's spread. In our model, individuals can occupy one of four states: susceptible, infected (an individual is contingent on proximity to an existing infection), detected (an infected individual has a small probability of being detected), or recovered (an infected individual receives necessary treatment; after recovery, they will not be reinfected).

5.2   Experiment Setting

Simulator Setting: Following Sect.3, we set the simulation within a $N \times N$ grid where $N = 10$, providing a defined space for the individuals and the spread of the disease. A population size of 500 individuals is used in the simulation. To account for the inherent variability in the epidemic process, we conduct a total of 30 independent simulations during our testing period and each simulation lasts for 50 timesteps. Following existing work [5], within each simulation, we set the distance exponent α and transmission rate β in Eq. (1) to 1 and 1.1e-2 respectively. The detection parameter γ in Eq. (2) is 1.4e−3.

The grid size and population size, along with the parameter values for α, β, and γ, are chosen for illustrative purposes, and are not based on any specific infection. Rather, they are chosen such that transmission is fast enough to spread from an initial case, but not so fast that all individuals are infected in a very short time frame. Hence management strategies have the ability to have a measurable impact on the progress of the epidemic, and strategies can be compared.

A common control practice when dealing with infectious diseases is to inspect neighbours of infected individuals within an inspection radius. In our simulation, the inspection radius is set to 1, defining the range within which individuals are considered for inspection (i.e., added to the inspection queue). Commonly during an epidemic, only limited resources (e.g., medical professionals, medical supplies, and logistic resources) will be available. To account for different resource restrictions, different daily inspection limits are simulated. We evaluate multiple scenarios with varying inspection capacities, setting daily inspection limits at 50, 100, 200, 300, 400, and 500 inspections per timestep.

Algorithm Settings: In this work, we evaluate 3 different algorithms to decide the number of daily inspections:

Reinforcement Learning-based policy (denoted by "RL"): This is our proposed management strategy using an RL-trained neural network to decide the number of daily inspections. We set the Gaussian noise standard deviation in Eq. (7) $\sigma = 5 \times 10^{-3}$ and the learning rate in Eq. (8) $\delta = 0.01$. Following [10], we set the ES population size $M = 40$ and the maximum number of generations is 3000. Each candidate is evaluated for 1 episode. The management rule is a multi-layer perceptron with two hidden layers of 64 tanh units.

Random Inspection (denoted by "Random"): Although Random is not often carried out in practice, it is usually used for illustrating the effects of management strategies. In Random, the number of inspections conducted daily is randomly selected, falling within the prescribed inspection limit. This strategy serves to examine the effectiveness of inspections that are conducted without following any specific policy or guidance.

Genetic Programming (denoted by "GP"): We employ a GP tree representation as a substitute for the management rule due to its high interpretability, expressiveness, and flexibility [20]. The tree is initialized with functional and terminal sets, and trained via canonical GP algorithm. The functional set encompasses arithmetic operations $(+, -, \times)$ and a protected division $(/)$ operation that returns 1 if the denominator is zero. The terminal set includes all features from s_t in Eq. 5. The parameter settings of GP follow the settings reported in some recent works [20], with a population size of 40, an elite size of 5, a tournament selection size of 2, a crossover rate of 0.7, and a mutation rate of 0.1. The minimum and maximum depth of the evolved trees are set to 2 and 17, respectively. To maintain a consistent number of evaluations for comparison, we set the number of generations to 300. Additionally, each tree-based management rule is evaluated 10 times across 10 different simulations.

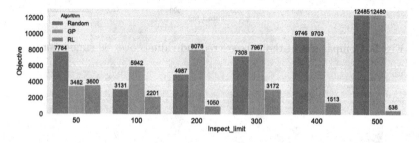

Fig. 3. Comparison of the average objective function value over 30 simulations.

5.3 Experiment Results

Figure 3 shows the values of our objective function in Eq. (3) achieved by our proposed RL-based approach and two baselines. It illustrates that apart from the scenario when the daily inspection limit is 50, our RL-based approach (depicted by the green bars) consistently achieves lower objective values measured by Eq. (3), compared to both Random and GP, indicating more efficient epidemic management. We also noticed that when the daily inspection limit is 50, RL achieved very competitive performance with slightly 3% higher than the optimal value obtained by GP. Furthermore, for the other scenarios, RL can dramatically reduce the objective value by up to 95%. In comparison, GP did not perform well when the inspection limit is more than 50. This observation indicates that learning an effective tree-based management rule becomes increasingly challenging with a larger terminal set, especially when the training task grows in complexity due to an increased inspection limit. This increase significantly enlarges the search space. These performance results suggest GP does not scale well, demanding more effective design on GP trees and training methods.

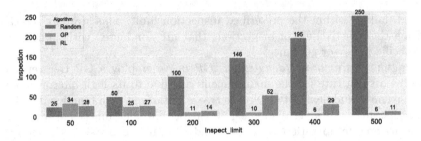

Fig. 4. Comparison of the number of daily inspections over 30 simulations.

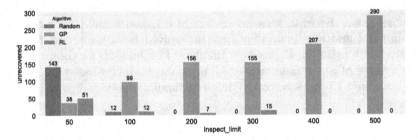

Fig. 5. Comparison of the unrecovered individuals over 30 simulations.

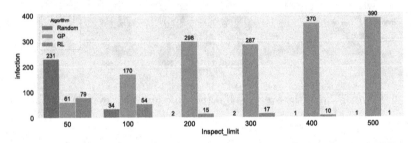

Fig. 6. Comparison of the infected individuals over 30 simulations.

Figure 4 reveals the average number of daily inspections under different management strategies. Management rules trained by RL often require fewer inspections compared to Random, except when the daily inspection limit is set at 50. Despite conducting marginally more inspections in this specific scenario (a 50-daily inspection limit), the benefits are evident in Fig. 5 and Fig. 6. These figures show that the number of unrecovered and infected individuals under RL is significantly less than under Random. To illustrate, under the 50 daily inspection limit, with an increment of just three daily inspections, RL reduces the number of unrecovered individuals from 143 to 51, and the number of infected individuals from 231 to 79.

On the contrary, Fig. 4 also shows that RL produces more inspections compared to GP, except for the 50 inspection limit scenario. With more inspections (no more than 20% of the inspection limit), RL can clearly reduce the number of unrecovered individuals as shown in Fig. 5, and the number of infected individuals as shown in Fig. 6. Therefore, compared to GP, RL can more effectively control the epidemic size while minimizing the inspection cost, especially when the search space is large.

In addition to the previous comparisons, we have visualized results for two specific scenarios where the daily inspection limit is set to 50 and 100 respectively. Figure 7 compares RL with Random and GP in terms of changes in the number of inspections over the simulated 50-day period over 30 simulations.

When the inspection limit is 50 as shown in Fig. 7(a), both RL and Random exhibit similar trends and operate within a comparable range while GP sets a higher and relatively stable daily inspection number. On the other hand, when the inspection limit increases to 100, noticeable differences can be observed from Fig. 7(b). Random fluctuates between 40 to 60 inspections for the whole simulation period. GP remains relatively stable during the 50-day simulation period with a slight increase after 35 days. This increase is likely due to the increase of the epidemic size. More inspections need to be carried out to recover the infected individuals to control the epidemic. On the other hand, RL adopts a very different approach. It first increases the number of inspections to more than 60 at the beginning of the epidemics and then reduces the inspection frequency to the level between 20 to 30 daily inspections after 15 days. This management strategy is "smart" as a higher inspection rate at the beginning can quickly identify the infected individuals and treat them to dampen the future spreading of the epidemics. When the epidemic is considered under control, the daily inspection number can be set lower to reduce the inspection cost without the risk of increasing the epidemic size.

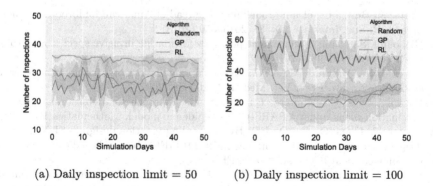

(a) Daily inspection limit = 50   (b) Daily inspection limit = 100

Fig. 7. Comparison of the number of inspections changes within the simulated 50 days over 30 simulations.

6   Conclusions

In this paper, a novel Reinforcement Learning (RL)-based approach is proposed to automatically design a neural network-based epidemic management rule. This rule can determine optimal daily inspection numbers during an outbreak while jointly minimizing both the inspection cost and the epidemic size. Our rule can be effectively trained using evolutionary strategy, a deep neuroevolution RL algorithm. To facilitate experimentation, an epidemic simulator is developed to accurately simulate the dynamics of an epidemic outbreak. Through extensive experiments, we have demonstrated that our RL-based management rule can significantly outperform existing approaches.

Acknowledgements. This work is supported by The Strategic Science Investment Fund (SSIF) funds under Grant CDFP2303, administrated by the National Institute of Water and Atmospheric Research, New Zealand.

References

1. Abdin, A.F., Fang, Y.P., Caunhye, A., Alem, D., Barros, A., Zio, E.: An optimization model for planning testing and control strategies to limit the spread of a pandemic - the case of COVID-19. Divers. Distrib. **304**, 308–324 (2023)
2. Bolzoni, L., Bonacini, E., et al.: Optimal control of epidemic size and duration with limited resources. Math. Biosci. **315**, 108232 (2019)
3. Bushaj, S., Yin, X., et al.: A simulation-deep reinforcement learning (SiRL) approach for epidemic control optimization. Ann. Oper. Res. 1–33 (2022)
4. Chen, G., Huang, V.: Ensemble reinforcement learning in continuous spaces-a hierarchical multi-step approach for policy training. arXiv preprint arXiv:2209.14488v2 (2022)
5. Datta, S., Bull, J.C., et al.: Modelling the spread of American foulbrood in honeybees. J. R. Soc. Interface **10**(88), 20130650 (2013)
6. Giakoumi, S., et al.: Management priorities for marine invasive species. Sci. Total Environ. **688**, 976–982 (2019)
7. Hayes, B.H., Andraud, M., et al.: Mechanistic modelling of African swine fever: a systematic review. Prev. Vet. Med. **191**, 105358 (2021)
8. Hazard-Valdés, C., Montero, E.: A heuristic approach for determining efficient vaccination plans under a SARS-COV-2 epidemic model. Mathematics **13** (2023)
9. Huang, V., Chen, G., Fu, Q.: Effective scheduling function design in SDN through deep reinforcement learning. In: ICC 2019–2019 IEEE International Conference on Communications (ICC), pp. 1–7. IEEE (2019)
10. Huang, V., Wang, C., Ma, H., Chen, G., Christopher, K.: Cost-aware dynamic multi-workflow scheduling in cloud data center using evolutionary reinforcement learning. In: Troya, J., Medjahed, B., Piattini, M., Yao, L., Fernández, P., Ruiz-Cortés, A. (eds.) ICSOC 2022. LNCS, vol. 13740, pp. 449–464. Springer, Cham (2022)
11. Hunter, E., Namee, B.M., Kelleher, J.D.: A hybrid agent-based and equation based model for the spread of infectious diseases (2020)
12. Leung, B., Cacho, O.J., et al.: Searching for non-indigenous species: rapidly delimiting the invasion boundary. Divers. Distrib. **16**(3), 451–460 (2010)

13. Liessner, R., Schmitt, J., et al.: Hyperparameter optimization for deep reinforcement learning in vehicle energy management. In: ICAART (2), pp. 134–144 (2019)
14. Lin, H.: Optimal design of cordon sanitaire for regular epidemic control. Adv. Civil Eng. **2021** (2021)
15. M, X., Bottcher, L., Chou, T.: Controlling epidemics through optimal allocation of test kits and vaccine doses across networks. IEEE Trans. Netw. Sci. Eng. **9**(3), 1422–1436 (2022)
16. Miguel Navascués, Costantino Budroni, Y.G.: Disease control as an optimization problem. PLoS ONE **16** (2021)
17. Milner, F.A., Zhao, R.: Sir model with directed spatial diffusion. Math. Popul. Stud. **15**(3), 160–181 (2008)
18. Salimans, T., Ho, J., Chen, X., et al.: Evolution strategies as a scalable alternative to reinforcement learning. arXiv preprint arXiv:1703.03864 (2017)
19. Sepulveda-Salcedo, L.S., et al.: Optimal control of dengue epidemic outbreaks under limited resources. Stud. Appl. Math. **144**(2), 185–212 (2020)
20. Tan, B.: An evolutionary computation approach to resource allocation in container-based clouds. Ph.D. thesis, Open Access Te Herenga Waka-Victoria University of Wellington (2020)

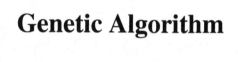

Genetic Algorithm

A Semantic Genetic Programming Approach to Evolving Heuristics for Multi-objective Dynamic Scheduling

Meng Xu[ID], Yi Mei[ID], Fangfang Zhang[(✉)][ID], and Mengjie Zhang[ID]

Centre for Data Science and Artificial Intelligence & School of Engineering and Computer Science, Victoria University of Wellington, Wellington 6140, New Zealand
{meng.xu,yi.mei,fangfang.zhang,mengjie.zhang}@ecs.vuw.ac.nz

Abstract. Multi-objective dynamic flexible job shop scheduling (MO-DFJSS) is a challenging problem that requires finding high-quality schedules for jobs in a dynamic and flexible manufacturing environment, considering multiple potentially conflicting objectives simultaneously. A good approach to MO-DFJSS is to combine Genetic Programming (GP) with Non-dominated Sorting Genetic Algorithm II (NSGA-II), namely NSGP-II, to evolve a set of non-dominated scheduling heuristics. However, a limitation of NSGPII is that individuals with different genotypes can exhibit the same behaviour, resulting in a loss of population diversity. Semantic genetic programming (SGP) considers individual semantics during the evolutionary process and can enhance population diversity in various domains. However, its application in the domain of MO-DFJSS remains unexplored. Therefore, it is worthy to incorporate semantic information with NSGPII for MO-DFJSS. This study focuses on semantic diversity and semantic similarity. The results demonstrate that NSG-PII considering semantic diversity yields better performance compared with the original NSGPII. Moreover, NSGPII incorporating semantic similarity achieves even better performance, highlighting the importance of maintaining a reasonable semantic distance between offspring and their parents. Further analysis reveals that the improved performance achieved by the proposed methods is attributed to the attainment of a more semantically diverse population through effective control of semantic distances between individuals.

Keywords: Heuristic learning · Multi-objective genetic programming · Semantic · Multi-objective dynamic scheduling

1 Introduction

Multi-objective dynamic flexible job shop scheduling (MO-DFJSS) is a complex combinational optimisation problem that involves scheduling multiple jobs on multiple machines in a flexible manufacturing environment, considering dynamic job arrivals and conflicting objectives. Solving the MO-DFJSS problem is challenging due to its combinatorial nature, the presence of multiple conflicting

T. Liu et al. (Eds.): AI 2023, LNAI 14472, pp. 403–415, 2024.
https://doi.org/10.1007/978-981-99-8391-9_32

objectives, and its dynamic and flexible characteristics. Scheduling heuristics [14] can quickly adapt and make real-time decisions based on the most up-to-date information, allowing for practical applications in real-world dynamic environments. However, manually designing scheduling heuristics can be particularly challenging, time-consuming, labor-intensive, and require deep domain knowledge [15]. Further, balancing multiple objectives and finding a Pareto front of high-quality scheduling heuristics require specialised algorithms and techniques, which might be difficult to incorporate into the manual design process.

Genetic programming (GP) methods have been widely used to automatically evolve scheduling heuristics for solving the DFJSS problem. There are some studies incorporating well-known Pareto dominance-based methods (i.e., non-dominated sorting genetic algorithm II [5] and strength Pareto evolutionary algorithm 2 [23]) and scalarising function-based methods (i.e., multi-objective evolutionary algorithm based on decomposition [22]) into GP, named NSGPII [19], SPGP2 [19], and MOGP/D [16], for MO-DFJSS. Among them, NSGPII showed the best performance in terms of hypervolume (HV) [24] and inverted generational distance (IGD) [7], which are two important performance indicators used to measure multi-objective algorithms [9]. However, NSGPII acts on the genotype of individuals and does not consider semantic information, which reflects the behaviour of the genotype. Semantic GP [13] has been proposed to enhance population diversity by integrating semantic information into the evolutionary process. Its effectiveness has been demonstrated across diverse domains, including symbolic regression [11], classification [1,10], and feature selection [8]. However, to the best of our knowledge, semantic information has not been incorporated into NSGPII for solving the MO-DFJSS problem. Given these promising results, it becomes particularly intriguing to explore how to improve the performance of NSGPII for MO-DFJSS by incorporating semantic information.

For this purpose, the objectives of this paper are as follows:

1. Define the *semantic* and *semantic distance* concepts in the context of MO-DFJSS domain, and design strategies to measure the semantic information derived from MO-DFJSS.
2. Design a semantic NSGPII algorithm to evolve a Pareto front of scheduling heuristics by considering the semantic information during the evolution process for solving the MO-DFJSS problem.
3. Study the effects of incorporating semantic information into NSGPII on the performance of the evolved scheduling heuristics for MO-DFJSS in terms of HV and IGD.
4. Analyse how semantic information affects the performance of evolved scheduling heuristics by NSGPII on solving the MO-DFJSS problem.

2   Background

2.1   Multi-objective Dynamic Flexible Job Shop Scheduling

In MO-DFJSS, a set of jobs $\mathcal{J} = \{J_1, J_2, ..., J_n\}$ needs to be processed by a set of machines $\mathcal{M} = \{M_1, ..., M_m\}$. Each job J_i is characterised by its arrival time r_i, due date d_i, weight w_i, and a sequence of operations $[O_{i,1}, O_{i,2}, ..., O_{i,p_i}]$ that must be performed in order. Each operation $O_{i,j}$ has a workload $\pi_{i,j}$ and can be processed by a machine from a set of optional machines $\mathcal{M}_{i,j} \subseteq \mathcal{M}$. The processing time $t_{i,j,k} = \pi_{i,j}/\gamma_k$ for operation $O_{i,j}$ on machine $M_k \in \mathcal{M}_{i,j}$ depends on the processing speed γ_k of the machine. Additionally, the machines are distributed, and there is a transport time τ_{k_1,k_2} required to move a job between two machines M_{k_1} and M_{k_2}.

The problem assumptions are as follows:

1. An operation cannot start until its preceding operation in the sequence has been completed and the job has been transported to the designated machine.
2. Each machine can handle only one operation at a time.
3. Each operation can be processed by only one of its optional machines.
4. The scheduling is non-preemptive, meaning once an operation starts, it must be completed without interruption.

This paper considers six common scheduling objectives: max-flowtime ($Fmax$), mean-flowtime ($Fmean$), max-weighted-flowtime ($WFmax$), max-tardiness ($Tmax$), max-weighted-tardiness ($WTmax$), and mean-weighted-tardiness ($WTmean$). The definitions of these objectives are as follows:

$$Fmax = \max_{i=1}^n\{C_i - r_i\}, \quad Fmean = \frac{1}{n}\sum_{i=1}^n(C_i - r_i)$$
$$WFmax = \max_{i=1}^n\{w_i(C_i - r_i)\}, \quad Tmax = \max_{i=1}^n\{T_i\}$$
$$WTmax = \max_{i=1}^n\{w_iT_i\}, \quad WTmean = \frac{1}{n}\sum_{i=1}^n(w_iT_i)$$

where C_i is the completion time of the job J_i in the schedule, and $T_i = \max\{C_i - d_i, 0\}$ is the tardiness of the job J_i.

To facilitate analysis, we focus on bi-objective scenarios in this paper, which consider two out of the above six objectives for each scenario. More details regarding the objective selection will be provided in Sect. 4.

2.2   Related Work

MOGP for MO-DFJSS: In [19], GP is combined with two well-known Pareto dominance-based multi-objective optimisation algorithms (i.e., non-dominated sorting genetic algorithm II [5] and strength Pareto evolutionary algorithm 2 [23]) to form NSGPII and SPGP2 to evolve scheduling heuristics to address the MO-DFJSS problem. Experimental results demonstrate that NSGPII outperforms the SPGP2 in terms of both training and test HV and IGD values. Except for Pareto dominance-based methods, in [16], a multi-objective GP method based on decomposition (MOGP/D) that incorporates the advantages of multi-objective evolutionary algorithm based on decomposition [22] and GP to learn scheduling heuristics for MO-DFJSS is proposed. Among the aforementioned three classical multi-objective GP algorithms, NSGPII performs the best in terms of HV and IGD performance. Following this, some further studies were carried out on the basis of NSGPII. In [17], a novel NSGPII approach is presented for MO-DFJSS by incorporating surrogate technique and brood recombination

technique. By leveraging the surrogate and brood recombination-assisted approach, the improved NSGPII obtains high-quality scheduling heuristics compared
to the original NSGPII within the same training time. In [20], the influence of
terminal settings on NSGPII for solving MO-DFJSS is studied. Some studies
focus on interpretability [21] or multitask [18] topics in MO-DFJSS. In a word,
MO-DFJSS has become a popular problem and NSGPII has become a widely
used algorithm for solving it.

Semantic GP: SGP [13] has recently gained significant attention in the field
of GP. It represents a valuable approach for incorporating semantic information into the evolution process, thereby improving the performance of evolved
solutions. One of the key advantages of SGP is its ability to consider the
behaviours/semantics rather than the genotype of individuals [13]. By considering the behaviour of individuals, SGP introduces semantic information, enabling
a more nuanced understanding of the evolved solutions. The semantic analysis
facilitates the discovery of individual relationships and population composition,
making SGP particularly valuable in domains where different genotypes can give
the same behaviour, such as MO-DFJSS.

Most SGP methods are based on the usage of genetic operators that act on
the genotype to produce offspring, and then accept offspring that satisfy some
semantic criteria into the next population [13]. The semantic criteria can be
semantic diversity [2–4,6] and semantic similarity [11,12]. The consideration of
semantic information enhances the exploration of different dimensions of search
space. SGP has demonstrated its effectiveness across various problem domains,
including symbolic regression [11], classification [1,10], and feature selection [8].

However, the impact of semantic information on NSGPII for MO-DFJSS has
not been investigated. By investigating this, we will be able to open new avenues
for solving the MO-DFJSS problem by extracting meaningful knowledge from
the evolutionary process, which will be explored in this paper.

3   Methods

3.1   Overall Framework

The proposed method uses the NSGPII parent selection, crossover, and mutation to generate offspring for the next generation. On top of that, it designs
novel strategies to decide which kind of offspring is allowed to be added to the
next generation by considering semantic diversity and semantic similarity. In
this section, we begin by providing the definitions of *semantic* and *semantic
distance* in MO-DFJSS, then describe the proposed strategies. The flowchart of
the improved NSGPII with the proposed strategies is shown in Fig. 1.

3.2   Semantic in MO-DFJSS

In the research domain of DFJSS, phenotypic characterisation (PC) [17] is usually used to describe the behaviour of an individual. This paper defines the

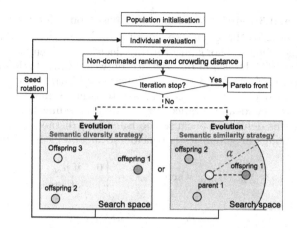

Fig. 1. The flowchart of the proposed NSGPII with the semantic diversity strategy or the semantic similarity strategy for evolution.

Table 1. An example of calculating the PC of an individual.

Sequencing				Routing			
Decision points	Reference rule	Sequencing rule	Decision	Decision points	Reference rule	Routing rule	Decision
$1(O_1)$	1	3	3	$1(M_1)$	2	1	2
$1(O_2)$	3	1		$1(M_2)$	1	2	
$1(O_3)$	2	2		$1(M_3)$	3	3	
$2(O_1)$	3	2	1	$2(M_1)$	2	2	1
$2(O_2)$	1	1		$2(M_2)$	3	3	
$2(O_3)$	2	3		$2(M_3)$	1	1	
$3(O_1)$	1	2	2	$3(M_1)$	1	2	3
$3(O_2)$	2	1		$3(M_2)$	3	1	
$3(O_3)$	3	3		$3(M_3)$	2	3	

semantic in MO-DFJSS as the PC, which is a list of decisions given by an individual on a given number of decision points. These decision points are derived by applying a reference scheduling heuristic to a given DFJSS instance. Specifically, this paper employs the weighted shortest processing time (WSPT) as the reference sequencing rule and working remaining in the queue (WIQ) as the reference routing rule. Considering that each instance often contains thousands of decision points, to save time and ease of use, we randomly select 20 sequencing decision points and 20 routing decision points, each involving a set of 7 candidates (operations for sequencing rule and machines for routing rule). Then, to calculate the PC of an individual, the sequencing/routing rule in an individual is applied to these decision points, and the ranks of the selected operations/machines across these decision points are utilised to construct the PC. An illustrative example of calculating the PC for an individual is shown in Table 1, considering 3 sequencing decision points and 3 routing decision points based on the reference scheduling heuristic. According to the given description, the PC of this example is a combination of sequencing decisions and routing decisions, which is $[3, 1, 2, 2, 1, 3]$.

Based on the PC, the *semantic distance* between individuals ind_a and ind_b is defined as the number of different decisions between their semantics and can be calculated based on Eq. (1). In contrast to other semantic methods that typically rely on Euclidean distance for calculating semantic differences, this paper proposes this definition because the Euclidean distance between machine or operation rankings in semantics is considered meaningless in DFJSS.

$$dis_{a,b} = \sum_{i=1}^{40} d_{a,b,i} \text{ where } d_{a,b,i} = \begin{cases} 0 & \text{if } pc_{a,i} = pc_{b,i} \\ 1 & \text{otherwise} \end{cases} \tag{1}$$

3.3 Semantic Diversity Strategy

This strategy aims to enhance the semantic diversity of the population by only accepting offspring that are semantically different from each other. Two individuals are considered semantically different if their semantic distance is greater than 0. This strategy is used whenever an offspring is generated by crossover or mutation. To be specific, when an offspring ind_o is generated, we compare it with all the other offspring already generated at the current generation. If ind_o is found to be semantically different from all the existing ones, it is accepted as a new offspring. However, if any duplicates are found, indicating that it is not semantically different, the offspring ind_o is discarded. This process is repeated iteratively until the population is filled with semantically different offspring.

3.4 Semantic Similarity Strategy

This strategy builds upon the aforementioned semantic diversity approach and introduces an additional constraint to control the *semantic similarity* between generated offspring and their parents. The degree of semantic similarity between individuals is restricted by a threshold α. Specifically, when an offspring ind_o is generated, we first compare it with all the previously generated offspring in the current generation. If ind_o is found to be semantically different from all existing ones, we further examine whether it is semantically similar to at least one of its parents. This is done by assessing whether the semantic distance between ind_o and its parent ind_p is smaller than the threshold α. If ind_o is similar to any of its parents, then it is accepted as a new offspring; otherwise, it is discarded. This process is iterated until the offspring population is filled. The idea behind this strategy is that by limiting the similarity between offspring with their parents, we expect evolution to be smooth, without losing convergence, and at the same time maintain diversity. To achieve this goal, the key point is to determine an appropriate α value.

4   Experiment Design

4.1   Dataset

This paper utilises the DFJSS simulation model [14] as an experimental tool to do the investigation. For each instance, we assume the presence of 6000 jobs (the first 1000 are warm-up jobs) that need to be processed by 10 heterogeneous machines with varying processing rates. The processing rates are randomly generated within the range of $[10, 15]$. The distances between machines and the entry/exit point are assigned using a uniform discrete distribution between 35 and 500. The transportation speed is set to 5. New jobs arrive over time according to a Poisson process. Each job consists of a random number of operations, generated from a uniform discrete distribution between 2 and 10. Jobs have different importance, represented by weights. Specifically, 20%, 60%, and 20% of jobs have weights of 1, 2, and 4, respectively. The workload for each operation is assigned using a uniform discrete distribution within the range of $[100, 1000]$. The due date for each job is determined by adding 1.5 times its processing time to its arrival time. The utilisation level plays a significant role in simulating different scenarios. A higher utilisation level indicates a busier job shop.

In this paper, six scenarios are examined by considering different combinations of objectives and different utilisation levels (0.85 and 0.95), which are: Scenario 1: $Fmax$ and $WTmax$ with 0.85; Scenario 2: $Fmax$ and $WTmax$ with 0.95; Scenario 3: $WFmax$ and $Tmax$ with 0.85; Scenario 4: $WFmax$ and $Tmax$ with 0.95; Scenario 5: $Fmean$ and $WTmean$ with 0.85; and Scenario 6: $Fmean$ and $WTmean$ with 0.95. For each scenario, 50 instances are used for training, while a separate set of 50 unseen instances is used for test.

4.2   Parameter Setting

The terminal set and function set for GP are displayed in Table 2. The terminal set comprises features associated with machines (e.g., NIQ, WIQ, and MWT), operations (e.g., PT, NPT, and OWT), jobs (e.g., WKR, NOR, rDD, SLACK, W, and TIS), and transport (e.g., TRANT). The function set consists of arithmetic operators that require two arguments. The division operator ("/") is protected and returns 1 if divided by 0. The "max" and "min" functions take two arguments and return the maximum and minimum values, respectively. Regarding the parameter configurations, the population size is set as 1000, and the Pareto front is output after 50 generations. The Ramped-half-and-half method is employed for population initialisation. The crossover and mutation rates are set to 0.85 and 0.15, respectively. Parent selection is performed using tournament selection with a size of 7.

5   Experimental Results

To simplify the algorithm description, we name NSGPII with semantic diversity strategy as $NSGPII^d$, and NSGPII with semantic similarity strategy as $NSGPII^s$.

Table 2. The GP terminal and function set for DFJSS.

Notation	Description
NIQ	the number of operations in the queue
WIQ	the work in the queue
MWT	the waiting time of the machine = t^* - MRT^*
PT	the processing time of the operation
NPT	the median processing time for next operation
OWT	the waiting time of the operation = t - ORT^*
WKR	the work remaining
NOR	the number of operations remaining
rDD	the relative due date = DD^* - t
SLACK	the slack
W	the job weight
TIS	time in system = t - releaseTime*
TRANT	the transportation time
Function	$+, -, \times, /, max, min$

* t: current time; MRT: machine ready time; DD: due date;
ORT: operation ready time; releaseTime: release time.

For NSGPII^s, different α values are tested, including 6, 8, 10, 12, and 14. To measure and compare the performance of algorithms, we conducted 30 independent runs for each algorithm and employed Friedman's test and Wilcoxon rank-sum test for comparison. If Friedman's test yielded significant results, we proceed with the Wilcoxon rank-sum test for pairwise comparisons between the improved NSGPII considering the semantic diversity strategy or the semantic similarity strategy and the classical NSGPII, using a significance level of 0.05.

In the subsequent results, we use the symbols "↑", "↓", and "=" to indicate statistical significance, denoting better, worse, or similar results compared to their counterparts, respectively. We use two widely used measurement indicators, HV [24] and IGD [7], to assess algorithms. A higher HV value (or a smaller IGD value) represents superior performance.

5.1   Test Performance

Tables 3 and 4 present the mean and standard deviation of the HV and IGD results for different algorithms across 30 independent runs on the test instances of the six scenarios. The bottom of the tables shows the results of the Wilcoxon comparison and Friedman's test.

For NSGPII^s, we expect evolution to be smooth, without losing convergence, and at the same time maintain diversity. To achieve this goal, the key point is to determine an appropriate α value. Therefore, we first analyse the effect of α on NSGPII^s. From the tables, we can see that NSGPII^s with $\alpha = 6$ shows significantly better HV and IGD performance across all the 6 scenarios. NSGPII^s

Table 3. The mean (standard deviation) test **HV** of 30 independent runs of NSGPII, NSGPII^d and NSGPII^s with different α for 6 scenarios.

Scenario	NSGPII	NSGPII^d	NSGPII^s				
			$\alpha = 6$	$\alpha = 8$	$\alpha = 10$	$\alpha = 12$	$\alpha = 14$
1	0.82(0.04)	0.86(0.04)	0.85(0.03)	0.86(0.04)	0.86(0.04)	0.86(0.03)	0.86(0.04)
2	0.79(0.04)	0.82(0.04)	0.84(0.03)	0.84(0.03)	0.83(0.03)	0.84(0.02)	0.82(0.03)
3	0.87(0.03)	0.88(0.03)	0.89(0.04)	0.89(0.03)	0.89(0.03)	0.89(0.03)	0.89(0.02)
4	0.95(0.01)	0.95(0.01)	0.96(0.01)	0.96(0.02)	0.96(0.01)	0.96(0.01)	0.96(0.01)
5	0.61(0.20)	0.59(0.18)	0.73(0.09)	0.66(0.12)	0.70(0.14)	0.60(0.18)	0.69(0.14)
6	0.98(0.01)	0.98(0.01)	0.98(0.01)	0.98(0.01)	0.98(0.01)	0.98(0.01)	0.98(0.01)
↑/=/↓	-	2/4/0	6/0/0	4/2/0	5/1/0	4/2/0	4/2/0
rank	6.67	5.0	2.33	3.0	3.17	4.0	3.83

Table 4. The mean (standard deviation) test **IGD** of 30 independent runs of NSGPII, NSGPII^d and NSGPII^s with different α for 6 scenarios.

Scenario	NSGPII	NSGPII^d	NSGPII^s				
			$\alpha = 6$	$\alpha = 8$	$\alpha = 10$	$\alpha = 12$	$\alpha = 14$
1	0.12(0.03)	0.10(0.03)	0.10(0.02)	0.10(0.03)	0.10(0.02)	0.09(0.02)	0.10(0.03)
2	0.12(0.03)	0.10(0.02)	0.10(0.02)	0.10(0.02)	0.10(0.02)	0.10(0.01)	0.10(0.02)
3	0.07(0.03)	0.07(0.02)	0.06(0.02)	0.06(0.02)	0.06(0.02)	0.06(0.02)	0.06(0.01)
4	0.03(0.01)	0.03(0.01)	0.02(0.01)	0.02(0.01)	0.02(0.00)	0.03(0.01)	0.02(0.01)
5	0.28(0.19)	0.29(0.17)	0.16(0.07)	0.21(0.11)	0.19(0.11)	0.28(0.18)	0.20(0.13)
6	0.01(0.01)	0.01(0.01)	0.01(0.00)	0.01(0.01)	0.01(0.01)	0.01(0.01)	0.01(0.01)
↑/=/↓	-	2/4/0	6/0/0	3/3/0	5/1/0	3/3/0	5/1/0
rank	6.5	5.17	2.67	2.67	3.17	4.17	3.67

with $\alpha = 8$, $\alpha = 12$, and $\alpha = 14$ show significantly better HV performance than NSGPII on 4 scenarios and show similar HV performance as NSGPII on the other 2 scenarios. NSGPII^s with $\alpha = 10$ shows significantly better HV performance on 5 scenarios and shows similar HV performance as NSGPII on the remaining 1 scenario. Also, NSGPII^s with $\alpha = 8$ and $\alpha = 12$ both show significantly better HV performance than NSGPII on 3 scenarios and show similar HV performance as NSGPII on the other 3 scenarios. NSGPII^s with $\alpha = 10$ and $\alpha = 14$ both show significantly better HV performance on 5 scenarios and show similar HV performance as NSGPII on the remaining 1 scenario. Based on the Friedman's test results, in terms of HV performance, NSGPII^s with $\alpha = 6$ achieves the highest rank, followed by NSGPII^s with $\alpha = 8$, NSGPII^s with $\alpha = 10$, NSGPII^s with $\alpha = 14$, NSGPII^s with $\alpha = 12$ in order. In terms of test IGD performance, both NSGPII^s with $\alpha = 6$ and $\alpha = 8$ secure the first position, NSGPII^s with $\alpha = 10$ ranks second, followed by NSGPII^s with $\alpha = 14$, NSGPII^s with $\alpha = 12$ in order. Since NSGPII^s with $\alpha = 6$ performs the best among all the NSGPII^s

with different α values, it is used for further analysis. For simplicity, we refer to NSGPII^s with $\alpha = 6$ as NSGPII^s.

Then we compare NSGPII, NSGPII^d, and NSGPII^s. We can see that, NSGPII^d shows significantly better HV and IGD performance than NSGPII on 2 scenarios and obtains similar HV and IGD performance as NSGPII on the remaining 4 scenarios. NSGPII^s gives even better performance. NSGPII^s shows significantly better HV and IGD performance than NSGPII across all the 6 scenarios. Based on the Friedman's test results, NSGPII^s ranks the first among these three methods, followed by NSGPII^d and NSGPII in order.

Through the above analysis, we can see that semantic information plays an important role in improving the performance of NSGPII on the MO-DFJSS problem. The HV and IGD performance of NSGPII can be improved by increasing the diversity of the behaviours of the individuals in the population. Moreover, requiring the offspring to have similar semantic behaviour with their parents can further improve the HV and IGD performance of NSGPII in solving the MO-DFJSS problem. This finding highlights the positive impact of considering semantic diversity and semantic similarity in NSGPII on addressing the MO-DFJSS problem.

5.2   Population Distribution

The proposed semantic diversity and semantic similarity strategies aim to limit the semantic distance between individuals in the population. It is interesting to study the semantic distribution of individuals in the population. The semantic represents the behaviour of the individual, which is a 40-dimensional vector. To visualise the semantics of individuals in the population, we employ t-SNE to reduce the dimensions to a 2-dimensional space.

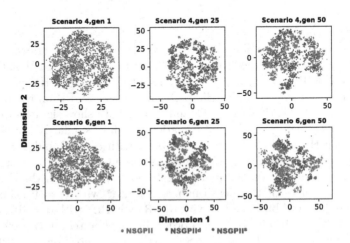

Fig. 2. Visualisations of the dimensionality reduced semantic of individuals of NSGPII, NSGPII^d, and NSGPII^s of one run in the scenario 4 and scenario 6 during the start (generation 1), middle (generation 25), and late (generation 50) stages of evolution.

Specifically, Fig. 2 visualises the dimensionality reduced semantic of individuals in the population of NSGPII, $NSGPII^d$, and $NSGPII^s$ across different generations (1, 25, and 50) in scenarios 4 and 6. From Fig. 2, we can clearly see that NSGPII has several regions of more concentrated semantic distribution in each subfigure. This aligns with our expectations, as NSGPII does not impose limitations on semantic distance between individuals. Compared to NSGPII, the semantic distributions obtained by $NSGPII^s$, on the other hand, are relatively widespread and do not have as clearly concentrated areas as NSGPII. Compared to NSGPII and $NSGPII^s$, $NSGPII^d$ gives the most diverse semantic distributions. This finding highlights the significance of restricting the semantic distance between individuals, as it allows to achieve a population with a relatively more uniform semantic distribution, avoiding losing diversity, and potentially leading to better final scheduling heuristics. These insights emphasise the importance of controlling semantic distances between individuals during the evolutionary process of NSGPII. Furthermore, it reveals that the improved final performance achieved by the inclusion of semantic information in NSGPII is attributed to its ability to evolve a more semantically diverse population.

6   Conclusions

This study has successfully demonstrated the advantages of integrating semantic information into NSGPII for addressing the MO-DFJSS problem. Firstly, this study contributes to giving the definitions of the semantic and semantic distance of scheduling heuristics for DFJSS. Then, by incorporating semantic diversity and semantic similarity within NSGPII, this study contributes to evolving better scheduling heuristics than using the original NSGPII. The results highlight the benefits of considering semantically diverse individuals for achieving high-quality scheduling heuristics. Moreover, NSGPII, considering semantic similarity, achieves the best overall performance, offering valuable insight into the importance of maintaining a reasonable semantic distance between offspring and their parents to further enhance the quality of scheduling heuristics. This emphasises the trade-off between semantic diversity and semantic similarity. Furthermore, the analysis of the population semantic distribution reveals that by controlling semantic distances between individuals, we are able to achieve a more semantically diverse population. This is the key factor contributing to the enhanced performance achieved by the proposed methods.

Overall, this paper demonstrates the potential of incorporating semantic information into the evolution process of NSGPII for MO-DFJSS, providing valuable insights into the benefits and considerations of utilising semantic information in solving complex scheduling problems. Further deeper studies are needed to explore and optimise the integration of semantic information in GP to achieve even better results for solving the MO-DFJSS problem. Some other research techniques (e.g., feature selection and surrogate) can also be combined with GP for solving the MO-DFJSS problem. In addition, although the study here is conducted on the MO-DFJSS problem, we believe that the techniques and results presented here are transferable to other complex problems. We expect the semantic information used in this work to be easily extendable to other different problems.

References

1. Bakurov, I., Castelli, M., Fontanella, F., di Freca, A.S., Vanneschi, L.: A novel binary classification approach based on geometric semantic genetic programming. Swarm Evol. Comput. **69**, 101028 (2022). https://doi.org/10.1016/j.swevo.2021.101028
2. Beadle, L., Johnson, C.G.: Semantically driven crossover in genetic programming. In: Proceedings of the IEEE Congress on Evolutionary Computation, pp. 111–116 (2008)
3. Beadle, L., Johnson, C.G.: Semantic analysis of program initialisation in genetic programming. Genet. Program Evolvable Mach. **10**, 307–337 (2009)
4. Beadle, L., Johnson, C.G.: Semantically driven mutation in genetic programming. In: Proceedings of the IEEE Congress on Evolutionary Computation, pp. 1336–1342 (2009)
5. Deb, K., Pratap, A., Agarwal, S., Meyarivan, T.: A fast and elitist multiobjective genetic algorithm: NSGA-II. IEEE Trans. Evol. Comput. **6**(2), 182–197 (2002)
6. Galván, E., Schoenauer, M.: Promoting semantic diversity in multi-objective genetic programming. In: Proceedings of the Genetic and Evolutionary Computation Conference, pp. 1021–1029 (2019)
7. Liu, H.L., Gu, F., Zhang, Q.: Decomposition of a multiobjective optimization problem into a number of simple multiobjective subproblems. IEEE Trans. Evol. Comput. **18**(3), 450–455 (2013)
8. Papa, J.P., Rosa, G.H., Papa, L.P.: A binary-constrained geometric semantic genetic programming for feature selection purposes. Pattern Recogn. Lett. **100**, 59–66 (2017)
9. Riquelme, N., Von Lücken, C., Baran, B.: Performance metrics in multi-objective optimization. In: Proceedings of the Latin American Computing Conference, pp. 1–11 (2015)
10. Sánchez, C.N., Graff, M.: Selection heuristics on semantic genetic programming for classification problems. Evol. Comput. **30**(2), 253–289 (2022)
11. Uy, N.Q., Hoai, N.X., O'Neill, M., McKay, R.I., Galván-López, E.: Semantically-based crossover in genetic programming: application to real-valued symbolic regression. Genet. Program Evolvable Mach. **12**, 91–119 (2011)
12. Uy, N.Q., McKay, B., O'Neill, M., Hoai, N.X.: Self-adapting semantic sensitivities for semantic similarity based crossover. In: Proceedings of the IEEE Congress on Evolutionary Computation, pp. 1–7 (2010)
13. Vanneschi, L., Castelli, M., Silva, S.: A survey of semantic methods in genetic programming. Genet. Program Evolvable Mach. **15**(2), 195–214 (2014). https://doi.org/10.1007/s10710-013-9210-0
14. Xu, M., Mei, Y., Zhang, F., Zhang, M.: Genetic programming with cluster selection for dynamic flexible job shop scheduling. In: Proceedings of the IEEE Congress on Evolutionary Computation, pp. 1–8 (2022)
15. Xu, M., Mei, Y., Zhang, F., Zhang, M.: Genetic programming with lexicase selection for large-scale dynamic flexible job shop scheduling. IEEE Trans. Evol. Comput. (2023). https://doi.org/10.1109/TEVC.2023.3244607

16. Xu, M., Mei, Y., Zhang, F., Zhang, M.: Multi-objective genetic programming based on decomposition on evolving scheduling heuristics for dynamic scheduling. In: Proceedings of the Companion Conference on Genetic and Evolutionary Computation, pp. 427–430 (2023)

17. Zhang, F., Mei, Y., Nguyen, S., Zhang, M.: Phenotype based surrogate-assisted multi-objective genetic programming with brood recombination for dynamic flexible job shop scheduling. In: Proceedings of the IEEE Symposium Series on Computational Intelligence, pp. 1218–1225 (2022)

18. Zhang, F., Mei, Y., Nguyen, S., Zhang, M.: Multitask multiobjective genetic programming for automated scheduling heuristic learning in dynamic flexible job shop scheduling. IEEE Trans. Cybern. 53(7), 4473–4486 (2023)

19. Zhang, F., Mei, Y., Zhang, M.: Evolving dispatching rules for multi-objective dynamic flexible job shop scheduling via genetic programming hyper-heuristics. In: Proceedings of the IEEE Congress on Evolutionary Computation, pp. 1366–1373 (2019)

20. Zhang, F., Mei, Y., Zhang, M.: An investigation of terminal settings on multitask multi-objective dynamic flexible job shop scheduling with genetic programming. In: Proceedings of the Companion Conference on Genetic and Evolutionary Computation, pp. 259–262 (2023)

21. Zhang, F., Shi, G., Mei, Y.: Interpretability-aware multi-objective genetic programming for scheduling heuristics learning in dynamic flexible job shop scheduling. In: Proceedings of the IEEE Congress on Evolutionary Computation (2023)

22. Zhang, Q., Li, H.: MOEA/D: a multiobjective evolutionary algorithm based on decomposition. IEEE Trans. Evol. Comput. 11(6), 712–731 (2007)

23. Zitzler, E., Laumanns, M., Thiele, L.: SPEA 2: improving the strength pareto evolutionary algorithm. TIK Rep. 103 (2001). https://doi.org/10.3929/ethz-a-004284029

24. Zitzler, E., Thiele, L., Laumanns, M., Fonseca, C.M., Da Fonseca, V.G.: Performance assessment of multiobjective optimizers: an analysis and review. IEEE Trans. Evol. Comput. 7(2), 117–132 (2003)

XC-NAS: A New Cellular Encoding Approach for Neural Architecture Search of Multi-path Convolutional Neural Networks

Trevor Londt[(✉)] [ID], Xiaoying Gao [ID], Peter Andreae [ID], and Yi Mei [ID]

Centre for Data Science and Artificial Intelligence and School of Engineering and Computer Science, Victoria University of Wellington, PO Box 600, Wellington 6140, New Zealand

{trevor.londt,xiaoying.gao,peter.andreae,yi.mei}@ecs.vuw.ac.nz

Abstract. Convolutional Neural Networks (CNNs) continue to achieve great success in classification tasks as innovative techniques and complex multi-path architecture topologies are introduced. Neural Architecture Search (NAS) aims to automate the design of these complex architectures, reducing the need for costly manual design work by human experts. Cellular Encoding (CE) is an evolutionary computation technique which excels in constructing novel multi-path topologies of varying complexity and has recently been applied with NAS to evolve CNN architectures for various classification tasks. However, existing CE approaches have severe limitations. They are restricted to only one domain, only partially implement the theme of CE, or only focus on the micro-architecture search space. This paper introduces a new CE representation and algorithm capable of evolving novel multi-path CNN architectures of varying depth, width, and complexity for image and text classification tasks. The algorithm explicitly focuses on the macro-architecture search space. Furthermore, by using a surrogate model approach, we show that the algorithm can evolve a performant CNN architecture in less than one GPU day, thereby allowing a sufficient number of experiment runs to be conducted to achieve scientific robustness. Experiment results show that the approach is highly competitive, defeating several state-of-the-art methods, and is generalisable to both the image and text domains.

Keywords: Neural architecture search · Convolutional neural networks · Cellular Encoding

1 Introduction

Designing state-of-the-art CNNs for a classification task is not trivial; it requires expert skills and costly development time. Furthermore, training times have progressively increased as CNN architectures have become more complex.

T. Liu et al. (Eds.): AI 2023, LNAI 14472, pp. 416–428, 2024.
https://doi.org/10.1007/978-981-99-8391-9_33

Researchers have focused on developing Neural Architecture Search (NAS) algorithms that automatically design CNNs without human intervention to mitigate these problems. One such variant approach, Evolutionary Computation-based Neural Architecture Search (ECNAS) [20], has shown success in automatically designing CNN architectures for various classification tasks. However, many ECNAS algorithms are typically restricted to only one problem domain and task, such as image or text. Due to the high computational cost, time, and intractable size of the search space involved [22], ECNAS algorithms typically evolve the micro-architecture and treat the macro-architecture exclusively as layers of micro-architectures stacked together in series to form a complete CNN architecture. The theory is that training a micro-architecture representing a simple computational block takes less time than training an entire CNN macro-architecture at once. The assumption is that this evolved micro-architecture can be stacked to produce a single-path CNN macro-architecture. Such works have succeeded in evolving highly performant CNNs, however, this approach prevents the possibility of exploring multi-path CNN macro-architecture topologies. Multi-path CNN macro-architectures can allow multiple feature reuse between various layers of the network or provide different processing paths through the network representing network ensembles, which may boost classification performance. Inspired by the capabilities of Cellular Encoding (CE) [5] to evolve multi-pathed topologies, this work proposes a new, configurable, and extensible CE representation (XC) with the ability to represent novel multi-pathed CNN architectures. Furthermore, the proposed encoding incorporates the ability to dynamically modify the depth and width of a CNN architecture during the evolutionary process to allow the creation of architectures of varying widths and depths. Since CE is involved with representing a sequence of operations to be executed that are encapsulated in a tree structure, Genetic Programming (GP) [13] is used to facilitate the evolution of CNN architectures using the proposed CE representation because GP itself is designed for running genotypes as programs in a tree data structure. The proposed CE representation forms the backbone of evolving multi-pathed macro-architectures for the proposed algorithm in this work. Experiments on commonly used image and text classification benchmark datasets show competitive results compared to the state-of-the-art models across all datasets. The contributions of this work are:

1. A new configurable CE representation (XC), including supporting network structures, that is designed to represent multi-pathed CNN architectures for image or text classification tasks, with the ability to dynamically adjust the width and depth of the architecture during its construction.
2. Three new ECNAS algorithms using the XC representation to evolve multi-path CNN architectures, each making use of a different handcrafted micro-architecture, namely a simple convolutional block, a ResNet block, or an Inception module. The proposed algorithms can evolve multi-path CNN architectures for image or text classification tasks in less than one GPU day.

2   Background

Cellular Encoding [5], inspired by the division of biological cells to produce complex organisms, is a generative encoding strategy to synthesise artificial neural network topologies. CE genotypes, represented as tree structures, are encoded as a sequence of instructions applied on an initial *ancestor network* containing a singular *cell*. Each execution of an instruction transforms the ancestor network into a different topology by adding, modifying, or removing cells in the ancestor network. This approach allows the neural network to grow organically until a suitably sized network topology with enough capacity for the problem under consideration has been generated. The instruction set of cellular encoding is extensive, however, the essential instructions are related to the duplication of a cell in the network. Sequential (SEQ), parallel (PAR), and recursion (REC) are the main duplication instructions. Sequential duplication indicates that a cell being operated on will divide into two cells, each connected in series. Parallel duplication is the division of a cell into two cells that are connected in parallel. REC operations are only executed once and repeat a sequence of instructions in the genotype. Terminal operations are the END operation, a leaf node in the genotype. The original CE encoding was used in a Genetic Algorithm [7] (GA) to facilitate evolutionary operations. However, modern variants of CE [1,12] are used in Genetic Programming [10] (GP) algorithms because GP is a population-based EC method that operates directly on tree-based genotypes representing executable programs, making it a perfect and convenient fit for CE-based genotypes.

EC-based NAS (ECNAS) [20] is an evolutionary-based approach for searching performant network architectures. ECNAS uses an evolutionary algorithm with appropriate representation to explore the space of possible network architectures. The representation aims to constrain the search space size. The training of each potential network architecture under consideration typically uses backpropagation. The fitness of a trained network influences its survivability for the next generation in the evolutionary process. ECNAS algorithms have succeeded in evolving network architectures for many domains, including complex tasks such as image and text classification. Various EC-based search methods have been used in proposed ECNAS algorithms, including GA [7], Particle Swarm Optimisation (PSO) [9], and CE [5].

The approaches taken by GA- [19] and PSO-based [4] algorithms typically search for CNNs containing single-path macro-architectures, missing the opportunity to explore novel multi-path architectures - which may offer added benefits of feature reuse and ensemble qualities to boost performance further. CE-based approaches can potentially explore unknown multi-path architecture search spaces. Londt et al. [12] have successfully used a CE-based approach to evolve multi-path CNN architecture for text classification tasks achieving competitive state-of-the-art results. However, their algorithm has not been designed or implemented for image classification tasks and does not consider modifying the width or depth of the network during evolution. Broni-Bediako et al.'s [1] work focused on a CE-based approach to perform evolution in the local space, evolving micro-architectures that are stacked to create a single path macro-architecture, achieving impressive performance results against peer-competitors, again showing the effectiveness and utility of CE. However, their approach does not consider the global search space to evolve multi-path macro-architectures for CNNs, leaving a gap in the literature that needs to be investigated. To address these limitations, a new CE encoding (XC) is proposed and implemented in a new ECNAS algorithm, XC-NAS, capable of evolving multi-path CNN architectures for image or text classification tasks of varying widths and depths.

3   XC-NAS: Cellular Encoding for NAS

3.1   Framework

The general process of the proposed algorithm is presented in Fig. 1, and a detailed listing is provided in Algorithm 1. A subset of the training set is generated based on a predetermined percentage. A surrogate model approach is taken where only a subset of the training dataset is used during the evolutionary process as is done in [1,12], thereby evolving low-resolution models, after which the best-evolved model is retrained using the entire training set to produce a higher resolution model for conducting inference. This approach is taken to decrease the evolutionary training time significantly. Each individual in the population is generated based on a randomly selected depth between a predefined minimum and maximum value.

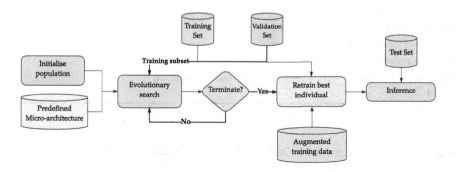

Fig. 1. Framework of XC-NAS.

Each individual in the population is decoded to a phenotype (CNN) and uploaded to the graphics processing unit (GPU). If the phenotype is too large to fit into the memory of the GPU, then the phenotype is considered unfit for the environment and assigned a fitness value of -1.0. This approach ensures that the phenotype will not survive into the next generation. The phenotype is trained using the generated training set and validation set, with the resulting fitness being returned when the training process terminates. The evolutionary operations of elitism, selection, crossover, and mutation are then conducted. A new population is generated, and the evolutionary process repeats until the required generations are reached. Finally, the fittest genotype in the population is decoded into a phenotype and trained using the entire training set, validation set, and data augmentation.

Algorithm 1: Pseudocode of XC-NAS.

---

1 **begin**;
2      $seed \leftarrow random()$;
3      $\mathcal{D}_{train} \leftarrow subset(seed, \mathcal{D}_{full\_train})$;
4      $\mathcal{P} \leftarrow init(min, max)$;
5 **while** *not maximum generations* **do**
6   **foreach** $\kappa \in \mathcal{P}$ **do**
7          $\rho \leftarrow decode(\kappa, GPU)$;
8   **if** $\rho \notin GPU$ **then**
9              $\mid \quad val_{acc} \leftarrow -1.0$;
10   **else**
11             $\mid \quad val_{acc} \leftarrow evaluate(\rho, \mathcal{D}_{train}, \mathcal{D}_{val})$;
12   **end if**
13   **end foreach**
14     $\varepsilon \leftarrow elite \subset \mathcal{P}$;
15     $\varsigma \leftarrow tournament(\mathcal{P})$;
16     $\phi \leftarrow crossover(\varsigma)$;
17     $\lambda \leftarrow mutate(\phi)$;
18     $\mathcal{P} \leftarrow limit(\lambda \cup \varepsilon)$;
19 **end while**
20 $\beta \leftarrow fittest \in \mathcal{P}$;
21 $\rho \leftarrow decode(\beta, GPU)$;
22 $test_{acc} \leftarrow evaluate(\rho, \mathcal{D}_{train}, \mathcal{D}_{val}, \mathcal{D}_{test}, \mathcal{D}_{augment})$;
23 **end**;

3.2 Ancestor Network, Cell, and Micro-architecture

The network cell is the fundamental unit in a phenotype produced by a CE-based algorithm. We adopt a similar approach proposed by Londt et al. [12], where a cell encapsulates a predefined micro-architecture. Figure 2 presents the proposed cell configuration. The cell configuration comprises three components: an input section, a *block* section containing a micro-architecture, and an output section. The input section includes operations that transform the incoming connections and data from preceding cells as required by CE operations, for example, increasing or decreasing filter counts. Similarly, the output section of the cell also allows specified operations to be executed during run-time to transform the cell's outgoing connections and output data as required. The middle or block section contains the micro-architecture of the cell.

In this paper, three different micro-architectures are used and evaluated, namely Simple, ResNet-like, and InceptionNet-like micro-architectures. The Simple micro-architecture serves as a baseline micro-architecture, containing a batch normalisation layer (BN) followed by a ReLU activation layer and a convolutional layer.

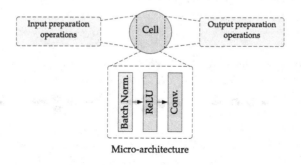

Fig. 2. Proposed cell structure.

ResNet [6] and InceptionNet [17] are highly performant CNN architectures, each making a significant contribution to the literature, and are therefore leveraged in this work. The ResNet-like micro-architecture is the same as the pre-activation block in [6]. The Inception-like micro-architecture replicates the Inception module proposed by Szegedy et al. [17]. The convolution type will be temporal for text classification tasks or spatial for image classification tasks.

All genotypes in the population have a corresponding phenotype representing a CNN architecture. The initial phenotypes of each genotype are based on an *ancestor network*. We define the ancestor network in Eq. 1.

$$\Phi(\varsigma, \varepsilon, \psi) : \psi(\varepsilon(\varsigma(\beta))) \to \mathbb{Z}_1^n \tag{1}$$

ς is the stem, ε is the feature extractor, ψ is the classifier, and β is a batch of input data which is mapped to a class $i \in \mathbb{Z}_1^n$.

The feature extractor, ε, is defined in Eq. 2

$$\tau(C, E) \to \varepsilon \tag{2}$$

C is the set of cells in the network, E is a set of edges between the cells, and τ is a mapping of the edges and cells to produce a topology representing the feature extractor. The initial feature extractor contains only one cell. The evolutionary process primarily involves the feature extractor part of the ancestor network, creating more cells and interconnections between them to form the macro-architecture. The architecture of the stem in the ancestor network depends on the domain in which the ancestor network will operate.

3.3   Encoding Scheme

The proposed algorithm uses a depth-first traversal approach. Figure 3 presents an example of the program execution of a genotype, represented by a program tree structure, operating directly on an ancestor network to construct a final CNN architecture. Two key CE operations are demonstrated namely the sequential (**S**) and parallel (**P**) split operations.

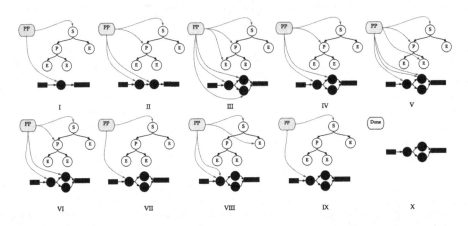

Fig. 3. Growing a neural network from an ancestor network. **PP**=Program Pointers, **S**=Sequential Split, **P**=Parallel Split, **E**=End Terminal.

In step **I**, the program pointer (PP), displayed in blue, points to the first program symbol, an **S** symbol, and the cell that will be operated on in the ancestor network. Step **II** represents the program's state after executing the **S** symbol. The cell in the ancestor network is divided into two cells connected in series. The PP retains the original pointers, in blue, to the **S** symbol and its corresponding cell it operated on. These pointers are maintained to facilitate the back tracing of the tree structure. A new green pointer points to the next program symbol, **P**, including the newly added cell on which the P symbol will operate. Step **III** displays the results of the execution of the **P** symbol. The cell has divided, and the resulting two cells are connected in parallel. The remainder of the program, from steps IV to IX, represents the results of executing the E symbols, which do not affect the ancestor neural network. In step **X**, the final architecture of the CNN is presented.

Two critically important hyperparameters of a CNN are its depth and width. The depth refers to the longest path of consecutive convolutional layers in the network from the input to the output. The width refers to the number of filters at a specific layer in the network. A new Cellular Encoding representation (XC) is proposed to encapsulate these two critical hyperparameters to allow the evolution of CNNs of varying depth and width. A cell that is being operated on is referred to as a mother cell. Any cell produced from a mother cell is called a child cell. An explanation of the proposed cellular encoding operations is provided below:

1. **S**: the child cell inherits the mother cell's number of filters and depth. The child cell is inserted into the network after the mother cell.
2. **SID**: the child cell inherits the mother cell's number of filters but increases the depth of the child cell by one block compared to the mother cell's depth.
3. **SDD**: the child cell inherits the mother cell's number of filters but decreases the depth of the child cell by one block compared to the mother cell's depth. If the mother cell's depth is one block, then there will be no decrease, and the child cell will have a depth of one block.
4. **SIW**: the child cell inherits double the number of filters of the mother cell and the same depth value as the mother cell.
5. **SDW**: the child cell inherits half the number of filters of the mother cell and the same depth value. A minimum filter count is predefined; if it is breached, the child cell will inherit the minimum filter count value.
6. **S*D*W**: the sequential split operation will inherit the mother cell's attributes and either increase or decrease the width and depth accordingly, as discussed in the previous operations. For example, SIDDW would perform a sequential split, increase the child cell's depth and also decrease the width of the child cell.
7. **P***: the equivalent P operations behave the same as the S operations except that the mother and child cells are connected in parallel, and the outputs from both cells are concatenated channel-wise. For example, PID would connect the mother and child cell in parallel and increase the child cell's depth.

Table 1 lists the new encoding operations and associated configurable hyperparameters. Each operation has two hyperparameters that can be adjusted to alter the magnitude of the effect an operation has on a cell. The first hyperparameter is the *depth adder*. The value "1" adds an extra block inside the cell. The value "−1" will remove a block from the cell unless only one block is left, in which case nothing will be done. The second hyperparameter is the *width multiplier*. A value of "2" indicates a doubling of the number of inherited filters, and a value of "0.5" will halve the number of inherited filters.

Table 1. ID = Increase Depth, **DD** = Decrease Depth, **IW** = Increase Width, **DW** = Decrease Width.

Operation	Depth +	Operation	Width ×
ID	1	IW	2
DD	−1	DW	0.5

3.4   Architecture Evolution

A uniform single-point mutation operation is chosen for the proposed algorithms. The operation involves randomly selecting a node in the genotype and replacing the sub-tree at that node with a randomly generated subtree of a specified size. The single-point crossover operation is chosen for its simplicity of operation and implementation. This operation works by randomly selecting a point in each parent genotype and crossing over the sub-tress located at these points in each parent to form two offspring genotypes. Tournament Selection is an appropriate initial choice for the proposed algorithms and provides a reasonable balance between exploration and exploitation, which the tournament size can control. After training a candidate phenotype using stochastic gradient descent (SGD), the phenotype's best-recorded validation accuracy is used to determine its corresponding genotype's fitness. Early stopping is implemented to perform regularisation and prevent wasting compute time. The validation accuracy at the end of each training epoch is monitored, and if the accuracy has not improved beyond a predefined delta for a specified number of epochs, training is stopped.

4   Experiment Design

4.1   Datasets and Peer Competitors

To determine the proposed approach's effectiveness and generalisability, we consider classification tasks from two disparate domains, namely the image and text domains. Five commonly used datasets are selected. Two datasets are from the text domain, and three are from the image domain. The AG's News [21] dataset is considered a medium-difficulty dataset, whereas the Yelp Reviews Full [3] dataset is considered high-difficulty. KMNIST [2] is an image dataset containing handwritten characters and bridges the text and image domains in this work. The Fashion-MNIST [18] dataset is chosen as a more challenging dataset than KMNIST. Finally, CIFAR-10 [11] is a popular benchmark dataset which is considered to be of medium difficulty to model. Several state-of-the-art peer competitors are chosen for benchmarking, including both manually designed architectures and NAS algorithms. For the text domains, Zhang et al.'s [21] original manually designed character-level model, Conneau et al.'s [3] manually designed VDCNN model, and the GP-Dense [12] NAS algorithm are compared against. For the image domain, ResNet-110 [6] and InceptionNet [17] architectures are chosen as the expert-designed manual models and numerous NAS algorithms, including the CE-based CE-GeneExpr [1], FPSO [8], EvoCNN [16], CGP-CNN [15], and EIGEN [14] are chosen for benchmarking.

4.2   Parameter Settings

The parameter settings used for the experiment are based on those used in [1, 12]. The number of generations and the population size are set to 20. Genotypes are randomly initialised to depths between 2 and 17, inclusively (Table 2).

Table 2. Parameter settings for CNN training process.

Parameter	Setting	Parameter	Setting
Number of generations	20	Population size	20
Genotype depth	[2,17]	Elitism	0.1
Crossover probability	0.5	Mutation probability	0.01
Mutation tree growth size	[1,3]	Tournament size	3
Epochs	Evol = 15, Full = 300	Batch size	128
Initial learning rate	0.01	Momentum	0.9
Learning schedule	Evol. = [12], Full= [6]	Weight Init	Kaiming. [6]
Training data usage	0.25	Experiment runs	30

The evolutionary and network training settings are those commonly used in the community. The training set used during the evolutionary process is set at 25% of the entire training set. Three different micro-architectures are tested with XC-NAS, and thirty independent experiment runs are conducted for each configuration.

5   Experimental Results and Discussions

The best test results recorded for each of the 30 independent runs, and the peer competitors, are listed in Table 3. On the text datasets, it can be seen that XC-NAS (Inception) has outperformed the current state-of-the-art CE-based competitor, GP-Dense. XC-NAS (Inception) has only slightly underperformed VDCNN-Convolution on the AG's News dataset. On the Yelp Reviews Full dataset, XC-NAS (ResNet) appears to have performed the best of the XC-NAS variants. Regardless, XC-NAS (ResNet) has outperformed the CE-based GP-Dense competitor but underperformed the manually designed architectures, demonstrating that the Yelp Reviews Full dataset is challenging to model automatically.

Table 3. Test accuracies (%) compared to peer competitors. (M = Manual, A = NAS, n/a = Not Applicable, '-' = Not Available, best = Blue, second best = Orange)

Algorithm	Type	AG's News	Yelp Reviews Full	KMNIST	Fashion-MNIST	CIFAR-10
XC-NAS (Simple)	A	90.71	60.92	98.42	94.31	93.24
XC-NAS (ResNet)	A	89.94	61.26	98.68	94.72	93.74
XC-NAS (Inception)	A	91.15	61.01	99.13	95.39	94.85
Zhang et al. Lg. Full Conv. [21]	M	90.15	61.60	n/a	n/a	n/a
VDCNN-Convolution [3]	M	91.27	64.72	n/a	n/a	n/a
GP-Dense [12]	A	89.58	61.05	n/a	n/a	n/a
ResNet-110. [6]	M	n/a	n/a	97.82	93.40	93.57
InceptionNet (GoogLeNet) [17]	M	n/a	n/a	97.95	92.74	93.64
EvoCNN [16]	A	n/a	n/a	-	94.53	-
FPSO [8]	A	n/a	n/a	-	95.07	93.72
CGP-CNN-ConvSet [15]	A	n/a	n/a	-	-	93.25
EIGEN [14]	A	n/a	n/a	-	-	94.60
CE-GeneExpr [1]	A	n/a	n/a	-	-	96.26

On the KMNIST and Fashion-MNIST datasets, XC-NAS (Inception) has outperformed all peer competitors. On CIFAR-10, XC-NAS (Inception) has surpassed all peer competitors except for the CE-GeneExpr algorithm. The results show the XC-NAS (Inception) is highly competitive and generalises well across both the image and text domains.

The number of GPU days required to evolve the best model is presented in Table 4. Only the peer competitors with available GPU day information available have been shown. XC-NAS has taken less than one GPU day to find the best architecture for any dataset. It is clear that the approach of XC-NAS is efficient in terms of computational costs.

Table 4. Number of GPU days required to run the algorithm and evolve the best model compared to peer competitors.

Algorithm	GPU Days
XC-NAS (Each dataset)	< 1
EvoCNN [16] (Fashion-MNIST)	4
CGP-CNN-ConvSet [15] (CIFAR-10)	29.8
EIGEN [14] (CIFAR-10)	2

6   Conclusions

This paper introduced a new Cellular Encoding representation (XC) and a new ECNAS algorithm (XC-NAS), capable of evolving multi-path CNN architectures for image and text classification tasks. The XC representation is configurable and includes operations that can automatically modify a CNN architecture's width and depth during construction. This ability results in performant multi-path CNN architectures of varying complexity and capacity. XC-NAS implements a surrogate approach where only a subset of the training dataset is used during the evolutionary process, thereby evolving low-resolution models to save on compute costs and training time, after which the best-evolved model is retrained using the entire training set to produce a higher-resolution model for conducting inference. This approach has significantly reduced the required evolutionary run time. Experiments across five datasets show that the XC encoding is capable of representing a multitude of performant multi-path CNN architectures and that XC-NAS generalises well across both the image and text domains, demonstrating promising performance by outperforming several state-of-the-art approaches both in terms of classification performance and GPU days required to evolve the best architectures. Currently, XC-NAS can use a predefined micro-architecture of any configuration; however, future work will focus on integrating the ability to automatically evolve micro-architectures in conjunction with evolving the macro-architecture of a CNN architecture.

References

1. Broni-Bediako, C., Murata, Y., Mormille, L.H.B., Atsumi, M.: Evolutionary NAS with gene expression programming of cellular encoding. In: 2020 IEEE Symposium Series on Computational Intelligence (SSCI), pp. 2670–2676 (2020)
2. Clanuwat, T., Bober-Irizar, M., Kitamoto, A., Lamb, A., Yamamoto, K., Ha, D.: Deep learning for classical Japanese literature (2018)
3. Conneau, A., Schwenk, H., Cun, Y.L., Barrault, L.: Very deep convolutional networks for text classification. In: 15th Conference of the European Chapter of the Association for Computational Linguistics, EACL 2017 - Proceedings of Conference, vol. 2, pp. 1107–1116. Association for Computational Linguistics (ACL) (2017)
4. Fernandes, J.F.E., Yen, G.G.: Particle swarm optimization of deep neural networks architectures for image classification. Swarm Evol. Comput. **49**, 62–74 (2019)
5. Gruau, F., et al.: Neural network synthesis using cellular encoding and the genetic algorithm (1994)
6. He, K., Zhang, X., Ren, S., Sun, J.: Deep residual learning for image recognition. In: Proceedings of the IEEE Computer Society Conference on Computer Vision and Pattern Recognition, pp. 770–778 (2016)
7. Holland, J.H.: Genetic algorithms. Sci. Am. **267**(1), 66–73 (1992)
8. Huang, J., Xue, B., Sun, Y., Zhang, M.: A flexible variable-length particle swarm optimization approach to convolutional neural network architecture design. In: 2021 IEEE Congress on Evolutionary Computation, CEC 2021 - Proceedings, pp. 934–941 (2021)
9. Kennedy, J., Eberhart, R.: Particle swarm optimization. In: Proceedings of ICNN'95 - International Conference on Neural Networks, vol. 4, pp. 1942–1948 (1995)
10. Koza, J.R.: Genetic programming as a means for programming computers by natural selection. Stat. Comput. **4**, 87–112 (1994)
11. Krizhevsky, A., Hinton, G.: Learning multiple layers of features from tiny images. Technical report (2009)
12. Londt, T., Gao, X., Andreae, P.: Evolving character-level DenseNet architectures using genetic programming. In: Castillo, P.A., Jiménez Laredo, J.L. (eds.) EvoApplications 2021. LNCS, vol. 12694, pp. 665–680. Springer, Cham (2021). https://doi.org/10.1007/978-3-030-72699-7_42
13. Poli, R., Langdon, W.B., McPhee, N.F.: A Field Guide to Genetic Programming. Lulu Enterprises, UK Ltd (2008)
14. Ren, J., Li, Z., Yang, J., Xu, N., Yang, T., Foran, D.J.: Eigen: ecologically-inspired genetic approach for neural network structure searching from scratch. In: Proceedings of the IEEE Computer Society Conference on Computer Vision and Pattern Recognition, pp. 9051–9060 (2019)
15. Suganuma, M., Shirakawa, S., Nagao, T.: A genetic programming approach to designing convolutional neural network architectures. In: Proceedings of the Genetic and Evolutionary Computation Conference, GECCO 2017, pp. 497–504 (2017)
16. Sun, Y., Xue, B., Zhang, M., Yen, G.G.: Evolving deep convolutional neural networks for image classification. IEEE Trans. Evol. Comput. **24**(2), 394–407 (2020)
17. Szegedy, C., et al.: Going deeper with convolutions. In: Proceedings of the IEEE Computer Society Conference on Computer Vision and Pattern Recognition, pp. 1–9 (2015)

18. Xiao, H., Rasul, K., Vollgraf, R.: Fashion-MNIST: a novel image dataset for benchmarking machine learning algorithms. CoRR (2017)
19. Xie, L., Yuille, A.: Genetic CNN. In: 2017 IEEE International Conference on Computer Vision (ICCV), pp. 1388–1397 (2017)
20. Zhan, Z.H., Li, J.Y., Zhang, J.: Evolutionary deep learning: a survey. Neurocomputing **483**, 42–58 (2022)
21. Zhang, X., Zhao, J., LeCun, Y.: Character-level convolutional networks for text classification. In: Proceedings of the 28th International Conference on Neural Information Processing Systems, NIPS 2015, vol. 1, pp. 649–657. MIT Press, Cambridge (2015)
22. Zhou, X., Qin, A.K., Sun, Y., Tan, K.C.: A survey of advances in evolutionary neural architecture search. In: 2021 IEEE Congress on Evolutionary Computation, CEC 2021 - Proceedings, pp. 950–957 (2021)

Bloating Reduction in Symbolic Regression Through Function Frequency-Based Tree Substitution in Genetic Programming

Mohamad Rimas$^{(\boxtimes)}$ (ID), Qi Chen (ID), and Mengjie Zhang (ID)

Centre for Data Science and Artificial Intelligence and School of Engineering and Computer Science, Victoria University of Wellington, Wellington, New Zealand
{mhdrimas,Qi.Chen,MengjieZhang}@ecs.vuw.ac.nz

Abstract. Genetic programming (GP) is an evolutionary machine learning method that can be used to address a wide range of both classification and regression conundrums. However, traditional GP algorithms can lead to unnecessary code growth known as bloating. This can slow down the convergence time, lead to over-fitting, and increase the computational cost required by the algorithm. The main focus of this paper is to control bloating caused by symbolic regression in GP trees. To address the bloating issue, this paper introduces a novel tree substitution method to reduce the tree size while increasing the exploring ability of the GP algorithm. The proposed method incorporates a comprehensive analysis to detect bloating in parent trees. When a bloated tree is detected, a new, smaller tree is generated, leveraging the function frequency of the identified bloated tree. A set of regression experiments have been conducted on six real-world datasets. Results showed that the proposed GP method obtains a reduction in the size of the best individual while maintaining similar performance as standard GP with a tree height limit.

Keywords: Genetic programming · Bloating · Symbolic regression · Tree substitution

1 Introduction

Genetic programming (GP) [11] is an evolutionary computation technique that enables the solution of complex problems through the evolution of computer programs. Inspired by the Darwinian principles of natural selection and genetics, GP begins by initializing a population of individuals. It then evaluates their fitness and selects individuals for the mating process based on the fitness scores. Next applies crossover and mutation operations to generate new program variations. Through iterative evolution, GP aims to discover effective solutions for complex problems. GP has been applied to solve various complex conundrums in real-world applications. One of the predominant fields of GP is known as symbolic regression (SR), which evolves a set of mathematical expressions to model

T. Liu et al. (Eds.): AI 2023, LNAI 14472, pp. 429–440, 2024.
https://doi.org/10.1007/978-981-99-8391-9_34

a given dataset. GP enables the search for accurate symbolic models without making prior assumptions about the distribution of the data and the structure of the model.

Though GP can evolve an SR model automatically, traditional GP algorithms can lead to a phenomenon called bloating [9] which refers to continuous tree growth without a corresponding fitness improvement. Thus bloat can lead to unnecessarily large and complex trees which can cause several issues to GP such as high computation time, and low interpretability, and could lead towards over-fitting. In general, bloating can hamper the overall performance of GP algorithms. Researchers have proposed many methods to address the bloating issue. These methods include implementing fixed depth limit [8], dynamic depth limit [13], punishing the largest individual during fitness calculation [7], adjusting population size [15], numerical simplification [5] and various designing specific genetic operators [17] for GP. Furthermore, there are other algorithms such as improving generalization [12] which indirectly address bloating. Despite all the efforts, there is not much research done addressing GP bloating by controlling only the individual that caused bloating. As the complexity of the modern world problem increases, finding complex models with good performance progress becomes difficult. Hence bloating remains an open question.

In this paper, a novel bloat control algorithm is developed to avoid bloating functions while reducing the tree size. The aim is to change both the function and size distribution of each generation, thereby reducing the likelihood of encountering the same bloating frequency in newly generated trees.

The objectives of this work can be summarised as follows:

- introduces an individual bloat detection method based on fitness scores and tree sizes,
- develop a new function frequency based tree generation method to control bloating, and
- investigate the effectiveness of the new bloat controlling method on real-world regression tasks

The remainder of the paper is organized as follows: Sect. 2 discusses related works, while Sect. 3 provides a detailed description of the methodology. In Sect. 4, the experiment settings are discussed. The results of the algorithm on six symbolic regression datasets are presented in Sect. 5. Finally, Sect. 6 concludes the paper and highlights some future directions.

2   Related Work

Numerous researchers have made efforts to address bloating issues. In this section, we aim to present the most relevant literature we have encountered, categorizing the related work into two main areas: bloat detection and bloat control.

2.1   Bloat Detection

One of the earlier techniques to detect bloating by simply setting an individual threshold size. Koza [6] proposed to substitute individuals that exceed the threshold tree size by its root node parent. However, setting a static tree size might not be ideal because in some situations a larger tree size may be needed to model the given dataset. Hence setting a tree size without prior knowledge or considering fitness score may be detrimental to the evolving model. The negative effects of size limit have been studied by Dignum et al. [2].

Gardner et al. [4] used a histogram envelop method to control bloating. This research uses the size of the best individual in the previous generation to decide the bloating histogram bin (cut off bin) to reduce the number of bloated trees in the population of large bloated individuals. This bloat detection method assumes that all the GP tree sizes higher than the cut-off histogram bin are bloated without considering individual fitness scores. This might lead to detecting trees with good fitness improvement but with larger tree sizes as bloated trees.

Vanneschi et al. [18] introduced a mathematical formula to detect bloating. The main idea is to measure the ratio of the increase in size to the fitness score of the whole generation with respect to the initial generation. Any generation that gets a positive value is considered a bloated generation. However, the authors admit that the suggested bloat detection equation can produce negative values for bloated trees in certain situations when conducting symbolic regression. Furthermore, the proposed method measures the bloating of the whole population instead of each individual. hence fails to identify individual bloating as the equation is for a whole generation.

In summary, there has been a small number of research conducted on bloat detection for individual GP trees and especially linking fitness scores for individual bloat detection. In this paper, we fill this research gap by developing an individual bloat detection method by considering both the tree size and the fitness score.

2.2   Bloat Correction

In addition to exploring bloat detection, researchers have explored various methods to rectify bloated trees. One typical method is known as Operator Equalization (OpEq) proposed by Dignum et al. [3]. This method assigns a histogram on the distribution of tree size to each generation to prevent bloating. In OpEq, a number of bins that contain trees with a range of tree sizes are used. The heights of bins determine the number of trees belonging to each bin. OpEq achieves the given target distribution pattern by accepting or rejecting each newly created individual to the corresponding bin. Hence this algorithm control bloating by setting a smaller distribution for larger trees that are more likely potentially bloat. The authors later extend OpEq to dynamic OpEq [14] which allows individuals to create their own bin if fitness is better than the best-so-far individual, mutation OpEq [15] algorithm which mutates the rejected individuals in order to force it to fit into the adjacent bin, and flat OpEq [16] algorithm which maintains a flat distribution for each generation. The main issue of these methods arises when the bin is full, the algorithm is forced to lose some potentially good individuals or to adjust them to fit into a norther bin.

Alfaro-Cid et al. [1] developed the Prune and Plant method to address the bloating. The Prune and Plant operator splits a tree into two trees to reduce bloating. The operator selects a random subtree in a parent and replaces this subtree with a terminal (Pruning). Then the leftover subtree is added to the population as a new tree (Planting). By doing this Prune and Plant method introduces smaller trees to the population, it increases the probability of selecting smaller trees for crossover which eventually leads to smaller offspring. Furthermore, pruning reduces complexity by removing a subtree from a large tree, while planting encourages exploring unexplored regions of the search space by introducing the removed tree as a new individual, potentially leading to the discovery of better solutions. The algorithm outperformed a number of state-of-the-art bloat control methods.

Another popular way of bloat controlling is parsimony pressure. Panait et al. [10] propose biased multi-objective parsimony pressure to control bloating. This method includes minimizing the size of the tree as a second objective in addition to reducing the error score.

In summary, the Prune and Plant method performs well compared to other bloat controlling algorithms. However, the Prune and Plant method reintroduces both trees back into the population, thereby failing to change the bloated function distribution. Consequently, there is a possibility that one of the trees may revert to its previously controlled bloated state. To address this concern, this research explores the concept of bloat-avoiding trees.

3   Methodology

This section introduces a new method for controlling bloating in GP for symbolic regression, known as the function frequency-based probabilistic tree generation method (FFBPTG). The proposed algorithm examines each parent selected for crossover to identify bloated trees. The selected parents are then analyzed to determine if they exhibit minimal fitness progress coupled with rapid growth. If the analyzed parent is detected to be bloating, the proposed algorithm replaces it by generating a smaller new tree based on the function frequency of the original parent tree. Hence, the proposed bloat controlling can be divided into two components of bloat detection and bloat correction.

3.1   Bloat Detection

Bloating refers to the phenomenon where the tree size continuously increases without any corresponding improvement in the fitness score. The bloat detection algorithm employed in this research focuses on identifying individual bloating rather than population bloating. To accomplish this, the FFBPTG algorithm evaluates individual bloating by considering the rolling average tree size which refers to the average tree size, and the rolling average fitness which is the average fitness of the selected individual's ancestors in the previous three generations. An individual is considered bloated if an individual is grown 30% more compared to the rolling average tree size and the fitness gain is less than 0.1% compared with the rolling average fitness. The thresholds of 30% and 0.1% are empirically chosen for this research. The algorithmic representation of this process is shown in Algorithm 1. Increasing the tree size growth threshold to more than 30% would reduce the number of newly generated trees which would limit the exploration ability of GPSR. Conversely, setting it to less than 30% would lead to generating an excessive number of new trees, resulting in the loss of already evolved good individuals. Similarly, increasing or decreasing the fitness threshold would have the same concern.

It is known that the average tree size of the algorithm is continuously growing along with the increase of generations. When too many generations are taken to average then the algorithm might become over-responsive to small tree growth while if too few generations are considered to average, the algorithm will become over-responsive to tree size reduction due to crossover. In some situations, the crossover operation can cause a big reduction in tree size. Since the average is sensitive to outliers this would reduce the average significantly if too few generations are considered. This can lead to detecting smaller trees as bloated. Thus the number of generations considered for the rolling average is decided empirically.

Algorithm 1. Bloat Detection

1: **Input** I ▷ Individual tree
2: compute the rolling average of tree size A_t ▷ average tree size of I's ancestors in the previous three generations
3: compute the rolling average of fitness score A_f ▷ average fitness of I's ancestors in the previous three generations
4: **if** $(A_f - \text{I's fitness}) \leq (0.1\% \text{ of } A_f)$ and I's tree size $> (30\% \text{ of } A_t) + A_t$ **then**
5:      do bloat correction   ▷ Given individual is bloated

3.2 Bloat Correction

Once an individual is identified as bloated, the FFBPTG algorithm replaces it with a newly generated smaller tree to correct bloat. The primary objective of generating new trees is to reduce the occurrence of frequently used functions within the bloated individual. This leads to a change in the function distribution of the next generation which will yield less chance of occurring the same bloated function tree due to crossover and mutation in future generations. To mitigate the bloated functions, this algorithm assigns selection probability (p_i) to each function when generating a new tree. These selection probabilities are calculated based on the bloated tree function histogram and probabilities are calculated in a way to minimize the utilization of functions that are predominantly used by the bloated individual. The calculation of p_i probability is accomplished using Eq. (1):

$$p_i = \frac{(1 - \frac{f_i}{M_f})}{\sum_{i=1}^{n}(1 - \frac{f_i}{M_f})} \tag{1}$$

where n is the number of functions, and f_i represents the function frequency of each function. M_f is the maximum value of f_i.

p_i is used to determine the probability to include the corresponding functions in a new tree that replaces the bloated one. The suggested function selection probability p_i is calculated in a way that frequent functions selected by the bloated tree would have a minimum selection probability to be included in the new tree generation and less selected functions would have a higher probability. Consequently, the new tree would search for a solution that is different from the bloated tree. Ideally, we anticipate the new tree to discover an alternative minimum in the solution space.

Algorithm 2. Modified Crossover Operator

1: **Input** I_1, I_2 ▷ Selected parents
2: **for** selected individual (I) **do**
3:       compute the rolling average of tree size A_t
4:       compute the rolling average of fitness score A_f
5:       **if** $(A_f - $ I's fitness$) \leq (0.1\%$ of $A_f)$ and I's tree size $> (30\%$ of $A_t) + A_t$ **then**
6:             compute P_i using Equation 1   ▷ Selection probabilities
7:             compute T_h using Equation 2   ▷ Maximum tree height
8:             $N_t = $ generate new tree (I , P_i , T_h) ▷ Using dynamic probabilities
9:             $I = N_t$ ▷ Replace I with new tree N_t
10: offspring1 , offspring2 = Crossover (I_1, I_2) ▷ Perform crossover
11: **return** offspring1 , offspring2

However, using a static p_i vector can be harmful in certain situations. For example, let's consider a GP algorithm that has three functions namely plus (+), minus (−), and multiplication (∗). If a tree was detected bloated with function frequency of 20, 20, and 0 respectively. The p_i vector will force the regeneration algorithm to choose only multiplications (∗) during the regeneration process. Hence this can lead to generating a tree that contains only one function. Hence it is important to use a dynamic p_i array. Thus in the proposed algorithm, the p_i array is recalculated after every function selection during the tree regeneration process. Considering the previous example, once the first function selection occurs, which in this case is multiplication (∗), the function frequencies undergo an update, resulting in values of 20, 20, and 1 respectively. As a consequence, during the next selection process, plus (+) and minus (−) functions are expected to have a greater probability of being selected compared to their previous probabilities.

After the initial calculation of p_i, the algorithm proceeds to regenerate a new tree. The tree generation process is guided by the p_i values. The ramped half-and-half method is used for the regeneration process. Inspired by the simple depth limiting algorithms [6], the maximum depth limit for generating new trees in FFBPTG is determined based on the depth of the bloated tree according to Eq. (2). Higher bloated tree depths are generated in later generations due to the mean tree size growth of the population. Hence it is highly likely that a bloat corrected tree would crossover with a larger tree. Thus it is important to reduce the newly generated tree depth more for bloated trees that contain higher depths compared to lower tree depths.

$$T_h = \begin{cases} 2 & B_h \leq 3 \\ B_h * 0.5 & 3 < B_h \leq 5 \\ B_h * 0.3 & 5 < B_h \end{cases} \qquad (2)$$

In the given conditions, B_h represents the depth of the bloated tree, while T_h denotes the maximum depth limit for tree regeneration. Tree height ratios (0.3 and 0.5) and threshold values of 3 and 5 were empirically determined.

It is flexible to integrate the bloat detection and control algorithm into different components of GP, e.g., tournament selection, fitness evaluation, crossover, etc. In this research, bloat detection and correction are applied during the crossover as shown in Algorithm 2. Applying at the stage of crossover forces the algorithm to perform crossover with a smaller tree, thus leading the algorithm to produce smaller offspring. Furthermore, correcting the individuals that are only selected for crossover instead of searching the whole population helps to improve the efficiency of the algorithm.

4   Experimental Settings

To evaluate the performance of the FFBPTG algorithm, a set of experiments is conducted under different settings. Table 1 provides a summary of the datasets used in this work. Test and training sets are separated by using a 25% and 75% ratio for all the datasets except for DLBCL where the training and the test sets are provided.

Table 1. Data-set Information

Name	Abbreviation	Features	Testing	Training
CCN	CCN	122	499	1,495
DLBCL	DLBCL	7399	180	240
Bike Sharing	BSharing	11	183	548
Appliance Energy	AppEng	26	375	1125
Concrete	Concrete	8	258	772
Boston House	BstHouse	13	127	379

Table 2. Experimental GP Settings

GP Setting	Value
Population size	512
Generations	60
Tournament size	3
Crossover, mutation probability	0.9, 0.1
Function set	$+, -, *, \%$protected, sin, cos, tanh, exp, log
Elitism	10 individuals
Initial Min-Max tree depth	1–3
Max tree depth	10
Initialisation	Ramped half-and-half

The FFBPTG algorithm is compared with standard GP to evaluate the performance in this research. The experiment settings for GP algorithms are shown in Table 2. The normalized root mean square error (NRMSE) Eq. (3) is used as the fitness score for all the experiments. Algorithms are tested 30 times for each dataset using 30 different random seeds.

In Eq. (3), $\hat{y}_i$ represents the predicted value, y_i represents the target value, and n denotes the total number of samples. The variables max(y) and min(y) represent the maximum and minimum values of the actual values, respectively.

$$\text{NRMSE} = \frac{\sqrt{\frac{1}{n}\sum_{i=1}^{n}\left(\hat{y}_i - y_i\right)^2}}{\max(y) - \min(y)} \tag{3}$$

5   Results and Analysis

This section shows the experiment results of the two GP methods on six real-world datasets. Algorithms are compared under different metrics including best individual training error, best individual test error, best individual tree size, and execution time. Furthermore, the Wilcoxon statistical significant test with a significance level of 0.05 is conducted to confirm the difference in the performance of the FFBPTG algorithm. Finally, an analysis of the simplified version of these example programs is presented to confirm if the FFBPTG algorithm has found shorter programs.

5.1   Performance Analysis

Figure 1 shows the best performance on the training datasets and test sets. In general, comparing the performance of the two GP methods, there is no improvement or deterioration in the training or test performance. FFBPTG obtain similar training and test performance on all of the six datasets except for BstHouse test performance. On BstHouse, the FFBPTG algorithm performs slightly worse than standard GP on the test set. Furthermore, both algorithms perform almost identically on DLBCL, Concrete, and AppEng datasets.

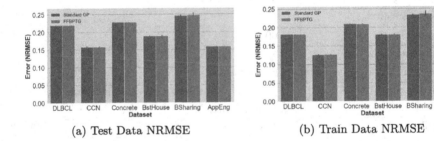

(a) Test Data NRMSE   (b) Train Data NRMSE

Fig. 1. Performance Measured in Normalized Root Mean Square Error (NRMSE)

Table 3 shows the results of the Wilcoxon statistical significance test conducted for the training and the test performance of the two GP methods. where "=" means no significant difference. As it shows, there is no significant difference in the performance between standard GP and the FFBPTG algorithm.

Table 3. Statistical Significance

Benchmark	Method	Training NRMSE (Mean ± SD)	Test NRMSE (Mean ± SD)	Significant Test (with FFBPTG Algo.) (Training, Test)
CCN	Standard GP	0.1261 ± 0.0009	0.1577 ± 0.0014	(=, =)
	FFBPTG Algo.	0.1266 ± 0.0011	0.1584 ± 0.0016	N/A
DLBCL	Standard GP	0.1821 ± 0.0	0.2211 ± 0.0	(=, =)
	FFBPTG Algo.	0.1821 ± 0.0	0.2211 ± 0.0	N/A
BSharing	Standard GP	0.2339 ± 0.0045	0.2457 ± 0.0046	(=, =)
	FFBPTG Algo.	0.2368 ± 0.0168	0.248 ± 0.015	N/A
AppEng	Standard GP	0.1235 ± 0.0001	0.159 ± 0.0001	(=, =)
	FFBPTG Algo.	0.1235 ± 0.0	0.159 ± 0.0001	N/A
Concrete	Standard GP	0.2091 ± 0.0005	0.2281 ± 0.0006	(=, =)
	FFBPTG Algo.	0.2091 ± 0.0003	0.228 ± 0.0004	N/A
BstHouse	Standard GP	0.1807 ± 0.0009	0.1885 ± 0.0011	(=, =)
	FFBPTG Algo.	0.1809 ± 0.0013	0.1892 ± 0.0018	N/A

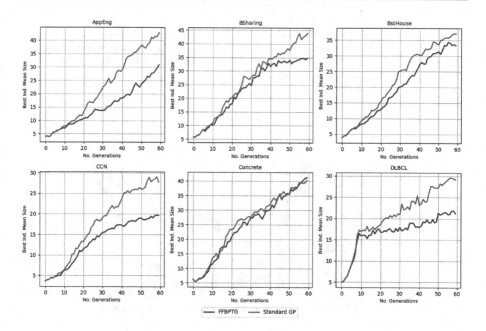

Fig. 2. Evolutionary Plots on Average Tree Size of Best Individuals

5.2   Tree Size Analysis

Figure 2 shows how the tree size changes over 60 generations for 30 different runs. It is clear that the FFBPTG algorithm produces shorter tree sizes compared to standard GP. The FFBPTG algorithm obtains a smaller program size than standard GP on the AppEng, CCN, and DLBCL datasets. At the end of 60 generations, the FFBPTG algorithm produces trees that are 28%, 29%, and 27% shorter compared to standard GP. The difference between the mean tree size of standard GP and FFBPTG increases except for Concrete. Hence, better results could be expected as the number of generations increases.

Table 4. Simplified Final Best Models of CCN Data-set

Seed number	Standard GP	FFBPTG algorithm
842	sin(sin(cos(sin(tanh(sin(sin(ARG119)))))))- tanh(sin(tanh(cos(sin(tanh(sin(sin(ARG119))))))))	sin(ARG119)- sin(sin(sin(ARG119)))
600	log(2*ARG110+exp(ARG106) -3*tanh(ARG110)+ tanh(tanh(tanh(tanh (tanh(tanh(ARG110))))))) *tanh(tanh(tanh(tanh(tanh(tanh (tanh(tanh(ARG110)))))))))	ARG106*tanh(tanh(tanh(tanh(tanh (tanh(tanh(tanh(tanh (ARG110)))))))))
519	ARG103*exp(-ARG103* tanh(ARG103)-tanh(4*(ARG103 +tanh(ARG103))*tanh(ARG103)) -tanh(tanh(tanh(tanh(tanh (ARG103))))))	ARG103*exp(-ARG103 - sin(ARG103)-tanh(tanh(tanh(tanh (tanh(tanh(ARG103)))))))

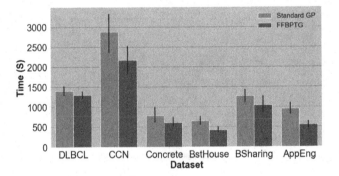

Fig. 3. Execution Time

5.3 Time Analysis

Figure 3 shows the execution time comparison for each dataset. The mean time consumed by the FFBPTG algorithm is shorter compared to standard GP in general. However, the improvement in time is not significant. The FFBPTG algorithm shows the highest time reduction for CCN and AppEng datasets. This might be due to the tree size reduction on these two datasets. Although, the mean tree size of the Concrete dataset is similar to standard GP, the execution time of the FFBPTG algorithm is shorter compared to standard GP. In general FFBPTG algorithm consumes a shorter time for all the examined datasets.

5.4 Evolved Tree Analysis

Table 4 shows a simplified version of the three examples of the best-evolved program in FFBPTG and standard GP for the same random seed. It is clear that the FFBPTG algorithm evolves less complicated programs.

6 Conclusions

This work proposes a new bloat control algorithm. The main goal of the research is to develop an algorithm that can generate smaller new trees based on the function frequency of bloated trees during the crossover stage. The algorithm was tested on six real-world data-sets and results showed that the FFBPTG algorithm achieves the goal of bloat control and evolves compact models with similar performance to standard GP. Execution time analysis showed a slight improvement compared to the standard GP algorithm. Simplified best individual results proved that the FFBPTG algorithm is capable to find more compact solutions.

In this research, we generated smaller trees and replaced the bloated trees with the generated trees. This shows that bloating can be controlled by tree replacement methods which is an area that has not been studied much. Further experiments in this area are encouraged. In this work, fitness improvement was neglected during tree generation which leads to no fitness improvement. This issue can be addressed in future work by considering the fitness scores of the substituting tree.

Acknowledgement. This work is supported in part by the Marsden Fund of New Zealand Government under Contract MFP-VUW2016 and MFP-VUW1913.

References

1. Alfaro-Cid, E., Esparcia-Alcázar, A., Sharman, K., Vega, F.F.D.: Prune and plant: a new bloat control method for genetic programming. In: 2008 Eighth International Conference on Hybrid Intelligent Systems, pp. 31–35 (2008)

2. Dignum, S., Poli, R.: Crossover, sampling, bloat and the harmful effects of size limits. In: O'Neill, M., et al. (eds.) EuroGP 2008. LNCS, vol. 4971, pp. 158–169. Springer, Heidelberg (2008). https://doi.org/10.1007/978-3-540-78671-9_14

3. Dignum, S., Poli, R.: Operator equalisation and bloat free GP. In: O'Neill, M., et al. (eds.) EuroGP 2008. LNCS, vol. 4971, pp. 110–121. Springer, Heidelberg (2008). https://doi.org/10.1007/978-3-540-78671-9_10

4. Gardner, M.A., Gagné, C., Parizeau, M.: Bloat control in genetic programming with a histogram-based accept-reject method. In: Proceedings of the 13th Annual Conference Companion on Genetic and Evolutionary Computation, New York, NY, USA, pp. 187–188 (2011)

5. Kinzett, D., Johnston, M., Zhang, M.: Numerical simplification for bloat control and analysis of building blocks in genetic programming. Evol. Intell. 2(4), 151–168 (2009)

6. Koza, J.R.: Genetic Programming: On the Programming of Computers by Means of Natural Selection. MIT Press, Cambridge (1992)

7. Koza, J.R.: Genetic programming as a means for programming computers by natural selection. Stat. Comput. 4, 87–112 (1994)

8. Luke, S., Panait, L.: A comparison of bloat control methods for genetic programming. Evol. Comput. 14(3), 309–344 (2006)

9. O'Neill, M.: Riccardo Poli, William B. Langdon, Nicholas F. Mcphee: a field guide to genetic programming. Genetic Program. Evol. Mach. 10(2), 229–230 (2009)

10. Panait, L., Luke, S.: Alternative bloat control methods. In: Deb, K. (ed.) GECCO 2004. LNCS, vol. 3103, pp. 630–641. Springer, Heidelberg (2004). https://doi.org/10.1007/978-3-540-24855-2_71

11. Poli, R., Langdon, W., Mcphee, N.: A field guide to genetic programming (2008)

12. Raymond, C., Chen, Q., Xue, B., Zhang, M.: Genetic programming with rademacher complexity for symbolic regression. In: 2019 IEEE Congress on Evolutionary Computation (CEC), pp. 2657–2664 (2019)

13. Silva, S., Costa, E.: Dynamic limits for bloat control in genetic programming and a review of past and current bloat theories. Genet. Program. Evol. Mach. 10(2), 141–179 (2009)

14. Silva, S., Dignum, S.: Extending operator equalisation: fitness based self adaptive length distribution for bloat free GP. In: Vanneschi, L., Gustafson, S., Moraglio, A., De Falco, I., Ebner, M. (eds.) EuroGP 2009. LNCS, vol. 5481, pp. 159–170. Springer, Heidelberg (2009). https://doi.org/10.1007/978-3-642-01181-8_14

15. Silva, S., Dignum, S., Vanneschi, L.: Operator equalisation for bloat free genetic programming and a survey of bloat control methods. Genet. Program Evolvable Mach. 13, 197–238 (2011)

16. Silva, S., Vanneschi, L.: The importance of being flat-studying the program length distributions of operator equalisation. In: Riolo, R., Vladislavleva, E., Moore, J. (eds.) Genetic Programming Theory and Practice IX. Genetic and Evolutionary Computation, pp. 211–233. Springer, New York (2011). https://doi.org/10.1007/978-1-4614-1770-5_12

17. Uy, N.Q., Chu, T.H.: Semantic approximation for reducing code bloat in genetic programming. Swarm Evol. Comput. 58, 100729 (2020)

18. Vanneschi, L., Castelli, M., Silva, S.: Measuring bloat, overfitting and functional complexity in genetic programming. In: Proceedings of the 12th Annual Conference on Genetic and Evolutionary Computation, New York, NY, USA, pp. 877–884 (2010)

Generating Collective Motion Behaviour Libraries Using Developmental Evolution

Md Khan, Kathryn Kasmarik, Michael Barlow, Shadi Abpeikar,
Huanneng Qiu, Essam Debie(✉)(iD), and Matt Garratt

School of Engineering and Information Technology, University of New South Wales,
Canberra, Australia
{e.debie,abc}@unsw.edu.au

Abstract. This paper presents an evolutionary framework for generating diverse libraries of collective motion behaviours. It builds upon recent advancements in machine recognition of collective motion and the transformation of random motions into structured collective behaviours. The paper describes the design of the framework, including the use of a fitness function and diversity metrics specifically tailored for this purpose. The proposed framework generates diverse behaviours with distinct collective motion characteristics. Analysing the relationship between genotypic and behavioural diversity, we observed that greater diversity emerges after a moderate number of evolutionary generations. Our findings highlight the effectiveness of task non-specific fitness functions in distinguishing structured collective behaviours in an evolutionary setting.

1 Introduction

Structured collective motion behaviours, such as movement in a line, movement in a group, aggregation or dispersion, offer the advantages of efficient movement, a level of protection for the group, and potential for human guidance of the single swarm organism. However, current approaches to swarm robotics still require significant human input to configure such behaviours, either in the form of rule weights or fitness objectives. This inhibits rapid, ad hoc formation of swarms from available units for new missions. In recent years, hardware for large numbers of autonomous ground and aerial vehicles is becoming increasingly accessible. It is feasible that a modern company might own an eclectic mix of different autonomous vehicle platforms. Tuning and re-tuning the parameters of collective motion behaviours for each platform is complex and time-consuming. As a result, recent work has been developing methods to recognise collective motion behaviour when it is occurring [10] and to tune random motion into structured collective motions [1]. This paper considers the next necessary step— generation of libraries of collective motion behaviours for humans to employ during human-swarm interaction (HSI) missions.

The remainder of the paper is organised as follows. Section 2 reviews the literature most closely related to this work. In Sect. 3, we present our developmental evolutionary framework and describe the components in detail. In Sect.

T. Liu et al. (Eds.): AI 2023, LNAI 14472, pp. 441–452, 2024.
https://doi.org/10.1007/978-981-99-8391-9_35

4, we present the results of the experiments and provide a detailed analysis in comparison to other recent work in this area. Section 5 concludes the paper and examines directions for future work.

2   Related Work

Structured collective motion includes movements in recognisable patterns such as lines, grouping, circles and so on. Early work in collective motion behaviour recognition used indicators such as 'group' and 'order' [4] to identify collective behaviour. However, it is now understood that these approaches can only detect subsets of structured behaviour [9]. More recently, Hamann [6] presented an evolutionary approach to generate collective behaviours by minimising surprise. Their work shows that 1-dimensional collective behaviours of point-mass agents can be evolved as a by-product of minimising prediction errors. In our current paper we consider a 2-dimensional environment (a flat arena) and simulate ground robots as point masses. Gomes et al. [5] uses novelty search for evolution of a swarm of robots. They used aggregation and sharing a charging station as representatives of simple and complex swarm robotics tasks. Our current paper focuses on movement in a group configuration, as well as aggregation, and presents an architecture for generating a library of generic structured behaviours that could be used by an operator for completing future missions.

Finally, some existing work has considered unsupervised learning algorithms as a means to identify whether there is structure in the behavioural data of a group of agents [10]. This work has assumed an evolutionary framework to generate behaviours, but focused on the design of the value system fitness function did not implement or test a full evolutionary system. In this paper we address this gap. In doing so, we extend other work that has used reinforcement learning to tuning individual collective behaviours [1] because our approach in this paper can generate libraries of collective motion behaviours through simultaneous evolution of a population of diverse behaviours. The following sections introduce the evolutionary computation processes required to do this.

3   Proposed Evolutionary Developmental Framework

The computational processes underlying our proposed approach consists of two processes: 1) Robot-Centric Cycle; and 2) Value System. We describe these two processes in the coming sections, see Fig. 1.

3.1   Robot-Centric Cycle

The robot-centric cycle in our framework consists of boids [12], representing a group of agents interacting with each other. The basic boid model consists of the three simple steering forces–cohesion, alignment and separation. "Cohesion" means that an agent moves towards the average position of the neighbours. "Alignment" signifies that an agent should steer to align itself with

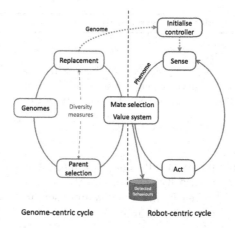

Fig. 1. The conceptual framework, which assumes parallel genome-centric (left) and robot-centric (right) processes. Adapted from the general architecture presented in [10].

the average heading of its neighbours. Lastly, an agent moves to avoid collision with the neighbours, producing "Separation". This paper will focus on an extended boid model [2] that includes additional parameters that affect the 'situational awareness' of the boids such as vision angle. Suppose there are N agents $A^1, A^2, A^3...A^N$ in the group. At time t each agent A^i has a position, p_t^i, and a velocity vector v_t^i. The velocity of each agent is updated step-wise as shown in Eq. 1:

$$v_{t+1}^i = v_t^i + W_c c_t^i + W_a a_t^i + W_s s_t^i \qquad (1)$$

where c_t^i, a_t^i, and s_t^i are the vectors representing the cohesion, alignment and separation forces. c_t^i is defined as the direction of the average position of agents within the neighbourhood; a_t^i is defined as the average direction of agents in the neighbourhood; and s_t^i is defined as the direction away from the average position of agents in the neighbourhood. W_c, W_a, and W_s are weights that strengthen or weaken the corresponding force for cohesion, alignment, and separation, respectively. Once velocity is calculated, agent position is updated by Eq. 2.

$$p_{t+1}^i = p_t^i + v_{t+1}^i \qquad (2)$$

It is quickly apparent that parameters such as the weights in Eq. 1 will need to be tuned for specific hardware platforms. The extent of tuning required depends on the quality of the collective motion resulting from a given parameter set. This quality is evaluated by a value system fitness function.

3.2 Value System

The value system in an embodied evolutionary framework changes over time based on experiences, unlike a static fitness function. It identifies new behaviours

---

Algorithm 1: Construct value system inputs

---
1 **for** *each time step* **do**
2   **for** *each boid B* **do**
3   Identify the k-nearest neighbours using position values
4   Calculate the relative velocity/position of the nearest neighbours
5   Create data line using the position/velocity of the *k* neighbours

---

for the library. In this paper, a self-organising map (SOM) serves as the value function and represents curiosity [11,13]. Observations of the boid's behaviour are passed to the SOM at each time step as an observation vector. The closest neuron to the input becomes the winning neuron, and its weights are adjusted along with its neighbouring neurons. Two types of inputs are proposed in this paper to assess the structured quality of collective motion: velocity of neighbouring agents and relative positions of neighbouring agents. For calculating the nearest neighbours, Euclidian distance is used. Algorithm 1 creates one data line for each boid for each time step. For a neighbourhood of k boids, each data line has the form of either Eq. 3 (when using velocity as input) or Eq. 4 (when using position as input). V_x, V_y denotes the x and y components for velocity and P_x, P_y denotes the x and y components for the position.

$$O_t^i = [V_{xt}^1, V_{yt}^1, V_{xt}^2, V_{yt}^2, ..., V_{xt}^k, V_{yt}^k] \tag{3}$$

$$O_t^i = [P_{xt}^1, P_{yt}^1, P_{xt}^2, P_{yt}^2, ..., P_{xt}^k, P_{yt}^k] \tag{4}$$

The quantization error (QE) that is computed at each timestep is referred to as the curiosity value, which is produced by the SOM in response to each input. Calculating QE involves determining the average distance that exists between the input vectors coming from each boid and the best matching unit that corresponds to them. The formula for average quantisation error is calculated using Eq. 5.

$$QE_t = \sum_{i=1}^{N} O_t^i - D_t^i \tag{5}$$

where D_t^i is the winning neuron for the corresponding input O_t^i. The value system runs within the evolutionary framework as new genome gets generated and the value system is provided with the newly generated data to get the QE value.

3.3 Genome-Centric Cycle

In this paper, we use a multi-modal genetic algorithm as the evolutionary algorithm.

We initialise the population with μ genomes. The set of parameters that makes up each genome is as follows: Separation Weight (W_s) range in [0–4], Alignment Weight (W_a) range in [0–4], Cohesion Weight (W_c) range in [0–4],

Maximum Speed (V_{max}) range in [5–20], Minimum Speed (V_{min}) range in [2–10], Avoid Distance (R_s) range in [2–62], Vision Range (R_c/R_a) range in [5–1005], Vision Angle (θ) range in [60°–360°], Rule likelihood (P_{rule}) range in [0–1], SA likelihood (P_{sa}) range in [0–1], Full scan likelihood ($P_{fullscan}$) range in [0–1], Separation likelihood (P_s) range in [0–1], Alignment likelihood (P_a) range in [0–1], Cohesion likelihood (P_c) range in [0–1]. The first eight parameters directly affect agent behaviour while the six "likelihood" parameters are probabilities for various agent rules. These features, and the corresponding ranges are adopted from [2]. The additional six likelihood parameters can be utilised to vary the corresponding features. The rule likelihood probability determines the probability that an agent will update the velocity each tick. Hence, each of the genomes G is realised as:

$$G = [W_s, W_a, W_c, V_{max}, V_{min}, R_s, R_{(c/a)}, \theta_{vision},$$
$$P_{rule}, P_{sa}, P_{fullscan}, P_s, P_a, P_c,] \quad (6)$$

Situational Awareness (SA) is defined as updating the information about position and velocity of other agents, and the SA likelihood parameter is used to set that probability. Fullscan likelihood represents the probability that the agent will look all around by doing a 360-degree scan rather than using the current vision angle. The remaining three parameters control the probability of applying the three basic rules. These additional parameters ensure that a varied set of structured and random behaviours can be generated through this framework.

A fitness function embodies the requirements to which the population should adapt. In our case, fitness is generated by the value system discussed in the previous section. As our aim is to generate structured behaviours, by our definition, a lower QE value will signify that the respective genome has higher fitness. Thus, we can set our fitness score as $f = \frac{1}{QE}$. Restricted Tournament Selection (RTS) is used in our framework for individual selection. Unlike fitness proportional or ranking based selection, tournament selection is conducted by comparing two individuals according to their fitness. Thus, it is conceptually simple and faster to implement. In Tournament selection, m individuals are chosen randomly and the best individual among them (as per fitness) is selected to be in the mating pool. This is conducted over λ times, which denotes the number of individuals to be selected. m is denoted as "tournament size" and increasing m means increasing the selection pressure. The crossover operator resembles biological crossing over, and in practice takes two genomes and swaps their sequences to create two offspring. Crossover combines features of parents to generate offspring which will probably perform better in the fitness landscape. A crossover probability, $P_{crossover}$ is defined, which determines the probability of crossover to be applied. As the parameters in our case have real values, we chose to do a blend crossover [3]. Blend crossover is also known as BLX-α crossover. It works as follows. If $G_1 = (g_1^1, ..., g_n^1)$ and $G_2 = (g_1^2, ..., g_n^2)$ are two genomes to be crossed, then BLX-α generates two offspring, $H_k = (h_1^k, ..., h_i^k, ..., h_n^k)$, $k = 1, 2$ where h_i^k is a randomly chosen number from the interval $[g_{min} - I\alpha, g_{max} + I\alpha]$, where $g_{max} = max(g_i^1, c_i^2)$, $g_{min} = min(g_i^1 g c_i^2)$ and $I = g_{max} - g_{min}$. Mutation intro-

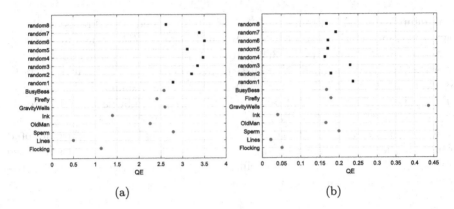

(a)　　　　　　　　　　　　　　　　(b)

Fig. 2. Mean QE for a set of known structured and random behaviours using velocity and position of the five nearest neighbours as input to the value system.

duces a random, unbiased change in the population. It introduces an arbitrary alteration in the gene, increasing the structural diversity of the population. Each feature of a genome goes through this change of value with a mutation probability, $P_{mutation}$. In our case, we apply uniform mutation, i.e. a new feature value x_i is drawn uniformly randomly from the domain of allowed values for that particular feature. The next section describes a series of experiments with this system.

4　Experiments

We begin this section by describing the experimental environment in which the robot-centric cycle is situated. We then describe three experiments in this setting.

4.1　Experimental Environment

We use $N = 200$ boids in a 1400×1000 arena to simulate the robot-centric cycle. The agents are homogeneous with respect to ability and control. That is, the parameter ranges are same for all the agents in the swarm. The value of k is set to 5, meaning we consider the 5 nearest neighbours. The SOM toolbox [14] is used for implementing the value system. The SOM toolbox is an efficient implementation of SOM in MATLAB. A heuristic formula was used to determine the number of neurons for the map: No. of neurons = $5\sqrt{M}$, where M is the number of observations. In our experiments we run SOM with 300,000 observations. Thus, the number of neurons used is 2756. The learning rate was fixed to 0.25 and the neighbourhood size was fixed to 1. These fixed values capture the property of a value system that it is always learning.

　　The parameters used for implementing the evolutionary algorithm are as follow: No. of generations = 100, No. of individuals = 50, Crossover rate = 0.9, Mutation rate = 0.1, Number of peaks = 5, Window size = 20. The first four parameters are generic evolutionary algorithm parameters, while the number of peaks and window size are particular to RTS [7], the strategy used here to incorporate diversity and encourage multimodality in generating the solutions. Unlike simpler functions, the number of peaks present in the fitness landscape of our value function is unknown. Thus, we have selected five peaks as a starting point and set the RTS window size as four times that (i.e., 20) as per the recommendation in the original paper [7]. Thus, we are allowing our system to attempt to identify a library of five collective motion behaviours.

In Fig. 2, we show one of the demonstrated performances, which shows the mean QEs when using the velocity or position of the five nearest neighbours as input to the value system. In Fig. 2, the structured behaviours, given representative names, are located at the bottom half whereas the random behaviours are located on the top. When using velocity, We can see that most of the structured behaviours have a QE lower than 3. Lines has the lowest QE at 0.5. Using position, structured behaviours (except for GravityWells) have a QE lower than 0.2. Lines has the lowest QE at 0.023.

4.2   Experiment 1: Establishing a Performance Baseline

The mean QE, along with the 95% confidence interval, is displayed in Fig. 4a. The QE value is inversely proportional to fitness, so a lower QE means higher fitness in the context of the optimisation algorithm. Generation 1 is the randomly initialised population, while generation 100 is the final iteration of the optimisation. The 50 genomes present in each iteration were averaged over 10 runs to give the mean. We can see there is an increasing trend of fitness (shown as lowering QE) across generations. As additional measures of collective motion, we also calculated the group and order values of the boids across generations. Group measures the cohesion of the swarm while order is a measure of alignment. We can see that Group values remain similar throughout generations (Fig. 3b) whereas the order values gradually decrease (Fig. 3c). This reflects the fact that the input to the value system in this experiment was boid velocity, so the evolutionary system is tuning velocity to create structured collective motions.

Behavioural Diversity. We analyse the behavioural diversity of individuals in earlier generations. Iteration 35, with the highest diversity, is selected. It has a mean QE of 0.6, falling within the range of the structured behaviour cluster in Fig. 2a. We choose five individuals from the 35th generation:

- **Individual 1:** This behaviour is similar to those described for iteration 100: the boids move together in the same general direction.
- **Individual 25:** This is similar to the behaviour produced by Individual 1 where boids generally move in the same direction. However, the separation weight is higher and cohesion weight is lower.
- **Individual 35:** Boids make clusters and move in the same direction.
- **Individual 30:** boids generally move in the same direction but with high speed range (6.5, 9.8).

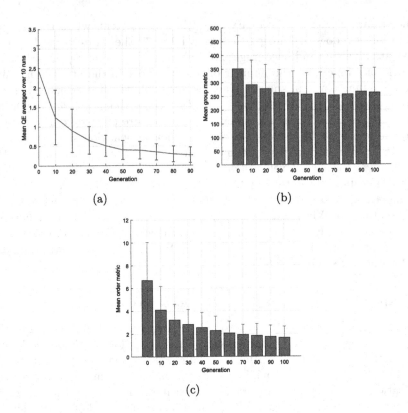

Fig. 3. Fitness, mean group, and order values across generations (velocity input). Averaged over 10 runs.

– **Individual 6:** The boids move either in clusters or in line formation with low speed range (9.2, 9.5). Low separation weight (0.93) and cohesion weight (0.9) compared to other individuals. The relatively low separation, cohesion weights, and small speed range (9.2, 9.5) make the boids behaviour more stable.

As shown in our analysis, each individual demonstrates some degree of structured behaviour. At this earlier iteration, there is more diversity among the genotypes of the individuals, and even those behaviours with a considerably higher QE show structured properties. The group and order results in Fig. 3 provide some insight into why this is, as we can see that around generations 40–50 that changes to grouping metric cease.

4.3 Experiment 2: Impact of Alternative Value System Inputs

In this experiment, we examine the change in performance of the evolutionary framework when using the position of the neighbouring boids as input to the

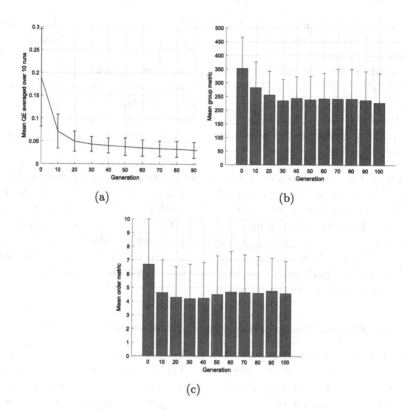

(a)

(b)

(c)

Fig. 4. Fitness, mean group, and order values across generations (position input). Averaged over 10 runs.

value system, rather than velocity. As can be observed, the framework is able to converge to populations with lower QEs (i.e., higher fitness). In Fig. 4, we show the fitness, group and order values across generations for using neighbourhood position.

Behavioural Diversity. We analyse the behavioural diversity of individuals in earlier generations. Iteration 50, with the highest diversity, is selected. It has a mean QE of 0.04, falling within the range of the structured behaviour cluster in Fig. 2b. We choose five individuals from the 50th generation:

- **Individual 5:** This behaviour is substantially different from those in the 100th generation. In this case, the boids make clusters and then keep moving across the landscape in the same direction. The clusters are looser than those in the final iteration, and the movement is fluid.
- **Individual 7:** This behaviour also makes multiple clusters, with a stronger separation compared to Individual 5.

- **Individual 9:** This behaviour is similar to that of Individual 5. Multiple clusters move around the grid at very high speed range (7, 13) compared to other individuals.
- **Individual 50:** This behaviour is different from the others in this set, with the boids making central cluster and moving together in the same direction with lower speed range compared to other individuals (6.5, 7.5).
- **Individual 28:** This behaviour is substantially different from other behaviours. Boids tend to move in line formation like a stream with a stronger separation weight.

Most of the individuals demonstrate some degree of structured behaviour. During this iteration, there is more diversity among the genotypes of the individuals. Even the behaviours with a considerably higher QE show structured properties. The group and order statistics in Fig. 4 suggest that this is an area of trade-off between grouping and ordering. However, in this case it is clear from the error bars on these charts that the change in ordering over generations is not statistically significant. Thus, it would not be possible to use these metrics directly as an effective fitness function.

4.4   Experiment 3: Comparison with Reinforcement Learning

We compare the evolutionary approach for behaviour library generation with a reinforcement learning (RL) approach [1]. The RL approach utilises a learned policy trained with an actor-critic deep neural network to tune structured collective motion based on random behaviour input. It employs a reward signal generator that rewards behaviours recognised as collective motion. In this comparison, the generations 35 and 50 of evolution, which correspond to velocity and position as inputs respectively, exhibit the highest behavioural diversity. Therefore, the RL-tuned behaviour is compared to the outputs at those generations. To ensure fair comparison, 10 genomes from generation 0 are randomly selected as input for the RL tuning. Figure 5 presents the average values of Repel, Attract+Repel, and Attract+Align+Repel forces for both evolved and tuned behaviours. The distributions of Repel values generated by the four approaches have similar ranges, although the average values differ. The Attract+Repel and Attract+Align+Repel metrics exhibit similar ranges and average values across both the evolutionary and RL approaches. The experiment concludes that both evolutionary and RL approaches possess a similar ability to generate behaviours recognisable as collective motion, as defined by the metrics in [8]. Figure 6 illustrates the percentage of behaviours generated by either the evolutionary or RL method that resemble each of the eight baseline structured behaviours. The results indicate that both approaches can generate behaviours with characteristics of each of the eight structured behaviours in the baseline set. However, when using position data as input, the RL approach produces fewer behaviours similar to Flocking, Lines, and Ink. On the other hand, both the evolutionary and RL approaches generate a higher number of behaviours similar to Old Man River and Gravity. In conclusion, while both methods generate relatively diverse

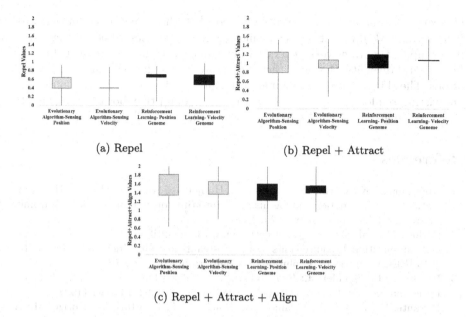

(a) Repel

(b) Repel + Attract

(c) Repel + Attract + Align

Fig. 5. Repel, Align, and Attract metrics for evolved and tuned behaviours. Averages are based on 10 runs and measured after 50/35 generations (position/velocity sensing) or 10 tuning steps (RL).

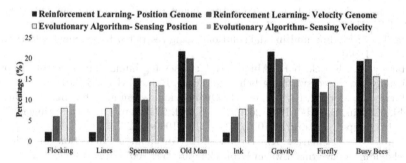

Fig. 6. Experiment 3: The similarity between the 10 runs of generated behaviours by each algorithm with the eight structured behaviours

sets of structured behaviours, they do not achieve the same level of diversity as manual tuning.

5 Conclusions

In this paper, we proposed an evolutionary framework with a value system as a fitness evaluator for generating structured collective behaviours. We conducted three experiments to validate the effectiveness of the framework. The experiments used two types of input: velocity of nearest neighbours and position of nearest neighbours. The results showed that the framework can generate behaviours with different collective motion characteristics. When using velocity as input, the framework generated behaviours with high alignment, but low behavioural diversity. However, there was more visible diversity in earlier generations. The third experiment compared the quality of generated behaviours with an existing reinforcement learning method and found that both could generate a diverse range of behaviours recognised as structured collective motion.

References

1. Abpeikar, S., Kasmarik, K., Garratt, M., Hunjet, R., Khan, M.M., Qiu, H.: Automatic collective motion tuning using actor-critic deep reinforcement learning. Swarm Evol. Comput. 101085 (2022)
2. Barlow, M., Lakshika, E.: What cost teamwork: quantifying situational awareness and computational requirements in a proto-team via multi-objective evolution. In: 2016 IEEE Congress on Evolutionary Computation (CEC) (2016)
3. Eshelman, L., Schaffer, J.: Real-coded genetic algorithms and interval-schemata. In: Foundations of Genetic Algorithms, vol. 2, p. 187–202. Elsevier (1993)
4. Ferrante, E., Turgut, A., Stranieri, A., Pinciroli, C., Birattari, M., Dorigo, M.: A self-adaptive communication strategy for flocking in stationary and non-stationary environments. Natural Comput. 13(2), 225–245 (2014)
5. Gomes, J., Urbano, P., Christensen, A.: Evolution of swarm robotics systems with novelty search. Swarm Intell. 7(2–3), 115–144 (2013)
6. Hamann, H.: Evolution of collective behaviours by minimizing surprise. In: ALIFE2014 (2014)
7. Harik, G.: Finding multimodal solutions using restricted tournament selection. In: ICGA (1995)
8. Harvey, J., Merrick, K.E., Abbass, H.A.: Assessing human judgment of computationally generated swarming behavior. Front. Robot. AI 5, 13 (2018)
9. Harvey, J., Merrick, K., Abbass, H.: Quantifying swarming behaviour. In: Tan, Y., Shi, Y., Niu, B. (eds.) ICSI 2016. LNCS, vol. 9712, pp. 119–130. Springer, Cham (2016). https://doi.org/10.1007/978-3-319-41000-5_12
10. Khan, M., Kasmarik, K., Barlow, M.: Autonomous detection of collective behaviours in swarms. Swarm Evol. Comput. 57, 100715 (2020)
11. Merrick, K., Maher, M.: Motivated Reinforcement Learning: Curious Characters for Multiuser Games. Springer, Berlin (2009). https://doi.org/10.1007/978-3-540-89187-1
12. Reynolds, C.: Flocks, herds and schools: a distributed behavioral model. In: Computer Graphics (SIGGRAPH 1987) Conference Proceedings, vol. 21, no. 4, pp. 25–34 (1987)
13. Shafi, K., Merrick, K.E., Debie, E.: Evolution of intrinsic motives in multi-agent simulations. In: Bui, L.T., Ong, Y.S., Hoai, N.X., Ishibuchi, H., Suganthan, P.N. (eds.) SEAL 2012. LNCS, vol. 7673, pp. 198–207. Springer, Heidelberg (2012). https://doi.org/10.1007/978-3-642-34859-4_20
14. Vesanto, J., Himberg, J., Alhoniemi, E., Parhankangas, J.: Self-organizing map in matlab: the SOM toolbox. In: Proceedings of the Matlab DSP Conference (1999)

A Group Genetic Algorithm
for Energy-Efficient Resource Allocation
in Container-Based Clouds
with Heterogeneous Physical Machines

Zhengxin Fang[1]($\boxtimes$), Hui Ma[1], Gang Chen[1], and Sven Hartmann[2]

[1] Centre for Data Science and Artificial Intelligence and School of Engineering and
Computer Science, Victoria University of Wellington, Wellington, New Zealand
{zhengxin.fang,hui.ma,aaron.chen}@ecs.vuw.ac.nz
[2] Department of Informatics, Clausthal University of Technology,
Clausthal-Zellerfeld, Germany
sven.hartmann@tu-clausthal.de

Abstract. Containers are quickly gaining popularity in cloud computing environments due to their scalable and lightweight characteristics.
However, the problem of Resource Allocation in Container-based clouds
(RAC) is much more challenging than the Virtual Machines (VMs)-
based clouds because RAC includes two levels of allocation problems:
allocating containers to VMs and allocating VMs to Physical Machine
(PMs). In this paper, we proposed a novel Group Genetic Algorithm
(GGA) with energy-aware crossover, Best-Fit-Decreasing Insert (BFDI),
and Local Search based Unpack (LSU) operator to solve RAC problems.
Meanwhile, we apply an energy model with heterogeneous PMs that
accurately captures the energy consumption of cloud data centers. Compared to state-of-the-art methods, experiments show that our method
can significantly reduce the energy consumption on a wide range of test
datasets.

Keywords: Cloud Resource Allocation · Group Genetic Algorithm ·
Container-based Cloud · Physical Machine · Cloud Computing

1 Introduction

Container-based cloud is rapidly gaining attention in cloud computing since it
encourages resource pooling and therefore can significantly reduce computation
overheads compared to Virtual Machines (VMs) based cloud [7]. To reduce the
energy consumption of a data center, cloud service providers need to effectively
allocate resource requests to cloud resources. This gives rise to the *Resource
Allocation problem in Container-based cloud* (RAC). RAC is more difficult
than resource allocation problems in VM-based clouds, as it is a two-level two-
dimensional bin packing problem, which is NP-hard [15]. As shown in Fig. 1,

container-based clouds involve resource allocations on two levels (container-VM level and VM-PM level). At each level, we need to make resource selection and creation decisions. Meanwhile, for each level of resource allocation, both the CPU and memory capacity must be considered jointly because a VM/PM instance is full when one of the resource types runs out of capacity. Different selections or creations of VM/PM could lead to different energy consumption. The resource selection and creation decisions at the two levels are strongly interdependent. Specifically, VM selection and VM creation at the first level have a huge impact on the PM selection and PM creation at the second level [12]. A solution of a RAC problems involves a combination of containers, VM types and PM types.

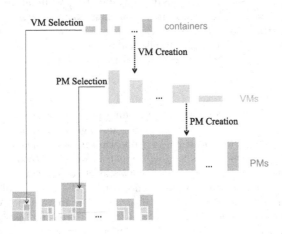

Fig. 1. Resource Allocation in Container-based Cloud Problem

Several heuristic and meta-heuristic approaches have been proposed to solve the cloud resource allocation problem [5,9,15]. Most of the heuristic approaches, such as [15], are based on greedy strategies that can easily get stuck at local optima. Due to their global search capacity, genetic algorithms (GAs) have been applied to solve the RAC problems. A GA based approach [11] is proposed to solve RAC problems based on indirect solution representation. Using indirect representation relies on decoding processes to evaluate evolved solutions. The chosen decoding process has negative impacts on the performance of the algorithm.

To directly represent RAC solutions, Tan et al. [12] propose a Group Genetic Algorithm (GGA) based approach to tackle RAC problems and achieves better performance than GA based approach in [11]. The GGA is an algorithm for solving grouping problems (e.g., bin packing problems) where each gene in GGA is a group. In the RAC problem, the GGA group consists of PMs where VMs are deployed. The GGA based approaches in RAC problem employ *gene-level crossover*, *rearrangement* and *mutation* operators that are based on CPU and memory utilization separately to evolve solutions with minimal energy consumption. However, [12] assumes that the data center uses homogeneous PMs and does not need to make PM creation decisions. In data centers with heterogeneous PMs, evolving solutions based on resource utilization are not effective because the energy consumption of PMs depends not only on their resource utilization but also on their sizes. That is PMs with different types (sizes) and the same CPU utilization consume significantly different energy due to their varied resource capacity and energy consumption profiles. As a result, if an algorithm just aims to improve the resource utilization, it may fail to produce good solutions by completely ignoring the differences in PM energy profiles. At the same time, improving CPU and memory separately may lead to unbalanced usage of these two resources' utilization, which causes resource wastage since one dimension of the resource is full but the other dimension is still in low utilization. Therefore, effective methods are needed for resource allocation in container-based clouds with heterogeneous PMs.

Motivated by the above, in this paper, we aim to propose a new GGA based algorithm for solving RAC problems with heterogeneous PMs. Firstly, we design a new energy-aware crossover operator, which considers not only different energy profiles but also normalized CPU and memory utilization to properly select VM and PM for allocation. Meanwhile, a Best-Fit-Decreasing Insert method is proposed for re-arranging *free containers* (will be introduced in Sect. 4) that are generated by crossover and mutation according to both CPU and memory usage. Further, we propose a local search operator that explicitly utilizes information regarding energy consumption to select and create PMs. Our proposed algorithm is named as Energy-aware Group Genetic Algorithm with Local Search-based Unpacking (EGG-LSU).

In summary, the contributions of this paper are as follows:

- We propose a newly designed energy-aware crossover operator to consider different energy profiles and normalized CPU and memory utilization to better select of VM and PM. We propose a Best-Fit-Decreasing Insert (BFDI) rearrangement operator to rearrangement for the un-allocated containers after crossover;
- We propose a local search operator with a novel unpack operator to refine PM selection and further improve the RAC solutions;
- Through extensive experiments, we thoroughly examine the performance of EGGA-LSU, which is shown to significantly outperform several state-of-the-art heuristic and meta-heuristic algorithms, including FF&BF/FF and GGA.

2 Related Work

Various algorithms have been proposed to solve the RAC problem. Zhang et al. proposed a hybrid heuristic algorithm for RAC problem [15]. First-Fit and Best-Fit is applied to allocate containers and VMs and for selecting VM types, respectively. Sengupta et al. apply Next-Fit heuristic for VM placement in cloud computing environment [10].

An improved genetic algorithm based VM placement algorithm was proposed in [1] to reduce the energy consumption and resource wastage. For two level resource allocation problem, a vector-based GA method with a dual-chromosome representation is proposed in [11] to solve the RAC problem. To get the allocation solutions on both levels and to evaluate their fitness, the chromosomes in [11] need to be decoded using the Next-Fit heuristic. That is, the list of containers contained in a chromosome on the first level, are allocated one by one to VMs using the Next-Fit heuristic. All of the above-mentioned EC algorithms employ indirect representation and require a decoding method to get the final solutions.

Since the decoding process has a negative impact on the performance [11], to directly search allocation solutions, Tan et al. [12] proposed a genetic algorithm method with a direct representation. Using the group-based representation, authors in [12] further introduced several operators (i.e., gene-level crossover, rearrangement, and merge) to search for the best solution of the RAC problem. Another method that applies group genetic algorithm to solve the RAC problem is proposed in [2]. It constrains the length of the group representation of the solution, ensuring that a small number of PMs will be used to reduce the total energy consumption. For multi-objectives optimization, authors in [13] propose a Pareto based genetic algorithm for the purposes of reducing energy consumed in cloud computing environment and improving reliability of application.

The GGA based algorithms mentioned above can directly find the solutions of RAC problems without decoding any evolved solutions. However, these algorithms assume that all PMs are of the same type, which is not be valid for many real-world cloud environments with heterogeneous PMs (as discussed in Sect. 1). Therefore, this paper aims to propose a novel GGA algorithm, namely EGGA-LSU, to solve the RAC problem with heterogeneous PMs.

3   Preliminaries

In a container-based cloud, given a set of containers $C = \{c_1, ..., c_i\}$, a set of VM types, and a set of PM types, RAC allocates containers to VM instances, which are further allocated to PMs. Each container c_i has a CPU occupation $\zeta^{cpu}(c_i)$ and a memory occupation $\zeta^{mem}(c_i)$. A VM type γ_i is defined by $\gamma_j = \{\Omega^{cpu}(\gamma_i), \Omega^{mem}(\gamma_i), \pi^{cpu}(\gamma_i), \pi^{mem}(\gamma_i)\}$, where Ω is the capacity and π is the overhead. A VM instance v_i is of a VM type γ_i. The PM type is defined as a tuple $\tau_i = (\Omega^{cpu}(\tau_i), \Omega^{mem}(\tau_i), E^{idle}(\tau_i), E^{full}(\tau_i))$, capturing its CPU and memory capacities, as well as the energy consumption when it is idle and under full workload respectively. A PM instance with PM type τ_i is denoted as p_i.

Equation (1) quantifies the energy consumption for a given PM instance p_i, based on a popular non-linear energy model proposed in [3].

$$E(p_i) = E^{idle}(p_i) + (E^{full}(p_i) - E^{idle}(p_i)) \times (2\mu^{cpu}(p_i) - (\mu^{cpu}(p_i))^{1.4}) \quad (1)$$

where $E^{idle}(p_i)$ and $E^{full}(p_i)$ are the energy consumption of the PM instance with PM type τ_i when it is idle or fully loaded, respectively. $\mu^{cpu}(p_i)$ is the CPU utilization level of the PM instance p_i, which is calculated by:

$$\mu^{cpu}(p_i) = \frac{\sum_{l=1}^{L}(\sum_{j=1}^{m} \pi^{cpu}(\gamma_j) * z_{j,l} + \sum_{i=1}^{n} \zeta^{cpu}(c_i) * x_{i,l}) * y_{l,p}}{\sum_{k=1}^{|\Pi|} \Omega^{cpu}(\tau_i) * w_{p,k}} \quad (2)$$

where $x_{i,l}$, $y_{l,p}$, $z_{j,l}$ and $w_{p,k}$ are binary decision variables, and L is the number of created VMs. $x_{i,l}$ takes 1 if c_i is allocated to the l-th created VM, and 0 otherwise. $y_{l,p}$ takes 1 if the l-th created VM instance is allocated to the p-th PM, and 0 otherwise. $z_{j,l}$ is 1 if the l-th created VM is of type j, and 0 otherwise. $w_{p,k}$ is 1 if the p-th created PM is of type k, 0 otherwise.

Let $P = \{p_1, p_2, ..., p_i\}$ denote a solution, which is a list of PM instances. The total energy consumption with respect to a solution P is calculated as follows:

$$TEC(P) = \sum_{p_i \in P} E(p_i) \quad (3)$$

where P is the number of PMs that have been created and used in the problem solution.

Let $S = \{P_1, P_2, ..., P_i\}$ denote the set of all possible solutions. The objective of the RAC problem is to find the resource allocation solution with minimal overall energy consumption, as shown below:

$$\min_{P \in S} TEC(P) \quad (4)$$

4   Proposed Algorithm

This section describes our proposed Energy-aware Group Genetic Algorithm with Local Search-based Unpacking (EGG-LSU). The flowchart of our proposed algorithm is shown in Fig. 2 (a), where the main contributions of the algorithm are highlighted. Our algorithm is based on the group genetic algorithm with newly designed gene-level crossover [8], rearrangement method and unpack operators. Our new developments realize several key advantages, allowing EGGA-LSU to explore energy-aware solutions and achieve balanced utilization of different resources.

Representation. EGGA-LSU employs the GGA's direct representation that represents the candidate RAC solutions as a combination of PMs, VMs and containers. Figure 2 (b) shows an example solution that includes a list of PMs, while every item in the list of PMs is a list of VMs. Similarly, every item in the list of VMs is a list of containers. Clearly, this direct representation is a variable-length representation. The length of a solution depends on the number of used PMs. The variable length representation makes it more flexible to explore different solutions using well-designed evolution operators.

Initialization. To create initial solutions for the first population, we first shuffle the container list randomly. Then, we allocate the shuffled list to VMs by using

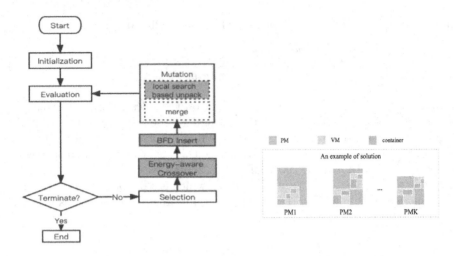

(a) The overall algorithm   (b) Representation of a chromosome

Fig. 2. Overall algorithm and representation of a chromosome in GGA

the First-Fit heuristic [4]. If there are no suitable existing VM instances for a given container, EGGA-LSU creates a new VM randomly that has sufficient capacity to host the container. Similarly, upon allocating VMs to PMs, EGGA-LSU first allocates the VMs using the First-Fit heuristic, and creates a new PM instance randomly if existing PM instances do not have sufficient remaining capacity to host the VM instance to be allocated. By using the First-Fit heuristic to allocate containers and VMs, EGGA-LSU ensures that majority of VMs and PMs can enjoy high utilization in the initial solutions, reducing the total number of PMs used in these solutions.

Energy-Aware Crossover. Our proposed energy-aware crossover is illustrated in Fig. 3. To maintain the population diversity, we propose to use two different preservation criteria, *energy consumption*, see Eq. (1), and *normalized resource*, see Eq. (5), respectively, to generate two offspring individuals from two parent individuals.

$$NOR = \frac{\zeta^{cpu}(c_i))}{\Omega^{cpu}(p_k)} \times \frac{\zeta^{mem}(c_i)}{\Omega^{mem}(p_k)} \tag{5}$$

According to each of the criteria, it first sorts the genes, i.e., PMs, in each parent individual. The genes in two parent individuals are then compared in a pairwise manner. An offspring is created by preserving the better gene of the two parents. That is, an offspring is created by preserving genes with lower energy consumption while the other offspring is generated by preserving genes with higher normalized resource values.

As shown in Fig. 3, every pair of genes in two parents are compared, e.g., PM1 is compared with PM1', PM2 is compared with PM2', and so on. The winning PMs in the comparison are preserved to the offspring individuals by copying

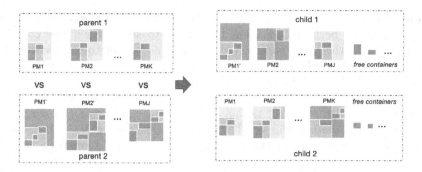

Fig. 3. An example of energy-aware crossover

the structure of the PM and all of the VM instances in the PM. Afterward, the crossover operator checks whether each container in VMs has been allocated to the previous PMs. If so, the allocated containers will not be reallocated; otherwise, containers will be allocated to the corresponding VMs. Finally, the crossover operator will further remove PMs and VMs that do not have any workload. As shown in Fig. 3, some containers may not be allocated after the crossover operator. We refer to these unallocated containers as *free containers*. The *free containers* will be allocated subsequently through a new rearrangement operator, which will be introduced below. Note that, our proposed energy-aware crossover operator is different from the crossover operator in [12], which is based on the utilization of parents.

The energy consumption profile with respect to CPU utilization varies significantly for different PMs. As a result, our proposed crossover operator prefers to preserve PMs with low energy consumption in the first offspring solution so that it can explore the possible use of PMs that consumes less energy. Since RAC is a two-dimensional bin packing problem, therefore we should consider both CPU and memory capacity utilization when preserving PMs for the next generation. Otherwise, the resource utilization VMs and PMs may not be balanced. For example, only considering memory utilization may lead to some PMs with memory utilization much higher than CPU utilization, which leads to CPU resource wastage, because a lot of CPU resource have not been used but the memory resource is full, as in [12]. Instead, in our crossover method, we preserve PMs for the next generation according to the normalized resource, which considers CPU and memory utilization together. It generates offspring with more balanced resource utilization so as to reduce resource wastage and overhead energy consumption.

Best-Fit-Decreasing Insert (BFDI). To reallocate *free containers* to suitable PMs, we propose a Best-Fit-Decreasing Insert (BFDI) method. Our BFDI rearrangement method is developed based on the Best-Fit-Decreasing (BFD) heuristic [6], proposed for the bin packing problem. BFD inserts the items from the largest to smallest into the bins from the most packed to the least packed. However, BFD cannot be directly used to allocate containers to VMs since we need to consider two dimensions. To allocate *free containers* resulted from energy-aware crossover, BFDI performs insertion from large to small. Each free container is inserted into the most packed VM and PM according to their normalized resource. All PMs in the solution are sorted by the normalized resource and all VMs in each PM are also sorted by the normalized resource. After that, BFDI scans each VM instance in each PM instance one by one and checks whether there are VMs with sufficient resource capacity for the target *free containers*. If so, the *free container* will be allocated to corresponding VMs. If there are no suitable existing VMs for the target *free container* BFDI creates a new VM instance randomly to host the container.

BFDI inserts the largest container first to the most packed VM and PM to improve resource utilization. Subsequently, smaller containers are allocated to reduce resource fragmentation. It considers the CPU and memory together when re-arranging *free containers*, in order to balance the CPU and memory utilization and reduce resource wastage. The VM and PM will be created randomly if there are no suitable VMs/PMs to allocate.

Mutation. The mutation process in EGGA-LSU includes two operations: *Local Search Unpack ()LSU* and *Merge*.

Algorithm 1: Local Search Unpack (LSU)

    Input   : a list of PMs
    Output: a list of PMs with the least energy consumption

1 Calculate the unpack possibility of each PM according to Eq. (6);
2 bestPmList ← None;
3 **while** there are neighbors that consume less energy **do**
4   **for** i = 0 to 9 **do**
5   chosenPM ← chose PMs to unpack according to unpack possibility;
6   newPmlist ← Unpack(chosenPM);
7   newPmlist ← **BFDI(free containers)**;
8   calculate the new list of PMs according to Eq. (1);
9   **if** newPmlist consume less than bestPmList **do**
10   bestPmList ← new list of PMs;
11   **end**
12   **end**
13 **end**

Local Search Unpack (LSU), see Algorithm 1, aims to remove PMs with high resource wastage and also explore allocation solutions with less energy consumption. It combines the unpack method with local search. Specifically, LSU unpacks PMs according to the unpack probability which is calculated by Eq. (6), according to which PMs with lower resource utilization have higher probabilities to be unpacked. The containers in the unpacked PMs will be reallocated by BFDI. After unpacking and BFDI, a new solution is generated. As in Algorithm

1, the *bestPMList* is defined as a solution that consumes the least energy. So, the energy consumption of the new solution is compared with *bestPMList*. If the new solution consumes less energy, then the new solution will become the *bestPMList*. The LSU repeats for multiple iterations until no better solutions can be found to further reduce energy consumption.

$$probabiliy = \frac{1 - \Omega^{cpu}(p_k) * \Omega^{memory}(p_k)}{\sum_{k=1}^{K} 1 - \Omega^{cpu}(p_k) * \Omega^{memory}(p_k)} \qquad (6)$$

Merge aims to reduce the overhead consumption of VMs. It repeatedly checks whether the two smallest VMs can be replaced by a larger randomly selected VM. If so, all the containers in the two small VMs are reallocated to the larger VM so that the two small VMs can be released.

5 Evaluation

This section presents the details of the experiment design and the experiment results to evaluate the performance of our proposed EGGA-LSU. We will evaluate the performance of all competing algorithms in terms of the overall energy consumption of the resulting allocations.

5.1 Experiment Design

Experiment Setting. There are eight test instances in our experiments that cover four different numbers of containers (500, 1000, 1500, 5000) with two sets of VM types (real-world VM types and synthetic VM types) [12]. The 10 synthetic VM types are generated randomly. These irregular VM types bring additional challenges for the problem [2]. Additionally, we consider 20 different real-world VM types from Amazon EC2[1]. Furthermore, there are 12 types of PMs with different CPU and memory capacities [14]. Finally, the parameter settings of EGGA-LSU are as follows: the crossover rate is 70%, the mutation rate is 20%, the elite size is 5, the tournament size is 7, and the population size is 100, which follows the setting in [12].

Baseline Algorithms. We compare EGGA-LSU with two state-of-the-art algorithms for RAC problem: FF&BF/FF [15] and GGA [12], which have been introduced in Sect. 2.

5.2 Experiment Results

All the experiments are run for 30 times. The results are verified by the Wilcoxon rank-sum test with a significance level of 0.05. Table 1 compares the energy consumption of the baseline algorithms (introduced above) with EGGA-LSU.

We can observe from the results in Table 1 that GGA performs better than FF&BF/FF, while EGGA-LSU performs the best when compared with each

---

[1] https://aws.amazon.com/ec2/pricing/on-demand/.

Table 1. Energy consumption (in kWh) of 8 test instances with different numbers of containers (size) and two sets of VM types for the baseline algorithms FF&BF/FF and GGA, and for our proposed algorithm EGGA-LSU (*Note: the lower energy consumption values the better*)

Instance	Size	VM Type	FF&BF/FF [15]	GGA [12]	EGGA-LSU
1	500	Real-world	548.0585	466.6287 ± 30	**388.9988 ± 4**
2	1000		1213.2225	957.5432 ± 27	**844.3967 ± 18**
3	1500		1850.3094	1413.6897 ± 12	**1267.1683 ± 33**
4	5000		6733.7918	4481.4496 ± 49	**4011.5749 ± 42**
5	500	Synthetic	576.0420	463.6227 ± 3	**387.8889 ± 5**
6	1000		1209.0082	962.5782 ± 25	**891.8277 ± 29**
7	1500		1972.0720	1386.5216 ± 32	**1285.0675 ± 43**
8	5000		7421.2896	4545.0675 ± 62	**4199.0209 ± 52**

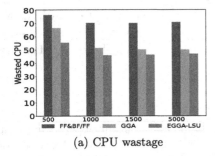

(a) CPU wastage

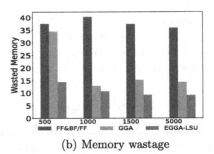

(b) Memory wastage

Fig. 4. CPU and memory wastage in real-world VM types test instances with different numbers of containers (i.e., test instances 1 to 4) for the baseline algorithms FF&BF/FF and GGA, and our proposed algorithm EGGA-LSU (*Note: the lower resource wastage the better*)

baseline algorithm, achieving substantially less energy consumption. EGGA-LSU reduces about 20% energy consumption for test instances 1 and 5, and reduces at least 10% for test instances 4 and 8, when compared with GGA.

The experiment results in Table 1 show that EGGA-LSU outperforms all baseline algorithms for all test instances. To continue with, we also analyze the solutions generated by EGGA-LSU with respect to further quality criteria.

Resource Wastage. Fig. 4 presents the comparison of the baseline algorithms, and our proposed algorithm EGGA-LSU regarding the CPU and memory wastage for four test instances with real-world VM types. As evidenced in this figure, EGGA-LSU significantly reduces the resource wastage.

Number of VM. Fig. 5 presents the number of used VMs for each algorithm. We can see that the number of used VMs was reduced significantly by EGGA-LSU, which can effectively save energy consumption caused by the VMs overhead.

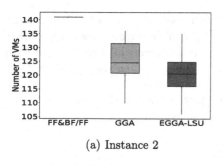
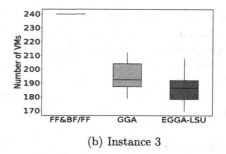

(a) Instance 2   (b) Instance 3

Fig. 5. Average number of VMs (over 30 runs) used by the baseline algorithms FF&BF/FF and GGA, and our proposed algorithm EGGA-LSU (*Note: the lower VM numbers the better*)

Ablation Study. To investigate the effectiveness of proposed new operators for EGGA-LSU, including *Energy-aware Crossover* (EC), *Best-Fit-Decreasing Insert* (BFDI), and *Local Search Unpack* (LSU), we conducted an ablation study. The EGGA-LSU is compared with the original GGA and GGA with newly proposed operators; while GGA with newly proposed operators is compared with the original GGA. Table 2 shows the effectiveness of all of our proposed new operators, where "+" indicates significantly better while "-" means significantly worse performance, and "=" indicates that there are no significantly differences.

Table 2. Ablation experiment for our proposed new operators

Instance	GGA [12]	GGA with EC	GGA with BFDI	GGA with LSU	EGGA-LSU
1	466.6287 ± 30	$464.6300 \pm 28(+)$	$466.5480 \pm 10(=)$	$398.7692 \pm 23(+)$	$388.9988 \pm 4(+)(+)(+)(+)$
2	957.5432 ± 27	$950.7127 \pm 12(+)$	$957.8832 \pm 15(-)$	$909.2499 \pm 20(+)$	$844.3967 \pm 18(+)(+)(+)(+)$
3	1413.6897 ± 12	$1407.8338 \pm 5(+)$	$1407.5753 \pm 14(+)$	$1346.1233 \pm 54(+)$	$1267.1683 \pm 33(+)(+)(+)(+)$
4	4481.4496 ± 49	$4434.0011 \pm 41(+)$	$4402.7946 \pm 20(+)$	$4341.0828 \pm 91(+)$	$4011.5749 \pm 42(+)(+)(+)(+)$

The ablation study confirms that all of our proposed new operators are effective, since they can improve the performance of GGA in some of the test instances. In particular, GGA with the proposed energy-aware crossover, and GGA with the proposed BFDI outperform the basic GGA in all the test instances, while GGA with BFDI outperforms the basic GGA in two test instances. Meanwhile, the EGGA-LSU outperforms the GGA with EC, GGA with BFDI and GGA with LSU.

6 Conclusions

In this paper, we proposed a novel Group Genetic Algorithm to solve the resource allocation problem in container-based clouds with heterogeneous PMs. For EGGA-LSU, we proposed three novel operators, *energy-aware crossover, best-fit-decreasing insert* and *local search-based unpack*. The EGGA-LSU algorithm can explore different PM energy profiles to produce better combinations of containers, VM types and PM types effectively. Meanwhile, EGGA-LSU considers normalized CPU and memory utilization to better select VM and PM for allocation so as to reduce resource wastage. The experiments show that EGGA-LSU significantly outperforms two state-of-the-art algorithms, including a heuristic algorithm FF&BF/FF [15] and a meta-heuristic algorithm GGA [12].

References

1. Abohamama, A.S., Hamouda, E.: A hybrid energy-aware virtual machine placement algorithm for cloud environments. Expert Syst. Appl. **150**, 113306 (2020)
2. Akindele, T., Tan, B., Mei, Y., Ma, H.: Hybrid grouping genetic algorithm for large-scale two-level resource allocation of containers in the cloud. In: Long, G., Yu, X., Wang, S. (eds.) AI 2022. LNCS (LNAI), vol. 13151, pp. 519–530. Springer, Cham (2022). https://doi.org/10.1007/978-3-030-97546-3_42
3. Dayarathna, M., Wen, Y., Fan, R.: Data center energy consumption modeling: a survey. IEEE Commun. Surv. Tutorials **18**(1), 732–794 (2015)
4. Dósa, G., Sgall, J.: First fit bin packing: a tight analysis. In: International Symposium on Theoretical Aspects of Computer Science (STACS) (2013)
5. Gawali, M.B., Shinde, S.K.: Task scheduling and resource allocation in cloud computing using a heuristic approach. J. Cloud Comput. **7**(1), 1–16 (2018). https://doi.org/10.1186/s13677-018-0105-8
6. Kaaouache, M.A., Bouamama, S.: Solving bin packing problem with a hybrid genetic algorithm for VM placement in cloud. Procedia Comput. Sci. **60**, 1061–1069 (2015)
7. Piraghaj, S.F., Dastjerdi, A.V., Calheiros, R.N., Buyya, R.: A framework and algorithm for energy efficient container consolidation in cloud data centers. In: IEEE International Conference on Data Science and Data Intensive Systems, pp. 368–375. IEEE (2015)
8. Ramos-Figueroa, O., Quiroz-Castellanos, M., Mezura-Montes, E., Kharel, R.: Variation operators for grouping genetic algorithms: a review. Swarm Evol. Comput. **60**, 100796 (2021)
9. Saidi, K., Bardou, D.: Task scheduling and VM placement to resource allocation in cloud computing: challenges and opportunities. Cluster Comput. 1–19 (2023)
10. Sengupta, J., Singh, P., Suri, P.K.: Energy aware next fit allocation approach for placement of VMs in cloud computing environment. In: Arai, K., Kapoor, S., Bhatia, R. (eds.) FICC 2020. AISC, vol. 1130, pp. 436–453. Springer, Cham (2020). https://doi.org/10.1007/978-3-030-39442-4_33
11. Tan, B., Ma, H., Mei, Y.: Novel genetic algorithm with dual chromosome representation for resource allocation in container-based clouds. In: IEEE International Conference on Cloud Computing (CLOUD), pp. 452–456. IEEE (2019)
12. Tan, B., Ma, H., Mei, Y.: A group genetic algorithm for resource allocation in container-based clouds. In: Paquete, L., Zarges, C. (eds.) EvoCOP 2020. LNCS, vol. 12102, pp. 180–196. Springer, Cham (2020). https://doi.org/10.1007/978-3-030-43680-3_12
13. Tan, B., Ma, H., Mei, Y.: A NSGA-II-based approach for multi-objective microservice allocation in container-based clouds. In: 2020 20th IEEE/ACM International Symposium on Cluster, Cloud and Internet Computing (CCGRID), pp. 282–289. IEEE (2020)

14. Wang, C., Ma, H., Chen, G., Huang, V., Yu, Y., Christopher, K.: Energy-aware dynamic resource allocation in container-based clouds via cooperative coevolution genetic programming. In: Correia, J., Smith, S., Qaddoura, R. (eds.) EvoApplications 2023. LNCS, vol. 13989, pp. 539–555. Springer, Cham (2023). https://doi.org/10.1007/978-3-031-30229-9_35
15. Zhang, R., Zhong, A., Dong, B., Tian, F., Li, R.: Container-VM-PM architecture: a novel architecture for docker container placement. In: Luo, M., Zhang, L.-J. (eds.) CLOUD 2018. LNCS, vol. 10967, pp. 128–140. Springer, Cham (2018). https://doi.org/10.1007/978-3-319-94295-7_9

Genetic Programming with Adaptive Reference Points for Pareto Local Search in Many-Objective Job Shop Scheduling

Atiya Masood[1]([⊠])([iD]), Gang Chen[2], Yi Mei[2], Harith Al-Sahaf[2], and Mengjie Zhang[2]

[1] Iqra University, Karachi, Pakistan
atiya.masood@iqra.edu.pk
[2] Victoria University of Wellington, Wellington, New Zealand
{aaron.chen,yi.mei,harith.al-sahaf,mengjie.zhang}@ecs.vuw.ac.nz

Abstract. Genetic Programming (GP) is a well-known technique for generating dispatching rules for scheduling problems. A simple and cost-effective local search technique for many-objective combinatorial optimization problems is Pareto Local Search (PLS). With some success, researchers have looked at how PLS can be applied to many-objective evolutionary algorithms (MOEAs). Many MOEAs'performance can be considerably enhanced by combining local and global searches. Despite initial success, PLS's practical application in GP still needs to be improved. The PLS is employed in the literature that uniformly distributes reference points. It is essential to maintain solution diversity when using evolutionary algorithms to solve many-objective optimization problems with disconnected and irregular Pareto-fronts. This study aims to improve the quality of developed dispatching rules for many-objective Job Shop Scheduling (JSS) by combining GP with PLS and adaptive reference point approaches. In this research, we propose a new GP-PLS-II-A (adaptive) method that verifies the hypothesis that PLS's fitness-based solution selection mechanism can increase the probability of finding extremely effective dispatching rules for many-objective JSS. The effectiveness of our new algorithm is assessed by comparing GP-PLS-II-A to the many-objective JSS algorithms that used PLS. The experimental findings show that the proposed method outperforms the four compared algorithms because of the effective use of local search strategies with adaptive reference points.

Keywords: many-objective optimization · adaptive reference points · genetic programming · Pareto local search · job shop scheduling

1 Introduction

Job shop scheduling (JSS) [20] is a significant combinatorial optimization problem in the scheduling and artificial intelligence fields. It captures many practical

T. Liu et al. (Eds.): AI 2023, LNAI 14472, pp. 466–478, 2024.
https://doi.org/10.1007/978-981-99-8391-9_37

issues in real-world scheduling problems, such as cloud computing. JSS is to process a group of tasks or jobs by a set of machines. The goal of a JSS problem is to design an optimal schedule to achieve the objectives, such as minimizing the makespan, flowtime, and tardiness.

Additionally, industries are interested in solving problems with many objectives [4,10]. It is widely evidenced in the literature that JSS, by nature, presents several potentially conflicting objectives, including maximal flowtime, mean flowtime, and mean tardiness [16]. Finding a set of non-dominating solutions is known as the Pareto front. Many objective optimization algorithms aim to approximate the Pareto front effectively.

JSS has been proven to be NP-hard [1], and an exact optimization method can hardly tackle JSS. Dispatching rule is one of the popular scheduling heuristics to solve such NP-hard problems efficiently. Conceptually, a dispatching rule can be represented as a priority function that calculates the priority value of each waiting operation in a queue, and the operation with the highest priority value at the decision is selected to be processed next by the idle machine.

There are two main issues in designing a dispatching rule manually for JSS. First, dispatching rules are time-consuming to design manually, especially for optimizing multiple potentially conflicting objectives that are frequently demanded in a manufacturing environment. Second, the behavior of a dispatching rule can vary from one scenario to another. For example, the rule good at minimizing the mean flowtime may perform poorly in minimizing the maximal tardiness. Therefore, researchers have proposed automated design approaches that solve the issues related to manually designing dispatching rules [19]. Genetic Programming (GP)-based hyper-heuristic (GP-HH) has attracted many researchers because it has been a promising approach for generating rules automatically [17,19]. It was shown in [17] that GP-HH can evolve much more effective rules than manually designed rules on many JSS problems. The ability of GP-HH to be integrated with many-objective optimization methodologies and automatically generate dispatching rules to optimize several conflicting objectives simultaneously is a further benefit. However, GP is typically ineffective in fine-tuning a solution locally [19]. This limitation is often mitigated by introducing local search with GP [19].

For many-objective combinatorial optimization problems like JSS [2], Pareto local search (PLS) is an effective local search technique. With some success, researchers have looked into using PLS with MOEAs [2]. In fact, many MOEAs' performance can be improved considerably by combining local and global search [2,7]. Memetic algorithms are a common name for such a hybrid algorithm. Despite PLS's initial success, its application to GP for many-objective JSS still needs to be improved. To the best of our knowledge, just two studies employ PLS in GP-HH for many-objective JSS. PLS solutions for the study in [13] were chosen randomly. Although random sampling is simple, it might not be able to identify the solutions with the greatest potential to be improved by PLS. The other study [14] divided the entire objective space into several subregions using uniformly distributed reference points. However, JSS problems commonly have irregular Pareto-front [18]. As a result, many uniformly distributed reference points may be found in areas with no Pareto-optimal solutions. They are never associated with any solutions on the evolving Pareto-front. In this study, we used adaptive reference points to decompose the objective space, establishing the appropriate search direction for each solution. [9] shows that the distribution of reference points can be closely matched with the distribution of possible solutions in order to generate a high-quality dispatching rule and minimize the number of useless points.

Our goal in this research is to propose an effective selection technique for choosing the PLS's initial solutions. In order to accomplish this, we proposed a new fitness-based selection mechanism that used the decomposition-based method, which balances convergence and diversity requirements. Furthermore, to enhance the solutions' diversity, the distribution of reference points is expected to match closely with the distribution of the candidate solutions; therefore, this study used adaptive reference points.

The objectives of this study are to: 1) develop a new fitness-based selection for GP-PLS with adaptive reference points called GP-PLS-II-Adaptive, 2) investigate whether the inclusion of adaptive reference points in GP-PLS-A can increase the likelihood of finding highly effective dispatching rules for many-objective JSS, and 3) assess the performance of GP-PLS-II-A by contrasting it with four other many-objective algorithms.

The remainder of this paper is structured as follows. The background is provided in Sect. 2, along with a description of the JSS problem and relevant research. The proposed algorithm is described in Section 3 (GP-PLS-II-A). The experimental design is presented in Sect. 4. Section 5 presents the results and discussions. Finally, the conclusions and further research are presented in Sect. 6.

2   Research Background

In this section, the JSS problem will be described first. Then we will discuss some related works.

2.1   Problem Description

Job Shop Scheduling. In a JSS problem, there are a set of N jobs and a set of M machines. Each job j_i has a sequence of K operations to be performed, i.e., $\{o_i^1, o_i^2, \ldots, o_i^k\}$. Each operation o_i^k has a fixed processing time $t_i^k > 0$ and has to be processed on a specific machine $m_i, 1 \leq i \leq M$. Any solution to a JSS problem must satisfies a few requirements, as described below.

- Consider $m \in M$ machines in the job shop are *persistently available* to process new and at most one operation whenever the machine m is idle.
- Every operation o_i^k is *non-preemptive*. This means that once an operation on a machine begins processing, other operations cannot interrupt it during this time.

2.2 Related Works

GP is considered the most popular method for discovering and constructing dispatching rules for scheduling problems [17,19]. GP shows its effectiveness not only in single-objective JSS problems but can also evolve useful rules for multi-objective and many-objective JSS problems [11,16]. Previous studies have shown that in later generations, the exploitation ability of GP is also limited by the population size, but applying local search can improve its exploitation ability [13,19]. PLS is very effective for tackling *NP-hard* multi-objective JSS problems. In particular, [7] showed that suitable candidates for local search should be carefully selected based on specific scalarization mechanisms. [5] applied PLS and improved the overall quality of the evolved Pareto-front. However, only two research works [13] has been studied PLS in GP-HH for many-objective JSS. The study in [13] used a random selection mechanism for PLS solutions. Furthermore, one of the research [14] used the fitness selection mechanisms with uniformly distributed reference points. However, the efficiency of many-objective algorithms has been proven to be intensely dependent on the curvature of the Pareto-front [8]. As a result, when the Pareto-front has irregular shapes, many algorithms that use uniformly distributed reference points cannot perform well on the many-objective optimization problems (MaOPs). JSS problems often have irregular and discontinuous Pareto-front [18]. To address this limitation, we investigate the effectiveness of the adaptive reference points and PLS in GP-HH in our current study. The investigation in this paper is expected to inspire many future studies on PLS in GP-HH for many-objective JSS.

3   Adaptive Reference Method for Pareto Local Search

3.1 Representation of Rules

Consider the popular manually-designed 2PT+WINQ+NPT rule [6], which represents as a GP tree in Fig. 1. In the GP tree representation, terminals in the tree are in Table 1 for a summary of all terminal types used in this paper, and the functions are $\{+, *, -, \div, min, max, if\}$.

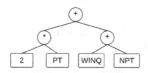

Fig. 1. The GP tree representation of the 2PT+WINQ+NPT rule.

3.2   General Framework of GP-PLS-II-A

GP-PLS-II-A (the second work that used reference points for GP-PLS; therefore, we named it GP-PLS-II). the proposed algorithm contains the following steps.

Selection of Diverse Set of Solutions. In this study, the decomposition-based approach is used to divide the objective space into a number of independent sub-spaces according to a set of reference points. This decomposition of the objective space determines the appropriate search direction of each solution. This component has been demonstrated to be very effective in simultaneously locating multiple local optimal solutions in the search space. We used the following steps to select the diverse set of solutions.

1. Generate the offspring population Q_g using the crossover, mutation and reproduction of GP
2. Combine the parent (P_g) and offspring (Q_g) population and obtain the combined population (R_g).
3. Use a decomposition-based approach to split the objective space into a number of independent sub-regions according to a set of adaptive reference points.
4. For generating the adaptive reference points, we select the MARP-NSGA-III algorithm [15]. In the literature, MARP-NSGA-III outperformed other adaptive reference point approaches (A-NSGA-III, NSGA-III-DRA).
5. This decomposition of the objective space determines the appropriate search direction of each solution. Solutions associated with similar reference points have an identical search direction.
6. In our algorithm, the vector angle reflects the similarity of search directions between two individuals, and later, the angle information between two individuals in the objective space is used to maintain the diversity. The acute angle can be calculated as follows:

$$cos\theta_{i,j} = \frac{r_{i,j} \cdot w_{i,j}}{|r_{i,j}|}. \tag{1}$$

where $r_{i,j}$ is an individual from the combined population of the size 2N, and $w_{i,j}$ is a reference point. If an individual $r_{i,j}$ and reference point $w_{i,j}$ have a minimal acute angle among all the reference points, $r_{i,j}$ becomes the member of the subpopulation $R_{g,k}$.

Fitness-Based Selection. In GP-PLS-II-A, we used the following steps for the selection of K initial solutions.

1. Once the population R_g is partitioned into $2N$ subpopulations, associated solutions are selected from each subspace using Eq. (2). So, K_r best solutions are selected according to their fitness values from each subspace.

$$(K_r) = \left(\frac{number\ of\ solutions\ from\ each\ subgroup}{Total\ population} \right) \times N, \tag{2}$$

where $r = 1, 2 \ldots, 2N$.

The selection criteria based on the fitness value (FV) are designed based on two sub-criteria: (1) the convergence criterion ($C(r)$) (d_1 in Fig. 2), and (2) the diversity criterion ($D(r)$) (d_2 in Fig. 2).

The ($C(r) = d_1$) is represented by the distance from the solution ($r_{i,j}$) to the ideal point (Z^*) ,i.e., $\| r_{i,j} - Z^* \|$. Similarly, ($D(r) = d_2$) is represented by the inverse of the acute angle between ($r_{i,j}$) and $w_{i,j}$, i.e., $\theta_{i,j}$. In order to balance between the $C(r)$ and the $D(r)$, the total fitness value (FV) of each individual can be formulated as a scalarization function:

$$FV = d_1 + \frac{g}{g_{max}} \times \frac{d_2}{\theta_m}. \tag{3}$$

where θ_m is used to normalize $\theta_{i,j}$ in the Eq. (3). The proposed algorithm adopted an angle normalization procedure [3] and g is the current generation and g_{max} is the maximum number of generations. $\frac{g}{g_{max}}$ is a penalty parameter which can better regulate the proportion of convergence and diversity information. During the exploration stage, it is a good idea to exert significant selection pressure on convergence and move the population toward the Pareto-front of the search process. Therefore we add the penalty parameter $\frac{g}{g_{max}}$ in Eq. (3), which can effectively control the ratio of convergence and diversity information In the early stage, FV determines the convergence value (d_1) because $g << g_{max}$, therefore the diversity criterion (d_2) ≈ 0. However, when g approaches g_{max}, the penalty parameter gradually increases to emphasize the importance of the diversity criterion $\theta_{i,j}$.

2. After getting the fitness values of each individual, then N solutions are selected and saved in the *archive*. From the *archive*, K solutions are selected based on their fitness values and saved in P_k. These solutions are selected for neighborhood exploration.

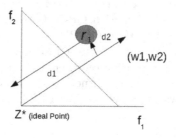

Fig. 2. Distance measurement is used in the context of minimization with respect to a reference direction.

3.3   Neighborhood Solution

In GP-PL-II-A, a neighboring rule of any given rule p is obtained using the restricted mutation operator [13]. When the restricted mutation is applied to rule p, we randomly select a node in p whose corresponding sub-tree has a depth of 2. A randomly generated depth-2 sub-tree then replaces the selected node and its sub-tree. With the help of this restricted mutation, GP-PL-II-A can effectively prevent a new neighboring rule discovered during the local search process from being significantly different from the original rule.

3.4   Exploration and Comparison

In the local search, the neighborhood exploration strategies govern the size of the neighborhood for exploring neighboring solutions. GP-PLS-II-A used a partial exploration of the neighborhood since the entire neighborhood is infinitely large. Moreover, the partial exploration strategy requires less computation than the best improvement approach (exploring the neighborhood entirely), particularly for problems with a large number of features. Specifically, we select a neighbor based on the dominance relation from the neighborhood repetitively until the maximum number of steps ($step_{max}$) is reached and return the best neighbor sampled so far.

For the comparison of rules, we consider the replacement strategy [2], which is based on the dominance relation while comparing two rules p' and p_{new} during PLS. Three outcomes are possible when comparing two rules $pnew$ and p': If p_{new} dominates p', choose p_{new}; if p' dominates p_{new} select p' ; and if p_{new} and p' are incomparable, choose one at random. It is obvious that replacing the current rule p' with any neighborhood rule p_{new} that dominates it in the PLS archive (P_k) imposes selection pressure on the P_k and pushes it towards the Pareto-front. When the new population P_{best} and $P_{N/best}$ are combined, a new population P_{g+1} is produced.

4   Experiment Design

The experimental studies will be undertaken in this section and compare the proposed algorithm GP-PLS-II-A with GP-NSGA-III, GP-PLS-II-U, GP-PLS-I-s [13], and GP-PLS-I-r [13]. Here GP-PLS-s refers to the variation of GP-PLS where the scalarization approach is used for selection in [13]. On the other hand, GP-PLS-r represents the variation where the replacement strategy is used for selection in [13]. In addition, the fitness selection technique with uniformly spaced reference points was employed by GP-PLS-II-U [14].

4.1   Dataset for JSS

Taillard static (TA) job shop is a widely used static JSS benchmark set [21] and is selected for our experiments. TA consists of 80 JSS problem instances. We group the problem instances with the same number of jobs and machines into the same group. As a result, TA is divided into 8 groups (denoted as TA-1, ..., TA-8). The number of jobs varies from 15 to 100, and the number of machines varies from 15 to 20 across these groups. In the experiments, the total 80 instances were further divided into the *training set* and the *test set*, where each set consists of 40 instances.

4.2   Parameter Settings

The terminals are summarized in Table 1. Based on previous studies [16], each competing algorithm's crossover, mutation, and reproduction rates are set to 85%, 10%, and 5%, respectively. The tournament selection method is used to choose the parents in each generation, with a tournament size of seven. The population size for GP-PLS-II-A, GP-NSGA-III, and GP-PLS-II-U is set to 1000. The maximum number of generations is set to 50. To compare the HV and IGD results obtained by each algorithm, the Wilcoxon rank-sum test [22] with a significance level of 0.05 has been applied independently.

Table 1. Terminal set of GP for JSS.

Attribute	Notation	Attribute	Notation	Attribute	Notation
Processing time of the operation	PT	Ready time of the operation	ORT	Flow due date	FDD
Inverse processing time of the operation	IPT	Ready time of the next machine	NMRT	Work Remaining	WKR
Number of operations in the next queue	NOINQ	Work in the next queue	WINQ	Due Date	DD
Processing time of the next operation	NOPT	Number of operation remaining	NOR	Weight	W
Number of operations in the queue	NOIQ	Ready time of the next machine	MRT	Work in the queue	WIQ

Table 2. The mean and standard deviation over the average HV and IGD values on training and test instances of the compared methods in the four-objective experiment. The significantly better results are shown in bold.

	Training		Test	
	HV ($\bar{x} \pm s$)	IGD ($\bar{x} \pm s$)	HV ($\bar{x} \pm s$)	IGD ($\bar{x} \pm s$)
GP-NSGA-III	0.65550±0.0121	0.00132±0.00019	0.50322±0.01423	0.00180±0.00029
GP-PLS-I-s	0.66820±0.0222	0.00130±0.00016	0.51422±0.03223	0.00178±0.00010
GP-PLS-I-r	0.68850±0.0221	0.00128±0.00015	0.52522±0.02423	0.00177±0.00020
GP-PLS-II-U	0.70513±0.0130	0.00125±0.00013	0.5564±0.02433	0.00166±0.00017
GP-PLS-II-A	**0.7113±0.0010**	**0.00120±0.00005**	**0.5994±0.02117**	**0.00160±0.00016**

GP-PLS-II-A has two additional parameters, the size of the archive (K) and the maximum number of local search steps ($step_{max}$) are set to 250 and 4 respectively as same as the GP-PLS-II-U [13]. The parameter settings for GP-PLS-I-s and GP-PLS-I-r are taken from [13]. In our experiment, we aim to minimize four objectives, i.e., the mean flowtime (Obj1), maximal flowtime (Obj2), mean weighted tardiness (Obj3), and maximal weighted tardiness (Obj4). Existing work [16] showed that the four objectives are mutually conflicting.

4.3   Performance Measures

Several performance measures are proposed in the literature for evaluating many-objective optimization algorithms. However, the two main performance measures *Hyper-Volume* (HV) [23] and the *Inverted Generational Distance* (IGD) [23] are used in this study. Theoretically, a set of tradeoff rules should have a *higher* HV values and a *smaller* IGD values.

5   Results and Discussions

For each algorithm in the experiment, 30 GP runs were conducted to obtain 30 sets of dispatching rules. Then, the rules were tested on the 40 test instances.

5.1   Performance of Obtained Dispatching Rules

Table 2 reveals that the mean and standard deviation of the training and test performance in terms of HV and IGD of the rules obtained by GP-NSGA-III, GP-PLS-I-r, GP-PLS-I-r, GP-PLS-II-U, and GP-PLS-II-A. The Wilcoxon rank-sum test with a significance level of 0.05 is applied to the HV and IGD of the Pareto front evolved by the three compared algorithms. Table 2 shows that GP-PLS-II-A performs significantly better than GP-NSGA-III GP-PLS-I-r, GP-PLS-I-r, and GP-PLS-II-U in terms of IGD and HV. This is because GP-PLS-II-A selects rules from each subregion based on their convergence and diversity, which can evolve a well-distributed set of Pareto-optimal solutions because of using adaptive reference points. Further, the fitness-based solution selection from each search direction explores promising rules from evolved Pareto front.

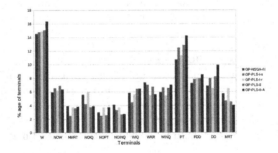

Fig. 3. Terminals Frequency of GP-NSGA-III, GP-PLS-I-s, GP-PLS-I-r, GP-PLS-II-U, and GP-PLS-II-A.

In fact, GP-PLS-II-A and GP-PLS-II-U algorithms performed significantly better than GP-PLS-I-r and GP-PLS-I-s. The results indicate the effectiveness of selecting solutions based on the fitness value with adaptive reference points for neighborhood exploration. The results also demonstrate that GP-PLS-II-A and GP-PLS-II-U, GP-PLS-I-r, and GP-PLS-I-s algorithms performed significantly better overall than GP-NSGA-III. This is because the local search enhances the exploitation ability of the algorithms integrated with PLS.

5.2   Further Analysis

Analysis of Dispatching Rules. The percentage of terminals in evolved rules from each algorithm is displayed as a bar chart in Fig. 3. We can further examine the relevant and irrelevant terminals based on the frequency of terminals. For example, more than 10% of the algorithms' developed rules have W and PT terminals, as shown in Fig 3. The research [6,12] shows that the three most important terminals for maximizing the flowtime objective are PT, WINQ, and WKR. On the other hand, the relevant terminals for maximizing tardiness objectives are WINQ, NOINQ, NOPT, W, PT, MRT, and DD. Therefore, PT, DD, and W are the most relevant terminals for minimizing tardiness objectives. The number of occurrences of W, WINQ, PT, DD, MRT, NOPT, and FDD terminals is higher in the local search algorithms (GP-PLS-I-s, GP-PLS-I-r, GP-PLS-II-U, and GP-PLS-II-A) than in the GP-NSGA-III method, as can be seen in Fig. 3. This is because all the GP-PLS algorithms improved the exploitation ability and evolved noticeably better rules than the GP-NSGA-III. As a result, these algorithms choose well-optimized rules. However, more relevant terminals to appear in the GP-PLS-II-A algorithm. This can show the usefulness of adaptive reference points with PLS.

Convergence Curves. Figure 4(a) and Fig. 4(b) reveal that GP-PLS-II-A has better convergence curves in terms of both HV and IGD than other compared algorithms. These also show that both algorithms of GP-PLS-II-U and GP-PLS-II-A achieved better performance than the GP-PLS-I-r, GP-PLS-I-s, and GP-NSGA-III in terms of HV and IGD. These results reveal that selecting the solutions based on fitness value (convergence and diversity) can improve the GP-PLS algorithm's overall performance.

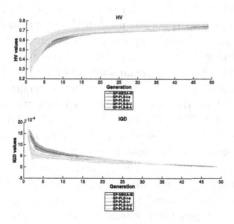

Fig. 4. The curves of the average number of (a) HV value, and (b) IGD value of the non-dominated solutions on the training set during the 30 independent GP runs.

Parallel Coordinate Plots. The parallel coordinate plots in Fig. 5(a) to Fig. 5(e) depict the non-dominated set of dispatching rules obtained by all the competing algorithms. As shown in Fig. 5, the rules evolved by GP-PLS-II-A are more diversified to cover a much wider range of the objective space than other compared algorithms. We can also observe that GP-PLS-II-U and GP-PLS-I-r are well-diversified compared to GP-NSGA-III and GP-PLS-I-s. Figure 5 also reveals that GP-PLS-II-A obtained better coverage for the third and fourth objectives (i.e., mean weighted tardiness and maximum weighted tardiness) than GP-PLS-II-U. In this study, our experiment results reveal that selecting solutions based on the fitness value (convergence and diversity) with adaptive reference points for neighborhood exploration will improve the solutions' quality. GP-PLS-II-A performed significantly better than GP-PLS-II-U and has widely distributed optimal solutions than GP-PLS-II-U, GP-PLS-I-r, GP-PLS-I-s, and GP-NSGA-III.

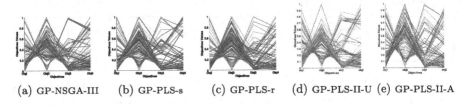

(a) GP-NSGA-III   (b) GP-PLS-s   (c) GP-PLS-r   (d) GP-PLS-II-U   (e) GP-PLS-II-A

Fig. 5. Parallel coordinates of non-dominated front obtained by each algorithm.

6   Conclusions

In this study, we successfully proposed a new algorithm with a new fitness-based selection mechanism with adaptive reference points and verified its effectiveness. The decomposition-based method was used in the proposed GP-PLS-II-A. This method divides the entire objective space adaptively into a number of small subspaces using a set of adaptive reference points. Then a fitness-based selection criterion was proposed for selecting initial solutions for neighborhood exploration - the selection criteria based on convergence and diversity. Using the Taillard static job shop benchmark dataset, extensive experiments have been conducted to verify the proposed GP-PLS-II-A's effectiveness. The adaptive distributed reference points enhance solution diversity during evolution and impact algorithm performance for problems with irregular Pareto-fronts. The experiment's findings demonstrated that GP-PLS-II-A outperformed the other evaluated algorithms in HV and IGD. In addition, the diverse preferences for using terminals have been revealed through further analysis of the proposed algorithm.

By creating an intelligent local search operator to direct exploitation based on recently assessed rules and adaptive selection techniques, we will improve the performance of our proposed PLS in future investigations. However, more research is needed to fully understand the potential of local search approaches on many-objective GP-HH.

References

1. Błażewicz, J., Domschke, W., Pesch, E.: The job shop scheduling problem: conventional and new solution techniques. Eur. J. Oper. Res. **93**(1), 1–33 (1996)
2. Chen, B., Zeng, W., Lin, Y., Zhang, D.: A new local search-based multiobjective optimization algorithm. IEEE Trans. Evol. Comput. **19**(1), 50–73 (2015)
3. Cheng, R., Jin, Y., Olhofer, M., Sendhoff, B.: A reference vector guided evolutionary algorithm for many-objective optimization. IEEE Trans. Evol. Comput. **20**(5), 773–791 (2016)
4. Deb, K., Jain, H.: An evolutionary many-objective optimization algorithm using reference-point-based nondominated sorting approach, part I: solving problems with box constraints. IEEE Trans. Evol. Comput. **18**(4), 577–601 (2014)
5. Dubois-Lacoste, J., López-Ibáñez, M., Stützle, T.: Anytime pareto local search. Eur. J. Oper. Res. **243**(2), 369–385 (2015)
6. Holthaus, O., Rajendran, C.: Efficient jobshop dispatching rules: further developments. Prod. Plann. Control **11**(2), 171–178 (2000)
7. Ishibuchi, H., Murata, T.: A multi-objective genetic local search algorithm and its application to flowshop scheduling. IEEE Trans. Syst. Man Cybern. Part C Appl. Rev. **28**(3), 392–403 (1998)
8. Ishibuchi, H., Setoguchi, Y., Masuda, H., Nojima, Y.: Performance of decomposition-based many-objective algorithms strongly depends on Pareto front shapes. IEEE Trans. Evol. Comput. **21**(2), 169–190 (2016)
9. Jain, H., Deb, K.: An improved adaptive approach for elitist nondominated sorting genetic algorithm for many-objective optimization. In: Purshouse, R.C., Fleming, P.J., Fonseca, C.M., Greco, S., Shaw, J. (eds.) EMO 2013. LNCS, vol. 7811, pp. 307–321. Springer, Heidelberg (2013). https://doi.org/10.1007/978-3-642-37140-0_25
10. Jain, H., Deb, K.: An evolutionary many-objective optimization algorithm using reference-point based nondominated sorting approach, Part II: handling constraints and extending to an adaptive approach. IEEE Trans. Evol. Comput. **18**(4), 602–622 (2014)
11. Koza, J.R.: A genetic approach to the truck backer upper problem and the intertwined spiral problem. In: Proceedings of IJCNN International Joint Conference on Neural Networks, vol. IV, pp. 310–318. IEEE Press (1992)
12. Lee, Y.H., Bhaskaran, K., Pinedo, M.: A heuristic to minimize the total weighted tardiness with sequence-dependent setups. IIE Trans. **29**(1), 45–52 (1997)
13. Masood, A., Chen, G., Mei, Y., Al-Sahaf, H., Zhang, M.: Genetic programming with pareto local search for many-objective job shop scheduling. In: Liu, J., Bailey, J. (eds.) AI 2019. LNCS (LNAI), vol. 11919, pp. 536–548. Springer, Cham (2019). https://doi.org/10.1007/978-3-030-35288-2_43
14. Masood, A., Chen, G., Mei, Y., Al-Sahaf, H., Zhang, M.: A fitness-based selection method for pareto local search for many-objective job shop scheduling. In: 2020 IEEE Congress on Evolutionary Computation (CEC), pp. 1–8. IEEE (2020)
15. Masood, A., Chen, G., Mei, Y., Al-Sahaf, H., Zhang, M.: Genetic programming hyper-heuristic with gaussian process-based reference point adaption for many-objective job shop scheduling. In: 2022 IEEE Congress on Evolutionary Computation (CEC), pp. 1–8. IEEE (2022)
16. Masood, A., Mei, Y., Chen, G., Zhang, M.: Many-objective genetic programming for job-shop scheduling. In: IEEE WCCI 2016 Conference Proceedings, IEEE (2016)

17. Nguyen, S.: Automatic Design of Dispatching Rules for Job Shop Scheduling with Genetic Programming. Ph.D. thesis (2013)
18. Nguyen, S., Mei, Y., Ma, H., Chen, A., Zhang, M.: Evolutionary scheduling and combinatorial optimisation: applications, challenges, and future directions. In: 2016 IEEE Congress on Evolutionary Computation (CEC), pp. 3053–3060. IEEE (2016)
19. Nguyen, S., Zhang, M., Johnston, M., Tan, K.C.: Automatic programming via iterated local search for dynamic job shop scheduling. IEEE Trans. Cybern. **45**(1), 1–14 (2015)
20. Pinedo, M.L.: Scheduling: Theory, Algorithms, and Systems. Springer, Berlin (2012)
21. Taillard, E.: Benchmarks for basic scheduling problems. Eur. J. Oper. Res. **64**(2), 278–285 (1993)
22. Wilcoxon, F.: Individual comparisons by ranking methods. In: Kotz, S., Johnson, N.L. (eds.) Breakthroughs in Statistics. Springer Series in Statistics. Springer, New York (1992). https://doi.org/10.1007/978-1-4612-4380-9_16
23. Zhang, Q., Zhou, A., Zhao, S., Suganthan, P.N., Liu, W., Tiwari, S.: Multiobjective optimization test instances for the CEC 2009 special session and competition, pp. 1–30 (2008)

A Study of Fitness Gains in Evolving Finite State Machines

Gábor Zoltai[1]([✉])[iD], Yue Xie[2][iD], and Frank Neumann[1][iD]

[1] Optimisation and Logistics, School of Computer and Mathematical Sciences,
University of Adelaide, Adelaide, Australia
{gabor.zoltai,frank.neumann}@adelaide.edu.au
[2] Bio-Inspired Robotics Laboratory, Department of Engineering,
University of Cambridge, Cambridge, UK
yx388@cam.ac.uk

Abstract. Among the wide variety of evolutionary computing models, Finite State Machines (FSMs) have several attractions for fundamental research. They are easy to understand in concept and can be visualised clearly in simple cases. They have a ready fitness criterion through their relationship with Regular Languages. They have also been shown to be tractably evolvable, even up to exhibiting evidence of open-ended evolution in specific scenarios. In addition to theoretical attraction, they also have industrial applications, as a paradigm of both automated and user-initiated control. Improving the understanding of the factors affecting FSM evolution has relevance to both computer science and practical optimisation of control. We investigate an evolutionary scenario of FSMs adapting to recognise one of a family of Regular Languages by categorising positive and negative samples, while also being under a counteracting selection pressure that favours fewer states. The results appear to indicate that longer strings provided as samples reduce the speed of fitness gain, when fitness is measured against a fixed number of sample strings. We draw the inference that additional information from longer strings is not sufficient to compensate for sparser coverage of the combinatorial space of positive and negative sample strings.

Keywords: evolutionary computation · finite state machines · regular languages

1 Introduction

In all areas of computing inspired by biological evolution, the relationship between the parameters of selection and increases of fitness measures over time is an important factor. This applies to genetic programming, evolutionary computation, evolutionary artificial intelligence, or artificial life. In genetic programming, this relationship has been studied with respect to the speed at which a desired solution can be derived [1]. In artificial life, this relationship is connected

© The Author(s), under exclusive license to Springer Nature Singapore Pte Ltd. 2024
T. Liu et al. (Eds.): AI 2023, LNAI 14472, pp. 479–490, 2024.
https://doi.org/10.1007/978-981-99-8391-9_38

to the arising of complexity (being an enabler of fitness) [2]. For constrained opti-misation problems, the rate of change of selection parameters over time has been seen to enhance the performance of genetic algorithms [3]. Diversity, which can be seen as another aspect of selection (as a result of the severity of selection) has been extensively studied for its effects on fitness gains [4]. We investigate an example of the relationship between selection parameters and a fitness measure's increase over generations.

1.1 Background

Understanding the relationship between selection function and adaptive perfor-mance benefits from simplified models, such as the constructs of automata the-ory. For the levels of automata from Finite State Machines (FSMs) up to deter-ministic Turing Machines, successful evolution to recognise formal languages has been reported [5].

Finite State Machines (FSMs, also known as deterministic finite automata or DFAs) are one of the simplest such classes of entities. They have several advantages as a model of study:

- They are simple in concept;
- their structure can be communicated using clear diagrams (for low numbers of states);
- their complexity can be measured unambiguously; and
- measures of how well an FSM recognises a regular language provide readily comprehensible fitness metrics.

For these reasons, FSMs have been a viable research tool in constructing simplified analogies reflecting aspects of biological evolution. For example, Rasek et al. [6] have proposed a definition of FSM-species and studied their formation. In another study, Moran and Pollack [7] have created simple ecosystem models of lineages of FSMs interacting in competitive and cooperative modes. They found that some results exhibit indications of open-endedness, i.e. the absence of an apparent upper bound to evolving complexity.

However, seemingly unbounded increases in complexity of evolving computa-tional entities is the exception, rather than the rule. The review of the literature by Packard et al. [8] states the realisation that evolutionary simulations and genetic algorithms tend to approach plateaus of complexity. In other words, the complexity of evolved solutions to a problem posed tends to level out as gener-ations succeed each other. Most commonly, a state is reached where few novel solutions of higher complexity are generated by such algorithms.

1.2 Our Contribution

On the one hand, the literature documents plateaus of evolution under fixed selection parameters. On the other, there are examples of (at least an impression of) open-endedness with more complex regimes such as co-evolution of three or more interacting lineages. This suggests a need for better understanding of the influence of the selection regime on the evolution of FSMs, especially in the case of several simultaneous selection pressures.

In order to investigate this, we constructed an experimental setup using two simple selection pressures acting in opposite directions:

- maximise the correct recognition of strings in or out of a language, and
- minimise the number of states in the FSMs.

We use this setup to study the effect of an aspect of selection on fitness gain.

A good metric for the complexity of regular languages is their "state complexity" [9]. Unlike other proposed measures (see [10]), state complexity relates directly to the resources needed for recognition. The family of languages used in this study has one language for each state complexity value. This provides control over the selection pressure promoting the addition of states over generations.

The remainder of the paper is organised as follows: Sect. 2 provides definitions for the key concepts we have combined in our study, and the way these experiments have applied these ideas. Section 3 presents the algorithms for creation of sample sets, managing the process of evolution, and the mutation operator. Section 4 describes our experiments and their results. Section 5 includes discussion and interpretation of the results. Finally, Sect. 6 presents our conclusions and proposes future work.

2   Preliminaries

We summarise the elements we have drawn on in the experimental configuration: FSMs, the *universal witness* family of regular languages, and the way we have combined conflicting selection pressures.

2.1   Finite State Machines

A Finite State Machine is a 5-tuple $(Q, \Sigma, \delta, q_0, F)$, where

- Q is a set of states;
- Σ is a set of symbols used as an input alphabet;
- $\delta : Q \times \Sigma \to Q$ is a transition function from one state to another depending on the next input symbol;
- q_0 is the starting state; and
- $F \subseteq Q$ is the subset of accepting states

The transition function δ can be extended to describe what state the machine transitions to if given each symbol of a string in succession. Using ϵ for the empty string, and cs to denote the symbol c followed by the trailing sub-string s, we can define the extended transition function $\hat{\delta} : Q \times \Sigma^* \to Q$:

$$\hat{\delta}(q, \epsilon) = q \tag{1}$$

$$\hat{\delta}(q, cs) = \hat{\delta}(\delta(q, c), s) \tag{2}$$

An FSM "accepts" a string s if the extended transition function maps it from the starting state to an accepting state, i.e. the following is true:

$$\hat{\delta}(q_0, s) \in F \tag{3}$$

The set of strings accepted by an FSM is termed its language; the machine is said to "recognise" the language. For a machine $\mathcal{M} = (Q, \Sigma, \delta, q_0, F)$, we define its language $L_{\mathcal{M}}$ as:

$$L_{\mathcal{M}} = \{s \in \Sigma^* \mid \hat{\delta}(q_0, s) \in F\} \tag{4}$$

The languages recognisable by FSMs are the "regular" languages, i.e. those that can be described by a regular expression [11]. Each regular language is recognised by a set of FSMs which are equivalent in that for any string, each of them produces the same result of acceptance or rejection. The lowest number of states of any FSM recognising a regular language is termed the "state complexity" of the language, as explained by Yu [9]. For an algorithm to compute the state complexity from any equivalent FSM, see [12].

2.2   The U_n Languages as Drivers of Complexity

To be able to run experiments with a range of language complexities, we looked for a family of languages that differed primarily in their complexity only, and preferably only included one member language for each state complexity value. Both or these criteria are met by the "universal witness" languages described by Brzozowski [13], which we use here. This is a family (or "stream") of languages U_n, which has one member for any $n <= 3$, which are examples ("witnesses") of the upper bounds of state complexity in a number of theorems concerning the complexity of FSMs. In particular, it has the property that U_n cannot be recognised by an FSM of less than n states.

Each of the U_n languages has a corresponding FSM $\mathcal{U}_n$ that accepts it. The general structure of these FSMs is shown in Fig. 1. The formal definition is as follows:

Definition of $\mathcal{U}_n$ - for $n \geq 3$, the FSM $\mathcal{U}_n$ is a quintuple $(Q, \Sigma, \delta, q_0, F)$, where $Q = \{q_0, q_1, ..., q_{n-1}\}$ is the set of states, $\Sigma = \{a, b, c\}$ is the alphabet, q_0 is the initial state, $F = \{q_{n-1}\}$ is the set containing the one accepting state, and δ is:
$\delta(q_i, a) = q_{(i+1) \bmod n}$;
$\delta(q_0, b) = q_1$; $\delta(q_1, b) = q_0$; $\delta(q_i, b) = q_i$ for $i \notin \{0, 1\}$;
$\delta(q_{n-1}, c) = q_0$; $\delta(q_i, c) = q_i$ for $i \neq n - 1$.

The proportion of strings from the ternary alphabet that are members of the U_n languages decreases with increasing n, as shown in Table 1.

2.3   Conflicting Objectives: Sample Set Recognition vs State Count

The two objectives we combine in the selection applied in this study are:

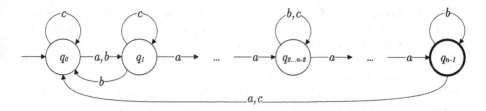

Fig. 1. FSM structure for $\mathcal{U}_n$ (based on [13])

Table 1. Member strings of some U_n languages as percentage of all ternary strings up to a length

Up to length	All strings #	%	U_3 #	%	U_4 #	%	U_5 #	%
7	3,280	100%	656	20%	490	15%	320	10%
8	9,841	100%	1,968	20%	1,452	15%	1,122	11%
9	29,524	100%	5,904	20%	4,280	15%	3,616	12%
10	88,573	100%	17,714	20%	12,688	14%	11,040	13%
11	265,720	100%	53,144	20%	37,874	14%	32,640	12%
12	797,161	100%	159,432	20%	113,548	14%	95,042	11%
13	2,391,484	100%	478,296	20%	340,992	14%	276,016	12%
14	7,174,453	100%	1,434,890	20%	1,024,128	14%	806,000	11%
15	21,523,360	100%	4,304,672	20%	3,074,490	14%	2,375,360	11%
16	64,570,081	100%	12,914,016	20%	9,225,836	14%	7,065,762	11%

1. "linguistic fitness" - maximising the correct classification by FSMs of strings as being in or out of a language; and
2. minimising the number of states of the FSMs.

The first objective favours more states, as more states can embody more complex information about the target language. The second objective selects for fewer states. In this way, the two objectives counteract each other.

To quantify the fitness of an FSM as an acceptor of a language, we use its performance in accepting or rejecting strings in two sample sets, positive (strings in the language) and negative (strings not in the language).

In this study, the "linguistic fitness" $F_{A,R}(\mathcal{M})$ of an FSM $\mathcal{M}$ against a set A of strings to be accepted and a set R of strings to be rejected, is the total of correctly accepted and correctly rejected examples:

$$F_{A,R}(\mathcal{M}) = |\{a \in A \mid \mathcal{M}(a)\}| \\ + |\{r \in R \mid \neg\mathcal{M}(r)\}| \tag{5}$$

The maximum value of $F_{A,R}$ for an FSM is the total size of the positive and negative reference sets A and R. The fitness gains we study are specifically the increases of this linguistic fitness over time.

A genetic algorithm selecting for multiple objectives needs to decide which of a set of candidates to favour in the reproductive pool, based on how well they satisfy each of the objectives. As described by Neumann [14], this can be expressed in a "dominance" relationship combining the two objectives.

In this study, we combine the objectives of maximisation of fitness and minimisation of states. The former is measured by a function $F_{A,R}$ (A and R being the positive and negative sample sets), the latter by the number of states, C.

This paper uses the two relations "weakly dominates" (denoted by the symbol $\succeq$) and "dominates" (denoted by $\succ$) between FSMs X and Y:

$$X \succeq Y \iff F_{A,R}(X) \geq F_{A,R}(Y) \wedge C(X) \leq C(Y)$$
$$X \succ Y \iff (X \succeq Y) \wedge (F_{A,R}(X) > F_{A,R}(Y) \vee C(X) < C(Y))$$

Informally, X weakly dominates Y if it is no worse than Y on either fitness or complexity, and X dominates Y if it is no worse on either fitness or number of states, but better on at least one of the two.

3   Algorithms

3.1   Sample Set Generation Methods

We use two methods of generating sample sets of positive and negative examples. In both methods, the positive and negative sets contain 500 strings each, for a total of 1,000 strings. The two methods labelled Bss and Rle_n, are defined below.

The reason for equal numbers of positive and negative examples (regardless of the percentage of strings of a given length that are in versus out of the language) is technical. An FSM of a single state can either accept all strings or rejects all strings. Its fitness score therefore would be the number of positive and negative samples respectively, but its complexity is the same in either case. Using the same number of positive and negative examples provides a common baseline of performance for minimal complexity.

Bss Method. The Bss method (for "balanced short strings") generates positive and negative sets that have short strings, with the lengths of positive and negative examples being balanced, i.e. neither the positive nor the negative subset favours longer strings. The sample sets are generated by Algorithm 1.

Rle_n Method. To investigate the effect of longer sample strings, we devised the Rle_n method (for "Random, less than or equal to n"). This constructs positive and negative sample sets of equal size, from strings drawn uniformly at random from all strings over the alphabet of the target language, of length zero through n. The Rle_n method is implemented by by Algorithm 2.

Algorithm 1: Sample set generation: Bss method

Input : $\mathcal{U}$, FSM recognising the target language

 D, total size of positive and negative sample sets to generate

Output: A, set of positive example strings

 R, set of negative example strings

$A \leftarrow \{\}$;

$R \leftarrow \{\}$;

for *all strings s on the alphabet of $\mathcal{U}$, in alphabetic order* **do**

 if $\mathcal{U}(s) \wedge |A| \leq |R|$ **then**

  | $A \leftarrow A \cup \{s\}$;

 else if $\neg(\mathcal{U}(s)) \wedge |A| \geq |R|$ **then**

  | $R \leftarrow R \cup \{s\}$;

 end

 if $|A| + |R| \geq D$ **then**

  | break;

 end

end

3.2   The Evolution Algorithm

The study of complexity in multi-objective genetic programming by Neumann [14] describes the SMO-GP algorithm (Simple Multi-Objective Genetic Programming). It starts with a population of one individual. In each iteration, SMO-GP selects one member of the population, applies a mutation operation, and then either replaces all dominated individuals with the (single) new mutant, or drops the mutant if it does not dominate any existing members.

For this study, we adapted this algorithm is adapted to evolve FSMs starting with the simplest possible FSM, which we call $\mathcal{N}$, see Algorithm 3. $\mathcal{N}$ is an FSM of a single non-accepting state, with the transition function mapping all inputs back to that state. This is in a sense the simplest possible FSM, and corresponds to the empty language, as it rejects all strings. (An FSM accepting all strings, i.e. accepting $*\Sigma$, would be equally simple.) The formal definition of such an FSM for a ternary alphabet is:

$$\mathcal{N} = (\{q_0\}, \{a, b, c\}, \{((q_0, a), q_0), ((q_0, b), q_0), ((q_0, c), q_0)\}, q_0, \emptyset) \qquad (6)$$

Note that half the 1,000 sample strings will not be elements of the language, so the FSM $\mathcal{N}$ by rejecting them will achieve a linguistic fitness score of 500. The only way for a new mutant to enter the population is to weakly dominate an existing FSM in the population. This can come about by having higher fitness or fewer states, but no FSM can have less states than $\mathcal{N}$. Therefore, all other FSMs that will become part of the population will score at least 500.

In some preliminary work, we also experimented with starting evolution towards U_n not with $\mathcal{N}$, but instead with the result of evolution over a set number of generations toward U_{n-1}. This approach did not produce any notable

---

Algorithm 2: Sample set generation: Rle_n method

Input : n, maximum length of strings to include in the sample sets
             $\mathcal{U}$, FSM recognising the target language
             D, total size of positive and negative sample sets to generate
Output: A, set of positive example strings
             R, set of negative example strings
$A \leftarrow \{\}$;
$R \leftarrow \{\}$;
$C \leftarrow |\text{alphabet of } \mathcal{U}|$;
$p \leftarrow$ distribution of 0 to C, weighted by number of strings of that length;
while $|A| + |R| < D$ **do**
    $L \leftarrow$ random choice of length from the weighted distribution p;
    $s \leftarrow$ random string of length L from the alphabet of $\mathcal{U}$;
    if $s \notin A \cup R$ **then**
        if $\mathcal{U}(s) \wedge |A| < D/2$ **then**
            $A \leftarrow A \cup \{s\}$;
        else if $\neg\mathcal{U}(s) \wedge |R| < D/2$ **then**
            $R \leftarrow R \cup \{s\}$;
        end
    end
end

---

patterns correlating the starting point with improved speed of adaptation to higher fitness, and we did not pursue it further.

3.3 The Mutation Operation

The mutation operation used on FSMs selects one random state for the source of an arc, and a random input symbol from the alphabet for the label of the arc. A target state is then selected; half the time, this is an existing state of the FSM, otherwise it is a new state which has all outgoing arcs leading back to itself.

The FSM is then modified so that the arc from the source state, labelled with the chosen input symbol, is set to lead to the chosen target state.

Lastly, a random state is chosen and changed from accepting to non-accepting and vice versa. For details, see Algorithm 4.

4 Experimental Results

Table 2 shows the mean fitness scores observed after given numbers of generations, under four different methods of generating sample sets for the U_3 language, across 30 runs with different randomisation seeds. Figure 2 shows a more detailed set of observations for the four methods.

The relative position of the data series in Fig. 2 made us curious about whether a general trend might exist for using longer strings in sample sets to lead to slower evolution of better recognisers. We repeated the experiment for

---

Algorithm 3: Multi-objective Evolution Algorithm for FSMs

Input : G, then number of iterations
Output: P, the final set of FSMs in the population after G generations

$q_0 \leftarrow$ an initial FSM state;
$\mathcal{N} \leftarrow (\{q_0\}, \{a, b, c\}, \{((q_0, a), q_0), ((q_0, b), q_0), ((q_0, c), q_0)\}, q_0, \emptyset)$;
$P \leftarrow \{\mathcal{N}\}$;

for G *iterations* **do**
    $M \leftarrow$ uniformly random choice from P;
    // Pois denotes the Poisson distribution with parameter $\lambda = 1$.
    for $1 + Pois(1)$ *iterations* **do**
        $\mid$ $M \leftarrow M$ modified by mutation operation;
    end
    if $\not\exists i \in P | i \succ M$ **then**
        $\mid$ $P \leftarrow (P \cup \{M\}) \setminus \{j \in P | M \succcurlyeq j\}$;
    end
end

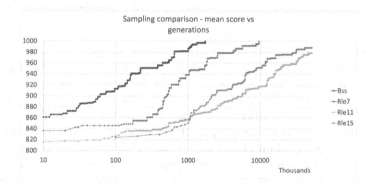

Fig. 2. Evolving to U_3: Comparison of sampling styles

U_4 and U_5; results are shown in Fig. 3. Note that the Rle sampling was adjusted for U_4 and U_5. In the U_4 experiment, we generated 500 positive and 500 negative sample strings, for each of Rle_7, Rle_{11}, and Rle_{15}. However, the languages U_4 and U_5 do not contain 500 strings of at most 7 characters. In order to create comparable sample sets for U_4 and U_5, we used Rle_8, Rle_{12}, and Rle_{16}.

5 Discussion

A pattern of higher n used in Rle_n sampling leading to slower improvement of recognition fitness is seen in all three Figures summarising the generated data. This observation suggests that selection by using longer strings provides less relevant information towards recognising the language than selection by shorter

---

Algorithm 4: Mutation operation

Input : $M = (Q, \Sigma, \delta, q_0, F)$, the FSM to be mutated
Output: M, the input FSM modified
$q \leftarrow$ uniformly random choice from Q;
$c \leftarrow$ uniformly random choice from Σ;
if *uniformly random choice of {existing, new}* $=$ *existing* **then**
 | $t \leftarrow$ uniformly random choice from Q;
else
 | $t \leftarrow$ a new FSM state $\notin Q$;
 | $Q \leftarrow Q \cup \{t\}$;
end
$\delta(q, c) \leftarrow t$;
$r \leftarrow$ uniformly random choice from Q;
if *uniformly random choice of {accepting, rejecting}* $=$ *rejecting* **then**
 | $F \leftarrow F \setminus \{r\}$;
else
 | $F \leftarrow F \cup \{r\}$;
end

---

strings. This is perhaps counter-intuitive. Longer sample strings of necessity embed more information about the language in question, on a per string basis.

A possible explanation may be that as we increase n in Rle_n, a fixed-size sample set (e.g. 500 positive and 500 negative strings) is a smaller percentage of all strings of length up to n. For a long string, a larger set of its leading substrings is likely to be missing from the sample set. So, a mutant FSM may randomly be correct in its categorisation of the long string, while not correctly categorising its prefixes. This implies that a gain of fitness under these circumstances is less likely to lead to further gains of fitness, compared to sample sets that are a more comprehensive coverage of shorter strings in the language.

Table 2. Score comparison for evolving U_3 from $\mathcal{N}$ (mean of 30 runs with different randomisation seeds)

Generation	Mean score			
	Bss	Rle_7	Rle_{11}	Rle_{15}
100,000	912.67	845.47	826.37	823.83
200,000	941.03	854.90	836.57	826.33
500,000	960.63	898.60	841.90	833.87
1,000,000	986.80	938.73	856.80	851.03
2,000,000	1000.00	969.60	871.70	894.30
5,000,000	1000.00	986.97	898.97	923.57
10,000,000	1000.00	1000.00	917.87	951.63

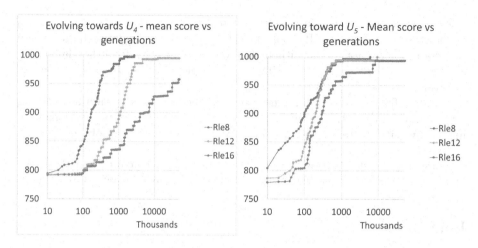

Fig. 3. Evolving to U_4 and U_5: Comparison of sampling styles

In other words, the information about the language that is embedded in the sample set is less representative at higher values of n in Rle_n. With less comprehensive coverage of examples for higher n in Rle_n, the positive and negative sets of examples may become more akin to random strings rather than providing selection conducive to higher linguistic fitness in the next generation.

Although these experiments only investigated evolution against sample sets of U_3 through U_5, for further members of the series, growth in the number of FSM states may be an important factor. As it is known that a perfect recogniser for U_n needs to have at least n states, we would expect that the growth in the number of states would be an important factor in enabling the evolution of better recognition for higher values of n in U_n.

Whether this finding transfers to other languages, and if so, whether it may be able to be generalised to other automata or evolutionary scenarios, is not answered by our study. Even within the U_n languages, we have only investigated low values of n. Further, languages with significantly different structures may behave differently. Algorithm 3 uses very small populations, so the effect of diversity on our observation is also an open question.

A potential application of our observation, if generalised, may be to find better solutions in fewer generations in evolutionary optimisation.

6   Conclusion

Under the experimental scenario we used, extending the sample sets of training strings to longer string lengths tends to slow down the achievement of higher fitness scores. A fixed size of sample set, drawn from pools of strings that increase in size as maximum length is extended provides less directed selection pressure. It appears that this is not counteracted by the additional information embedded in longer sample strings, possibly because gains of fitness may occur with a lower likelihood of leading to further gains of fitness, using the same fitness function.

This work was based on observation of the U_n languages, for low values of the state complexity n. It would be interesting to investigate whether these observations hold for: - U_n with higher values of n, or - for other languages arranged in a sequence of increasing state complexity, or - when the linguistic fitness function shifts from U_n to U_{n+1}.

Another fruitful avenue of further work may be to relax the criteria governing retention of FSMs in the population, and observe the effect of populations with higher diversity interacting with rising string lengths in the sample sets.

Acknowledgements. This work was supported by the Commonwealth of Australia. Frank Neumann was supported by the Australian Research Council through grant FT200100536.

References

1. Roostapour, V., Neumann, A., Neumann, F.: Single- and multi-objective evolutionary algorithms for the knapsack problem with dynamically changing constraints. Theoret. Comput. Sci. **924**, 129–147 (2022)
2. Standish, R.: Open-ended artificial evolution. Int. J. Comput. Intell. Appl. **3**(02), 167–175 (2003)
3. Kazarlis, S., Petridis, V.: Varying fitness functions in genetic algorithms: studying the rate of increase of the dynamic penalty terms. In: Eiben, A.E., Bäck, T., Schoenauer, M., Schwefel, H.-P. (eds.) PPSN 1998. LNCS, vol. 1498, pp. 211–220. Springer, Heidelberg (1998). https://doi.org/10.1007/BFb0056864
4. Gabor, T., Phan, T., Linnhoff-Popien, C.: Productive fitness in diversity-aware evolutionary algorithms. Nat. Comput. **20**, 363–376 (2021)
5. Naidoo, A.: Evolving Automata Using Genetic Programming. Master's thesis, University of KwaZulu-Natal, Durban, South Africa (2008)
6. Rasek, A., Dörwald, W., Hauhs, M., Kastner-Maresch, A.: Species formation in evolving finite state machines. In: Floreano, D., Nicoud, J.-D., Mondada, F. (eds.) ECAL 1999. LNCS (LNAI), vol. 1674, pp. 139–148. Springer, Heidelberg (1999). https://doi.org/10.1007/3-540-48304-7_20
7. Moran, N., Pollack, J.: Evolving complexity in prediction games. Artif. Life **25**, 74–91 (2019)
8. Packard, N., et al.: An overview of open-ended evolution: editorial introduction to the open-ended evolution II special issue. Artif. Life **25**, 93–103 (2019)
9. Yu, S.: State complexity of regular languages. J. Autom. Lang. Comb. **6**(2), 221–234 (2001)
10. Ehrenfeucht, A., Zeiger, P.: Complexity measures for regular expressions. J. Comput. Syst. Sci. **12**, 134–146 (1976)
11. Sipser, M.: Introduction To The Theory Of Computation, pp. 55–76. PWS Publishing Company, Boston, MA (1997)
12. Hopcroft, J.: An $n \log n$ Algorithm For Minimizing States In A Finite Automaton. Stanford University, Stanford, CA (1971)
13. Brzozowski, J.: In search of most complex regular languages. In: Moreira, N., Reis, R. (eds.) CIAA 2012. LNCS, vol. 7381, pp. 5–24. Springer, Heidelberg (2012). https://doi.org/10.1007/978-3-642-31606-7_2
14. Neumann, F.: Computational complexity analysis of multi-objective genetic programming. In: GECCO 2012: Proceedings of the 14th Annual Conference on Genetic and Evolutionary Computation, pp. 799–806. Philadelphia, Pennsylvania, USA (2012)

Author Index

T. Liu et al. (Eds.): AI 2023, LNAI 14472, pp. 491–494, 2024.
https://doi.org/10.1007/978-981-99-8391-9

Printed in the United States
by Baker & Taylor Publisher Services